MW00608053

Introduction to Approximate Solution Techniques, Numerical Modeling, and Finite Element Methods

Civil and Environmental Engineering

A Series of Reference Books and Textbooks

Editor

Michael D. Meyer

Department of Civil and Environmental Engineering
Georgia Institute of Technology
Atlanta, Georgia

Additional Volumes in Production

Introduction to Approximate Solution Techniques, Numerical Modeling, and Finite Element Methods

Victor N. Kaliakin

University of Delaware
Newark, Delaware

MARCEL DEKKER, INC.　　　　　NEW YORK · BASEL

ISBN: 0-8247-0679-X

This book is printed on acid-free paper.

Headquarters
Marcel Dekker, Inc.
270 Madison Avenue, New York, NY 10016
tel: 212-696-9000; fax: 212-685-4540

Eastern Hemisphere Distribution
Marcel Dekker AG
Hutgasse 4, Postfach 812, CH-4001 Basel, Switzerland
tel: 41-61-261-8482; fax: 41-61-261-8896

World Wide Web
http://www.dekker.com

The publisher offers discounts on this book when ordered in bulk quantities. For more information, write to Special Sales/Professional Marketing at the headquarters address above.

Copyright © 2002 by Marcel Dekker, Inc. All Rights Reserved.

Neither this book nor any part may be reproduced or transmitted in any form or by any means, electronic or mechanical, including photocopying, microfilming, and recording, or by any information storage and retrieval system, without permission in writing from the publisher.

Current printing (last digit):
10 9 8 7 6 5 4 3 2 1

PRINTED IN THE UNITED STATES OF AMERICA

To Leeza, Yanni, Dimitri & Maria

Preface

This book has evolved from notes compiled while teaching the senior/first-year graduate level course CIEG 401/601 (*Introduction to the Finite Element Method*) offered at the University of Delaware. The goals of this course are to: (1) Introduce students to the general subject of approximate solutions; (2) Introduce and emphasize the finite element method; (3) Help students understand the strengths and limitations of various approximation solution techniques; and (4) Develop in the students the ability to assess the correctness of approximate solutions.

Quite often in the past, introductory finite element texts were written on the assumption that students possessed a background in structural mechanics. This is evidenced by the common use of simple bar and beam examples to illustrate the development of element equations. The present reality is, however, that a wider range of students is striving to learn the finite element method. For example, here at the University of Delaware, CIEG 401/601, though populated largely by students from Mechanical and Civil Engineering, is also taken by students from the College of Marine Studies, Water Resources, and Health and Exercise Sciences. Consequently, structural mechanics concepts, though well understood by engineers, represent potential stumbling blocks for students with non-engineering backgrounds.

In order to address the changing make-up of the student body enrolled in introductory finite element classes, this book emphasizes fairly generic forms of differential equations. The approximate solution techniques studied are applied to these generic problems; once introduced and "digested," these techniques are then applied to specific physical problems such as heat conduction, elastostatics, flow through porous media, etc.

Many of the chapters have been designed to be quite modular. As such, certain sections or chapters can be omitted from a course plan without detrimentally affecting the understanding of material in subsequent chapters. A brief overview of the respective chapters is presented at the end of Chapter 1.

Victor N. Kaliakin

Acknowledgements

Very rarely do individuals work successfully in a vacuum. The preparation and refinement of this book over the past ten years have been greatly aided by the comments of students who have taken the introductory course in approximate solution techniques. Their help is thankfully acknowledged.

The author is very fortunate to have studied under R. L. Taylor (University of California, Berkeley) and L. R. Herrmann (University of California, Davis). These pioneers in the finite element method have significantly influenced the structure and content of these notes.

The author's philosophy towards programming, however peculiar, has been influenced by past collaborations with Dr. K. J. Perano.

If one phrase is chosen to best summarize approximate solution techniques, it would be:

> "When considering a specific numerical procedure, you don't get something for nothing."

cheers,

V.N.K.

Glossary of Notations and Units

Units

E : denotes units of energy (e.g., Joules, cal, BTU, ft-lb)
F : denotes units of force (e.g., Newtons, dynes, pounds, etc.)
L : denotes units of length (e.g., meters, centimeters, inches, etc.)
m : denotes units of mass (e.g., kilograms, grams, slugs, etc.)
Q : denotes units of charge (e.g., Coulombs)
t : denotes units of time (e.g., seconds, minutes, etc.)
T : denotes units of temperature (e.g., degrees Kelvin, Fahrenheit, etc.)

Sets

$\cup$: union (.OR.)
$\cap$: intersection (.AND.)
$\emptyset$: empty set
$\in$: is member of (a set)
$\notin$: is not a member of (a set)
$\subset$: is subset of (contained in)
$\not\subset$: is not a subset of (not contained in)
$\Rightarrow$: implies
$\Leftrightarrow$: if and only if
$\forall$: for all
$\exists$: there exists
$\ni$: such that

Miscellaneous Integers

N_{eddof} = number of element displacement degrees of freedom.
N_{en} = total number of nodes in an element.
N_{pt} = total number of points used to describe the element geometry.
N_{rowb} = number of rows in the strain-displacement matrix $\mathbf{B}$.
N_{sdim} = spatial dimension of the analysis (1, 2 or 3).

Contents

Introduction to Approximate Solution Techniques, Numerical Modeling, and Finite Element Methods

Chapter 1

Governing Equations and Their Approximate Solution

1.1 Introductory Remarks

Analysts are routinely called upon to numerically simulate physical phenomena of varying complexity. To successfully perform such simulations, the phenomena must first be described mathematically. Such descriptions consists of the equations governing the phenomena and suitable boundary conditions, initial conditions, or both, that are imposed on the governing equations. The mathematical description is typically referred to as the *classical* or *strong form* of a problem.

In most undergraduate programs of study, students are exposed to rather diverse topics from applied mathematics, physics and engineering. The associated problems are typically solved exactly. Not surprisingly, the geometry of the associated solution domain, the boundary conditions and applied loadings tend to be relatively simple. However, in many real-world applications the complexity of problems precludes their solution exactly. Instead, such problems must be solved using *approximate* techniques.

When developing such techniques the diversity of the aforementioned physical problems represents a potential source of confusion, for it may not always be clear whether a particular technique is general in nature or is applicable only to a given class of problems. Fortunately, many quite different physical problems are governed by equations possessing similar forms.

1.2   Some Simple Governing Equations

In support of the above observation, and to introduce some physical problems that will subsequently be studied, the mathematical statements of a number of simple, yet diverse examples are now presented.

Example 1.1: One-Dimensional Steady State Heat Conduction

Consider the steady state heat conduction of the rod shown in Figure 1.1. In such problems the quantities of interest are the distribution of temperature T and the heat flux $\tilde{q}$. The heat flux is related to the source term S (i.e., internal heat production due to such things as dielectric or induction heating, radioactive decay, absorption from radiation, chemical reaction, etc.) through the balance of energy equation

$$\frac{d\tilde{q}_x}{dx} - AS(x) = 0 \tag{1.1}$$

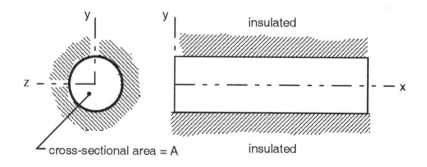

Figure 1.1: Schematic Illustration of an Insulated Rod

The heat conduction is governed by Fourier's law [169], viz.,

$$\tilde{q}_x = -k_x A \frac{dT}{dx} \tag{1.2}$$

where k_x is the thermal conductivity and A represents the cross-sectional area of the rod.

This constitutive relation is typically combined with the balance equation (1.1) to give the following single second-order differential equation for the primary dependent variable T:

$$\frac{d}{dx}\left(k_x A \frac{dT}{dx}\right) + AS(x) = 0 \qquad (1.3)$$

The possible boundary conditions for the problem are a prescribed temperature $T = \bar{T}$ or a prescribed value of heat flux $\tilde{q}_x = \bar{q}_x$ specified at a point (typically at one of the ends) along the rod.

Example 1.2: One-Dimensional Elastostatics

Consider a bar or rod of cross-sectional area A that can, in general, vary along its length. The y-axis is chosen to coincide with the bar's longitudinal axis, which passes through the centroid of the cross-section.

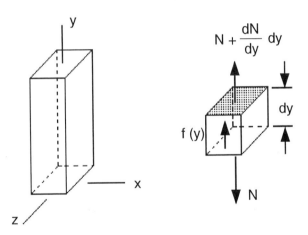

Figure 1.2: Schematic Illustration of a Simple Bar

Noting that $f(y)$ is a known quantity with units of force per unit length,[1] force equilibrium (balance of linear momentum) in the y-direction for the infinitesimal element of the bar shown in Figure 1.2 gives

---

[1] $f(y)$ is typically a force that acts on points interior to the bar. If it is produced by an external source $g(y)$ (e.g., a gravitational or centrifugal force) having units of force per volume, then $f(y) = g(y)A$.

$$\frac{dN}{dy} + f(y) = 0 \tag{1.4}$$

Assuming infinitesimal displacements and displacement gradients and an isotropic, linearly elastic constitutive relation for the material, the uniaxial stress and strain are related by Hooke's law [247],[2] viz.,

$$\sigma_y = E\frac{dv}{dy} = \frac{N}{A} \tag{1.5}$$

where E represents the elastic modulus, v is the longitudinal (y-direction) displacement, and dv/dy is the axial strain.

Solving equation (1.5) for the axial force N, and substituting the resulting expression into equation (1.4) gives the final governing differential equation

$$\frac{d}{dy}\left(EA\frac{dv}{dy}\right) + f(y) = 0 \tag{1.6}$$

The possible boundary conditions for the problem are a prescribed displacement $v = \bar{v}$ or a prescribed force $EA(dv/dy) = \bar{P}$ at a point in the rod. Typically the boundary conditions are specified at the ends of the bar.

Example 1.3: Steady State Flow Through Porous Geologic Media

Consider the one-dimensional flow of fluid through a confined porous aquifer of cross-sectional area A such as that shown in Figure 1.3. The confining layer prevents seepage of fluid into or out of the aquifer along its length. The governing differential equations for this problem are derived beginning with the continuity equation

$$\frac{dq_x}{dx} = e_x \tag{1.7}$$

In addition, the material is characterized by the constitutive relation proposed by Darcy [131], viz.,

$$q_x = -k_x A\frac{dh}{dx} \tag{1.8}$$

---

[2]Named in honor of Robert Hooke (1635-1703). Further references related to the theory of elasticity are given in Chapter 12.

where q_x represents the quantity of fluid flow through the porous medium, k_x represents the coefficient of permeability in the direction of flow, A denotes the cross-sectional area perpendicular to the direction of flow, e_x represents the seepage per unit length of medium, and h represents the excess hydrostatic (piezometric) head, or potential. The latter is a sum of the pressure head and the elevation head; i.e., $h = p/\gamma + z$, where p is the pressure, γ is the density of the fluid, and z is the elevation at some location along the aquifer.

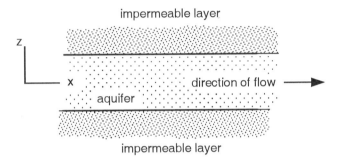

Figure 1.3: Schematic Illustration of One-Dimensional Flow Through Confined Aquifer

Equations (1.7) and (1.8) are typically combined to give the following single equation for h:

$$\frac{d}{dx}\left(k_x A \frac{dh}{dx}\right) + e_x = 0 \qquad (1.9)$$

The requisite boundary conditions, specified at two locations in the aquifer, are either prescribed piezometric head, $h = \bar{h}$ at a location, or prescribed fluid flow $q_x = \bar{q}_x$ at some other location.

From the previous examples it is evident that the equations governing three different physical processes are *identical in form*. Any solution procedure developed for one process is therefore applicable to the other two and, in fact, to any other physical problem governed by this particular second-order ordinary differential equation (see Table 1.1).

Table 1.1: Second-Order Governing Equations in One-Dimension

$$\frac{d}{dx}\left(k\frac{d\phi}{dx}\right) + S = 0 \quad ; \quad 0 < x < L$$

Physical Problem	ϕ	$q = k\left(\frac{d\phi}{dx}\right)$
	Material Coefficient (k)	Source Term (S)
Axial deformation of a bar	displacement elastic modulus $*$ area	force force/length
Cable deflection	transverse displacement cable tension	force transverse load
Channel flow	velocity viscosity	stress pressure gradient
Electrostatics	electrostatic potential permittivity	electric flux charge density
Flow through porous media	hydraulic head hydraulic conductivity $*$ area	flow velocity fluid flux
Heat flow	temperature thermal conductivity	heat flux heat source/sink

The similarities in form observed for one-dimensional problems also apply to certain two-dimensional problems. In support of this, we provide the following examples. These and other two-dimensional problems are summarized in Table 1.2.

Example 1.4: Two-Dimensional Steady State Heat Conduction

The steady state heat conduction problem of Example 1.1 is extended to heat flow in two dimensions. Such an analysis would, for example, apply to a plate with its top and bottom surfaces thermally insulated.

The balance of energy is again obtained using first law of thermodynamics, which states that

$$\frac{\partial q_x}{\partial x} + \frac{\partial q_y}{\partial y} - S = \nabla^T \mathbf{q} - S = 0 \tag{1.10}$$

In equation (1.10) $\mathbf{q} = \begin{Bmatrix} q_x & q_y \end{Bmatrix}^T$ is the vector of heat flux components, S is a source term, a superscript T denotes the operation of vector transposition, and ∇ is the gradient operator in two dimensions, viz.,

$$\nabla = \begin{Bmatrix} \dfrac{\partial}{\partial x} & \dfrac{\partial}{\partial y} \end{Bmatrix}^T \tag{1.11}$$

The relation between heat flux per unit area and the gradient of temperature is assumed to follow generalized Fourier's law [169], viz.,

$$\mathbf{q} = - \begin{bmatrix} k_{11} & k_{12} \\ k_{12} & k_{22} \end{bmatrix} \begin{Bmatrix} \dfrac{\partial T}{\partial x} \\ \dfrac{\partial T}{\partial y} \end{Bmatrix} = -\mathbf{K}\nabla T \tag{1.12}$$

where $\mathbf{K}$ is a symmetric matrix of thermal conductivity coefficients. Substituting equation (1.12) into equation (1.10) gives the equation governing steady-state heat conduction in two dimensions, viz.,

$$\nabla^T (\mathbf{K}\nabla T) + S = 0 \tag{1.13}$$

The boundary conditions for the problem are prescribed temperature $T = \bar{T}$ or prescribed heat flux $-(\mathbf{K}\nabla T)^T \mathbf{n} = \bar{q}$ normal to a portion of the boundary Γ.

Example 1.5: Two-Dimensional Electrostatics

Consider a dielectric plate with its top and bottom surfaces electrically insulated. Assuming the plate lies in the $x - y$ plane, the conservation of electric charge is described by Gauss' law [214], viz.,

$$\frac{\partial d_1}{\partial x_1} + \frac{\partial d_2}{\partial x_2} = \nabla^T \mathbf{d} = \rho \tag{1.14}$$

where $\mathbf{d} = \begin{Bmatrix} d_x & d_y \end{Bmatrix}^T$ is a vector of displacement components of electric charge in the x and y directions, and ρ is the charge density.

The constitutive relation with respect to principal material axes is

$$\mathbf{d} = - \begin{bmatrix} \varepsilon_{11} & 0 \\ 0 & \varepsilon_{22} \end{bmatrix} \begin{Bmatrix} \dfrac{\partial \phi}{\partial x} \\[2mm] \dfrac{\partial \phi}{\partial y} \end{Bmatrix} = -\varepsilon \nabla \phi \tag{1.15}$$

where ε_{11} and ε_{22} are permittivity coefficients in the x and y directions, respectively, and ϕ is the scalar electrostatic potential. Substituting equation (1.15) into (1.14) gives the governing differential equation

$$-\nabla^T \left(\varepsilon \nabla \phi \right) = \rho \tag{1.16}$$

or

$$-\frac{\partial}{\partial x_1} \left(\varepsilon_{11} \frac{\partial \phi}{\partial x_1} \right) - \frac{\partial}{\partial x_2} \left(\varepsilon_{22} \frac{\partial \phi}{\partial x_2} \right) = \rho \tag{1.17}$$

The boundary conditions for the problem are a specified electrostatic potential $\phi = \bar{\phi}$ or a specified normal displacement of electric charge $\left(\varepsilon \nabla \phi \right)^T \mathbf{n} = \bar{d}$.

Example 1.6: Two-Dimensional Seepage in Porous Media

Consider the problem of confined[3] seepage in porous geologic media. From the continuity of flow we obtain the following equation (compare to equation 1.10):

$$\frac{\partial q_x}{\partial x} + \frac{\partial q_y}{\partial y} - S = 0 \qquad (1.18)$$

where $\mathbf{q} = \{q_x \quad q_y\}^T$ is the vector of velocity components, and S is the flow generated per unit unit area.

As in Example 1.3, the relation between velocity and the gradient of the total hydraulic head ϕ is assumed to follow generalized Darcy's law [131],

$$\mathbf{q} = - \begin{bmatrix} k_{11} & k_{12} \\ k_{12} & k_{22} \end{bmatrix} \begin{Bmatrix} \dfrac{\partial \phi}{\partial x} \\ \dfrac{\partial \phi}{\partial y} \end{Bmatrix} \qquad (1.19)$$

Substituting equation (1.19) into equation (1.18) gives the governing equation for the problem, viz.,

$$\frac{\partial}{\partial x}\left(k_{11}\frac{\partial \phi}{\partial x} + k_{12}\frac{\partial \phi}{\partial y}\right) + \frac{\partial}{\partial y}\left(k_{12}\frac{\partial \phi}{\partial x} + k_{22}\frac{\partial \phi}{\partial y}\right) + S = 0 \qquad (1.20)$$

The boundary conditions for the problem are prescribed total head $\phi = \bar{\phi}$ or prescribed flow velocity $(\mathbf{K}\nabla\phi)^T \mathbf{n} = \bar{v}$ normal to a portion of the boundary Γ.

Based on the above observations for one- and two-dimensional problems, the importance of developing approximate solution techniques within a *mathematical* framework rather than from a physical argument applicable to a specific class of problems is evident. For this reason, it is timely to next discuss some mathematical concepts that will prove useful in this development.

---

[3]The unconfined flow problem is complicated by the need to calculate the position of the phreatic surface.

Table 1.2: Second-Order Governing Equations in Two-Dimensions

$$\nabla^T \left(\mathbf{K} \nabla \phi \right) + S = 0$$

Physical Problem	ϕ	$\left(\mathbf{K} \nabla \phi \right)^T \mathbf{n} = \bar{q}$
	Material Coefficients (**K**)	"Source" Term (S)
Displacement of elastic membrane	transverse displacement	normal force
	tension in membrane	applied transverse load
Electrostatics	electrostatic potential	displacement flux
	permittivity	charge density
Groundwater flow	hydraulic head	flow velocity
	hydraulic conductivity	recharge or pumping
Irrotational flow of ideal fluid	stream function	flow velocity
	identity matrix	mass production
Magnetostatics	magnetic potential	magnetic flux
	permeability	current density
Heat flow	temperature	heat flux
	thermal conductivity	heat source/sink
Torsion of prismatic bars	stress function	shear stress
	1/(shear modulus)	2 ∗ (angle of twist)

1.3 Mathematical Preliminaries

1.3.1 General Comments

Since the equations governing different physical phenomena often have similar forms, it is desirable to write these equations in a single general form and to subsequently develop approximate solution techniques for this form. Subsequent specializations to specific physical problems then simply involve the proper interpretation of the various parameters associated with the general form of the equations. Through such specializations additional insight into the approximate solution techniques can often be realized. Requisite to the proper development of the aforementioned general form is a bit of mathematical formalism, which is the subject of this section.

1.3.2 Classification of Physical Problems

Most problems in applied mathematics, physics and engineering can be classified as being either *discrete* or *continuous* [125]. A discrete system consists of a finite number of interconnected elements, whereas a continuous system involves a physical phenomenon distributed over a continuous region. Lumped masses interconnected by a system of springs is an example of a discrete system. The flow of heat through a solid is an example of a continuous system.

Discrete and continuous systems can each be further subdivided into the following three problem types:

Equilibrium Problems are those in which the state of the system remains constant with time. These are often referred to as *steady-state* problems. Examples of such problems are listed in Tables 1.1 and 1.2.

Eigenvalue Problems are typically considered as extensions of equilibrium problems. In eigenvalue problems, critical values of certain parameters must be determined in addition to the steady-state configuration. Examples include the determination of natural frequencies of vibration in mechanical systems, the buckling and stability of structures, and the determination of resonances in electric circuits.

Propagation Problems include those problems in which a subsequent state of the system must be related to an initially known state. These include *transient* and *unsteady-state* phenomena such as unsteady heat conduction, mechanical vibrations, structural dynamics, and stress waves in elastic media.

The approximate solution techniques developed for continuous systems reduce the number of unknowns in the system from *infinity* to a *finite* number. The *continuous* problem is thus reduced to the simpler case of a *discrete* one.

1.3.3   The Solution Domain and its Boundary

Whether discrete or continuous, the state of a physical problem is described by the values of unknown functions called *primary dependent variables.* The primary dependent variables satisfy the governing equation over a particular solution *domain,* as well as certain conditions (constraints) imposed along the *boundary* of the domain.

The domain is defined by the sets of all possible coordinate vectors $\mathbf{x}$. These are the *independent variables.* A particular set of allowable values of $\mathbf{x}$ defines a *point* in the domain. Each point is identified with its position vector, emanating from the origin of an appropriately defined coordinate system. If a given point is in the domain, then all points sufficiently close to this point also belong to the domain. This property implies that a domain consists only of internal points. A domain is *simply connected* if any simple closed contour in the domain can be shrunk continuously to a point without leaving the region; otherwise the domain is said to be *multiply connected* [184]. A hypothetical example of each type of domain is shown in Figure 1.4. In the subsequent development, an arbitrary domain is denoted by the symbol Ω.

In describing the material comprising Ω, the observed macroscopic behavior is often explained by disregarding molecular considerations. Instead, the material is assumed to be continuously distributed throughout its volume and to completely fill the space it occupies. Within the limitations for which such an assumption is valid, this *continuum concept* provides a suitable framework for studying the behavior of solids, liquids and gasses [379].

The boundary of Ω is the set of points such that in any neighborhood of each of these points there are points that belong to the domain as well as points that do not [461]. From the definition of a domain, it follows that the points on the boundary do not belong to the domain. The boundary of an arbitrary domain is assumed to be piecewise smooth; spatially it is one dimension less than the domain. In the subsequent development, the boundary is denoted by the symbol Γ.

In certain problems there is no upper or lower bound on one or more of the independent variables (e.g., time). In this case the boundary is said to be *open* with respect to the variable. When all the independent variables are bounded, the boundary is *closed.*

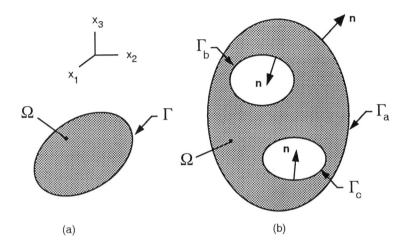

Figure 1.4: Schematic Illustration of a Typical Solution Domain and its Boundary: (a) Simply Connected, (b) Multiply Connected ($\Gamma = \Gamma_a + \Gamma_b + \Gamma_c$)

The unit outward vector normal to Γ is denoted by **n**. Representing the spatial dimension of the problem by the integer N_{sdim}, possible representations for **x** and **n** are shown in Table 1.3.

1.3.4 General Form of Governing Equations

It is convenient to denote an ordinary or partial differential equation describing a physical phenomenon over the domain Ω in the following general form

$$L(\phi) - f = 0 \qquad (1.21)$$

where L represents a differential operator, ϕ denotes the primary dependent variable, and f is a smooth scalar-valued function that is known throughout Ω and is independent of ϕ. For systems of partial differential equations, ϕ is a vector of unknowns and L is a matrix of operators.

The symbol $L(\phi)$ represents an expression containing ϕ and its derivatives (ordinary or partial) up to order[4] $2s$. Known functions of position

---

[4]The highest derivative is assumed to be even-ordered because this is representative of the classical field theories associated with the majority of engineering applications.

Table 1.3: Coordinate and Unit Outward Normal Vectors

N_{sdim}	Ω	Γ	$\mathbf{x}$	$\mathbf{n}$
1	line segment	endpoints	$\{x_1\} = \{x\}$	$\{n_1\} = \{n\}$
2	surface	curve	$\begin{Bmatrix} x_1 \\ x_2 \end{Bmatrix} = \begin{Bmatrix} x \\ y \end{Bmatrix}$	$\begin{Bmatrix} n_1 \\ n_2 \end{Bmatrix} = \begin{Bmatrix} n_x \\ n_y \end{Bmatrix}$
3	volume	surface	$\begin{Bmatrix} x_1 \\ x_2 \\ x_3 \end{Bmatrix} = \begin{Bmatrix} x \\ y \\ z \end{Bmatrix}$	$\begin{Bmatrix} n_1 \\ n_2 \\ n_3 \end{Bmatrix} = \begin{Bmatrix} n_x \\ n_y \\ n_z \end{Bmatrix}$

in Ω may appear in $L(\phi)$, but ϕ and its derivatives must enter in such a manner that $L(\phi)$ vanishes when ϕ and all its derivatives are set equal to zero [125].

If ϕ and all its derivatives appear linearly, L is said to be a *linear operator*. If some terms are non-linear but the derivatives of order $2s$ appear linearly, L is said to be *quasi-linear* [125]. If L is linear and $f = 0$ throughout Ω, then the differential equation is said to be *homogeneous*.

A differential equation is said to describe a *boundary-value problem* if the primary dependent variables and possibly their derivatives are required to take on specified values on the boundary. An *initial-value problem* is one in which the primary dependent variables and possibly their derivatives are specified initially. Initial-value problems typically describe time-dependent phenomena. Finally, time-dependent boundary-value problems are referred to as *boundary- and initial-value problems*.

1.3.5 General Form of Boundary Conditions

The proper solution of boundary-value problems requires the specification of known values of ϕ or its derivatives on the boundary Γ. These specifications are referred to as *boundary conditions*.

More precisely, a boundary-value problem of order $2s$ requires s boundary conditions to be specified at every point along Γ [125]. A boundary condition is an equation relating the values of ϕ, its derivatives from order 1 to $2s - 1$, or both, at points along Γ. Typically, two principal types of boundary conditions are associated with a boundary-value problem.[5]

The first type of boundary condition consists of an equation relating the values of ϕ, its derivatives up to order $s - 1$, or both, at points along a portion Γ_1 of the boundary. This type of boundary condition is referred to as *essential, forced* or *Dirichlet*.

The second type of boundary condition consists of an equation relating the values of any of the derivatives of ϕ from order s to $2s - 1$, at points along a portion Γ_2 of the boundary. This type of boundary condition is referred to as *natural, non-essential* or *Neumann*.

The classification of boundary conditions as essential or natural is determined solely by the order of governing differential equation [112]. It is independent of the method used to solve the differential equation.[6]

The boundary-value problems considered in Examples 1.1 to 1.3 involved second-order differential equations. Since $2s = 2$ for these problems, it follows that essential boundary conditions involve only the primary dependent variable ϕ and have the general form

$$\phi = \bar{\phi} \tag{1.22}$$

where $\bar{\phi}$ denotes a *known* value of ϕ. Natural boundary conditions for second-order differential equations involve only first derivatives of ϕ and have the general form

$$-k\frac{\partial \phi}{\partial n} = \bar{q} \tag{1.23}$$

where k is a material constant, n represents the unit outward vector normal to Γ (as discussed in Section 1.3.3), and $\bar{q}$ denotes a *known* value of the secondary dependent variable. The normal derivative appearing in equation (1.23) is defined as

$$\frac{\partial \phi}{\partial n} = n \bullet \nabla \phi = n_i \frac{\partial \phi}{\partial x_i} \tag{1.24}$$

where ∇ denotes the gradient operator, and indicial notation with summation over repeated indices is assumed.

---

[5] A third type of boundary condition also exists. It is commonly referred to as a *mixed* boundary condition, and involves ϕ and possibly its derivatives up to order $2s - 1$.

[6] The words "essential" and "natural" appear to have originated in the context of the calculus of variations during the 19th century [100]. They are, however, now applied to any boundary value problem, irrespective of whether it is in variational or differential equation form. Further details pertaining to variational methods are given in Chapter 5; the issue of boundary conditions is discussed in Section 5.4.

In the case of fourth-order differential equations, such as those associated with Bernoulli-Euler beam theory and certain plate and shell theories, $s = 2$. Consequently, essential boundary conditions involve specified values of ϕ and its first derivatives. Natural boundary conditions involve specified values of second and third derivatives of ϕ.

The boundary conditions can be written in a general form analogous to equation (1.21). That is, on Γ

$$B(\phi) - r = 0 \qquad (1.25)$$

where B represents expressions containing ϕ and its derivatives normal to the boundary up to order $2s - 1$, to be evaluated at each point along Γ. These expressions may be non-linear, but ϕ and its derivatives must enter in such a manner that $B(\phi)$ vanishes when ϕ and its derivatives vanish [125]. The symbol r represents prescribed values, independent of ϕ, and known at every point along Γ. If all the B are linear, the boundary conditions are said to be *homogeneous* when $r = 0$. Otherwise they are *non-homogeneous*.

Applying equation (1.25) to the essential boundary condition given by equation (1.22), it follows that on Γ_1

$$B(\phi) = \phi, \; r = \bar{\phi} \qquad (1.26)$$

The natural boundary condition given by equation (1.23) would be expressed on Γ_2 as

$$B(\phi) = -k\frac{\partial \phi}{\partial n}, \; r = \bar{q} \qquad (1.27)$$

It is not necessary for the $B(\phi)$, and r to be continuous along Γ. However, the portions of the boundary Γ_1 and Γ_2 are related in the following manner:

$$\Gamma_1 \cup \Gamma_2 = \Gamma \qquad (1.28)$$

$$\Gamma_1 \cap \Gamma_2 = \emptyset \qquad (1.29)$$

where the symbols $\cup$ and $\cap$ denote *set union* and *set intersection*, respectively, and $\emptyset$ denotes the *null set*.

In closing this section, it is timely to note that supplementary information concerning boundary conditions can be found in the thorough discussion presented by Burnett [100].

1.3.6 Defining a Well-Posed Problem

The boundary conditions must be chosen such that the mathematical problem is well-posed. For example, in the case of the second-order governing equation

$$\frac{d}{dx}\left(k\frac{d\phi}{dx}\right) + S = 0 \quad ; \quad 0 < x < L \tag{1.30}$$

at least one of the two requisite boundary conditions must be an essential (Dirichlet) boundary condition. Stated equivalently, we cannot specify natural (Neumann) boundary conditions at both ends of the domain.

Physically this can be explained by considering the one-dimensional steady-state heat conduction problem, applied to the domain (rod) shown in Figure 1.1. For simplicity, we assume the source term S to be zero. If we specify heat flux values at both ends of the rod (the same magnitude of heat flux, so that the heat balance equation is satisfied), it follows that the temperature distribution T along the rod in not unique. The only requirement is that the difference in temperature across the length of the rod yield the appropriate heat flux (through Fourier's law). Clearly this can be achieved by adding an arbitrary constant to the temperature at both ends of the rod. To render a unique solution, the temperature must be specified at either or both of the ends of the rod.

In the case of the elastostatic analysis of a simple bar (Figure 1.2), natural boundary conditions consist of specified values of axial force. Although such forces can be specified at both ends of the rod in such a way as to maintain equilibrium, the axial displacement v in the rod will not be unique. That is, the governing differential equation and (natural) boundary conditions will be satisfied by an infinite number of displacement solutions. Each solution differs from the others by an arbitrary constant. To preclude this ambiguity, the displacement must be specified at either or both of the ends of the bar.

Mathematically the above conditions involving only natural boundary conditions are explained as follows. Let ϕ_1 be a solution of equation (1.30), and let p be an arbitrary constant. Since only derivatives of ϕ appear in the governing equation and in the boundary conditions, it follows that $\phi_2 = \phi_1 + p$ is also a solution. This is easily seen since $dp/dx = 0$ and $d\phi_2/dx = d\phi_1/dx$. In fact, the problem has an *infinite* number of solutions, all differing from each other by p.

1.4   Comments on Approximate Solutions

Once the governing differential equations are derived and the boundary conditions and initial conditions prescribed, the mathematical statement of a given physical problem is complete. It is inherently assumed that in developing this statement the laws of physics have not been violated.

The process of going from a physical problem to its mathematical statement generally involves some, hopefully minor, degree of approximation. For example, displacements and displacement gradients may be assumed to be infinitesimal, thermal and mechanical loadings may be assumed to be applied quasi-statically, etc.

The governing equations must next be solved. At this point, difficulties arise, as only relatively simple equations within geometrically simple domains, can be solved exactly. Ordinary differential equations with constant coefficients represent one of the few problem classes for which standard solution procedures are readily available.

Once a problem cannot be solved exactly, it must be solved approximately. By their very definition, all approximation procedures introduce the notion of *error* into the solution. In a very general sense, "error" can be defined as the difference between the actual response (say measured experimentally) and that predicted by an approximate solution technique. Although such errors are unavoidable, it is imperative that they be assessed and *minimized* in order to economically obtain solutions within a given accuracy. Implicit to such a minimization is the availability of accurate error estimates associated with a given approximate solution technique.

One form of approximation involves the making of simplifying assumptions regarding the boundary conditions, the material constitution, etc. The resulting simplified problem can then be solved exactly, or if no exact solution exists, *numerically* in the manner described below.

A second form of approximation involves solving the actual problem in an *approximate* manner. Examples of this form of approximation include Fourier series approximations, polynomials, etc. Typically such problems are solved *numerically*.

From a formal mathematical point of view, the best devices for numerical solutions are the so-called *direct methods* [392]. According to Sobolev [503]:

> direct methods are those methods for the approximate solution of the problems of the theory of differential and integral equations which reduce these problems to finite systems of algebraic equations.

Perhaps the prime motivation for using direct methods is realization that numerical solutions typically enlist the aid of the digital computer. Digital computers can perform only standard arithmetic operations (addition, subtraction, multiplication, division, raising to powers, etc.) and certain logical operations. Thus recasting a problem in algebraic form as a result of using a direct method only facilitates the eventual solution of the problem on a digital computer.

The conversion of a problem to purely algebraic form requires the use of suitable *discretization* techniques. In such discretizations a *mathematical model* of the problem is created in which the infinite degrees of freedom representing the unknown primary dependent variables are replaced by a *finite* number of unknown parameters. A mathematical model of a physical problem is an attempt to give mathematical relationships between certain quantities of physical interest. Due to the complexity of physical reality, a variety of simplifying assumptions, such as those discussed above, are typically made to construct more tractable mathematical models. As a consequence of these assumptions, the resulting model has limitations on its accuracy. Depending on the use of the model, these limitations may or may not prove troublesome. If a model is not sufficiently accurate, the numerical solution of the model cannot improve upon this basic lack of accuracy.

In performing numerical simulations, it is of paramount importance to use only *accurate*, *reliable* and *computationally efficient* approximate solution techniques and numerical implementation schemes. Consequently, the student of approximate solution techniques must understand not only the theory underlying such techniques, but also their relative strengths and weaknesses and errors associated with their use. While certain errors are beyond the control of the analyst, others can be minimized by employing efficient principles and practices from the fields of Numerical Analysis and Computer Science.

1.5   The Role of Mathematical Modeling in Design

In recent years the reduction in cost and increase in computational speed of digital computers has an increasing number of engineers to rely more heavily on numerical analyses in the design process. This, in turn, has fueled a steady increase in the level of sophistication in the mathematical models developed. Consequently, the engineer is not constrained only to exact or analytical solutions.

The role of mathematical modeling and approximate solution techniques in the design process is schematically illustrated in Figure 1.5. In reference to this figure, it is important to point out that the mathematical model must be developed in full compliance with the laws of physics. Likewise, in the interpretation of results, a step whose importance cannot be overemphasized, the numerical solution is commonly compared to the results of simplified analyses that are based upon these same laws.

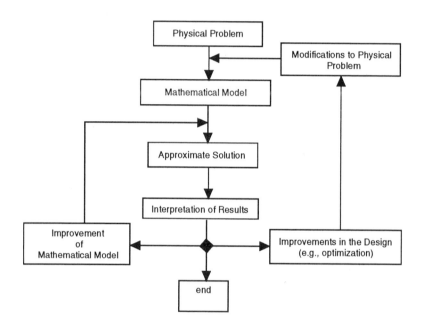

Figure 1.5: The Role of Mathematical Modeling and Approximate Solutions in the Design Process

In light of the above discussion, it is timely to comment on evolving trends in computer-aided design (CAD) and computer-aided engineering (CAE). Prior to the mid-seventies, design drafting, though increasingly computerized, was not linked (e.g., by a shared disk-resident database) to approximate solution techniques. In the mid-seventies and early eighties, this changed, as simple (linear) approximate solution techniques were increasingly being integrated with computer drafting applications. Since the mid-eighties, this trend has continued, only with more complicated (non-linear) approximate solution techniques employed. During this time, expert systems and artificial intelligence (AI) were increasingly used in the areas of automated design and evaluation. In addition, *numerical simulation* of manufacturing processes and prototype testing progressively became more widespread. Since this trend will surely continue in the future, it is important that students of approximate solution techniques understand the idea behind such simulation.

Traditionally, designs were verified by constructing prototypes and by testing these in the laboratory. This process was costly and time-consuming, for changes to the design typically meant fabrication of new specimens and then subsequent testing. Since realistic computer simulations were either unavailable or prohibitively expensive, designers had little choice but to follow this approach.

The availability of fast and relatively inexpensive computational resources has resulted in a transition towards more simulation-based designs. In such designs, a few prototypes are still fabricated and tested in order to get an insight into their behavior. The prototypes are then modeled mathematically. The boundary conditions and applied loadings are chosen so as to represent, as closely as possible, those employed in the actual tests. Using a suitable approximate solution technique, the models are then analyzed. The numerical results obtained are compared with the experimental results they sought to simulate.

If agreement between the two sets of results is deemed unacceptable, the mathematical model is appropriately (and hopefully rationally) modified and re-analyzed. Once the comparisons are acceptable, the mathematical models are deemed *verified,* and the analyst (and designer) has gained confidence in his or her ability to simulate the behavior of the prototype under specific experimental conditions.

The next step is to perform "numerical experiments" in which the designer stipulates possible changes in geometry, material properties, loading, etc., and the analyst performs further analyses to generate *predictions* of model behavior. In this manner, variations in the original design can be evaluated at a typically substantial savings in cost and time.

Closely related to numerical experiments are so-called "parametric studies" that are performed on the design prototypes. In such studies, one aspect of a mathematical model is changed (for example, an applied loading, the magnitude of a material parameter, a boundary condition, etc.), the model is re-analyzed, and the numerical results compared to the ones associated with the model without this change. The analyst thus attempts to see how sensitive the prototype's behavior is to variations in the specific parameter.

1.6   Concluding Remarks

As evident from the above discussion, the successful study of approximate solution techniques requires knowledge of physics, mathematics and some rudimentary aspects of computer science.

From physics come the principles of mechanics, thermodynamics, electromagnetics, etc. on which the governing equations and associated mathematical models are based. From the examples presented in this chapter, it is evident that many different physical problems are governed by differential equations that are identical in form. Consequently, approximate solution techniques should be developed within a mathematical framework rather than from a physical argument applicable to a specific class of problems.

The contributions of mathematics are not limited to the governing differential equations. The specific areas of numerical analysis and linear algebra are used extensively in the study of approximate solution techniques. From numerical analysis come the algorithms necessary for the efficient computer implementation of such techniques. The design of such algorithms is facilitated by the efficiency and elegance of matrices and linear algebra.

Finally, since approximate solution techniques require the power of the digital computer in order to be practical, rudimentary aspects of such computer science topics as efficient storage, finite precision arithmetic, solution errors, etc. must also be understood by students. These topics are investigated further in the next chapter.

Chapter 2

Computer Storage and Manipulation of Numbers

2.1 Introductory Remarks

For efficient application, approximate solution techniques require the use of a digital computer. Before discussing such methods and their implementation in detail, it is useful to briefly investigate the manner in which numbers are stored in, and manipulated by digital computers. Such an investigation brings to light certain sources of error that plague the numerical implementation and use of *all* approximate solution techniques.

Digital computers store information either as *characters* (text number representation) or as *numbers* (binary number representation) [168]. Both types of information are represented by binary numbers contained in a fixed number of binary digits or *bits*. Every bit has a value of either zero (0) or one (1); it is the smallest unit for storing data in a computer.

2.2 Storage of Characters

Characters consist of the letters of the alphabet, in both upper and lower case, numbers, and special characters (e.g., punctuation symbols). Combinations of characters form text strings. In general, each such string uses a variable amount of space, has a variable range of values, a variable precision, and is readable by humans [168].

Since the focus here is on the storage and manipulation of numbers, for further information on characters and character strings the reader is directed to books on the subject [289, 491, 168].

Table 2.1: Examples of Unsigned and Signed Numbers

Unsigned Numbers		Signed Numbers	
Decimal	Binary	Decimal	Binary
0	0000 0000	-128	1000 0000
5	0000 0101	-127	1000 0001
16	0001 0000	0	0000 0000
120	0111 1000	64	0100 0000
255	1111 1111	127	0111 1111

2.3   Storage of Numbers

Digital computers store numbers not with infinite precision but rather in some approximation that can be contained in a fixed number of bits. Such binary numbers fall into one of three types: *bytes*, *integers* or *fixed-point numbers*, and *floating-point numbers*. On most computers, the programmer is able to choose one of these three types of binary numbers. Each binary number is stored in a fixed amount of space with a fixed range of values and a fixed (finite) precision. Because of this, arithmetic on digital computers cannot, in general, be performed within the real number system as in pure mathematics.

2.3.1   Bytes

As noted at the outset, all data in digital computers is stored as bits. Computers typically work with combinations of eight bits called *bytes*. Since one bit can represent at most 2^1 values, it follows that a byte can represent the $2^8 = 256$ distinct values 0 to 255.

The simplest mapping of a decimal number to a binary (base 2) number involves *unsigned bytes*; i.e., all numbers are represented without regard for their sign. Some examples are given in Table 2.1.

Unsigned bytes cannot, however, be used to represent negative numbers. To do this, the 256 distinct values that can be stored by a byte are stored with the first bit serving as a "sign bit." Then, for all negative numbers this bit can be set to 1; for positive numbers it would be set to 0 (this convention can obviously be reversed). The remaining seven bits in the byte are then used to represent $2^7 = 128$ distinct values (including 0). Thus, the so-called *signed bytes* typically range in magnitude from -128 to 127. Some binary representations of positive and negative numbers are given in Table 2.1.

Remark

1. Most users have, at one time or another, exceeded their disk quota on a given computer. Disk quotas are typically represented in either *kilobytes* (K bytes) or *megabytes* (M bytes). The former is defined as $2^{10} = 1024$ bytes; the latter is defined as $2^{20} = 1,048,576$ bytes or 1024 kilobytes. Thus a disk quota (meager as it may be) of 5 megabytes equals 5,120 kilobytes, 5,242,880 bytes, or 41,943,040 bits.

2.3.2 Integers (Fixed–Point Numbers)

Since unsigned bytes can store no number larger than 255, they place rather severe restrictions on mathematical manipulations. These restrictions are overcome by combining bytes to form larger numbers.

The simplest approach is to combine two bytes. This gives a number with 2^{16} or 65,536 possible values (0 to 65,535). This two-byte (16 bit) number is commonly referred to as an *unsigned integer* or an *unsigned short integer*. To represent even larger numbers, four bytes (32 bits) are combined to form an unsigned long integer; such a number has 2^{32} possible values (0 to 4,294,967,295).

In order to handle negative numbers, the concept of a signed byte is extended to integers. In the case of a two-byte (16 bit) or *short* integer, the presence of a sign bit means that 2^{15} or 32,768 numbers can thus be represented. Extending the discussion to a signed four-byte (32 bit) or *long* integer, it follows that 2^{31} or 2,147,483,648 different numbers can be represented. The actual ranges of numbers associated with signed integers are summarized in Table 2.2. When performing calculations with integers, the analyst must always be aware of the maximum range of the integer type (signed or unsigned) employed. Storage of numbers outside this range will always yield erroneous, or at best unpredictable, results.

Concerning the topic of numerical precision, it is important to realize how and what information is lost when integers are used. If a decimal number is stored as an integer, the fractional part of the number is discarded; however, the non-fractional portion of the number stored is *exact*. Before being stored as an integer, a decimal number may also be *rounded* up to the next greatest integer. Thus, the number 73.25 would be stored as the integer 73; the number 98.67 would be typically stored as 99. Assuming that the result of an operation is not outside the range of a specific integer type, arithmetic between integers is exact.[1]

---

[1] A possible exception is integer division. Integer division produces an integer result; the remainder is discarded. However, the remainder can be extracted via a "modulo" command specific to the programming language used.

Table 2.2: Ranges of Signed Bytes and Integers

Type of Signed Number	Range
byte	-128 to 127
short integer (2 bytes)	-32,768 to 32,767
long integer (4 bytes)	-2,147,483,648 to 2,147,483,647

2.3.3   Floating–Point Numbers

Floating-point numbers overcome many of the shortcomings associated with integers. They get their name from the fact that the decimal point is not fixed at the outset; rather its position with respect to the first digit is indicated for each number separately. Schematically, floating-point numbers are represented internally in the form

$$N = s * m * b^e \qquad (2.1)$$

where s is a sign bit, m is an exact positive integer mantissa, e is an exact signed integer exponent, and b denotes the base of the numeric representation (usually $b = 2$, though sometimes $b = 16$). The exponent e indicates the position of the decimal point with respect to the mantissa. To facilitate the subsequent discussion, a decimal representation ($b = 10$) is adopted in lieu of the actual (binary) representation of numbers in digital computers.

Significant Digits

Any one of the digits 1, 2, 3, $\cdots$, 9 and 0 is a *significant digit* except when it is used to fix the decimal point or to fill the places of unknown or discarded digits [483]. In evaluating digits to see if they are significant, the following rules are typically used for numbers represented in conventional (as opposed to exponential) form [284]:

1. Leading zeros are not significant.

2. Following zeros that appear after the decimal point are significant.

3. Following zeros that appear before the decimal point may or may not be significant; more information is required.

These rules are illustrated in the following example.

Example 2.1: Significant Digits

- In the number 0.00137 the significant digits are 1, 3 and 7; the zeros are used merely to fix the decimal point and are therefore not significant.

- The number 0.001370 has four significant digits: 1, 3, 7 and 0.

- In the number 1024, however, all the digits including the zero are significant.

- The number 102,400 has at least four significant digits (1, 0, 2, and 4); depending on the situation, it may have five or six.

- Any ambiguity related to significant digits can be removed by writing a number in exponential form. For the above examples this would give 1.37 x 10^{-3}, 1.370 x 10^{-3}, 1.024 x 10^3, and 1.024 x 10^5 (or 1.0240 x 10^5 or 1.02400 x 10^5), respectively.

True, Absolute and Relative Errors

The *true error* of a number, measurement, or calculation is the numerical difference between the true value of the quantity (N) and its approximation ($\hat{N}$) as given, or obtained by measurement or calculation [483]. The *absolute error* is equal to the absolute value of the true error.

The *relative error* is the absolute error divided by the true value of the quantity. The *percentage error* is 100 times the relative error.

Example 2.2: Absolute and Relative Error

- Let $N = 0.10$ and $\hat{N} = 0.11$. The absolute error is

$$\left| N - \hat{N} \right| = 10^{-2} \tag{2.2}$$

and the relative error is

$$\frac{\left| N - \hat{N} \right|}{|N|} = 10^{-1} \tag{2.3}$$

- Next let $N = 0.10 \times 10^{-5}$ and $\hat{N} = 0.11 \times 10^{-5}$. The absolute error is 10^{-7}, and the relative error is 10^{-1}.

- Finally, let $N = 0.10 \times 10^{5}$ and $\hat{N} = 0.11 \times 10^{5}$. The absolute error is 10^{3}, and the relative error is 10^{-1}.

We conclude this brief discussion of absolute and relative error with the following remarks [483]:

Remarks

1. Since it does not involve the large or small exponents associated with absolute errors, the relative error is usually a more meaningful measure of accuracy.

2. If a floating-point number is correct to t significant digits, it is evident that its absolute error can not be greater than half a unit in the tth place.

3. It is often perceived that the accuracy of a measurement or of a computed result is indicated by the number of decimal places required to represent it. This notion is incorrect, for the accuracy of a result is indicated by the number of significant digits required to express it. The true index of the accuracy of a measurement or of a calculation is the relative error.

4. The absolute error is connected with the number of decimal places, whereas the relative error is connected with the number of significant digits.

Machine Numbers

Suppose the number $N \neq 0$ has the *exact* decimal representation

$$N = \pm (.d_1 d_2 \ \cdots) \times 10^q \tag{2.4}$$

where q is an integer and the $d_1 d_2 \ \cdots$ are digits. If d_1 is different from zero, the decimal floating-point representation is said to be *normalized*.

The number of significant digits (t), together with the base of the numeric representation b (in this case 10) and the exponent e of the floating-point number determine the subset of real numbers that can be represented exactly within a given computer [512]. The elements of this subset are called the *machine numbers*. On a given computer, this subset is finite.

The normalized t-digit machine number representation of the floating-point number N is obtained by terminating the mantissa of N at t decimal digits to give [263]

$$fl(N) = \pm (\delta_1 \delta_2 \cdots \delta_t) \times 10^q \tag{2.5}$$

where $\delta_1 \neq 0$ and $\delta_1 \delta_2 \cdots \delta_t$ are floating-point digits. The number $(\delta_1 \cdots \delta_t)$ is the t-digit representation of the true mantissa.

Machine number representations of floating-point numbers place the following restrictions on the exponent q

$$-L \leq q \leq H \tag{2.6}$$

for some large positive integers L and H [263]. If the number $N \neq 0$ has an exponent outside of this range, it cannot be represented in the form of equation (2.5). If, during the course of a calculation a machine number has an exponent $q > h$, the condition is termed *exponent overflow*. If $q < L$, *exponent underflow* occurs. Digital computers treat exponent overflow and underflow as irregularities in the calculation process; the results of this process are usually meaningless [512]. This brings to light the important observation that arithmetic operations on floating-point numbers do not necessarily produce machine numbers.

There are two ways in which the floating-point digits δ_j appearing in machine numbers are obtained from the exact digits d_j. The first consists of simply ignoring or *chopping* the digits $d_{t+1}, d_{t+2}, \cdots$ from the mantissa of the exact floating-point number. The resulting machine number is thus

$$fl(N) = chop(N) = \pm (\delta_1 \delta_2 \cdots \delta_t) \times 10^q \tag{2.7}$$

where $\delta_j = d_j$; $j = 1, 2, \cdots, t$.

In the second, and often preferable approach, the number $5 \times 10^{q-(t+1)}$ is added to N and this number is then "chopped." The resulting machine number is thus

$$fl(N) = chop \left(\pm (\delta_1 \delta_2 \cdots \delta_t \delta_{t+1} + 0.50) \times 10^q \right) \tag{2.8}$$

This approach is commonly referred to as *rounding* [263]. Rounding is typically preferred to chopping, as the latter results in the loss of significance of the final digit half the time and will tend to double the roundoff error. Furthermore, there is much greater danger of a significant non-random accumulation of roundoff error when chopping [284].

The error associated with the t-digit machine representation of the floating-point number $N \neq 0$ is bounded by [263]

$$\frac{|N - fl(N)|}{|N|} \leq 5 \times 10^{-t}p \qquad (2.9)$$

where $p = 1$ for rounding and $p = 2$ for chopping. Provided that 10^{-t} is replaced by 2^{-t}, equation (2.9) also holds for binary numbers.

Example 2.3: Chopping and Rounding of Floating-Point Numbers

Consider the real number 1.31449. Its exact, normalized decimal representation is

$$N = .131449 \times 10^1 \qquad (2.10)$$

For $t = 5$ *chopping* gives the machine number

$$fl(N) = chop(N) = .13144 \times 10^1 \qquad (2.11)$$

Rounding the same number gives

$$fl(N) = chop(.131449 \times 10^1 + 5 \times 10^{-6})$$

$$= chop(.131454 \times 10^1) = .13145 \times 10^1 \qquad (2.12)$$

The relative error associated with the "chopped" representation is

$$\frac{\left|.131449 \times 10^1 - .13144 \times 10^1\right|}{|.131449 \times 10^1|} = 6.8468 \times 10^{-5} \qquad (2.13)$$

which is less than $(5 \times 10^{-5}) * 2 = 1.0 \times 10^{-4}$.

For the "rounded" representation, the relative error is

$$\frac{\left|.131449 \,\text{x}\, 10^1 - .13145 \,\text{x}\, 10^1\right|}{\left|.131449 \,\text{x}\, 10^1\right|} = 7.6075 \,\text{x}\, 10^{-6} \qquad (2.14)$$

which is less than $5 \,\text{x}\, 10^{-5}$.

Both of the above relative errors are consistent with the error bounds given by equation (2.9).

Machine Accuracy

The smallest (in magnitude) floating-point number which, when added to the number 1.0, produces a floating-point result different from 1.0, is referred to as the machine accuracy *eps*. The machine accuracy is an estimate that is usually given for a specific computer. In light of the error bounds given by equation (2.9), it follows that if a particular computer allows t significant digits, then $eps \approx 5 \,\text{x}\, 10^{-t}$ or 2^{-t} for decimal and binary representations, respectively. The approximation $\hat{N}$ is thus related to the exact value N of a number by

$$\hat{N} \approx N(1 + eps) \qquad (2.15)$$

It is important to note that *eps* is not the smallest floating-point number that can be represented on a computer. That number depends upon the number of bits available for the exponent. Instead, *eps* depends upon how many bits comprise the mantissa [443].

2.3.4 Roundoff Error

All arithmetic operations involving floating-point numbers whose representation is restricted to a finite number of digits introduce an additional fractional error of at least *eps* in magnitude [443]. This type of error is referred to as *roundoff error*. Since the magnitude of *eps* is machine dependent, it follows that the roundoff error will likewise depend upon the specific digital computer used.

The estimation of roundoff errors is a non-trivial problem; no ideal solution has yet been found [565, 284]. With increasing numbers of arithmetic operations, roundoff errors typically accumulate [443]. Roundoff errors that are independently distributed in a statistical sense are usually not serious; however, if they become correlated, the effect on computations can be more severe [284]. If, in the course of obtaining a calculated value, P arithmetic

operations are performed, a lower bound on the total roundoff error would
be on the order of $\sqrt{P} * (eps)$. This assumes the roundoff errors enter
the computations randomly up or down. In reality, the above optimistic
estimate may not be realized for two reasons, namely [443]:

1. Regularities in the calculations performed or peculiarities of the particular computer cause the roundoff errors to accumulate preferen-
   tially in one direction. In this case the total roundoff error will be on
   the order of $P * (eps)$.

2. Some "especially unfavorable" occurrences increase the roundoff error
   of single operations. Typically these can be traced to the subtraction of two very nearly equal numbers, giving a result whose only
   significant bits are those low-order ones in which the operands differ.

In general, there is little an analyst can do about roundoff error other
than to choose stable algorithms that do not amplify it unnecessarily. A
comprehensive treatment of roundoff error in floating-point computations is
given in [564, 511]. Algorithmic stability is discussed further in Section 2.5.

Floating–Point Arithmetic

Even if floating-point numbers could be represented exactly, arithmetic
operations involving them will not be exact due to roundoff errors. For
example when two floating-point numbers are added, the mantissa of the
smaller (in magnitude) number is first right-shifted (divided by two and
simultaneously increasing its exponent) until the two operands have the
same exponent. Lower-order (least significant) bits in the smaller number
are lost by this shifting. If the two numbers differ too greatly in magnitude,
then the smaller one is effectively replaced by zero, since it is right-shifted
to oblivion [443].

To better understand the ramifications of floating-point arithmetic, define a set of substitute floating-point arithmetic operators for addition $\oplus$,
for subtraction $\ominus$, for multiplication $\otimes$, and for division $\oslash$. If x and y represent the *exact* values of floating-point numbers, the most basic floating-
point arithmetic operations involving x and y give [406]

$$x \oplus y = (x + y)(1 + \varepsilon_1) \tag{2.16}$$

$$x \ominus y = (x - y)(1 + \varepsilon_2) \tag{2.17}$$

$$x \otimes y = (x * y)(1 + \varepsilon_3) \tag{2.18}$$

$$x \oslash y = (x/y)(1 + \varepsilon_4) \tag{2.19}$$

where ε_1 to ε_4 represent roundoff errors with $|\varepsilon_i| \leq eps$; $i = 1, 2, 3, 4$.

The preceding discussion of floating-point arithmetic has, by design, been rather terse. Further details pertaining to this subject can be found in [406].

Example 2.4: Roundoff Error in Floating-Point Addition

We desire to investigate the roundoff error associated with the floating-point addition. As such, consider the following double sum

$$z \oplus (x \oplus y) = [z + (x \oplus y)] (1 + \varepsilon_2) = [z + (x + y)(1 + \varepsilon_1)] (1 + \varepsilon_2)$$
$$= (x + y + z) \left[1 + \frac{(x+y)\varepsilon_1}{x+y+z} \right] + (x + y)\varepsilon_1\varepsilon_2 \tag{2.20}$$

where x, y and z represent the *exact* values of floating-point numbers. The term involving the product $\varepsilon_1\varepsilon_2$ is extremely small and is thus neglected. The "true" error is thus

$$|z \oplus (x \oplus y) - (x + y + z)| = \left| (x + y + z) \left[\frac{(x+y)\varepsilon_1}{x+y+z} + \varepsilon_2 \right] \right|$$
$$= |(x + y)\varepsilon_1 + (x + y + z)\varepsilon_2| \tag{2.21}$$

To better interpret this error measure, assume that $x \approx -z$ (though x is not necessarily small in magnitude) and that y is arbitrary. It follows that the general expression for the error becomes

$$|z \oplus (x \oplus y) - (x + y + z)| = |(x + y)\varepsilon_1 + y\varepsilon_2| \tag{2.22}$$

Thus, depending on the particular magnitudes of x and y, the error could be quite large indeed.

In general, the subtraction of floating-point numbers of similar magnitudes can lead to a loss of significant digits. For example assume that $t = 6$, $x = 0.13456728519$ and $y = 0.134554928367$. Then $x \oplus y = 0.000013$.

The loss of significant digits is obvious. A similar loss of significant digits is possible when adding or subtracting a number that is several orders of magnitude smaller or larger than the other number involved in the operation. Taking this to an extreme, if a number is very small, it will have no effect on the floating-point operations of addition and/or subtraction. For example in base ten representation, let

$$|y| < \frac{eps}{10}|x| \tag{2.23}$$

Then $x \oplus y = x$. In general, $(x \oplus y) \oplus z \neq x \oplus (y \oplus z)$. In summary, floating-point operations do not always satisfy the well known arithmetic properties such as associativity or distributivity.

2.4   Approximation Error

There exists a second general category of error that differs in nature from roundoff error. This error is referred to as *approximation error* [512]. Approximation error depends upon the specific algorithms used in a numerical analysis and is independent of the particular digital computer used. It includes the so-called *truncation error* and *discretization error*.

Truncation error comes from terminating an infinite series at a finite number of terms. As such, truncation error arises when using: 1) Infinite geometric progressions, 2) Binomial expansions, 3) Taylor series, and any other series expansion.[2]

Many numerical methods are obtained by "discretizing" the original problem. For example, derivatives are approximated by difference quotients, definite integrals are approximated by finite sums, polynomials are used to approximate non-polynomials, etc. The difference between the exact solution of the discrete analogue and the exact solution of the original problem is called the *discretization error*.[3] A major problem of numerical analysis is to estimate such error, usually in terms of the data of the original problem together with parameters of the discretization [421].

Due to approximation error, numerical methods will typically not give the exact solution of a problem even if the computations were carried out on a hypothetical computer with an infinitely accurate representation of

---

[2] Roundoff error can also be viewed as a kind of truncation error, arising not from truncation of an infinite series, but from truncating the number system.

[3] Some authors extend the classification truncation error to include discretization error as well. Herein, both truncation error and discretization error are considered to be different subclasses of approximation error.

floating-point numbers and no roundoff error. Instead, they will give the solution of another, simpler problem that approximates the original one.

Unlike roundoff error, approximation error is entirely under the analyst's control. Indeed, the field of numerical analysis can be viewed primarily as consisting of the study of algorithms and techniques that minimize such error [443].

Roundoff and approximation error can be either positive or negative. As such, they may partially offset each other [284]. Typically, truncation error and roundoff error do not strongly interact with each other [512]. Thus, a calculation associated with a specific algorithm can be viewed as first having the truncation error that would exist if it was executed on a computer with infinite precision. To this is "added" the roundoff error associated with the number of arithmetic operations performed.

Example 2.5: Error in Repetitive Floating-Point Operations

In an effort to quantitatively demonstrate the effect of roundoff error, consider the following bit of Fortran 90 code.

```
integer i
real f
double precision ff

f   = 2.0e0   ;   ff = 2.0d0

do i=1,20
    f = SQRT (f)   ;   ff = SQRT (ff)
end do

do i=1,20
    f = f*f   ;   ff = ff*ff
end do
```

Clearly the expected result following the twenty square root and twenty squaring operations is the original value of 2.0. When compiled and executed on various computers, the above bit of code produced the results shown in Table 2.3.

The failure to return the original value of 2.0 is due primarily to roundoff error, though some truncation error is likely associated with the intrinsic square root function.

It is timely to note that the exact value, to the decimal places printed, was obtained only when using *quadruple* precision on a Cray J90 computer.

Table 2.3: Summary of Results

Computer	Single Precision	Double Precision
exact	0.200000000000e+01	0.200000000000d+01
Macintosh	0.186813199520E+01	0.200000000016E+01
PC compatible	0.186813200000E+01	0.200000000016E+01
SUN Sparc 1	0.186813199520E+01	0.200000000016E+01
IBM RS6000	0.186813199520E+01	0.200000000016E+01
SGI Power Challenge	0.186813199520E+01	0.200000000016E+01
SGI Origin 2000	0.186813199520E+01	0.200000000016E+01

2.5 Algorithmic Stability and Error Growth

Underlying roundoff and approximation error is the general notion of *algorithmic stability*. First, there is the issue of the stability of the solution of the problem to be solved: generally, a solution is said to be stable if "small" changes in the data of the problem produce only "small" changes in the solution [421]. If large changes in the solution occur, then it is unstable or *ill-conditioned*. The problem of ill-conditioning, as applied to the solution of systems of linear algebraic equations, is examined in Appendix E.

Secondly, there is the stability of the discrete analogue. It may be, for example, that the solution of the original problem is stable, but that of the discrete analogue is unstable.

Closely related to the notion of algorithmic stability is the topic of *error growth*. In general, one of the main objectives in numerical analysis is to use algorithms that minimize the roundoff and approximation error for an increasing number of arithmetic operations. Error that grows proportionally, or slower, is considered "normal" and often unavoidable. Error growth that is more rapid is usually disastrous [284]. The notion of error growth is quantified through the following definition.

- **Definition**

  Let E_n represent the roundoff error generated after n subsequent operations. Then, if

$$|E_n| \leq C(n)(eps) \qquad (2.24)$$

where C is a constant independent of n, the error is said to grow linearly. If, for some $m > 1$,

$$|E_n| \leq C(m^n)(eps) \qquad (2.25)$$

then the error growth is exponential, an undesirable condition.

Remarks

1. An algorithm with linear error growth is called stable ("small errors remain small"); an algorithm with exponential growth is called unstable.

2. If roundoff growth is exponential, but $m < 1$, the situation is actually desirable.

2.6 Concluding Remarks

To underscore the importance of the subject matter presented in this chapter, it is timely to recount the case of the European Ariane 5 launcher. On June 4, 1996, 40 seconds after takeoff, the maiden flight of this rocket ended in failure. Media reports indicated that the cost of the launcher was a half billion dollars [267].

Shortly after this disaster, the CNES (French National Center for Space Studies) and the European Space Agency appointed an international inquiry board, composed of respected experts from major European countries. The board concluded [317] that the crash of the Ariane 5 was the result of a software error. This error came from a piece of software that was not needed during the crash, but had to do with the Inertial Reference System (SRI) that measures the attitude of the launcher and its movements in space. Prior to lift-off, certain computations are performed to align the SRI. Normally these should be stopped nine seconds before lift-off. However, in the unlikely event of a hold in the countdown, the computation continues 50 seconds after the start of the launch mode. Although following lift-off this computation is useless, in the case of the Ariane 5 it caused a software exception.

The software exception was caused during the conversion from a 64-bit floating-point number to a 16-bit signed integer. The floating-point number that was converted had a value *greater* than what could be represented by a 16-bit signed integer. This resulted in an operand error. Since there was no explicit handler to catch the exception, this operation crashed the entire software, hence the on-board computers, and thus the Ariane 5 launcher.

Further details pertaining to the Ariane 5 failure are given in [317]. An insightful discussion of the software testing lessons learned from the failure is found in [267].

2.7   Exercises

2.1

To gain further insight into roundoff error, write a simple computer program to compute (in both *single* and *double* precision) the quantities

$$D = (A + B) + C$$

and

$$E = A + (B + C)$$

where

$$A = 0.23371258e + 01$$

$$B = 0.33678429e + 06$$

and

$$C = -0.33677811e + 06$$

Print the results to *twelve* significant digits. Using relative error as the measure, compare D and E to the exact result $A + B + C$. Submit a listing of the simple program written, along with a print out of the results obtained. Make sure to indicate which computer was used to perform the calculations. For the computer used, which of the quantities D or E approximates the exact sum best? Discuss your results.

Chapter 3

The Finite Difference Method

3.1 Introductory Remarks

This chapter is devoted to the finite difference method, which is classified as a *discrete* approximate solution technique. Rather than providing a continuous solution for the primary dependent variables defined throughout the domain Ω, discrete solution techniques yield approximations only at a specific number of locations. When necessary, intermediate values of the primary dependent variables and their derivatives can be computed from the discrete approximation by suitable interpolation techniques.

The most commonly used discrete solution technique is the finite difference method. The finite difference method makes use of the Calculus of Finite Differences, in which derivatives are replaced by suitable difference formulas involving simple algebraic operations.

3.2 Historical Note

Finite differences have long been used by engineers and mathematicians to approximately solve differential equations associated with boundary value problems. Although the exact origin of this method of approximation is debatable, certain historical observations are noteworthy. The collection of references presented herein is not meant to be exhaustive, but only representative. A far more detailed discussion of the development of the finite difference method prior to 1949 is given by Higgins [239].

The origin of the Calculus of Finite Differences has been ascribed to the work of Taylor [531] and Stirling [510] in the early 18th century [305, 271], though not in the context of a general theory [281]. The first formal treatise on this calculus was presented by Euler [153]. This was followed by the work of Lacroix [301] and Boole [88], the latter being one of the earliest books on the Calculus of Finite Differences that was well–suited for both students and instructors [271]. The notion of finite difference approximations to differential operators appears [506] to have been first systematized by Sheppard [493]. Beginning in the late 1890s, other books devoted to the Calculus of Finite Differences [373, 412, 395, 271, 167] and to related topics [507] were published. Texts devoted to the solution of boundary value problems using finite differences were, however, lacking [239].

In engineering calculations, finite differences have been used since the latter half of the 19th century. In the context of one-dimensional problems, Mohr's use of the graphical string polygon method for computing deflections of beams [399] is cited as an early example [125]. For the analysis of two-dimensional systems, the first application of the finite difference method was by Runge [474], who solved torsional problems in elasticity. Additional early applications of the method to problems in elasticity are discussed by Timoshenko and Goodier [538].

In an effort to extend and improve the basic finite difference method, various iteration techniques were developed in order to more efficiently solve the system of simultaneous equations generated by the method. The work of Richardson [463] represents the earliest application of iteration to the solution of equilibrium equations by finite differences, though his approach was not as simple to use or easy to grasp as subsequent techniques [239]. A different iteration approach, which received rather wide acceptance, was subsequently proposed by Liebmann [316]. The convergence of this approach, as applied to harmonic and bi-harmonic equations was discussed by Courant [120]. Courant also showed that if the original continuous system is positive definite, then the set of finite difference equations will also be positive definite and that iteration by single steps must always converge. However, for systems of appreciable size, the convergence was generally so slow as to preclude hand computation [125]. Useful iteration techniques were subsequently presented by Mikeladze [389] and Collatz [111].

Investigations of the convergence and stability of finite difference approximations were initiated in the late 1930s. Initial emphasis was on difference equations associated with second-order partial differential equations [333, 121, 432, 307]. A formal proof of convergence for the finite difference method was given by Kantorovich and Krylov [278].

Various techniques were proposed for accelerating the rate of convergence of iterative solutions [239]. Some of these techniques were well-suited for automatic programming [174], while others were better suited for hand computation. In the latter category, the so-called *relaxation* methods have been the most thoroughly developed. According to Sokolnikoff [504], the basic relaxation method appears to have been originated by Gauss [190], who used it and "many of the standard tricks" associated with the method. Convergence of the basic relaxation method when applied to linear systems was rigorously proved by Seidel [489]. Investigations of necessary and sufficient conditions ensuring that the approximations resulting from the method converge to the exact solutions were postulated by Black and Southwell [81] and by Temple [534]. In the 1930s, Southwell "rediscovered Gauss' method and named it" [504], apparently after a physical analogy used in the early development of the method [239]. He extensively developed relaxation methods and applied them to a wide variety of physics and engineering problems [505, 506].

The commercial introduction of high-speed digital computers in the 1950s lessened the limitations on the size of problems that could reasonably be solved using the finite difference method. As the method grew in popularity, studies of the errors associated with its use were initiated [96, 130, 125, 220]. Beginning in the mid-1950s [394, 311, 166], the number of books devoted primarily to the finite difference method its computer implementation, and application to physical problems such as fluid flow, heat transfer, electromagnetics and solid mechanics increased steadily. Somewhat more limited discussions of the method have also been included in numerous numerical analysis textbooks.

3.3   General Steps

The following steps are commonly associated with an approximate solution using the finite difference method.

1. In the governing differential equations $L(\phi) - f = 0$, derivatives are replaced by suitable difference expressions[1] involving only *algebraic operations*. This process, which uses the Calculus of Finite Differences, transforms the governing equations into the form $\hat{L}(\phi) - f = 0$. Here $\hat{L}$ denotes an approximate operator associated with the difference formulas used. These will be linear or non-linear depending upon whether the original operator L is linear or non-linear [125].

---

[1] A useful sampling of such expressions is given in Appendix A.

2. The domain Ω is divided into a set of *discrete* grid points.[2] It is at these grid points that the approximate values of the primary dependent variables will be solved for. The *minimum* number of grid points associated with a given independent variable is typically dictated by the finite difference formula chosen. The finite difference (discrete) equations are typically written in matrix form. Since the difference formulas in a given equation involve only a small number of unknowns, the coefficient matrix is typically sparse, possibly banded, and, in certain cases, symmetric.

3. The boundary conditions are next applied. Essential (Dirichlet) boundary conditions involve no further approximation. Since they require further differentiation, natural (Neumann) boundary conditions introduce further approximation through use of additional difference expressions.

4. The modified matrix equations are solved for the unknown values of the primary dependent variables. Taking advantage of the structure typically possessed by the coefficient matrix (see step 2), this solution process can be made to be computationally efficient.

5. Suitable difference formulas are used to approximate the secondary dependent variables. The fact that additional difference formulas are used implies that further approximation is introduced into the solution.

The application of the finite difference method to ordinary differential equations is discussed next.

3.4   Ordinary Differential Equations

When using the finite difference method to obtain approximate solutions to boundary value problems involving ordinary differential equations, the domain Ω is one-dimensional. Denoting the independent variable by x, a grid of $(P + 1)$ discrete, typically equally spaced points with coordinates x_j, $(j = 0, 1, \cdots, P)$ is specified on the range $0 \leq x \leq L$. This gives P equally-spaced segments, each of length $\Delta x = x_j - x_{j-1} = h$. The distance h is commonly referred to as the "grid spacing."

---

[2]In this chapter, for a given independent variable, these points are assumed to be *equally spaced*. For two-dimensional solution domains this leads to *rectangular* grids or "nets." It is, however, possible to use non-rectangular grids such as triangular and hexagonal lattices [506] and polynomial and curvilinear nets [426].

The solution process is best understood by studying the following examples.

Example 3.1: Second-Order Differential Equation

Consider the differential equation

$$\frac{d^2\phi}{dx^2} + \phi + x^2 = 0 \tag{3.1}$$

The solution domain Ω is the set of points x such that $0 < x < 1$. The boundary Γ is simply the points $x = 0$ and $x = 1$. The boundary conditions at both points are homogeneous and of the essential type, viz.,

$$\phi(0) = \phi(1) = 0 \tag{3.2}$$

The exact solutions for ϕ and its first derivative are

$$\phi = 2 - 2\cos x - \frac{1 - 2\cos(1)}{\sin(1)}\sin x - x^2 \tag{3.3}$$

$$\frac{d\phi}{dx} = 2\sin x - \frac{1 - 2\cos(1)}{\sin(1)}\cos x - 2x \tag{3.4}$$

We choose to approximate the second derivative of ϕ with a central difference $O(h^2)$, viz.,

$$\frac{d^2 f}{dx^2}\bigg|_{x=x_j} = \frac{f_{j+1} - 2f_j + f_{j-1}}{h^2} + O(h^2) \tag{3.5}$$

where $f_{j+1} \equiv f(x)|_{x=x_j+h}$, etc. The *discrete* form of equation (3.1) is thus

$$\frac{\hat{\phi}_{j+1} - 2\hat{\phi}_j + \hat{\phi}_{j-1}}{h^2} + \hat{\phi}_j + x^2 = 0 \quad ; \quad j = 1, 2, \cdots, (P-1) \tag{3.6}$$

where the total number of grid points is equal to $(P + 1)$. The associated boundary conditions, which are imposed *without approximation*, are

$$\hat{\phi}_0 = \hat{\phi}_P = 0 \tag{3.7}$$

We next discretize the domain. From equation (3.6) it follows that at least three grid points must be used; i.e., that $P = 2$. As a first try, we indeed assume such a coarse grid, which is shown in Figure 3.1(a). The grid spacing h is thus 0.50, and the total number of unknowns is three. The first of the requisite three equations comes from equation (3.6), evaluated for $j = 1$

$$\frac{\hat{\phi}_2 - 2\hat{\phi}_1 + \hat{\phi}_0}{h^2} + \hat{\phi}_1 + (x_1)^2 = 0 \tag{3.8}$$

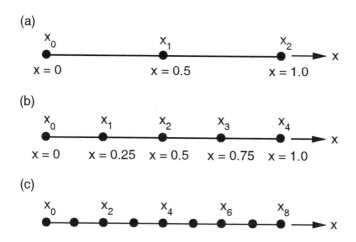

Figure 3.1: Finite Difference Grids for: (a) $P = 2$, (b) $P = 4$, and (c) $P = 8$

Substituting for h and x_1, and simplifying gives

$$4\hat{\phi}_2 - 7\hat{\phi}_1 + 4\hat{\phi}_0 = -0.25 \tag{3.9}$$

To this we add the two boundary conditions described in equation (3.7). Imposing these on equation (3.9) leads to the following solution:

$$\hat{\phi}_1 = \frac{1}{28} = 0.03571 \tag{3.10}$$

The exact solution at $x = 0.50$ is 0.04076, indicating that the relative error is about 12.4 percent. The only other grid points at which the solution is known are the ends of the domain, where the approximate solution is identical to the exact one as a result of enforcing the essential boundary conditions.

We next consider the secondary dependent variable, the slope $d\phi/dx$. To approximate the slope at $x = x_0$, a *forward* difference $O(h^2)$ is used. Accounting for the boundary conditions gives

$$\frac{d\phi}{dx}\bigg|_{x=x_0} = \frac{-\hat{\phi}_2 + 4\hat{\phi}_1 - 3\hat{\phi}_0}{2h} = \frac{1}{7} = 0.1429 \qquad (3.11)$$

The relative error associated with this approximation is 49.1 percent. To approximate the slope at $x = x_1$, the following *central* difference $O(h^2)$ is used:

$$\frac{d\phi}{dx}\bigg|_{x=x_1} = \frac{\hat{\phi}_2 - \hat{\phi}_0}{2h} = 0 \qquad (3.12)$$

The relative error is thus 100 percent. Finally, at $x = x_2$ a *backward* difference $O(h^2)$ is used, giving

$$\frac{d\phi}{dx}\bigg|_{x=x_2} = \frac{3\hat{\phi}_2 - 4\hat{\phi}_1 + \hat{\phi}_0}{2h} = -\frac{1}{7} = -0.1429 \qquad (3.13)$$

The relative error associated with this approximation is 46.2 percent.

The additional approximation related to the computation of secondary dependent variables is clearly quantified by the increase in the relative error. This trend is not restricted to the finite difference method.

In Figures 3.2 and 3.3, the finite difference approximations for ϕ and its derivative, respectively, are graphically compared to the exact values. Since the finite difference approximation is discrete, the approximate solution should, strictly speaking, be represented using discrete symbols and not continuous curves.

Since we typically do not know the exact solution, the approximate solution for $P = 2$, by itself, cannot be used to ascertain convergence. To begin to investigate the convergence of the finite difference approximation, we must *refine* the grid by halving the spacing to $h = 0.25$. The resulting grid is shown in Figure 3.1(b). The equations for the five unknowns $\hat{\phi}_0$ to $\hat{\phi}_4$ are

$$\hat{\phi}_0 = 0 \qquad (3.14)$$

$$\frac{\hat{\phi}_2 - 2\hat{\phi}_1 + \hat{\phi}_0}{h^2} + \hat{\phi}_1 + (x_1)^2 = 0 \qquad (j = 1) \tag{3.15}$$

$$\frac{\hat{\phi}_3 - 2\hat{\phi}_2 + \hat{\phi}_1}{h^2} + \hat{\phi}_2 + (x_2)^2 = 0 \qquad (j = 2) \tag{3.16}$$

$$\frac{\hat{\phi}_4 - 2\hat{\phi}_3 + \hat{\phi}_2}{h^2} + \hat{\phi}_3 + (x_3)^2 = 0 \qquad (j = 3) \tag{3.17}$$

$$\hat{\phi}_4 = 0 \tag{3.18}$$

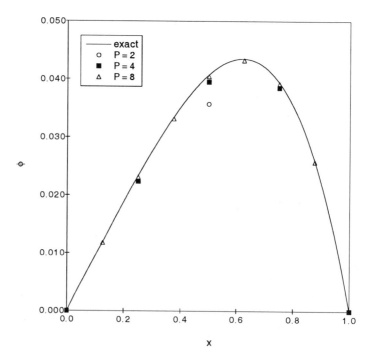

Figure 3.2: Comparison of Finite Difference Approximation and Exact Solution for ϕ

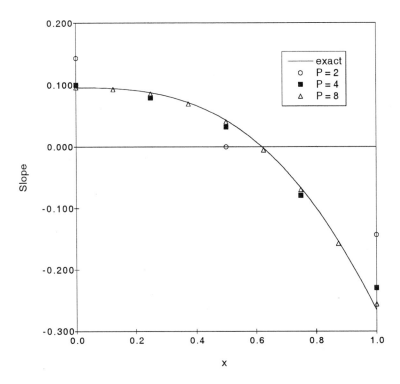

Figure 3.3: Comparison of Finite Difference Approximation and Exact Solution for $d\phi/dx$

For consistency with the previous analysis, central differences $O(h^2)$ have again been used. Substituting for h and the grid coordinates leads to the following system of equations:

$$
\begin{bmatrix}
1 & 0 & 0 & 0 & 0 \\
-16 & 31 & -16 & 0 & 0 \\
0 & -16 & 31 & -16 & 0 \\
0 & 0 & -16 & 31 & -16 \\
0 & 0 & 0 & 0 & 1
\end{bmatrix}
\begin{Bmatrix}
\hat{\phi}_0 \\
\hat{\phi}_1 \\
\hat{\phi}_2 \\
\hat{\phi}_3 \\
\hat{\phi}_4
\end{Bmatrix}
= \frac{1}{16}
\begin{Bmatrix}
0 \\
1 \\
4 \\
9 \\
0
\end{Bmatrix}
\tag{3.19}
$$

The solution to equation (3.19) and the associated relative error are listed in the second and third columns, respectively, of Table 3.1. In Figure 3.2 the finite difference approximation for ϕ is graphically compared to the exact solution. The former is seen to be converging to the latter.

Table 3.1: Summary of Results for Approximation with $P = 4$

x	$\hat{\phi}$	Relative Error (%)	$d\hat{\phi}/dx$	Relative Error (%)
0.000	0.0	0.0	0.1003	4.7
0.250	0.02242	4.1	0.0791	9.7
0.500	0.03953	3.0	0.0322	25.0
0.750	0.03854	2.2	-0.0791	18.7
1.000	0.0	0.0	-0.2293	13.6

The secondary dependent variable (slope) is next approximated. At $x = x_0$, we must use a forward difference $O(h^2)$. At the next three grid points, central differences $O(h^2)$ are used. Finally, at $x = x_4$, a backward difference $O(h^2)$ is used. The approximate slopes, along with the associated percent relative error, are listed in the fourth and fifth columns, respectively, of Table 3.1.

In Figure 3.3 the finite difference approximation for $d\phi/dx$ is graphically compared to the exact solution. We reiterate the previous observation concerning the computation of secondary dependent variables using the finite difference method. That is, since they require the use of additional difference formulas applied to already approximate results, secondary dependent variables are less accurate than the primary dependent variables.

To investigate the convergence of the finite difference approximation further, we *refine* the grid by again halving the spacing to $h = 0.125$. The result is the grid shown in Figure 3.1(c). The applicable equations for the nine unknowns $\hat{\phi}_0$ to $\hat{\phi}_8$ are: $\hat{\phi}_0 = 0$, equation (3.6) evaluated for $j = 1, 2, \cdots, 7$, and $\hat{\phi}_8 = 0$. The solution to this set of nine equations and the associated percent relative error are listed in the second and third columns, respectively, of Table 3.2.

The secondary dependent variable, the slope, is next approximated. Similar to the solutions for $P = 2$ and $P = 4$, at $x = x_0$ we must use a forward difference $O(h^2)$. At the next seven grid points, central differences $O(h^2)$ are used. Finally, at $x = x_8$, a backward difference $O(h^2)$ is used. The approximate slopes, along with the associated percent relative error, are listed in the fourth and fifth columns, respectively, of Table 3.2.

In Figures 3.2 and 3.3, the three finite difference approximations for ϕ and its derivative, respectively, are graphically compared to the exact values. The convergence of the approximate solution is evident.

Table 3.2: Summary of Results for Approximation with $P = 8$

x	$\hat{\phi}$	Relative Error (%)	$d\hat{\phi}/dx$	Relative Error (%)
0.000	0.0	0.0	0.09597	0.2
0.125	0.01178	1.2	0.09255	2.0
0.250	0.02314	1.0	0.08548	2.4
0.375	0.03315	0.9	0.06927	3.4
0.500	0.04045	0.8	0.04026	6.2
0.625	0.04322	0.6	-0.005014	136.2
0.750	0.03920	0.6	-0.06974	4.7
0.875	0.02578	0.5	-0.15680	2.1
1.000	0.0	0.0	-0.25573	3.6

Example 3.2: One-Dimensional Steady State Heat Conduction

We next consider the specific physical problem of steady state heat conduction of the rod shown in Figure 3.4. This example provides insight into the truncation error associated with finite difference approximations of derivatives.

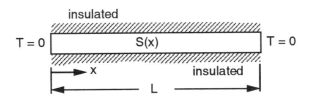

Figure 3.4: Schematic of an Insulated Rod

Assuming a constant thermal conductivity k and a linearly varying heat source $S(x) = S_o x$ where S_o is a constant, the governing differential equation for the problem is

$$k\frac{d^2T}{dx^2} + S_o x = 0 \tag{3.20}$$

Homogeneous essential boundary conditions $T(0) = T(L) = 0$ are assumed. The exact solution for this problem is

$$T = \frac{S_o}{6k} \left(L^2 - x^2\right) x \qquad (3.21)$$

The corresponding flux distribution (per unit area) is

$$q = \frac{S_o}{6} \left(3x^2 - L^2\right) \qquad (3.22)$$

Using central differences $O(h^2)$, the discrete form of equation (3.20) is

$$\frac{\hat{T}_{j+1} - 2\hat{T}_j + \hat{T}_{j-1}}{h^2} + \frac{S_o}{k} x_j = 0 \qquad (3.23)$$

where $\hat{T}$ denotes approximate values of temperature.

As a consequence of the choice of central differences, it follows that the simplest grid would contain three points. Since essential boundary conditions must be specified at the two ends of the domain, this would leave only one grid point with an unknown temperature. To slightly improve upon this discretization, the grid shown in Figure 3.5(a) with four points ($P = 3$ and $h = L/3$) is used instead.

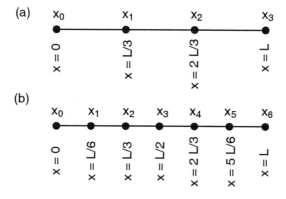

Figure 3.5: Finite Difference Grids for: (a) $P = 3$, (b) $P = 6$

Associated with this grid are four unknowns ($\hat{T}_0$ to $\hat{T}_3$). The requisite four equations come from equation (3.23), evaluated at $j = 1$ and $j = 2$, and from the two boundary conditions $\hat{T}_0 = \hat{T}_3 = 0$. Substituting the latter two equations into the former two, and writing the resulting equations in matrix form gives:

$$\frac{9}{L^2} \begin{bmatrix} 2 & -1 \\ -1 & 2 \end{bmatrix} \begin{Bmatrix} \hat{T}_1 \\ \hat{T}_2 \end{Bmatrix} = \frac{S_o L}{3k} \begin{Bmatrix} 1 \\ 2 \end{Bmatrix} \tag{3.24}$$

The solution of this set of linear equations is

$$\begin{Bmatrix} \hat{T}_1 \\ \hat{T}_2 \end{Bmatrix} = \frac{S_o L^3}{81k} \begin{Bmatrix} 4 \\ 5 \end{Bmatrix} \tag{3.25}$$

These values are *identical* to the exact solution at $x = L/3$ and $x = 2L/3$, respectively.

Approximate values for the heat flux per unit area are next computed. At $x = x_0$, we must use a *forward* difference. Furthermore, to maintain the order of difference formulas consistent with the central difference used in equation (3.23), we use a forward difference $O(h^2)$, viz.,

$$\hat{q} = \frac{-k\left(-\hat{T}_2 + 4\hat{T}_1 - 3\hat{T}_0\right)}{2h} = -\frac{33}{162} S_o L^2 \tag{3.26}$$

From equation (3.22), $q(0) = -S_o L^2/6$. Consequently, the relative error associated with this approximation is about 22 percent.

At $x = x_1$, we use the *central* difference $O(h^2)$, which leads to

$$\hat{q} = \frac{-k\left(\hat{T}_2 - \hat{T}_0\right)}{2h} = -\frac{5}{54} S_o L^2 \tag{3.27}$$

Since $q(L/3) = -S_o L^2/9$, the relative error associated with this approximation is about 16 percent.

At $x = x_2$ we likewise use the *central* difference $O(h^2)$, which leads to

$$\hat{q} = \frac{-k\left(\hat{T}_3 - \hat{T}_1\right)}{2h} = \frac{2}{27} S_o L^2 \tag{3.28}$$

Since $q(2L/3) = S_o L^2/18$, the relative error associated with this approximation is about 33 percent.

Finally, at $x = x_3$ we must use a *backward* difference $O(h^2)$, which leads to

$$\hat{q} = \frac{-k\left(3\hat{T}_2 - 4\hat{T}_1 + \hat{T}_0\right)}{2h} = \frac{8}{27}S_oL^2 \tag{3.29}$$

Since $q(L) = S_oL^2/3$, the relative error associated with this approximation is about 11 percent.

Remark

1. Although the approximate values of temperature (the primary dependent variable) are identical to the exact ones, the heat fluxes (the secondary dependent variable) are not. This trend of increased error with subsequent differentiation is true in general, and is not limited to the finite difference method.

We next investigate the effect of *refining* the finite difference grid. The new gird is shown in Figure 3.5(b). Since now $P = 7$ and $h = L/6$, the general discrete form of the governing equation (3.23) becomes

$$\frac{36\left(\hat{T}_{j+1} - 2\hat{T}_j + \hat{T}_{j-1}\right)}{L^2} + \frac{S_o}{k}x_j = 0 \quad ; \quad j = 1, 2, 3, 4, 5 \tag{3.30}$$

Specializing equation (3.30) for the respective values of j, and imposing the homogeneous essential boundary conditions at $x = 0$ and at $x = L$, leads to the following set of linear equations

$$\begin{bmatrix} 2 & -1 & 0 & 0 & 0 \\ -1 & 2 & -1 & 0 & 0 \\ 0 & -1 & 2 & -1 & 0 \\ 0 & 0 & -1 & 2 & -1 \\ 0 & 0 & 0 & -1 & 2 \end{bmatrix} \begin{Bmatrix} \hat{T}_1 \\ \hat{T}_2 \\ \hat{T}_3 \\ \hat{T}_4 \\ \hat{T}_5 \end{Bmatrix} = \frac{S_oL^3}{216k} \begin{Bmatrix} 1 \\ 2 \\ 3 \\ 4 \\ 5 \end{Bmatrix} \tag{3.31}$$

The coefficient matrix is seen to be *banded*, *sparse* and *symmetric*.[3] Solving the above system of equations gives

---

[3] Further details concerning these matrix attributes can be found in standard linear algebra textbooks such as [17, 272, 410, 514], to name a few.

$$\begin{Bmatrix} \hat{T}_1 \\ \hat{T}_2 \\ \hat{T}_3 \\ \hat{T}_4 \\ \hat{T}_5 \end{Bmatrix} = \frac{S_o L^3}{k} \begin{Bmatrix} 0.0270 \\ 0.0494 \\ 0.0625 \\ 0.0617 \\ 0.0424 \end{Bmatrix} \tag{3.32}$$

As expected, these approximate temperatures are identical to the exact ones. Using suitable forward, central, and backward differences as required, the approximate values for heat flux per unit area listed in Table 3.3 are obtained. The use of a finer grid has thus yielded more accurate flux approximations.

Since the exact solution to this problem is known — a luxury typically not afforded to the analyst, for if it were known, there would obviously be no need for an approximate solution — some insight into the errors associated with the approximation can be realized. In this example, the approximation for the primary dependent variable involved a central difference, viz.,

$$\left. \frac{d^2 T}{dx^2} \right|_{x=x_j} = \frac{\hat{T}_{j+1} - 2\hat{T}_j + \hat{T}_{j-1}}{h^2} + \frac{h^2}{12} T^{iv}(x_j) + \cdots \tag{3.33}$$

where the second term on the right-hand side of the equation represents the truncation error. Since the exact solution (equation 3.21) is cubic in x, it follows that $T^{iv} = 0$, indicating that the central difference approximation is *exact* for cubic (and lower order) functions.

Next consider the approximation for the heat flux per unit area. The forward and backward difference formulas used to approximate dT/dx have truncation errors of $T'''(x_j)h^2/3$; the central difference formulas have truncation errors that are approximately half of this value. Since the exact temperature solution is cubic in x, it follows that the truncation error associated with the calculation of heat flux will be non-zero.

Before leaving this example, it is timely to investigate the ramifications that the specification of a *natural* boundary condition would have on the solution. As such, change the boundary condition at $x = L$ to be a known value of heat flux, viz.,

$$-k \left. \frac{dT}{dx} \right|_{x=L} = \bar{q} \tag{3.34}$$

The finite difference grid shown in Figure 3.5(a) is once again used, as are central differences $O(h^2)$. The discrete form of the governing equation

Table 3.3: Approximate Heat Flux Values (per unit area)

Location (x)	$\hat{q}\ (/S_o L^2)$	Relative Error (%)
x_0	-0.1758	5.5
x_1	-0.1482	3.0
x_2	-0.1065	4.0
x_3	-0.0369	11.0
x_4	0.0603	8.5
x_5	0.1851	2.5
x_6	0.3237	2.9

is again given by equation (3.23), where $j = 1, 2$. Unlike the case with the essential boundary condition at $x = L$, the value of $\hat{T}_3$ is now *unknown*. The equation $T(L) = 0$ is thus replaced by a backward difference approximation $O(h^2)$ for the flux boundary condition at $x = L$, viz.,

$$-k \left.\frac{dT}{dx}\right|_{x=L} \approx -k \frac{\left(3\hat{T}_3 - 4\hat{T}_2 + \hat{T}_1\right)}{(2L/3)} = \bar{q} \qquad (3.35)$$

Accounting for the essential boundary condition at $x = 0$, the three required equations are thus

$$\begin{bmatrix} 2 & -1 & 0 \\ -1 & 2 & -1 \\ -1 & 4 & -3 \end{bmatrix} \begin{Bmatrix} \hat{T}_1 \\ \hat{T}_2 \\ \hat{T}_3 \end{Bmatrix} = \frac{L}{27k} \begin{Bmatrix} S_o L^2 \\ 2S_o L^2 \\ 18\bar{q} \end{Bmatrix} \qquad (3.36)$$

Remarks

1. Compared to the problem involving two essential boundary conditions, the coefficient matrix now has size (3 x 3) and has lost its symmetry

2. Whereas the essential boundary conditions are satisfied exactly, the natural boundary conditions involve additional approximation, introduced in the form of a suitable difference formula.

3. Due to the discrete nature of the finite difference method, *pointwise* error assessments have been used in this and the previous example. Although relative errors have been computed, other error measures could have likewise been used.

Example 3.3: Beam Resting on an Elastic Foundation

The equation governing the transverse deflection v of a Bernoulli-Euler beam resting on an elastic (Winkler) foundation[4] of stiffness k (units of FL^{-2}) is

$$EI\frac{d^4v}{dx^4} + kv - q = 0 \tag{3.37}$$

where EI represents the constant flexural rigidity (units of FL^2) of the beam, and q is the load per unit length (units of FL^{-1}) applied to the beam.

From basic strength of materials, we recall that in the analysis of beams, essential boundary conditions are those involving the displacement v and the slope dv/dx.

The particular case of a beam is of unit length $(L = 1)$ with $q/EI = k/EI = 1$ is analyzed using the finite difference method. The beam is assumed to be clamped at its left end and free on the right (Figure 3.6). As such, the boundary conditions at the left end of the beam are of the essential type, viz.,

$$v\big|_{x=0} = \frac{dv}{dx}\bigg|_{x=0} = 0 \tag{3.38}$$

At the right (free) end, the boundary conditions are both of the natural type, viz.,

$$\frac{d^2v}{dx^2}\bigg|_{x=1} = \frac{d^3v}{dx^3}\bigg|_{x=1} = 0 \tag{3.39}$$

The exact solution for the transverse deflection is [237]

$$v = \frac{q}{k}\left\{1 - \frac{v_1 - v_2}{(\cosh \lambda L)^2 + (\cos \lambda L)^2}\right\} \tag{3.40}$$

where

---

[4]The model of Winkler [572] assumes a continuous elastic foundation consisting of closely spaced, independent linear springs. Further details pertaining to basic elastic and viscoelastic foundation models are given by Kerr [287].

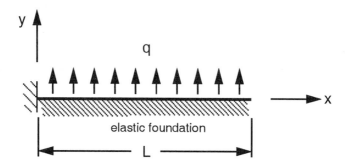

Figure 3.6: Uniformly Loaded Beam on Elastic Foundation, Clamped at Left End, Free on Right End

$$v_1 = \cosh \lambda L \left(\sin \lambda x \, \sinh \lambda x' + \cos \lambda x \, \cosh \lambda x' \right) \qquad (3.41)$$

$$v_2 = \cos \lambda L \left(\sinh \lambda x \, \sin \lambda x' - \cosh \lambda x \, \cos \lambda x' \right) \qquad (3.42)$$

Also of interest is the slope, the exact solution for which is

$$\frac{dv}{dx} = - \frac{2q\lambda}{k} \frac{\sinh \lambda x \, \cos \lambda x' \cos \lambda L - \sin \lambda x \cosh \lambda x' \cosh \lambda L}{(\cosh \lambda L)^2 + (\cos \lambda L)^2} \qquad (3.43)$$

where $x' = L - x$ and

$$\lambda \equiv \sqrt[4]{\frac{k}{4EI}} \qquad (3.44)$$

For the particular case under analysis, the governing differential equation simplifies to

$$\frac{d^4 v}{dx^4} + v - 1 = 0 \qquad (3.45)$$

Approximating the fourth derivative of the transverse displacement by a *central* difference $O(h^2)$, the discrete form of equation (3.45) is

$$\frac{\hat{v}_{j+2} - 4\hat{v}_{j+1} + 6\hat{v}_j - 4\hat{v}_{j-1} + \hat{v}_{j-2}}{h^4} + \hat{v}_j - 1 = 0 \quad ; \quad j = 2, 3, \cdots \quad (3.46)$$

The coarsest grid that can be employed thus contains five points ($P = 4$); the spacing between points is $h = 0.25$. The requisite five equations consist of equation (3.46), evaluated at $j = 2$, and the following four boundary conditions.

The homogeneous essential boundary condition involving transverse displacement at the left end of the beam is simply

$$\hat{v}\big|_{x=0} \equiv \hat{v}_0 = 0 \quad (3.47)$$

Using a *forward* difference $O(h^2)$, the slope boundary condition at the location is

$$\frac{d\hat{v}}{dx}\bigg|_{x=0} = \frac{-\hat{v}_2 + 4\hat{v}_1 - 3\hat{v}_0}{2h} = 0 \quad (3.48)$$

Finally, using *backward* differences $O(h^2)$, the remaining two boundary conditions are

$$\frac{d^2\hat{v}}{dx^2}\bigg|_{x=L} = \frac{2\hat{v}_4 - 5\hat{v}_3 + 4\hat{v}_2 - \hat{v}_1}{h^2} = 0 \quad (3.49)$$

$$\frac{d^3\hat{v}}{dx^3}\bigg|_{x=L} = \frac{5\hat{v}_4 - 18\hat{v}_3 + 24\hat{v}_2 - 14\hat{v}_1 + 3\hat{v}_0}{2h^3} = 0 \quad (3.50)$$

Writing equations (3.46) to (3.50) in matrix form gives

$$\begin{bmatrix} 1 & -4 & (6+h^4) & -4 & 1 \\ 1 & 0 & 0 & 0 & 0 \\ -3 & 4 & -1 & 0 & 0 \\ 0 & -1 & 4 & -5 & 2 \\ 3 & -14 & 24 & -18 & 5 \end{bmatrix} \begin{Bmatrix} \hat{v}_0 \\ \hat{v}_1 \\ \hat{v}_2 \\ \hat{v}_3 \\ \hat{v}_4 \end{Bmatrix} = \begin{Bmatrix} h^4 \\ 0 \\ 0 \\ 0 \\ 0 \end{Bmatrix} \quad (3.51)$$

Observe that, due to the boundary conditions, the coefficient matrix in equation (3.51) exhibits no special structure. Substituting for $h = 0.25$ and solving gives

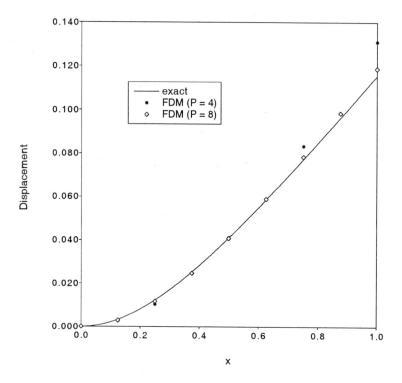

Figure 3.7: Comparison of Approximate and Exact Transverse Displacement for a Uniformly Loaded Beam Clamped at One End, Free at the Other, Resting on an Elastic Foundation

$$\begin{Bmatrix} \hat{v}_0 \\ \hat{v}_1 \\ \hat{v}_2 \\ \hat{v}_3 \\ \hat{v}_4 \end{Bmatrix} = \begin{Bmatrix} 0.0 \\ 1.030 \text{x} 10^{-2} \\ 4.120 \text{x} 10^{-2} \\ 8.333 \text{x} 10^{-2} \\ 1.311 \text{x} 10^{-1} \end{Bmatrix} \tag{3.52}$$

The relative error between the approximate tip displacement $\hat{v}_4$ and the exact value of $1.155 \text{x} 10^{-1}$ is approximately 13.5 percent. Using a backward difference $O(h^2)$, the approximate slope at the tip of the beam is

$$\left. \frac{d\hat{v}}{dx} \right|_{x=L} = \frac{3\hat{v}_4 - 4\hat{v}_3 + \hat{v}_2}{2h} = 2.022 \text{x} 10^{-1} \tag{3.53}$$

The relative error between this value and the exact value value of 1.536×10^{-1} is approximately 31.6 percent. In Figures 3.7 and 3.8 the approximate transverse displacement and slope for $P = 4$ are compared to exact values.

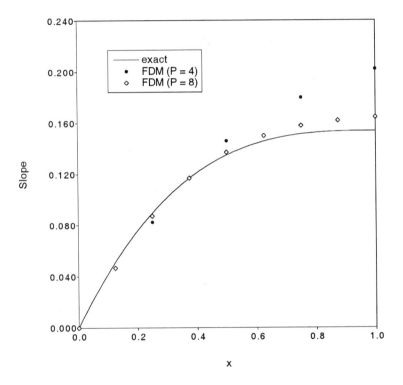

Figure 3.8: Comparison of Approximate and Exact Slope for a Uniformly Loaded Beam Clamped at One End, Free at the Other, Resting on an Elastic Foundation

The grid is next *refined* by halving the spacing to $h = 0.125$. Now $P = 8$, and there are nine unknowns. Five of the nine requisite equations consist of equation (3.46), evaluated at $j = 1$ to 5. Two additional equations are the boundary conditions at $x = 0$ given by equations (3.47) and (3.48). The final two equations are the boundary conditions at $x = L$, given above in equations (3.49) and (3.50). These must be modified to account for the larger number of grid points. In particular,

$$\frac{d^2\hat{v}}{dx^2}\bigg|_{x=L} = \frac{2\hat{v}_8 - 5\hat{v}_7 + 4\hat{v}_6 - \hat{v}_5}{h^2} = 0 \qquad (3.54)$$

and

$$\frac{d^3\hat{v}}{dx^3}\bigg|_{x=L} = \frac{5\hat{v}_8 - 18\hat{v}_7 + 24\hat{v}_6 - 14\hat{v}_5 + 3\hat{v}_4}{2h^3} = 0 \qquad (3.55)$$

The solution to the set of simultaneous equations is

$$\begin{Bmatrix} \hat{v}_0 \\ \hat{v}_1 \\ \hat{v}_2 \\ \hat{v}_3 \\ \hat{v}_4 \\ \hat{v}_5 \\ \hat{v}_6 \\ \hat{v}_7 \\ \hat{v}_8 \end{Bmatrix} = \begin{Bmatrix} 0.0 \\ 2.913\text{x}10^{-3} \\ 1.165\text{x}10^{-2} \\ 2.471\text{x}10^{-2} \\ 4.083\text{x}10^{-2} \\ 5.898\text{x}10^{-2} \\ 7.837\text{x}10^{-2} \\ 9.843\text{x}10^{-2} \\ 1.188\text{x}10^{-1} \end{Bmatrix} \qquad (3.56)$$

The relative error between the approximate tip displacement $\hat{v}_8$ and the exact value is approximately 2.9 percent. Again using a backward difference $O(h^2)$, the approximate slope at the tip of the beam is

$$\frac{d\hat{v}}{dx}\bigg|_{x=L} = \frac{3\hat{v}_8 - 4\hat{v}_7 + \hat{v}_6}{2h} = 1.645\text{x}10^{-1} \qquad (3.57)$$

The relative error between this value and the exact value is approximately 7.1 percent. In Figures 3.7 and 3.8 the approximate transverse displacement and slope for $P = 8$ are compared to exact values. From these figures it is evident that the approximate solution is converging to the exact one.

3.5   Partial Differential Equations

Physical phenomena involving more than one independent variable are typically expressed using equations involving partial derivatives. The Calculus of Finite Differences can be used to obtain approximate solutions to partial differential equations that characterize specific boundary/initial value problems.

In solving a given partial differential equation by the finite difference method, the independent variables (e.g., x, y, t, etc.) are discretized. This typically results in a two- or three-dimensional finite difference grid. Partial derivatives appearing in the governing differential equation are next replaced by approximations from the Calculus of Finite Differences. Thus, as was the case with ordinary differential equations, this process involves approximation and thus introduces error into the solution.

The specific approach followed in solving partial differential equations using the finite difference method is best understood by considering separately elliptic, parabolic and hyperbolic partial differential equations.[5]

3.5.1 Elliptic Partial Differential Equations

Elliptic partial differential equations possess the characteristic that a disturbance at a given point propagates in all directions. Consequently, the solution domain Ω for an elliptic partial differential equation is an enclosed one, with boundary conditions specified everywhere along the edges of Ω. The solution at each point is influenced by the solution at every other point in the region where the governing equation applies [265]. As it turns out, this characteristic of elliptic partial differential equations generally complicates their numerical solution as compared to that of parabolic partial differential equations (discussed in the next section) in which a "time marching" scheme is employed.

As an example of an elliptic partial differential equation, consider the two-dimensional Poisson's equation, which describes many different steady-state phenomena.

$$\frac{\partial^2 \phi}{\partial x^2} + \frac{\partial^2 \phi}{\partial y^2} - f(x,y) = 0 \quad \text{in} \quad \Omega \tag{3.58}$$

For simplicity, a rectangular domain Ω is assumed (Figure 3.9). Essential boundary conditions are prescribed along Γ, viz.,

$$\phi(x,y) = \bar{\phi}(x,y) \quad \text{on} \quad \Gamma \tag{3.59}$$

A set of equally spaced grid lines $x = x_i$, $i = 0,\ 1,\ \cdots,\ M$ is first constructed on the range $a \leq x \leq b$, such that $x_0 = a$ and $x_M = b$ in the manner shown in Figure 3.10. The spacing between these grid lines is denoted by $\Delta x = x_{i+1} - x_i \equiv h$. Thus $x_i = a + ih$ and $h = (b-a)/M$.

---

[5]The classification of partial differential equations is explained in Appendix A.

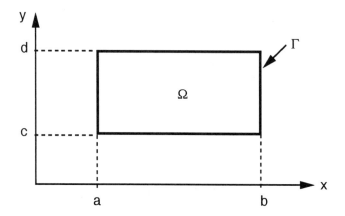

Figure 3.9: Two-Dimensional Solution Domain

Similarly, a set of equally spaced grid lines $y = y_j$, $j = 0, 1, \cdots, N$ is constructed such that $y_0 = c$ and $y_N = d$. The spacing between these grid lines is denoted by $\Delta y = y_{j+1} - y_j \equiv k$. Thus $y_j = c + jk$ and $k = (d-c)/N$.

The points of intersection (x_i, y_j) are the *grid* points. Rectangular grids such as that shown in Figure 3.10 represent a so-called *structured* mesh [194]. A two-dimensional structured mesh is one that is defined by only two parameters: the number of points along the x-axis $(M + 1)$ and the number of points $(N + 1)$ along the y-axis. Further details pertaining to structured as well as unstructured meshes are given in Chapter 13.

The derivatives appearing in equation (3.58) are next replaced by central differences $O(h^2)$, viz.,

$$\frac{\partial^2 \phi}{\partial x^2} = \frac{\hat{\phi}(x_{i+1}, y_j) - 2\hat{\phi}(x_i, y_j) + \hat{\phi}(x_{i-1}, y_j)}{h^2} + O\left(h^2\right) \qquad (3.60)$$

where y is fixed, and by central differences $O(k^2)$

$$\frac{\partial^2 \phi}{\partial y^2} = \frac{\hat{\phi}(x_i, y_{j+1}) - 2\hat{\phi}(x_i, y_j) + \hat{\phi}(x_i, y_{j-1})}{k^2} + O\left(k^2\right) \qquad (3.61)$$

where x is fixed.

Denoting the discrete approximate values of ϕ by $\hat{\phi}(x_i, y_j) \equiv \hat{\phi}_{i,j}$, etc., and substituting equations (3.60) and (3.61) into equation (3.58) gives the discrete form of the governing equations

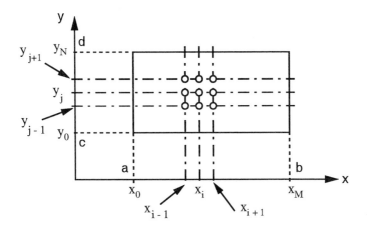

Figure 3.10: Typical Two-Dimensional Finite Difference Grid

$$\frac{\hat{\phi}_{i+1,j} - 2\hat{\phi}_{i,j} + \hat{\phi}_{i-1,j}}{h^2} + \frac{\hat{\phi}_{i,j+1} - 2\hat{\phi}_{i,j} + \phi_{i,j-1}}{k^2} - f(x_i, y_j) = 0 \quad (3.62)$$

This represents a system of $(M+1)*(N+1)$ equations in as many unknowns. If the spacing along the x and y-axes is identical, $h = k$ and equation (3.62) simplifies to

$$\hat{\phi}_{i+1,j} + \hat{\phi}_{i-1,j} - 4\hat{\phi}_{i,j} + \hat{\phi}_{i,j+1} + \hat{\phi}_{i,j-1} - h^2 f(x_i, y_j) = 0 \quad (3.63)$$

The boundary conditions associated with the finite difference approximation are specified in the following manner:

$$\bar{\phi}(x_0, y_j) = \bar{\phi}(a, y_j) = \hat{\phi}_{0,j} \quad (3.64)$$

$$\bar{\phi}(x_i, y_0) = \bar{\phi}(x_i, c) = \hat{\phi}_{i,0} \quad (3.65)$$

$$\bar{\phi}(x_M, y_j) = \bar{\phi}(b, y_j) = \hat{\phi}_{M,j} \quad (3.66)$$

$$\bar{\phi}(x_i, y_N) = \bar{\phi}(x_i, d) = \hat{\phi}_{i,N} \qquad (3.67)$$

The boundary conditions involve a total of $2(M + 1) + (N - 1) = 2(M + N)$ known values. The approximate solution thus involves a system of $(M - 1) * (N - 1)$ equations in as many unknowns.

The resulting linear system of simultaneous equations must be converted to the standard form $\mathbf{Au} = \mathbf{b}$, where u_p corresponds to some $\hat{\phi}_{i,j}$, where $p = 1, 2, \cdots, (M-1)*(N-1)$. More precisely, this conversion from a two-dimensional array of unknowns to a one-dimensional one is realized through the following mapping:

$$u_{i+(M-j-1)(N-1)} = \hat{\phi}_{i,j} \qquad (3.68)$$

where all parameters are as previously defined.

From either equation (3.62) or (3.63) it follows that the configuration shown in Figure 3.11 is realized at a typical *interior* grid point. It is thus evident that at each such interior grid point, only four surrounding points need to be considered in the finite difference approximation (stated another way, equations (3.62) and (3.63) each contain only five unknowns). Consequently, the coefficient matrix is *banded* and pentadiagonal. A typical computational molecule[6] associated with the approximate solution of Poisson's equation is shown in Figure 3.11.

The sparse nature of the coefficient matrix dictates that, for efficiency, special solution schemes need to be employed. In particular, if a *direct* equation solver based on Gauss elimination is employed, it should be one that is specialized for the banded (pentadiagonal) system under consideration. More commonly though, and particularly for large systems, equations (3.62) or (3.63) are solved using *iterative* techniques such as Successive Overrelaxation (SOR), the Multi-Grid method and Conjugate Gradient schemes.[7] An efficient method for solving two-dimensional steady-state diffusion problems is the so-called Alternating Direction Implicit (ADI) method. This method, which gives rise to a tridiagonal set of equations in each step, employs the unknown values of the dependent variable from the current iteration along one direction and the known values from the previous iteration along the other direction. In the following step, the directions are reversed. An acceleration parameter similar to that associated with the SOR method, is used to improve the rate of convergence [265].

---

[6]The "molecule" terminology is attributed to Bickley [75].

[7]Further details pertaining to equation solvers are given in Appendix E.

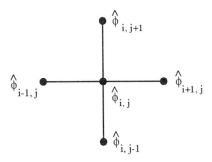

Figure 3.11: Finite Difference "Molecule" for Approximation of Poisson's Equation using Central Differences $O(h^2)$ and $O(k^2)$

3.5.2 Parabolic Partial Differential Equations

Unlike elliptic partial differential equations in which a disturbance at a given point propagates in all directions, in a parabolic partial differential equation there is a definite direction for the flow of information. The solution domain for parabolic partial differential equations thus stretches indefinitely in one coordinate direction, typically time t, from the given initial conditions. The governing equation is solved for the dependent variables ϕ by advancing, or "marching," in the independent variable t. The solution must satisfy the prescribed boundary conditions in the other independent variables. At a given time, the solution depends upon the values of ϕ at earlier times, but is independent of those at subsequent times.

As an example of a parabolic partial differential equation, consider the transient one-dimensional heat equation without source term[8]

$$\frac{\partial \phi(x,t)}{\partial t} = \alpha \frac{\partial^2 \phi(x,t)}{\partial x^2} \quad \text{in} \quad 0 < x < L \qquad (3.69)$$

where ϕ denotes the temperature, x is a spatial coordinate, t denotes time, and α is a positive constant known as the thermal diffusivity of the material.

The associated boundary conditions are

$$\phi(0,t) = \phi(L,t) = \bar{\phi} \quad \forall \quad t > 0 \qquad (3.70)$$

---

[8]This equation is derived in Appendix B.

The transient nature of the problem also requires the specification of a suitable initial condition

$$\phi(x,0) = \bar{\phi}_0 \quad \forall \quad 0 < x < L \tag{3.71}$$

For a finite difference solution, the domain is discretized in the manner shown in Figure 3.12. The grid point coordinates are thus $x_i = ih$ with $i = 0, 1, 2, \cdots, M$ and $t_j = jk$ with $j = 0, 1, 2, \cdots, N$.

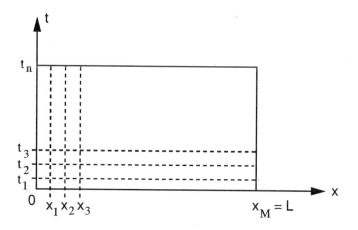

Figure 3.12: Finite Difference Grid For One-Dimensional Transient Heat Conduction Analysis

The spatial derivatives in equation (3.69) are replaced by central differences $O(h^2)$. Thus, as in the case of equation (3.60),

$$\frac{\partial^2 \phi}{\partial x^2} = \frac{\hat{\phi}(x_{i+1}, t_j) - 2\hat{\phi}(x_i, t_j) + \hat{\phi}(x_{i-1}, t_j)}{h^2} + O\left(h^2\right) \tag{3.72}$$

For the time derivative, note that the central difference approximation $O(k^2)$

$$\frac{\partial \phi}{\partial t} = \frac{\hat{\phi}(x_i, t_{j+1}) - \hat{\phi}(x_i, y_{j-1})}{k^2} + O\left(k^2\right) \tag{3.73}$$

leads to missing values (in particular, $\hat{\phi}(x_i, t_j)$ cannot be computed in this way). Consequently, a forward difference $O(k)$ is used instead, viz.,

$$\frac{\partial \phi}{\partial t} = \frac{\hat{\phi}(x_i, t_{j+1}) - \hat{\phi}(x_i, t_j)}{k} + O\left(k\right) \qquad (3.74)$$

Adopting the notation $\hat{\phi}(x_i, t_j) \equiv \hat{\phi}_{i,j}$, and substituting equations (3.72) and (3.74) into (3.69) gives

$$\frac{\hat{\phi}_{i,j+1} - \hat{\phi}_{i,j}}{k} - \alpha \frac{\hat{\phi}_{i+1,j} - 2\hat{\phi}_{i,j} + \hat{\phi}_{i-1,j}}{h^2} = 0 \qquad (3.75)$$

Solving for $\hat{\phi}_{i,j+1}$ gives

$$\begin{aligned}
\hat{\phi}_{i,j+1} &= \hat{\phi}_{i,j} + \frac{\alpha k}{h^2} \left(\hat{\phi}_{i+1,j} - 2\hat{\phi}_{i,j} + \hat{\phi}_{i-1,j} \right) \\
&= \left(1 - 2\frac{\alpha k}{h^2} \right) \hat{\phi}_{i,j} + \frac{\alpha k}{h^2} \left(\hat{\phi}_{i+1,j} + \hat{\phi}_{i-1,j} \right)
\end{aligned} \qquad (3.76)$$

The scheme described by equation (3.76) thus gives the solution for $\hat{\phi}_{i,j+1}$ (i.e., temperature) at time $t = (j + 1)k$ in terms of known values of $\hat{\phi}$ at $t = jk$. The values of $\hat{\phi}$ at a given time are thus determined explicitly from known values at the previous time step. Furthermore, the solution involves only multiplications; there is no need for factoring a coefficient matrix. Consequently, this scheme is classified as being *explicit*, and is commonly referred to as the *forward Euler* or the *forward time centered space* (FTCS) method [265]. The computational molecule associated with this method is shown in Figure 3.13.

For brevity, let

$$\hat{\phi}^{(j)} = \left\{ \begin{array}{c} \hat{\phi}_{1,j} \\ \vdots \\ \vdots \\ \hat{\phi}_{M-1,j} \end{array} \right\} \qquad (3.77)$$

The forward Euler method is thus written as

$$\hat{\phi}^{(j+1)} = A\hat{\phi}^{(j)} \qquad (3.78)$$

where only matrix multiplications are required to determine $\hat{\phi}^{(j+1)}$. In light of equation (3.76), the coefficient matrix is known to be symmetric and tridiagonal, viz.,

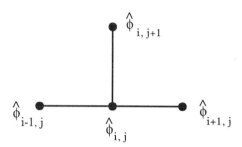

Figure 3.13: Finite Difference "Molecule" Associated with the Forward Euler Method

$$\mathbf{A} = \begin{bmatrix} (1-2\Lambda) & \Lambda & 0 & \cdots & & 0 \\ \Lambda & (1-2\Lambda) & \Lambda & 0 & \cdots & 0 \\ 0 & \Lambda & (1-2\Lambda) & \Lambda & 0 & \cdots & \vdots \\ 0 & \cdots & \Lambda & \ddots & \Lambda & \cdots & \vdots \\ 0 & \cdots & & \ddots & \ddots & \ddots & 0 \\ 0 & \cdots & & & \ddots & \ddots & \ddots & \Lambda \\ 0 & \cdots & & \cdots & \cdots & 0 & \Lambda & (1-2\Lambda) \end{bmatrix} \quad (3.79)$$

The parameter $\Lambda = \alpha k/h^2$ is sometimes referred to as the "grid Fourier number" [265].

As with any "time marching" scheme, the issue of stability is of utmost importance. To analyze the stability of the forward Euler method, recall that the spectral radius of $\mathbf{A}$, denoted by $\rho(\mathbf{A})$, is defined as the magnitude of the largest eigenvalue of $\mathbf{A}$. It is known that for the forward Euler method to be stable, $\rho(\mathbf{A})$ should be less than one [99]. The entries in $\mathbf{A}$ depend only upon Λ; the size of $\mathbf{A}$ is determined by the number of grid points in the spatial (x) direction (i.e., by the value of M). The eigenvalues of $\mathbf{A}$ are thus given by

$$\lambda_i = 1 - 4\Lambda \sin^2 \left(\frac{i\pi}{2M} \right) \quad , \quad i = 1, 2, \cdots, M-1 \qquad (3.80)$$

For stability, we require $|\lambda_i| < 1$ $\forall$ i. Noting that $\Lambda > 0$ and that $1 \ge \sin^2 \left(\frac{i\pi}{2M} \right) \ge 0$, it follows that

$$|\lambda_i| < 1 \quad \Leftrightarrow \quad 1 > 1 - 4\Lambda \sin^2\left(\frac{i\pi}{2M}\right) > -1 \tag{3.81}$$

or

$$\frac{1}{2} > \lambda > 0 \tag{3.82}$$

for $i = 0, 1, 2, \cdots, M - 1$. Satisfaction of the condition (3.82) guarantees $\rho(\mathbf{A}) < 1$. Recalling the definition of Λ, it follows that h and k cannot be chosen independently. That is, for stability, the time step

$$k < \frac{1}{2}\frac{h^2}{\alpha} \tag{3.83}$$

must be chosen. Typically such *conditional* stability necessitates the use of very small time steps.

To circumvent this restriction, *implicit* schemes are used. For example, if the forward difference of equation (3.74) is replaced by the backward difference $O(h)$, viz.,

$$\frac{\partial \phi}{\partial t} = \frac{\hat{\phi}(x_i, t_j) - \hat{\phi}(x_i, t_{j-1})}{k} + O(k) \tag{3.84}$$

the following implicit scheme is obtained:

$$\frac{\hat{\phi}_{i,j} - \hat{\phi}_{i,j-1}}{k} - \alpha \frac{\hat{\phi}_{i+1,j} - 2\hat{\phi}_{i,j} + \hat{\phi}_{i-1,j}}{h^2} = 0 \tag{3.85}$$

This scheme has the general form

$$B\hat{\phi}^{(j+1)} = \hat{\phi}^{(j)} \tag{3.86}$$

where

$$\mathbf{B} = \begin{bmatrix} (1 + 2\Lambda) & -\Lambda & 0 & \cdots & & 0 \\ -\Lambda & (1 + 2\Lambda) & -\Lambda & \ddots & & \vdots \\ 0 & -\Lambda & \ddots & \ddots & & 0 \\ \vdots & & \ddots & \ddots & \ddots & -\Lambda \\ 0 & & \cdots & 0 & -\Lambda & (1 + 2\Lambda) \end{bmatrix} \tag{3.87}$$

The coefficient matrix $\mathbf{B}$ is seen to be symmetric, strictly diagonally dominant, and tridiagonal.

For investigating stability, equation (3.86) is rewritten as

$$\hat{\phi}^{(j+1)} = \mathbf{B}^{-1}\hat{\phi}^{(j)} \tag{3.88}$$

The scheme will be stable provided that $\rho\left(\mathbf{B}^{-1}\right) < 1$. If λ is an eigenvalue of $\mathbf{B}$, then $\mathbf{B}\mathbf{x} = \lambda\mathbf{x}$ for some eigenvector $\mathbf{x}$. Since $\mathbf{x} = \lambda\mathbf{B}^{-1}\mathbf{x}$, and provided that $\lambda \neq 0$, it follows that $\mathbf{B}^{-1}\mathbf{x} = \frac{1}{\lambda}\mathbf{x}$, or $\frac{1}{\lambda}$ is an eigenvalue of $\mathbf{B}^{-1}$. Since the coefficient matrix of equation (3.87) is non-singular, all its eigenvalues will be *non-zero*. Thus,

$$\frac{1}{\mid \text{ smallest eigenvalue of } \mathbf{B} \mid} = \mid \text{ largest eigenvalue of } \mathbf{B}^{-1} \mid \tag{3.89}$$

The eigenvalues of $\mathbf{B}$ are thus

$$\lambda_i = 1 + 4\Lambda \sin^2\left(\frac{i\pi}{2M}\right) \quad , \quad i = 1, 2, \cdots, M-1 \tag{3.90}$$

Since $\sin^2\left(i\pi/2M\right) > 0$ and $\Lambda > 0$, it follows that $\lambda_i > 1 \;\forall\; i$. In particular, the smallest eigenvalue of $\mathbf{B}$ is greater than one, implying that $\rho\left(\mathbf{B}^{-1}\right) < 1$. Thus, for *any* choice of h and k, the backward difference scheme is stable. The order of this *unconditionally* stable scheme is $O(k) + O(h^2)$.

In an effort to improve upon the accuracy of the backward difference scheme, it is necessary to develop a high-order scheme. For example, retaining the error terms in a Taylor series expansion, we combine the following *forward* difference approximation at $t = jk$

$$\frac{\partial\phi(x_i, t_j)}{\partial t} = \frac{\phi(x_i, t_{j+1}) - \phi(x_i, t_j)}{k} + \frac{k}{2}\frac{\partial^2\phi(x_i, t_j)}{\partial t^2} + O(k^2) \tag{3.91}$$

with the following *backward* difference approximation at $t = (j+1)k$

$$\frac{\partial\phi(x_i, t_{j+1})}{\partial t} = \frac{\phi(x_i, t_{j+1}) - \phi(x_i, t_j)}{k} - \frac{k}{2}\frac{\partial^2\phi(x_i, t_{j+1})}{\partial t^2} + O(k^2) \tag{3.92}$$

Taking the average of equations (3.91) and (3.92) drops out the $O(k)$ terms. Combining the resulting expression with suitable central difference approximations for the spatial term leads to the so-called Crank-Nicolson method [126]:

$$\frac{\hat{\phi}_{i,j+1} - \hat{\phi}_{i,j}}{k} - \frac{\alpha}{2} \left[\frac{\hat{\phi}_{i+1,j} - 2\hat{\phi}_{i,j} + \hat{\phi}_{i-1,j}}{h^2} \right.$$

$$\left. + \frac{\hat{\phi}_{i+1,j+1} - 2\hat{\phi}_{i,j+1} + \hat{\phi}_{i-1,j+1}}{h^2} \right] = 0 \qquad (3.93)$$

This method has the general form

$$\mathbf{B}\hat{\phi}^{(j+1)} = \mathbf{A}\hat{\phi}^{(j)} \qquad (3.94)$$

where $\mathbf{A}$ and $\mathbf{B}$ are tridiagonal matrices. The Crank-Nicolson Method is *unconditionally stable*, and $O(h^2 + k^2)$. The computational molecule associated with this method is shown in Figure 3.14.

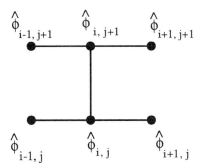

Figure 3.14: Finite Difference "Molecule" Associated with the Crank-Nicolson Method

In closing the discussion of solution schemes for parabolic partial differential equations, it is timely to point out that a family of implicit schemes may be obtained by approximating the second spatial derivative at a time between $t = jk$ and $t = (j+1)k$. The resulting finite difference equation is

$$\frac{\hat{\phi}_{i,j+1} - \hat{\phi}_{i,j}}{k} = \alpha \left[\Theta \frac{\hat{\phi}_{i+1,j+1} - 2\hat{\phi}_{i,j+1} + \hat{\phi}_{i-1,j+1}}{h^2} \right.$$

$$\left. + (\Theta - 1) \frac{\hat{\phi}_{i+1,j} - 2\hat{\phi}_{i,j} + \hat{\phi}_{i-1,j}}{h^2} \right] \qquad (3.95)$$

where $0 \leq \Theta \leq 1$. The second spatial derivative is thus written as a weighted average of the finite difference approximations corresponding to the times $t = jk$ and $t = (j + 1)k$.

The following values of Θ are noteworthy:

- If $\Theta = 0$, the explicit *forward Euler* method of equation (3.76) is obtained.

- If $\Theta = 0.5$, the second derivative is evaluated midway between the two time points, and the *Crank-Nicolson* method of equation (3.93) is obtained.[9]

- If $\Theta = 1.0$, the second spatial derivative is evaluated only at $t = (j + 1)k$ and the scheme is known as the fully implicit, *backward Euler* or *Laasonen* method [265].

The computational molecule associated with this method is shown in Figure 3.15.

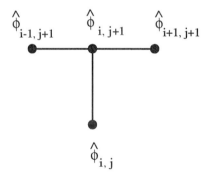

Figure 3.15: Finite Difference "Molecule" Associated with the Backward Euler Method

3.5.3   Hyperbolic Partial Differential Equations

Hyperbolic partial differential equations have two real and distinct characteristics. In regions defined by these characteristics, information travels at finite speeds. Hyperbolic partial differential equations arise in several

---

[9]Because of this fact, the Crank-Nicolson method is sometimes referred to as the "midpoint method."

problems of engineering interest, such as vibration of strings and rods, transmission of sound in air, and the flow of an object in a stationary fluid or of the fluid past stationary objects at speeds greater than the speed of sound in that fluid (i.e., supersonic flow).

Very effective numerical techniques for solving hyperbolic partial differential equations are based on the *method of characteristics*. Since the focus of this chapter is on the finite difference method, the subject of characteristics shall not be pursued further. The interested reader is referred to books on the subject such as [1].

To illustrate the use of finite difference solution of hyperbolic partial differential equations, consider the one-dimensional wave equation

$$\frac{\partial^2 \phi}{\partial t^2} - c^2 \frac{\partial^2 \phi}{\partial x^2} = 0 \tag{3.96}$$

where c represents the propagation velocity of the wave. In conjunction with the wave equation, the following homogeneous boundary conditions are assumed

$$\phi(0, t) = \phi(L, t) = 0 \quad \forall \quad t > 0 \tag{3.97}$$

To complete the statement of the problem, the following initial conditions are also required:

$$\phi(x, 0) = f(x) \quad \forall \quad 0 < x < L \tag{3.98}$$

$$\frac{\partial \phi(x, 0)}{\partial t} = g(x) \quad \forall \quad 0 < x < L \tag{3.99}$$

where $f(x)$ and $g(x)$ are known functions.

Using central differences $O(k^2)$ and $O(h^2)$, respectively, equation (3.96) is approximated by

$$\frac{\hat{\phi}_{i,j+1} - 2\hat{\phi}_{i,j} + \hat{\phi}_{i,j-1}}{k^2} - c^2 \frac{\hat{\phi}_{i+1,j} - 2\hat{\phi}_{i,j} + \hat{\phi}_{i-1,j}}{h^2} = 0 \tag{3.100}$$

where $x_i = ih$ and $t_j = jk$. Analyzing the initial conditions, we note that

$$\phi(x, 0) = f(x) \quad \rightarrow \quad \hat{\phi}_{i,0} = f(x_i) \tag{3.101}$$

Using a forward difference $O(k)$ in time,

$$\frac{\partial \phi(x,t)}{\partial t} = \frac{\hat{\phi}_{i,j+1} - \hat{\phi}_{i,j}}{k} + O(k) \qquad (3.102)$$

it follows that for $j = 0$

$$\frac{\hat{\phi}_{i,1} - \hat{\phi}_{i,0}}{k} = g(x_i) \qquad (3.103)$$

Using equation (3.101) in conjunction with equation (3.103), it is possible to determine $\hat{\phi}_{i,1}$. With $\hat{\phi}_{i,0}$ and $\hat{\phi}_{i,1}$ known, equation (3.100) can then be used to compute the remaining $\hat{\phi}_{i,j}$. This method is $O(k + h^2)$, and is stable provided that [263]

$$\lambda = \frac{\alpha k}{h} \leq 1 \qquad (3.104)$$

This stability condition is known as the *Courant condition* and the dimensionless parameter λ as the *Courant number* [265].

In closing this section, it is timely to note that unconditionally stable implicit schemes can be derived to solve hyperbolic partial differential equations. It is also possible to use a symmetric approximation for the first time derivative, and to arrive at a scheme that is $O(k^2 + h^2)$ [99].

In the area of explicit schemes, the *Lax-Wendroff* method is commonly used to solve linear hyperbolic partial differential equations. This explicit scheme is second-order accurate in both time and space, and is stable for Courant numbers less than unity [265].

3.6   Concluding Remarks

In this chapter, the finite difference method was introduced and used in the approximate solution of both ordinary and partial differential equations. The following summary of key points related to the finite difference method is thus in order.

- The finite difference method is simple to apply. In the governing differential equations $L(\phi) - f = 0$, derivatives are replaced by suitable difference equations containing only algebraic operations. This transforms the governing equations into the form $\hat{L}(\phi) - f = 0$.

- The domain Ω is discretized into a finite set of discrete grid points. The resulting grids are often regular; if this is not the case, the entire domain must be mapped.

- In applications of the finite difference method essential boundary conditions are represented exactly. Natural (Neumann) boundary conditions involve further finite difference approximations.

- The resulting set of linear simultaneous equations is frequently banded, though non-symmetric. Iterative solvers have traditionally been used to solve these equations.

- The finite difference method approximates the primary dependent variables only at the grid points; it gives no information about the function values between these points.

- Although it is possible to handle curved boundaries using the finite difference method, this is not a trivial undertaking and pretty much precludes writing general computer programs which use the method. Bodies of arbitrary geometry pose greater difficulties.

- The finite difference method is difficult to apply to bodies made of composite materials, for it is not clear what material properties should be assigned to grid points along material interfaces.

- The finite difference method can, without modification to the basic formulation, be applied to non-linear problems [11, 594]. The numerical implementation of the method for analysis of such problems involves *iteration.*

Although the finite difference method is simple to understand and apply, it approximates the primary dependent variables only at a finite number of discrete grid points in Ω. In an effort to overcome this shortcoming, a number of *continuous* solution techniques have, over the years, been developed. The discussion of such techniques commences in the next chapter.

3.7 Exercises

3.1

Consider the following boundary-value problem

$$\frac{d^2u}{dx^2} + 4u = \cos x \quad ; \quad 0 < x < \frac{\pi}{4} \quad ; \quad u(0) = u\left(\frac{\pi}{4}\right) = 0$$

Using the finite difference method with central differences $O(h^2)$, obtain the approximate solution for grid spacings of $\pi/12$, $\pi/24$ and $\pi/48$. At points common to all three grids (i.e., at $\Delta x = \pi/12$ and $\pi/6$), compare the approximate solution. Does the grid refinement significantly improve the results? Discuss your findings.

3.2

Consider the following boundary-value problem

$$\frac{d^2u}{dx^2} = -3\frac{du}{dx} + 2u + 2x + 3 \quad ; \quad 0 < x < 1 \quad ; \quad u(0) = 2, \, u(1) = 1$$

Using the finite difference method with central differences $O(h^2)$, obtain the approximate solution for grid spacings of 0.25, 0.125 and 0.0625. At points common to all three grids, compare the approximate solution. Does the grid refinement significantly improve the results? Discuss your findings.

3.3

Consider the following boundary-value problem

$$\frac{d^2u}{dx^2} = 64u \quad ; \quad 0 < x < 1 \quad ; \quad u(0) = 1, \, u(1) = e^{-8}$$

Using the finite difference method first obtain approximate solution using central differences $O(h^2)$ with grid spacings of 0.250 and 0.125. Compare these results with the exact solution $u = e^{-8x}$. Next repeat the analysis for a grid spacing of 0.250, but at $j = 2$ ($x = 0.5$) use a central difference $O(h^4)$. How do the results of this analysis with "localized enrichment" compare with those obtained using central differences $O(h^2)$ with a grid spacing of 0.125? Discuss your findings.

3.4

Consider the problem of one-dimensional steady state heat flow in a rod such as that shown in Figure 3.4. The rod has a unit cross-section and a length L equal to 8. The left end of the rod is held at a temperature of $50°$. At the right end of the rod heat is input at a rate of 200. The thermal

conductivity k is equal to 75, and heat is being generated within the rod at a rate of $S = 1.50x^2$ per unit length.
Using the finite difference method with mesh spacings of $h = \Delta x = 2.0$ and 1.0, calculate the approximate temperature distribution along the rod. Use central- and, where applicable, backward differences $O(h^2)$. Graphically compare the approximate solution with the exact one. Does the grid refinement drastically improve the results? Is the constraint of equally-spaced mesh points restrictive with respect to this problem? Ignoring this constraint, what would be an improved approach to re-meshing?

3.5
The distribution of bending moment M in a beam subjected to a distributed loading $q(x)$ per unit length (positive in the positive y-direction) satisfies the differential equation

$$\frac{d^2 M}{dx^2} = q(x)$$

A beam of *unit* length (Figure 3.16) is simply supported at both ends (i.e., $M(0) = M(1) = 0$). The beam carries the distributed load of $q(x) = \sin \pi x$ per unit length. Use the finite difference method with a mesh point spacing of $h = \Delta x = 0.25$. Concerning the order of approximation:

- First obtain an approximate solution using only *forward* differences $O(h)$.

- Next obtain an approximate solution using only *backward* differences $O(h)$.

- Finally, obtain a third approximate solution using *central* differences $O(h^2)$.

Compare your results to the exact solution. Does the use of central differences yield a substantial increase in accuracy of the approximate solution? If so, is this higher accuracy worth the added computational effort, if any, associated with central differences?

3.6
Consider the problem of one-dimensional steady state heat flow in a rod such as that shown in Figure 3.4. The rod has a unit cross-section and a length L equal to 8. The thermal conductivity k is equal to 0.20, and heat is generated within the rod at a rate of

$$S = 0.015 \left(\frac{x}{L}\right)^3$$

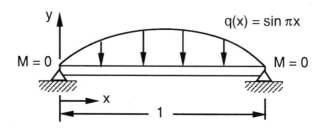

Figure 3.16: Beam of Unit Length

At the right end heat is lost to the environment through convection, that is,

$$\bar{q} = -k\frac{dT}{dx}\Big|_{x=L} = h(T_L - T_\infty)$$

where T_L represents the temperature at the right end of the rod, h is the convection coefficient and T_∞ denotes the ambient temperature. For the present problem, assume that $h = 0.10$ and that $T_\infty = -5^\circ$. Using the finite difference method with grid spacings of 2.0 and 1.0, calculate the approximate temperature distribution along the rod. Use central- and, where applicable, backward differences $O(h^2)$. Graphically compare the approximate solutions. Comment on what effect the mesh refinement has on the solution. What effect will increasing the coefficient h have on the solution ?

3.7

Consider the non-prismatic (i.e., one with varying cross-sectional area) rod shown in Figure 3.17. The rod is loaded by a distributed force of $f(x) = 0.50x^2$ kN/m, and by a concentrated 500 kN force applied at its end.

Using a suitable combination of central and backward differences $O(h^2)$, analyze the rod using the finite difference method with grid spacings of $h = 2.0, 1.0$ and 0.50 meters. Compare the end displacement for each of the three cases. There is no need to determine the exact solution to the problem, as this is a non-trivial undertaking. Does the approximate solution appear to be converging? Discuss your results.

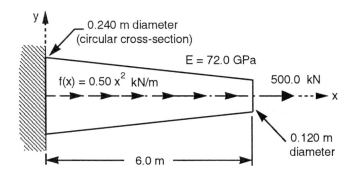

Figure 3.17: Non-Prismatic Bar

3.8

Repeat the analyses performed in Example 3.3, only for a beam of unit length that is *simply supported* at both ends. The associated boundary condition are

$$v = \frac{d^2v}{dx^2} = 0 \quad at \quad x = 0 \quad and \quad x = 1$$

The exact solutions for the transverse deflection and for the slope are [237]

$$v = \frac{q}{k} \left[1 - \frac{\cosh \lambda x \cos \lambda x' + \cosh \lambda x' \cos \lambda x}{\cosh \lambda L + \cos \lambda L} \right]$$

and

$$\frac{dv}{dx} = -\frac{q\lambda}{k} \frac{1}{\cosh \lambda L + \cos \lambda L} (\sinh \lambda x \cos \lambda x' + \cosh \lambda x \sin \lambda x'$$

$$- \sinh \lambda x' \cos \lambda x - \cosh \lambda x' \sin \lambda x)$$

3.9

Repeat the analyses performed in Example 3.3, only for a beam of unit length that is *clamped* at both ends. The associated boundary condition are

$$v = \frac{dv}{dx} = 0 \quad at \quad x = 0 \quad and \quad x = 1$$

The exact solutions for the transverse deflection and for the slope are [237]

$$v = \frac{q}{k}\left[1 - \frac{1}{\sinh \lambda L + \sin \lambda L}\left(\sinh \lambda x \cos \lambda x' + \sin \lambda x \cosh \lambda x'\right.\right.$$

$$+ \left.\sinh \lambda x' \cos \lambda x + \sin \lambda x' \cosh \lambda x\right)\Big]$$

and

$$\frac{dv}{dx} = -\frac{2q\lambda}{k}\frac{\sinh \lambda x \sin \lambda x' - \sin \lambda x \sinh \lambda x'}{\sinh \lambda L + \sin \lambda L}$$

3.10

Consider the one-dimensional bar shown in Figure 3.18. The boundary condition at $y = 0$ consists of a linear spring. That is, for the spring $F = kv$, where F is the force applied to the spring, k is the spring stiffness, and v is the vertical (axial) displacement of the spring.

Assuming a grid spacing equal to $L/3$, appropriately use central-, forward- and backward differences $O(h^2)$ to:

- Derive the general equations (in matrix form) for the values of displacement at the mesh points. Write these equations in terms of E, A, h, k and $\bar{P}$.

- Assuming $k = 1 \times 10^6$, 1×10^9 and 1×10^{12} lb/in., compute the approximate values of axial displacement and strain. Present these in a single table.

- Comment on the solutions obtained and their relation to the case where a rigid boundary exists at $y = 0$.

3.11

A fin is an appendage attached to a base structure in order to increase its surface area and thus increase the heat flux due to convection. Typical applications of fins include automotive radiators, electronic equipment, heat exchangers, etc.

In an idealized fin analysis, the Ω is assumed to be one-dimensional. The base temperature of the fin is assumed to be equal to the wall temperature T_b of the main structure. The fin temperature T is assumed to be a function

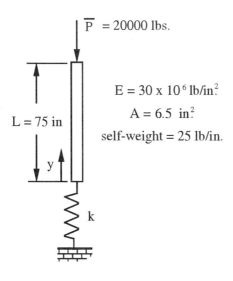

Figure 3.18: Bar With Elastic Support

only of the longitudinal coordinate of the fin (call this x); due to convection heat losses, the temperature must drop with x. The governing equation for the one-dimensional analysis of a fin of length L is [558]

$$\frac{d}{dx}\left(kA\frac{d\Theta}{dx}\right) - h_\infty p\Theta = 0 \quad , \quad 0 < x < L$$

where $\Theta = T - T_\infty$, T_∞ is the temperature of the surrounding fluid, k is the thermal conductivity coefficient, h_∞ is the external heat transfer coefficient, A is the cross-sectional area of the fin, and p is its perimeter. Although T_∞, k and h_∞ are typically assumed constant, A and p are, in general, functions of x.

Two boundary conditions are associated with the governing equation. The first is the known base temperature at the wall, viz.,

$$T(0) = T_b \quad \Rightarrow \quad \Theta(0) = T_b - T_\infty$$

The second boundary condition is the tip condition, for which there are *three* possibilities:

1. Convection loss at the tip :

$$-kA\frac{d\Theta}{dx} = h_\infty\Theta(L)$$

2. Insulated tip :

$$d\Theta dx = 0$$

3. Very long (and thus inefficient) fin :

$$\Theta \to 0 \quad \text{as} \quad x \to \infty$$

Solutions for various fin shape and boundary conditions are given in [286]. A finite difference approximation for a non-prismatic fin is considered herein. In particular, assume a linearly varying cross-sectional area and perimeter. Let $A_0 = 1$ and $P_0 = 4$, $\Theta_b = 80$, $k = 40$, $h_\infty = 10$, and $L = 20$.

Using the finite difference method with mesh spacings of $h = \Delta x = 4.0$, 2.0 and 1.0, calculate the approximate temperature distribution along the fin. Use central- and, where applicable, backward differences $O(h^2)$.

Chapter 4

The Method of Weighted Residuals

4.1 Introductory Remarks

The solutions generated using discrete approaches such as the finite difference method approximate the primary dependent variable ϕ only at a finite number of points in the domain Ω. The approximate solutions discussed in this and the following chapter are *continuous*, that is, they yield approximations to ϕ at all points within Ω and along its boundary Γ.

Given a system of governing differential equations

$$L(\phi) - f = 0 \tag{4.1}$$

along with suitable boundary conditions

$$B(\phi) - r = 0 \tag{4.2}$$

and possibly initial conditions, the general approach for generating continuous approximations is to assume a form for $\hat{\phi}$ whose functional dependence on position $\mathbf{x}$ is chosen a priori, but which includes undetermined functions of time t. Although this approach also applies to initial-value and eigenvalue problems, the present discussion is limited to boundary-value problems.

The following general form for $\hat{\phi}$ is assumed:

$$\hat{\phi}(\mathbf{x}) = \psi(\mathbf{x}) + \sum_{m=1}^{P} \alpha_m N_m(\mathbf{x}) \qquad (4.3)$$

Approximations of this form are sometimes referred to as "trial solutions with undetermined parameters" [125]. In equation (4.3) ψ represents a known function that exactly satisfies the boundary conditions on Γ. The N_m are a linearly independent set of *known* analytic *trial* or *interpolation functions* that satisfy

$$N_m(\mathbf{x}) = 0 \quad , \quad \mathbf{x} \text{ on } \Gamma \qquad (4.4)$$

Finally, the α_m represent *unknown* constants[1] whose values are computed in a manner that minimizes the error associated with the approximation. Although it is not necessary that $\hat{\phi}$ be linear in the α_m, such a choice is typically made for simplicity [159].

Since the summation in equation (4.3) has a finite upper bound P, it follows that the approximate solution is *finite-dimensional*. Stated another way, the continuous problem, possessing an infinite number of degrees of freedom, is reduced to an approximately equivalent problem with P degrees of freedom.

Clearly the trial functions N_m should be chosen to ensure that the approximation improves as P increases. Assuming the N_m are selected in some rational manner, the procedure for obtaining an approximate solution amounts to determining values for the α_m.

Two basic types of criteria are used in the determination of the α_m. In the first, the α_m are chosen so as to make weighted averages of residuals vanish. In the second, the α_m are chosen to give a stationary value to a functional $I(\phi)$ that is related to the system described by equations (4.1) and (4.2). The first criterion is discussed in this chapter; the second criterion is discussed in Chapter 5. The result of applying either of these criteria is a set of P simultaneous equations in the P unknowns α_m.

---

[1] For initial-value problems, ψ and the α_m would be functions of time; the N_m would be functions of time as well as of space.

4.2 Residuals

Requisite to the development of continuous approximate solution techniques is the existence of measures of the extent to which the function $\hat{\phi}$ in equation (4.3) satisfies the governing differential equation and boundary conditions. The specific measures employed are the *domain residual*

$$R_\Omega = L(\hat{\phi}) - f \quad \text{in} \quad \Omega \tag{4.5}$$

and the *boundary residual*

$$R_\Gamma = B(\hat{\phi}) - r \quad \text{on} \quad \Gamma \tag{4.6}$$

As the number P of trial functions in equation (4.3) is increased in successive approximations, it is hoped that the residuals R_Ω and R_Γ will become smaller. When both residuals are identically zero throughout Ω and along Γ, then equations (4.1) and (4.2) are obtained. The solution of this problem is the exact one. An approximation to this ideal is realized through the use of the method of weighted residuals.[2]

4.3 General Considerations

To simultaneously minimize R_Ω over Ω and R_Γ on Γ, the method of weighted residuals requires that the weighted integrals of the residuals be set equal to zero, viz.,

$$\int_\Omega w_i R_\Omega d\Omega + \int_\Gamma \bar{w}_i R_\Gamma d\Gamma = 0 \tag{4.7}$$

where $\{w_i(\mathbf{x}), \bar{w}_i(\mathbf{x}) \;\; ; \;\; i = 1, 2, \cdots, P\}$ constitute a *finite-dimensional* set of weighting functions.

Equation (4.7) thus requires that the spatial average or inner product of a finite set of weights and the residuals R_Ω and R_Γ be zero. The general convergence requirement that $\hat{\phi} \to \phi$ as $P \to \infty$ can likewise be viewed as requiring that equation (4.7) be satisfied for all i as $P \to \infty$. It is evident that this can only be true if $R_\Omega \to 0$ at all points within Ω and $R_\Gamma \to 0$ at all points along Γ.

---

[2]The name "method of weighted residuals" was coined by Crandall [125]. Other classical books on the subject are those by Collatz [113], Ames [11, 12], and Finlayson [158].

Since independent relationships are required in order to solve for the unknown coefficients α_m, it is clear that, provided they have an analytic form, the w_i and $\bar{w}_i$ must be *independent* functions. If the weighting functions are members of a complete set or "family" of functions,[3] then as $P \to \infty$, equation (4.7) indicates that R_Ω and R_Γ must be orthogonal to every member of a complete set of weighting functions. However, this implies that, as $P \to \infty$, the total residual converges to zero *in the mean*.[4] Consequently, the approximate solution would be expected to converge to the exact one in the mean, viz., [164]

$$\lim_{P \to \infty} \left\| \{\hat{\phi}\} - \{\phi\} \right\|_2 = \lim_{P \to \infty} \left\{ \sum_{i=1}^{P} \left(\hat{\phi}_i - \phi_i \right)^2 \right\}^{1/2} = 0 \qquad (4.8)$$

This is contrast with the notion of *uniform* convergence, which is typically defined by

$$\lim_{P \to \infty} \left\| \{\hat{\phi}\} - \{\phi\} \right\|_\infty = \lim_{P \to \infty} \left\{ \max_i \left| \hat{\phi}_i - \phi_i \right| \right\} = 0 \qquad (4.9)$$

where $i = 1, 2, \cdots, P$.

From the previous discussion, it is evident that, using the method of weighted residuals, approximate solutions can potentially be generated in which:

- The boundary conditions are satisfied exactly but the differential equation and initial conditions are approximated. Approximate solutions of this type are sometimes referred to as "interior methods" [113].

- The differential equation is satisfied exactly and the boundary conditions are approximated. Approximate solutions of this type lead to so-called "boundary methods" [113].

- The approximate solution satisfies neither the differential equation nor the boundary conditions. Approximate solutions of this type are referred to as "mixed methods" [159].

Although the emphasis herein is on "interior methods," a brief overview of the "mixed" and "boundary" methods is also presented.

---

[3] For example, a one-dimensional set of functions would be: $\sin x$, $\sin 2x$, $\sin 3x$, $\cdots$

[4] This is sometimes referred to as "weak convergence."

4.3.1 Interior Methods

In his book on the method of weighted residuals, Crandall [125] stated that the method could not be applied unless the approximate solution satisfied *all* of the boundary conditions (this was subsequently shown not to be a necessary condition for using the method). To this end, using the proposed approximation of equation (4.1), the function ψ and the trial functions N_m are chosen such that on Γ

$$B(\hat{\phi}) = r \tag{4.10}$$

where

$$N_m = 0 \quad ; \quad m = 1, 2, \cdots, P \tag{4.11}$$

In this manner, $\hat{\phi}$ automatically satisfies the boundary conditions for all values of the coefficients α_m, implying that $R_\Gamma = 0$.

For boundary conditions involving differentiation of ϕ, the approximation of derivatives of is straightforward. Provided that the functions N_m are continuous over Ω and that all their derivatives exist,[5] it follows that

$$\frac{\partial \hat{\phi}}{\partial x_i} = \frac{\partial \psi}{\partial x_i} + \sum_{m=1}^{P} \alpha_m \frac{\partial N_m}{\partial x_i} \quad ; \quad i = 1, 2, 3 \tag{4.12}$$

Substituting equation (4.3) into equation (4.5), and the resulting expression into equation (4.7) gives

$$\int_\Omega w_i R_\Omega d\Omega = \int_\Omega w_i \left\{ L(\psi) + L \left(\sum_{m=1}^{P} \alpha_m N_m \right) - f \right\} d\Omega = 0 \tag{4.13}$$

A total of P unknowns are associated with equation (4.13). The solution for the unknowns is obtained by solving the associated set of P linear algebraic equations that is produced. In matrix form this set of equations is written as

$$\mathbf{Ka} = \mathbf{b} \tag{4.14}$$

---

[5]Although in the present development the trial functions have been assumed to be continuously differentiable, such restrictions are quite often relaxed.

where (recall that L is a linear operator)

$$K_{im} = \int_\Omega w_i L(N_m) \, d\Omega \qquad (4.15)$$

$$b_i = \int_\Omega w_i f \, d\Omega - \int_\Omega w_i L(\psi) d\Omega \qquad (4.16)$$

and

$$\mathbf{a} = \left\{ \alpha_1 \quad \alpha_2 \quad \cdots \quad \alpha_P \right\}^T \qquad (4.17)$$

In general, the coefficient matrix $\mathbf{K}$ will be full and will not exhibit the banded structure characteristic of matrices produced, in certain cases, by the finite difference method. Once the coefficients in $\mathbf{K}$ are evaluated, the set of equations is solved for the unknowns α_m ; $i = 1, 2, \cdots, P$. The process of determining an approximate solution to the given differential equation is thus complete.

4.3.2   Boundary Methods

In the case of boundary methods,[6] the domain residual vanishes, leaving only the boundary residual given by equation (4.6). Since only the boundary needs to be considered, the dimensionality of the problem is reduced. Substituting equation (4.1) (without ψ) into equation (4.6) gives

$$R_\Gamma = B\left(\hat{\phi}\right) - r = B\left(\sum_{m=1}^{P} \alpha_m N_m\right) - r \qquad (4.18)$$

In the absence of a domain residual, equation (4.7) becomes

$$\int_\Gamma \bar{w}_i R_\Gamma \, d\Gamma = \int_\Gamma \bar{w}_i \left[B\left(\sum_{m=1}^{P} \alpha_m N_m\right) - r \right] d\Gamma = 0 \qquad (4.19)$$

---

[6]An early example of boundary solution methods was presented by Trefftz [543]. Subsequent work included the method of singularities [235] and the so-called panel method [472]. More recently, boundary element methods have become an important class of approximate solution techniques [51, 94]. In such methods, singular solutions (Green's functions) involving complex integration procedures are often used to satisfy the governing equations. Alternate approaches have, however, been proposed [593, 224, 225, 226] that accomplish the same task without introducing singularities.

Equation (4.19) can be written in matrix form as equation (4.14), with

$$K_{im} = \int_{\Gamma} \bar{w}_i B(N_m) d\Gamma \qquad (4.20)$$

and

$$b_i = \int_{\Gamma} \bar{w}_i r \, d\Gamma \qquad (4.21)$$

where $i, m = 1, 2, \cdots, P$. The vector of unknown coefficients is again given by equation (4.17). The matrix $\mathbf{K}$ will, in general, be non-symmetric, full, and will not exhibit a banded structure.

4.3.3 Mixed Methods

Although typically more difficult than interior and boundary methods, so-called *mixed methods* have also been successfully treated [494, 392, 85, 159]. Consider the expression

$$\hat{\phi} = \sum_{m=1}^{P} \alpha_m N_m \qquad (4.22)$$

which does not satisfy a priori some or all of the boundary conditions associated with the problem. The domain residual is thus supplemented by a boundary residual. Substituting R_Ω and R_Γ from equations (4.5) and (4.6), respectively, into equation (4.7) gives

$$\int_{\Omega} w_i \left[L(\hat{\phi}) - f \right] d\Omega + \int_{\Gamma} \bar{w}_i \left[B(\hat{\phi}) - r \right] d\Gamma = 0 \qquad (4.23)$$

Next, substituting equation (4.22) into equation (4.23) gives

$$\int_{\Omega} w_i L \left(\sum_{m=1}^{P} \alpha_m N_m \right) d\Omega + \int_{\Gamma} \bar{w}_i B \left(\sum_{m=1}^{P} \alpha_m N_m \right) d\Gamma$$

$$- \int_{\Omega} w_i f \, d\Omega - \int_{\Gamma} \bar{w}_i r \, d\Gamma = 0 \qquad (4.24)$$

Equation (4.24) can be written in matrix form as equation (4.14), with

$$K_{im} = \int_{\Omega} w_i L(N_m) d\Omega + \int_{\Gamma} \bar{w}_i B(N_m) d\Gamma \qquad (4.25)$$

and

$$b_i = \int_{\Omega} w_i f \, d\Omega + \int_{\Gamma} \bar{w}_i r \, d\Gamma \qquad (4.26)$$

where $i, m = 1, 2, \cdots, P$. The vector of unknown coefficients is again given by equation (4.17). As in the case of equation (4.14), the matrix $\mathbf{K}$ will, in general, be full and will not exhibit a banded structure.

In closing this general discussion of the method of weighted residuals some reflection is warranted. While continuous approximations represent an improvement over discrete ones, the "price" paid is the necessity of selecting a set of trial functions and a set of arbitrary independent weighting functions. The following two sections are provided to aid in this selection process.

4.4   Choice of Trial Functions

The importance of properly selecting the former set of functions may be summarized by Crandall's remark [125]:

> The selection of the trial family is, of course, the crucial point in any procedure of this type. It is up to the ingenuity of the analyst so to construct the trial solution that a maximum of information can be extracted with a minimum of computation. The more that is known about the expected behavior of a solution (symmetry, etc.), the more intelligently can the trial family be set up.

For a given problem, there is no fixed procedure by which a good, if not the best, choice of trial function can be realized. Nonetheless, the following general guidelines are noteworthy:

- As noted above by Crandall [125], in selecting trial functions any symmetry properties of a system should be exploited. However, there does not appear to be a systematic way to do this for all problems [159].

- In standard problems, it is typically convenient to have the trial functions satisfy the boundary conditions.[7] A method for constructing complete sets of trial functions that vanish on a boundary of complicated shape has been presented by Kantorovich and Krylov [278].

- In deciding upon the type of trial function to use, it is timely to note that *polynomials* and *trigonometric series* are quite popular. Orthonormalized trial functions have also been proposed [279], since their use avoids potential numerical difficulties that become manifest for large P.

Some more specific details pertaining to the choice of trial functions are given by Finlayson and Scriven [159], who note that

> Selecting approximating functions remains somewhat dependent on the user's intuition and experience, and this is often regarded as a major disadvantage of MWR.

Having addressed the choice of trial functions, we turn our attention to some specific weighting functions that have been proposed over the years.

4.5 Specific Weighting Functions

In theory, any set of independent weighting functions can be used in conjunction with the general weighted residual approximation given by equation (4.7). In practice, however, only a relatively small number of weighting functions have historically been used.[8] The specific choice of weighting function corresponds to a particular approximate solution technique that, though originally introduced as a separate method, was subsequently shown to be a specific application of the method of weighted residuals. Such a unification is attributed to Crandall [125], who gave the method its name [159].

In the sequel, only the boundary conditions are assumed to be satisfied exactly. Consequently, only *interior methods* are considered.

---

[7]This may require the choice of a suitable function ψ in equation (4.3).

[8]An excellent historical review of the method of weighted residuals, along with an exhaustive list of references related to specific applications of the method prior to 1967, is given by Finlayson and Scriven [159]. More current applications are discussed by Ames [12].

4.5.1   Collocation Method

In the collocation method[9] the weighting functions are Dirac delta functions, viz.,

$$w_i = \delta(\mathbf{x} - \mathbf{x}_i) \quad , \ i = 1, 2, \cdots , P \tag{4.27}$$

where

$$\delta(\mathbf{x} - \mathbf{x}_i) = \begin{cases} 0 & \text{for} \quad \mathbf{x} \neq \mathbf{x}_i \\ \infty & \text{for} \quad \mathbf{x} = \mathbf{x}_i \end{cases} \tag{4.28}$$

and

$$\int_{\mathbf{x} < \mathbf{x}_i}^{\mathbf{x} > \mathbf{x}_i} G(\mathbf{x})\, \delta(\mathbf{x} - \mathbf{x}_i) dx = G(\mathbf{x}_i) \tag{4.29}$$

Substituting these weighting functions of equation (4.27) into equation (4.13) gives

$$\int_{\Omega} w_i R_\Omega d\Omega = \int_{\Omega} \delta(\mathbf{x} - \mathbf{x}_i) R_\Omega d\Omega = R_\Omega\,(\mathbf{x}_i) = 0 \tag{4.30}$$

for $i = 1, 2, \cdots , P$. This choice of weighting function is thus equivalent to requiring that the residual equal to zero at the P arbitrarily chosen points $\mathbf{x}_i$ in Ω[10]. As P is increased, the residual vanishes at more and more points and presumably approaches zero throughout Ω.

Associated with the collocation method is the important issue of where to locate the collocation points. Some guidance on this matter is provided by Wright [574], who showed that for ordinary differential equations the residual is minimized if these points are given by the roots of the Chebyshev polynomials.

The application of the collocation method is illustrated in the following example.

---

[9]This method is also referred to as *point collocation* or *deterministic collocation*. It is attributed to Frazer et al. [175].

[10]Although the residual is set equal to zero at the P distinct collocation points, this does not imply that the approximate solution will be equal to the exact solution at these points.

Example 4.1: Application of the Collocation Method

The one-dimensional, steady-state heat conduction problem analyzed in Example 3.2 is again considered. The solution domain is shown in Figure 4.1.

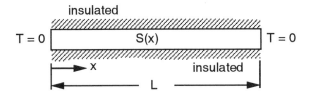

Figure 4.1: Schematic of an Insulated Rod

Recall that a constant thermal conductivity k and a linearly varying heat source $S = S_o x$ (where S_o is a constant) are assumed. The governing differential equation for the problem is

$$k\frac{d^2 T}{dx^2} + S_o x = 0 \qquad (4.31)$$

Homogeneous essential boundary conditions $T(0) = T(L) = 0$ are assumed. The exact solution for this problem is thus

$$T(x) = \frac{S_o}{6k}\left(L^2 - x^2\right)x \qquad (4.32)$$

The corresponding flux distribution per unit area is

$$q(x) = -k\frac{dT}{dx} = \frac{S_o}{6}\left(3x^2 - L^2\right) \qquad (4.33)$$

For the approximate solution, we use the following sine series:

$$\hat{T} = \sum_{m=1}^{P} \alpha_m \sin\frac{m\pi x}{L} \qquad (4.34)$$

Remarks

1. The form of equation (4.34) implies that the trial functions are

$$N_m = \sin \frac{m\pi x}{L} \qquad (4.35)$$

2. As mentioned in conjunction with equation (4.4), the chosen trial functions satisfy the condition that $N_m = 0$ on Γ. Since this happens to coincide with the boundary conditions, it follows that $\psi = 0$ in equation (4.3).

3. Since the approximate solution exactly satisfies the boundary conditions at $x = 0$ and at $x = L$, the boundary residual $R_\Gamma(x)$ is zero.

Differentiating equation (4.34) and substituting the resulting expression into equation (4.31) gives the following general expression for the total residual:

$$R(x) = R_\Omega = k\frac{d^2\hat{T}}{dx^2} + S_o x = -k \sum_{m=1}^{P} \left(\frac{m\pi}{L}\right)^2 \alpha_m \sin \frac{m\pi x}{L} + S_o x \quad (4.36)$$

We first consider the simplest approximation, that is $P = 1$. From equation (4.36) the total residual is thus:

$$R(x) = -k \left(\frac{\pi}{L}\right)^2 \alpha_1 \sin \frac{\pi x}{L} + S_o x \qquad (4.37)$$

For simplicity, the collocation point is located at $x = L/2$. Setting $R(L/2) = 0$ gives the requisite equation for α_1, viz.,

$$-k \left(\frac{\pi}{L}\right)^2 \alpha_1 \sin \frac{\pi}{2} + S_o \frac{L}{2} = 0 \qquad (4.38)$$

It follows that

$$\alpha_1 = \frac{S_o L^3}{2k\pi^2} \qquad (4.39)$$

Substituting for α_1 into equation (4.34) gives the following approximate temperature distribution:

$$\hat{T}(x) = \frac{S_o L^3}{2k\pi^2} \sin \frac{\pi x}{L} \qquad (4.40)$$

Differentiating equation (4.40) with respect to x leads to the corresponding approximate heat flux distribution per unit area:

$$\hat{q}(x) = -k\frac{d\hat{T}(x)}{dx} = -\frac{S_o L^2}{2\pi}\cos\frac{\pi x}{L} \qquad (4.41)$$

In Figures 4.2 and 4.3 the approximations given by equations (4.40) and (4.41) are compared to the exact solutions for $T(x)$ and $q(x)$, respectively. To facilitate the comparisons, values of $S_o = k = L = 1.0$ have been assumed.

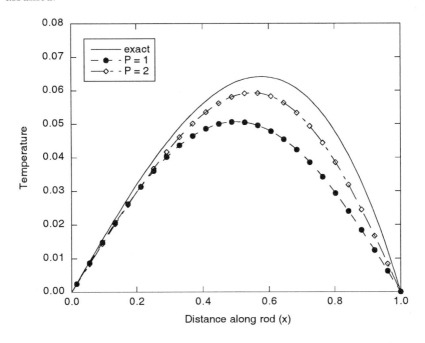

Figure 4.2: Assessment of Collocation Method: Temperature Distribution

From Figure 4.2 it is evident that the discrepancy between $\hat{T}(x)$ and $T(x)$ is rather large. This is not too surprising, since the approximate solution is symmetric about $x = 0.5$, while, due to the presence of the linearly varying source term, the exact solution is not. From Figure 4.3 the greatest discrepancy between $\hat{q}(x)$ and $q(x)$ occurs near the right end of the rod.

Seeking to improve the approximation, we take $P = 2$ in equation (4.34). From equation (4.36) the total residual is thus

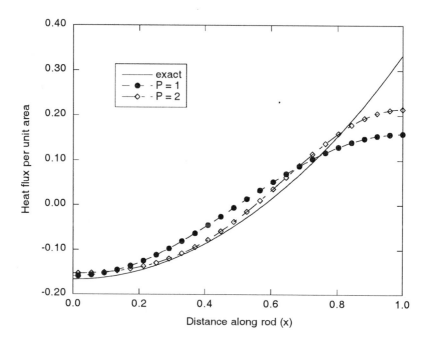

Figure 4.3: Assessment of Collocation Method: Heat Flux Distribution

$$R(x) = -k\left(\frac{\pi}{L}\right)^2 \left(\alpha_1 \sin\frac{\pi x}{L} + 4\alpha_2 \sin\frac{2\pi x}{L}\right) + S_o x \qquad (4.42)$$

The two collocation points are chosen to be $x = L/3$ and $x = 2L/3$. Setting $R(L/3) = R(2L/3) = 0$ gives the requisite two equations, viz.,

$$-k\left(\frac{\pi}{L}\right)^2 \left[\alpha_1 \sin\frac{\pi}{3} + 4\alpha_2 \sin\frac{2\pi}{3}\right] + \frac{S_o L}{3} = 0 \qquad (4.43)$$

$$-k\left(\frac{\pi}{L}\right)^2 \left[\alpha_1 \sin\frac{2\pi}{3} + 4\alpha_2 \sin\frac{4\pi}{3}\right] + \frac{2S_o L}{3} = 0 \qquad (4.44)$$

Simplifying these equations and writing them in matrix form gives

$$\frac{\sqrt{3}}{2}\begin{bmatrix} 1 & 4 \\ 1 & -4 \end{bmatrix} \begin{Bmatrix} \alpha_1 \\ \alpha_2 \end{Bmatrix} = \frac{S_o L^3}{3k\pi^2} \begin{Bmatrix} 1 \\ 2 \end{Bmatrix} \qquad (4.45)$$

The solution of equation (4.45) is

$$\begin{Bmatrix} \alpha_1 \\ \alpha_2 \end{Bmatrix} = \frac{S_o L^3}{\sqrt{3}k\pi^2} \begin{Bmatrix} 1 \\ -1/12 \end{Bmatrix} \tag{4.46}$$

Thus, from equation (4.34)

$$\hat{T}(x) = \langle N_1 \quad N_2 \rangle \begin{Bmatrix} \alpha_1 \\ \alpha_2 \end{Bmatrix} = \frac{S_o L^3}{\sqrt{3}k\pi^2} \left(\sin \frac{\pi x}{L} - \frac{1}{12} \sin \frac{2\pi x}{L} \right) \tag{4.47}$$

The corresponding approximate heat flux distribution per unit area is

$$\hat{q}(x) = -k \frac{d\hat{T}}{dx} = -k \left\langle \frac{dN_1}{dx} \quad \frac{dN_2}{dx} \right\rangle \begin{Bmatrix} \alpha_1 \\ \alpha_2 \end{Bmatrix}$$

$$= -\frac{S_o L^2}{\sqrt{3}\pi} \left(\cos \frac{\pi x}{L} - \frac{1}{6} \cos \frac{2\pi x}{L} \right) \tag{4.48}$$

In Figures 4.2 and 4.3 the approximations of equations (4.47) and (4.48) are compared to the exact solutions for $T(x)$ and $q(x)$, respectively. To facilitate the comparisons, values of $S_o = k = L = 1.0$ have again been assumed. From these figures it is evident that as P is increased from 1 to 2, the approximations converge to the exact solution.

4.5.2 Subdomain Method

In the subdomain method,[11] each weighting function is taken as unity over a specific portion of the domain Ω, viz.,

$$w_i = \begin{cases} 1 & \text{if } x \in \Omega_i \\ 0 & \text{otherwise} \end{cases} \tag{4.49}$$

where $i = 1, 2, \cdots, P$. This selection amounts to the differential equation being satisfied, on the average, in each of the P subdomains Ω_i. The application of the subdomain method is illustrated in the following example.

---

[11]This method is also referred to as *subdomain collocation*. It was introduced by Biezeno and Koch [77], who applied it to the analysis of plates and elastically supported beams and to the general problem of elastic equilibrium [76, 78].

Example 4.2: Application of the Subdomain Method

The problem solved in Example 4.1 is again considered. To facilitate the comparison of results obtained using the subdomain method with those obtained using collocation, the approximation given by equation (4.34) is once again used. The total residual is thus given by equation (4.36). Application of the subdomain method means that, over a specific interval $[a, b] \subset \Omega$, we must evaluate the integral

$$\int_a^b (1) R(x) \, dx = 0 \tag{4.50}$$

Substituting equation (4.36) into equation (4.50) and integrating gives the following general result:

$$\left[k \sum_{m=1}^{P} \left(\frac{m\pi}{L} \right) \alpha_m \cos \frac{m\pi x}{L} + \frac{S_o}{2} x^2 \right]_a^b = 0 \tag{4.51}$$

For $P = 1$ equation (4.51) becomes

$$\left[k \left(\frac{\pi}{L} \right) \alpha_1 \cos \frac{\pi x}{L} + \frac{S_o}{2} x^2 \right]_0^L = 0 \tag{4.52}$$

Evaluating the limits of integration and solving for α_1 gives

$$\alpha_1 = \frac{S_o L^3}{4k\pi} \tag{4.53}$$

Substituting for α_1 into equation (4.34) gives the following approximate temperature distribution:

$$\hat{T}(x) = \frac{S_o L^3}{4k\pi} \sin \frac{\pi x}{L} \tag{4.54}$$

The corresponding approximate heat flux distribution per unit area is

$$\hat{q}(x) = -k \frac{d\hat{T}(x)}{dx} = -\frac{S_o L^2}{4} \cos \frac{\pi x}{L} \tag{4.55}$$

In Figures 4.4 and 4.5 the approximations given by equations (4.54) and (4.55) are compared to the exact solutions for $T(x)$ and $q(x)$, respectively. To facilitate the comparisons, values of $S_o = k = L = 1.0$ have again been assumed. As evident from these figures, the discrepancy between the approximate solution corresponding to $P = 1$ and the exact one is rather large. We thus repeat the analysis, only for $P = 2$.

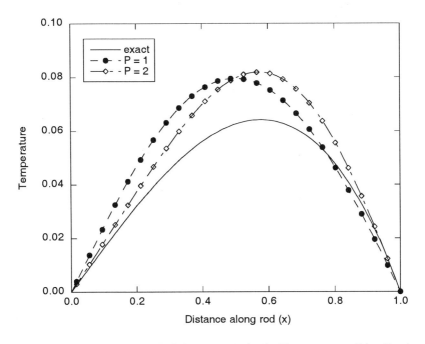

Figure 4.4: Assessment of Subdomain Method: Temperature Distribution

Since $P = 2$, we must subdivide Ω into two subdomains. Choosing subdomains of equal length and specializing equation (4.51) gives

$$\int_0^{L/2} R(x)\,dx = -\frac{k\pi}{L}\left(\alpha_1 + 4\alpha_2\right) + \frac{S_o L^2}{8} = 0 \qquad (4.56)$$

$$\int_{L/2}^{L} R(x)\,dx = -\frac{k\pi}{L}\left(\alpha_1 - 4\alpha_2\right) + \frac{3S_o L^2}{8} = 0 \qquad (4.57)$$

Writing equations (4.56) and (4.57) in matrix form gives

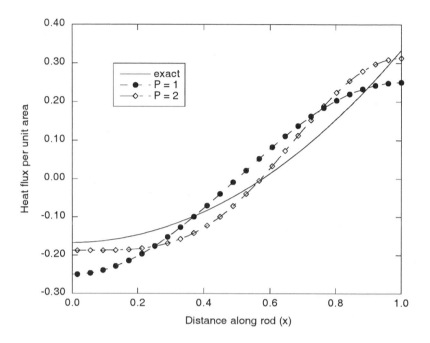

Figure 4.5: Assessment of Subdomain Method: Heat Flux Distribution

$$\begin{bmatrix} 1 & 4 \\ 1 & -4 \end{bmatrix} \begin{Bmatrix} \alpha_1 \\ \alpha_2 \end{Bmatrix} = \frac{S_o L^3}{8\pi k} \begin{Bmatrix} 1 \\ 3 \end{Bmatrix} \tag{4.58}$$

The solution to equation (4.58) is

$$\begin{Bmatrix} \alpha_1 \\ \alpha_2 \end{Bmatrix} = \frac{S_o L^3}{32 k\pi} \begin{Bmatrix} 8 \\ -1 \end{Bmatrix} \tag{4.59}$$

Thus,

$$\hat{T}(x) = \langle N_1 \quad N_2 \rangle \begin{Bmatrix} \alpha_1 \\ \alpha_2 \end{Bmatrix} = \frac{S_o L^3}{32 k\pi} \left(8 \sin \frac{\pi x}{L} - \sin \frac{2\pi x}{L} \right) \tag{4.60}$$

The corresponding approximate heat flux distribution per unit area is

$$\hat{q}(x) = -k \frac{d\hat{T}(x)}{dx} = -\frac{S_o L^2}{16} \left(4 \cos \frac{\pi x}{L} - \cos \frac{2\pi x}{L} \right) \tag{4.61}$$

In Figures 4.4 and 4.5 the approximations given by equations (4.60) and (4.61) are compared to the exact solutions for $T(x)$ and $q(x)$, respectively. As before, values of $S_o = k = L = 1.0$ have been assumed. From these figures it is evident that the approximate solution corresponding to $P = 2$ is significantly more accurate than the one for $P = 1$. This is particularly true for the primary dependent variable $T(x)$ which, as expected, converges more rapidly to the exact solution than does the secondary dependent variable $\hat{q}(x)$.

4.5.3 Method of Least Squares

In the method of least squares[12] the weighting functions are given by

$$w_i = \frac{\partial R_\Omega}{\partial \alpha_i} \quad , \quad i = 1, 2, \cdots P \tag{4.62}$$

Substituting equation (4.62) into equation (4.7) and recalling that an "interior" method has been assumed, gives

$$\int_\Omega \left(\frac{\partial R_\Omega}{\partial \alpha_i} \right) R_\Omega \, d\Omega = 0 \tag{4.63}$$

To better understand equation (4.63) consider an approximate solution that is obtained by minimizing the sum of the squares of the total residual, suitably weighted, at each point in Ω. This requires minimization of an error measure E, given by

$$E(\alpha_1, \alpha_2, \cdots, \alpha_P) = \int_\Omega w_1 \left(R_\Omega \right)^2 d\Omega \tag{4.64}$$

where the weighting function w_1 is independent of the α_i and is arbitrarily chosen. Although typically w_1 is set equal to one, it can be viewed as being a sort of "tuning parameter" [116].

---

[12] Karl Friedrich Gauss apparently discovered this method in 1794 [213], although he did not present it publicly until 1809 [125]. The method was independently discovered and published by Legendre [309] in France in 1805, and also by Adrain [3] in the United States in 1808. In 1928, Picone [436] presented a more specialized application of the method. Further historical details pertaining to the method, as well as an extensive list of applications and related references is given in the review paper by Eason [146].

To obtain the set of P equations in the unknowns α_i, E is minimized with respect to the α_i, viz.,

$$\frac{\partial E}{\partial \alpha_i} = 2 \int_\Omega w_1 \left(\frac{\partial R_\Omega}{\partial \alpha_i} \right) R_\Omega \, d\Omega = 0 \qquad (4.65)$$

for $i = 1, 2, \cdots, P$. If $w_1 = 1.0$, equation (4.65) reduces to equation (4.63).

The method of least squares always produces a *symmetric* coefficient matrix $\mathbf{K}$. However, $\mathbf{K}$ tends to be ill-conditioned [116]. The application of the method is illustrated in the following example.

Example 4.3: Application of the Method of Least Squares

The problem solved in the previous two examples is again considered. To facilitate the comparison of results obtained using the least squares method with those obtained using collocation and the subdomain method, the approximation given by equation (4.34) is once again used. The total residual is thus given by equation (4.36).

For $P = 1$, the weighted error of equation (4.64) becomes

$$E = \int_0^L (1) \left\{ -k \left(\frac{\pi}{L} \right)^2 \left[\alpha_1 \sin \frac{\pi x}{L} \right] + S_o x \right\}^2 dx \qquad (4.66)$$

Rather than expanding the integrand and then differentiating with respect to the α_1, it is easier to first differentiate and then integrate. This leads to

$$\frac{\partial E}{\partial \alpha_1} = 2 \int_0^L \left\{ -k \left(\frac{\pi}{L} \right)^2 \left[\alpha_1 \sin \frac{\pi x}{L} \right] + S_o x \right\}$$
$$\left[-k \left(\frac{\pi}{L} \right)^2 \sin \frac{\pi x}{L} \right] dx = 0 \qquad (4.67)$$

Using the orthogonality relation

$$\int_0^L \sin \frac{m\pi x}{L} \sin \frac{n\pi x}{L} dx = \begin{cases} \dfrac{L}{2} & \text{if } m = n \\[2ex] 0 & \text{otherwise} \end{cases} \qquad (4.68)$$

reduces equation (4.67) to

$$-k\left(\frac{\pi}{L}\right)^2 \frac{L}{2}\alpha_1 + S_o\left[\left(\frac{L}{\pi}\right)^2 \sin\frac{\pi x}{L} - \frac{Lx}{\pi}\cos\frac{\pi x}{L}\right]_0^L = 0 \qquad (4.69)$$

Evaluating the limits of integration gives

$$-k\left(\frac{\pi}{L}\right)^2 \frac{L}{2}\alpha_1 + S_o\frac{L^2}{\pi} = 0 \quad \Rightarrow \quad \alpha_1 = \frac{2S_o L^3}{k\pi^3} \qquad (4.70)$$

Substituting for α_1 into equation (4.34) gives the following approximate temperature distribution:

$$\hat{T}(x) = \frac{2S_o L^3}{k\pi^3}\sin\frac{\pi x}{L} \qquad (4.71)$$

The corresponding approximate heat flux distribution per unit area is

$$\hat{q}(x) = -k\frac{d\hat{T}(x)}{dx} = -\frac{2S_o L^2}{\pi^2}\cos\frac{\pi x}{L} \qquad (4.72)$$

In Figures 4.6 and 4.7 the approximations given by equations (4.71) and (4.72) are compared to the exact solutions for $T(x)$ and $q(x)$, respectively. To facilitate the comparisons, values of $S_o = k = L = 1.0$ have again been assumed. As evident from these figures, although the $\hat{T}(x)$ distribution corresponding to $P = 1$ is more accurate than corresponding solutions obtained using the collocation and subdomain methods, it still differs from the exact solution. Consequently, we repeat the analysis, only for $P = 2$. For $P = 2$ the weighted error of equation (4.64) becomes

$$E = \int_0^L (1)\left\{-k\left(\frac{\pi}{L}\right)^2\left[\alpha_1 \sin\frac{\pi x}{L} + 4\alpha_2 \sin\frac{2\pi x}{L}\right] + S_o x\right\}^2 dx \qquad (4.73)$$

This leads to the following two equations:

$$\frac{\partial E}{\partial \alpha_1} = 2\int_0^L \left\{-k\left(\frac{\pi}{L}\right)^2\left[\alpha_1 \sin\frac{\pi x}{L} + 4\alpha_2 \sin\frac{2\pi x}{L}\right] + S_o x\right\}$$
$$\left[-k\left(\frac{\pi}{L}\right)^2 \sin\frac{\pi x}{L}\right] dx = 0 \qquad (4.74)$$

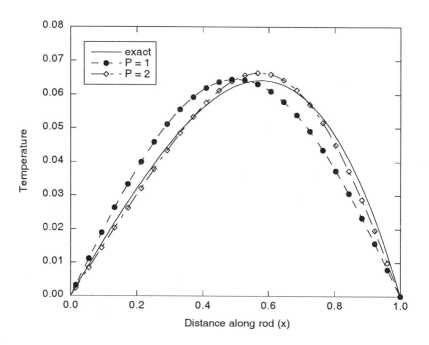

Figure 4.6: Assessment of Least Squares Method: Temperature Distribution

$$\frac{\partial E}{\partial \alpha_2} = 2 \int_0^L \left\{ -k \left(\frac{\pi}{L}\right)^2 \left[\alpha_1 \sin\frac{\pi x}{L} + 4\alpha_2 \sin\frac{2\pi x}{L}\right] + S_o x \right\}$$
$$\left[-4k \left(\frac{\pi}{L}\right)^2 \sin\frac{2\pi x}{L}\right] dx = 0 \qquad (4.75)$$

Using the orthogonality relation of equation (4.68), equations (4.74) and (4.75) reduce to

$$-k \left(\frac{\pi}{L}\right)^2 \frac{L}{2}\alpha_1 + S_o \left[\left(\frac{L}{\pi}\right)^2 \sin\frac{\pi x}{L} - \frac{Lx}{\pi} \cos\frac{\pi x}{L}\right]_0^L = 0 \qquad (4.76)$$

$$-k \left(\frac{\pi}{L}\right)^2 2L\alpha_2 + S_o \left[\left(\frac{L}{2\pi}\right)^2 \sin\frac{2\pi x}{L} - \frac{Lx}{2\pi} \cos\frac{2\pi x}{L}\right]_0^L = 0 \qquad (4.77)$$

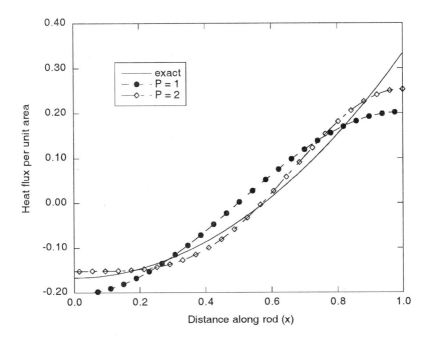

Figure 4.7: Assessment of Least Squares Method: Heat Flux Distribution

Evaluating the limits of integration leads to the following solution:

$$\begin{Bmatrix} \alpha_1 \\ \alpha_2 \end{Bmatrix} = \frac{S_o L^3}{4k\pi^3} \begin{Bmatrix} 8 \\ -1 \end{Bmatrix} \tag{4.78}$$

Substituting for α_1 and α_2 into equation (4.34) gives the following approximate temperature distribution:

$$\hat{T}(x) = \langle N_1 \quad N_2 \rangle \begin{Bmatrix} \alpha_1 \\ \alpha_2 \end{Bmatrix} = \frac{S_o L^3}{4k\pi^3} \left(8 \sin \frac{\pi x}{L} - \sin \frac{2\pi x}{L} \right) \tag{4.79}$$

The corresponding approximate heat flux distribution per unit area is

$$\hat{q}(x) = -k \frac{d\hat{T}}{dx} = -k \left\langle \frac{dN_1}{dx} \quad \frac{dN_2}{dx} \right\rangle \begin{Bmatrix} \alpha_1 \\ \alpha_2 \end{Bmatrix}$$

$$= -\frac{S_o L^2}{2\pi^2} \left(4 \cos \frac{\pi x}{L} - \cos \frac{2\pi x}{L} \right) \tag{4.80}$$

Remarks

   1. Unlike the present case involving functions that obey the orthogonality relationship of equation (4.68), the system of P equations in the unknowns α_1, α_2, $\cdots$, α_P does not generally uncouple.

   2. Due to the orthogonality of the sine functions, equation (4.76) is identical to equation (4.69), which was associated with the solution for $P = 1$. A portion of the computational effort is thus "reused" when progressively more terms are used in the sine series of equation (4.34).

   In Figures 4.6 and 4.7 the approximations given by equations (4.79) and (4.80) are compared to the exact solutions for $T(x)$ and $q(x)$, respectively. As before, values of $S_o = k = L = 1.0$ have been assumed. From these figures it is evident that the approximate solution corresponding to $P = 2$ is more accurate than the one for $P = 1$. In addition, the present solution represents an improvement over the comparable solutions generated using collocation and the subdomain method.

4.5.4   The Bubnov-Galerkin Method

The Bubnov-Galerkin method was the first example of a weighted residual method, and is probably the most popular such method.[13] Steady, unsteady, and eigenvalue problems have proved to be amenable to a Bubnov-Galerkin formulation [164]. This method, and its generalizations, have

---

[13]The method derives its name from the two men responsible for its development, namely I. G. Bubnov and B. G. Galerkin. As noted by Mikhlin [392], Bubnov [97] presented the formulation in conjunction with a variational formulation for solving eigenvalue problems. In this formulation, he required that the trial and weighting functions be orthogonal. Bubnov's work was prompted by a paper of Timoshenko [537] in which he studied the stability of rods and plates using the Ritz method (to be discussed in Chapter 5). The observations of Bubnov gave rise to a new approximate solution method, the development and application of which is closely associated with the Russian engineer and applied mechanician B. G. Galerkin. In 1915 Galerkin [185] presented a paper on the elastic equilibrium and stability of rods and thin plates. The method used was identical in form to that obtained by Bubnov. However, Galerkin's essentially new contribution was the fact that he did not connect his method with any variational statement, and did not require the orthogonality of the coordinate functions [392]. Following Galerkin's original paper, the method became very well-known in the Russian literature and was rather extensively applied to the solution of extremely diverse applied problems [392]. Only 25 years after Galerkin's publication was the Bubnov-Galerkin method substantiated in application to integral equations [462] and to certain special ordinary fourth-order differential equations [431]. A fairly general sufficient criterion for the convergence of the method was obtained by Mikhlin [390, 391] and applied to a number of problems. The Bubnov-Galerkin method appeared in the Western literature only in the late 1930s and early 1940s [164]. During the 1950s, use of the method increased rapidly.

been used to solve linear and non-linear problems in structural mechanics, dynamics, fluid flow, heat and mass transfer, hydrodynamic stability, acoustics, etc. Since the Bubnov-Galerkin method is commonly referred to as simply the Galerkin method, this terminology shall be adopted henceforth.

Application of the Galerkin method requires that the following conditions be satisfied [164]:

- The weighting functions are chosen from the *same* set as the trial functions, viz.,

$$w_i(\mathbf{x}) = N_i(\mathbf{x}) = \frac{\partial \hat{\phi}}{\partial \alpha_i} \quad ; \quad i = 1, 2, \cdots, P \qquad (4.81)$$

- The trial and weighting functions must be *linearly independent.*

- The trial and weighting functions should be chosen from the first P functions of a *complete set of functions* [278].

- The trial functions should *exactly* satisfy the boundary conditions and, if applicable, the initial conditions.

If the system of governing differential equations and corresponding boundary conditions is self-adjoint, the Galerkin method will yield a *symmetric* coefficient matrix $\mathbf{K}$ [125], whose entries are

$$K_{im} = \int_\Omega N_i L\left(N_m\right) d\Omega \quad ; \quad i, m = 1, 2, \cdots, P \qquad (4.82)$$

When $\mathbf{K}$ is symmetric, a significant computational advantage is realized.

From equation (4.82) it is evident that in the Galerkin method the trial functions are all orthogonal to the residual. Descriptively, this ensures that the residual is independent of, and thus as small as possible for, the given set of trial functions.

Remark

1. It is timely to mention the Petrov-Galerkin method [15], which is a generalization of the Galerkin method. In the Petrov-Galerkin method the weighting functions are

$$w_i\left(\mathbf{x}\right) = F_i\left(\mathbf{x}\right) \quad , \quad i = 1, 2, \cdots P \qquad (4.83)$$

where $F_i(\mathbf{x})$ represents an analytic function, similar to the trial functions N_m used by the Galerkin method, but with additional terms or factors included to impose some further requirement on the approximate solution. The motivation for such methods is the undesirable stability properties possessed by algebraic equations associated with conventional Galerkin based analyses of convection-dominated flows [218].

Example 4.4: Application of the Galerkin Method

Applying the Galerkin method to the same problem considered in Example 4.1 to 4.3, recall that the trial functions associated with the approximation of equation (4.35) are

$$N_m = \sin \frac{m\pi x}{L} \tag{4.84}$$

The Galerkin method requires that the weighting functions be identical to the trial functions, thus

$$w_i \equiv N_i = \sin \frac{i\pi x}{L} \quad ; \quad i = 1, 2, \cdots, P \tag{4.85}$$

Since the total residual is again given by equation (4.36), for $P = 1$ the weighted residual statement of the problem is

$$\int_0^L \sin \frac{\pi x}{L} \left[-k \left(\frac{\pi}{L}\right)^2 \alpha_1 \sin \frac{\pi x}{L} + S_o x \right] dx = 0 \tag{4.86}$$

Using the orthogonality relation given in equation (4.68), equation (4.86) becomes

$$-k \left(\frac{\pi}{L}\right)^2 \frac{L}{2} \alpha_1 + S_o \left[\left(\frac{L}{\pi}\right)^2 \sin \frac{\pi x}{L} - \frac{Lx}{\pi} \cos \frac{\pi x}{L} \right]_0^L = 0 \tag{4.87}$$

which is *identical* to equation (4.69) that was obtained for the method of least squares. It follows that

$$\alpha_1 = \frac{2 S_o L^3}{k \pi^3} \tag{4.88}$$

and

$$\hat{T}(x) = \frac{2S_o L^3}{k\pi^3} \sin \frac{\pi x}{L} \tag{4.89}$$

$$\hat{q}(x) = -k\frac{d\hat{T}(x)}{dx} = -\frac{2S_o L^2}{\pi^2} \cos \frac{\pi x}{L} \tag{4.90}$$

The graphical representations of the approximations given by equations (4.89) and (4.90) are given in Figures 4.6 and 4.7, respectively.

For $P = 2$, the weighted residual statement of the problem is

$$\int_0^L \sin \frac{i\pi x}{L} \left[-k\left(\frac{\pi}{L}\right)^2 \left(\alpha_1 \sin \frac{\pi x}{L} + 4\alpha_2 \sin \frac{2\pi x}{L} \right) + S_o x \right] dx = 0 \tag{4.91}$$

where $i = 1, 2$. Integrating equation (4.91) for $i = 1$ and for $i = 2$, and using the aforementioned orthogonality relations of equation (4.68) gives the following set of equations:

$$-\frac{kL}{2}\left(\frac{\pi}{L}\right)^2 \alpha_1 + \frac{S_o L^2}{\pi} = 0 \tag{4.92}$$

$$-2kL\left(\frac{\pi}{L}\right)^2 \alpha_2 - \frac{S_o L^2}{2\pi} = 0 \tag{4.93}$$

The solution of these equations is

$$\begin{Bmatrix} \alpha_1 \\ \alpha_2 \end{Bmatrix} = \frac{S_o L^3}{4k\pi^3} \begin{Bmatrix} 8 \\ -1 \end{Bmatrix} \tag{4.94}$$

which, not surprisingly, is identical to the solution obtained for the method of least squares (see equation 4.78). Consequently,

$$\hat{T}(x) = \frac{S_o L^3}{4k\pi^3} \left(8 \sin \frac{\pi x}{L} - \sin \frac{2\pi x}{L} \right) \tag{4.95}$$

and

$$\hat{q}(x) = -k\frac{d\hat{T}}{dx} = -\frac{S_o L^2}{2\pi^2} \left(4\cos \frac{\pi x}{L} - \cos \frac{2\pi x}{L} \right) \tag{4.96}$$

The graphical representations of $\hat{T}(x)$ and $\hat{q}$ are given in Figures 4.6 and 4.7, respectively.

4.5.5   Method of Moments

The method of moments[14] is similar to the Galerkin method except that the residual is made orthogonal to members of a system of weighting functions that are not the same as the trial functions. One popular choice for the weighting functions is [299]

$$w_i = x^i \quad , \quad i = 1, 2, \cdots P \tag{4.97}$$

regardless of the choice of trial functions. For the case of $i = 1$, the method of moments is equivalent to the subdomain method (with a single subdomain), and is usually referred to as the integral or Karman–Pohlhausen method [280, 437]. This method has proved to be quite effective when used to represent both laminar and turbulent boundary-layer flow [164]. Another potential choice for the weighting functions would be an arbitrary polynomial of degree $(i - 1)$.

   The application of the method of moments is illustrated in the following example.

Example 4.5: Application of the Method of Moments

   The problem solved in the previous four examples is again considered. To facilitate the comparison of results obtained using the previous four methods, the approximate solution given in equation (4.34) is once again selected; the total residual is thus again given by equation (4.36).

   For $P = 1$ the weighting function is $w_1(x) = x$. The weighted residual statement of the problem is thus

$$\int_\Omega w_1 R(x)dx = \int_0^L x \left\{ -k \left(\frac{\pi}{L} \right)^2 \left[\alpha_1 \sin \frac{\pi x}{L} \right] + S_o x \right\} dx = 0 \tag{4.98}$$

   Integrating, evaluating the limits of integration and solving for α_1 gives

$$\alpha_1 = \frac{S_o L^3}{3k\pi} \tag{4.99}$$

   The approximate temperature distribution is thus

---

[14]In the field of electromagnetics, the method of weighted residuals, as a whole, is commonly referred to as the *method of moments* [217, 430].

$$\hat{T}(x) = \frac{S_o L^3}{3k\pi} \sin \frac{\pi x}{L} \tag{4.100}$$

The corresponding heat flux distribution per unit area is

$$\hat{q}(x) = -\frac{S_o L^2}{3k} \cos \frac{\pi x}{L} \tag{4.101}$$

In Figures 4.8 and 4.9 the approximations given by equations (4.100) and (4.101) are compared to the exact solutions for $T(x)$ and $q(x)$, respectively. For simplicity in comparing the results, values of $S_o = k = L = 1.0$ have again been assumed. From these figure it is quite evident that the approximate solutions for $P = 1$ differs substantially from the exact solutions. As such, we seek to improve the approximation by taking $P = 2$.

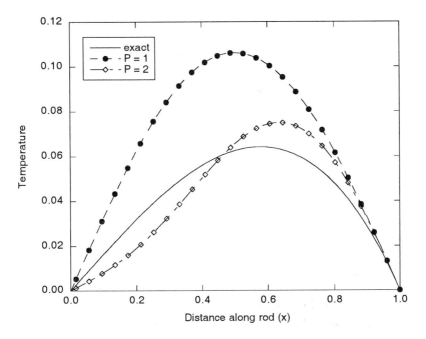

Figure 4.8: Assessment of Method of Moments: Temperature Distribution

For $P = 2$, the weights are $w_1(x) = x$ and $w_2(x) = x^2$. The weighted residual approximation is thus

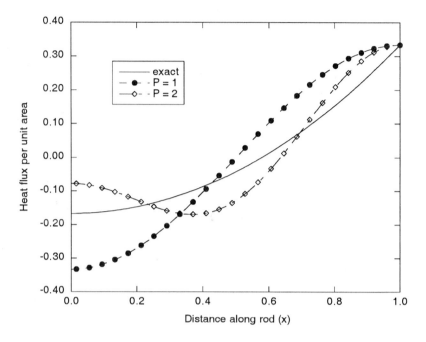

Figure 4.9: Assessment of Method of Moments: Heat Flux Distribution

$$\int_0^L x \left\{ -k \left(\frac{\pi}{L} \right)^2 \left[\alpha_1 \sin \frac{\pi x}{L} + 4\alpha_2 \sin \frac{2\pi x}{L} \right] + S_o x \right\} dx = 0 \qquad (4.102)$$

and

$$\int_0^L x^2 \left\{ -k \left(\frac{\pi}{L} \right)^2 \left[\alpha_1 \sin \frac{\pi x}{L} + 4\alpha_2 \sin \frac{2\pi x}{L} \right] + S_o x \right\} dx = 0 \qquad (4.103)$$

Integrating, evaluating the limits of integration and simplifying the resulting expressions gives the following system of equations:

$$\begin{bmatrix} 1 & -2 \\ (\pi^2 - 4) & -2\pi^2 \end{bmatrix} \begin{Bmatrix} \alpha_1 \\ \alpha_2 \end{Bmatrix} = \frac{S_o L^3}{k} \begin{Bmatrix} \dfrac{1}{3\pi} \\ \dfrac{\pi}{4} \end{Bmatrix} \qquad (4.104)$$

The solution of equation (4.104) is

$$\begin{Bmatrix} \alpha_1 \\ \alpha_2 \end{Bmatrix} = \frac{S_o L^3}{48k} \begin{Bmatrix} \pi \\ \dfrac{\pi^2 - 16}{2\pi} \end{Bmatrix} \tag{4.105}$$

The approximate temperature distribution is thus

$$\hat{T}(x) = \frac{S_o L^3}{48k} \left[\pi \sin \frac{\pi x}{L} + \left(\frac{\pi^2 - 16}{2\pi} \right) \sin \frac{2\pi x}{L} \right] \tag{4.106}$$

The associated approximation to the heat flux per unit area is

$$\hat{q}(x) = -\frac{S_o L^2}{48} \left[\pi^2 \cos \frac{\pi x}{L} + \left(\pi^2 - 16 \right) \cos \frac{2\pi x}{L} \right] \tag{4.107}$$

In Figures 4.8 and 4.9 the approximations given by equations (4.106) and (4.107) are compared to the exact solutions for $T(x)$ and $q(x)$, respectively. For simplicity in comparing the results, values of $S_o = k = L = 1.0$ have again been assumed.

From these figures it is evident that the solution for $P = 2$ is substantially more accurate than that for $P = 1$. It is also evident that the set of weighting functions used in this example leads to approximate solutions that are *less* accurate than those corresponding to the collocation, subdomain, least squares and Galerkin solutions. Recall that the same approximate solution for temperature (equation 4.34) has been used in all five cases.

4.5.6   Comparison of Results

The approximate solutions obtained in the previous five examples are now compared. Recall that in these examples the *same* trial solution for $\hat{T}(x)$ has been used. In addition, values of $S_o = k = L = 1.0$ have been assumed.

Values for the coefficients α_1 and α_2 corresponding to both $P = 1$ and $P = 2$ are summarized in Table 4.1. In progressing from $P = 1$ to $P = 2$, the biggest change in α_1 was realized in the method of moments.

In Tables 4.2 and 4.3 the approximations for temperature and heat flux per unit area corresponding to $P = 2$ are, respectively, compared to exact

Table 4.1: Summary of α_i Coefficients for $P = 1$ and $P = 2$

Method	α_1 $(\text{x } 10^{-2})$	α_1 $(\text{x } 10^{-2})$	α_2 $(\text{x } 10^{-3})$
Collocation	5.066	5.850	-4.875
Subdomain	7.958	7.958	-9.947
Least Squares and Galerkin	6.450	6.450	-8.063
Moments	10.61	6.545	-20.33

values at select sampling points in Ω. The relative accuracy of the various solutions is quantified through the discrete L_2 norm

$$\left\| \hat{\phi} - \phi \right\|_2 = \left\{ \sum_{i=1}^{N} \left(\hat{\phi}_i - \phi_i \right)^2 \right\}^{\frac{1}{2}} \qquad (4.108)$$

where N is the number of sampling points. For Tables 4.2 and 4.3, N equals nine and ten, respectively.

Based on the results shown in these tables, it is evident that the best approximation was realized using the least squares and Galerkin schemes. As expected, the primary dependent variable $T(x)$ was more accurately approximated than the secondary dependent variable $q(x)$.

It is timely to conclude this comparison of results with the following observations of Crandall [125]:

> The variation between results obtained by applying different criteria to the same trial family $\cdots$ is much less significant than the variations that can result from the choice of different trial families.

Table 4.2: Comparison of Temperature Results for $P = 2$

x	exact (x 10^{-2})	Collocation (x 10^{-2})	Subdomain (x 10^{-2})	Least Square and Galerkin (x 10^{-2})	Meth. of Moments (x 10^{-2})
0.10	1.650	1.521	1.874	1.519	0.828
0.20	3.200	2.975	3.731	3.025	1.914
0.30	4.550	4.269	5.492	4.452	3.362
0.40	5.600	5.277	6.984	5.661	5.030
0.50	6.250	5.850	7.958	6.450	6.545
0.60	6.400	5.850	8.153	6.609	7.419
0.70	5.950	5.196	7.384	5.985	7.228
0.80	4.800	3.902	5.623	4.558	5.780
0.90	2.850	2.094	3.044	2.467	3.217
Norm (x 10^{-2})	0.000	1.631	3.449	0.593	2.815

Table 4.3: Comparison of Heat Flux Results for $P = 2$

x	exact (x 10^{-1})	Collocation (x 10^{-1})	Subdomain (x 10^{-1})	Least Square and Galerkin (x 10^{-1})	Meth. of Moments (x 10^{-1})
0.10	-1.617	-1.500	-1.872	-1.517	-0.922
0.20	-1.467	-1.392	-1.829	-1.483	-1.269
0.30	-1.217	-1.175	-1.663	-1.348	-1.603
0.40	-0.867	-0.816	-1.278	-1.036	-1.669
0.50	-0.417	-0.306	-0.625	-0.507	-1.277
0.60	0.133	0.320	0.267	0.216	-0.398
0.70	0.783	0.986	1.276	1.035	0.814
0.80	1.533	1.581	2.216	1.796	2.058
0.90	2.383	1.996	2.883	2.337	2.989
1.00	3.333	2.144	3.125	2.533	3.333
Norm (x 10^{-1})	0.000	1.296	1.276	0.920	1.726

4.6   Continuity Requirements

Having discussed the specific weighting functions typically associated with the method of weighted residuals, we now also consider the trial functions N_m. In the previous sections of this chapter it was implicitly assumed that integrals associated with the method of weighted residuals can indeed be evaluated. This places certain restrictions on the possible families to which the weighting and trial functions and can belong.

In general, we seek to avoid functions that result in any term in the integrals becoming *infinite*. In assessing the behavior of the weighting and trial functions, the key characteristic is the *degree of continuity* exhibited by the functions. It is evident that, within Ω, these functions must be continuous. To investigate the issue of continuity, consider a hypothetical trial function N_m in the vicinity of a point $x = b$ (Figure 4.10).

The function shown in Figure 4.10a is discontinuous at the point $x = b$. It follows that at this location the derivative dN_m/dx will be *infinite*. The function shown in Figure 4.10(b) is continuous at $x = b$ but exhibits a discontinuity in the first derivative at this location. Consequently, at $x = b$ it is the second derivative that will be infinite.

From this simple example we conclude that in order to avoid functions that go to infinity, the following criterion must thus be satisfied: If s represents the highest order of differentiation of the weighting and trial functions, we must ensure that in the approximations for these functions the derivatives of order $(s-1)$ and lower are continuous. Using the notation of mathematics, we require that the w_i and N_m must be C^{s-1} continuous.

4.7   Weak Form

In Chapter 1, the general equation governing the steady-state response of several one-dimensional physical problems was found to be

$$\frac{d}{dx}\left(k\frac{d\phi}{dx}\right) + S = 0 \quad ; \quad 0 < x < L \tag{4.109}$$

where k and S are known constants or possibly known functions of x. The associated boundary conditions are

$$\phi = \bar{\phi} \quad \text{on} \quad \Gamma_1 \tag{4.110}$$

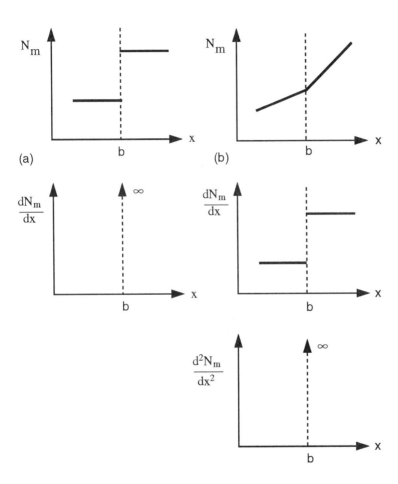

Figure 4.10: Continuity Requirements for Trial Functions : (a) Discontinuous Function, (b) Discontinuous First Derivative

and

$$-k\frac{d\phi}{dx} = \bar{q} \quad \text{on} \quad \Gamma_2 \tag{4.111}$$

Multiplying equation (4.109) by an arbitrary weighting function $w(x)$ and integrating over the domain Ω $(0 < x < L)$ gives

$$\int_0^L w\left[\frac{d}{dx}\left(k\frac{d\phi}{dx}\right) + S\right]dx = 0 \tag{4.112}$$

Clearly the solution of equation (4.112) will likewise satisfy equation (4.109). We note that equation (4.109) requires second derivatives of ϕ to exist in Ω.

If we integrate the first term in equation (4.112) by parts,[15] the resulting equation is

$$wk\frac{d\phi}{dx}\bigg|_\Gamma - \int_0^L \frac{dw}{dx}\left(k\frac{d\phi}{dx}\right)dx + \int_0^L wS\,dx = 0 \tag{4.113}$$

Expanding the boundary term gives

$$wk\frac{d\phi}{dx}\bigg|_\Gamma = wk\frac{d\phi}{dx}\bigg|_{\Gamma_1} + wk\frac{d\phi}{dx}\bigg|_{\Gamma_2} = wk\frac{d\phi}{dx}\bigg|_{\Gamma_1} - w\bar{q}|_{\Gamma_2} \tag{4.114}$$

where equation (4.111) has been used. Incorporating equation (4.114) into equation (4.113) and rearranging terms gives

$$\int_0^L \left(\frac{dw}{dx}k\frac{d\phi}{dx}\right)dx - \int_0^L wS\,dx + w\bar{q}|_{\Gamma_2} - wk\frac{d\phi}{dx}\bigg|_{\Gamma_1} = 0 \tag{4.115}$$

The solution of equation (4.115) likewise satisfies equation (4.112) [252]. However, equation (4.115) contains only first-order derivatives of ϕ and w. The differentiability requirements have thus been reduced or *weakened*. For this reason, equation (4.115) is referred to as the *weak form* of equation (4.109).[16]

---

[15] In general, a governing differential equation of order $2s$ would be integrated by parts s times.

[16] Recall that, as noted in Section 1.1, the governing differential equations comprise the so-called *strong form* of a problem. Whereas the strong form enforces conditions at every point in Ω and along Γ, the weak form enforces them in an average sense.

The weighting functions are not without restrictions. To investigate this issue further, consider equation (4.115). The essential boundary condition of equation (4.110) is assumed to be specified at $x = 0$. At $x = L$, the natural boundary condition of equation (4.111) is specified. Following the approach of Heinrich and Pepper [218], we consider a family of potential weighting functions $w_n(x)$ that does not vanish at $x = 0$ (i.e., on Γ_1). In particular,

$$w_n(x) = e^{-nx^2} \qquad (4.116)$$

Clearly $w_n(0) = 1$ for all values of n. Furthermore, as n increases, the w_n decay quite rapidly to zero. In the limit

$$w_\infty = \lim_{n \to \infty} w_n(x) = \begin{cases} 1 & \text{if } x = 0 \\ 0 & \text{if } x > 0 \end{cases} \qquad (4.117)$$

The derivatives of $w_n(x)$ are seen to be bounded and to rapidly approach zero with increasing n, viz.,

$$\lim_{n \to \infty} \frac{dw_n}{dx} = 0 \quad \forall\, x \geq 0 \qquad (4.118)$$

We choose the weighting function in equation (4.115) to be $w_n(x)$ for a value of n large enough that the first three terms in equation (4.115) can be made arbitrarily small. In the limit when $w_n(x) = w_\infty$, these terms are in fact zero and equation (4.115) reduces to

$$-(1.0)k\frac{d\phi}{dx}\bigg|_{\Gamma_1} = 0 \qquad (4.119)$$

Since the flux $d\phi/dx$ can, in general, assume any value at $x = 0$, equation (4.119) will not necessarily be satisfied. Furthermore, since ϕ is the exact solution to the problem, it follows that the left hand side of equation (4.115) must identically be zero. The only way in which these two constraints can be satisfied is to require the weighting function $w(x)$ to vanish at $x = 0$. Generalizing this for the one-dimensional case, we state that the weighting functions $w_n(x)$ must vanish on Γ_1. The further generalization to two- and three-dimensional problems is relatively straightforward.

If we next consider an *approximate* solution for ϕ of the form given by equation (4.3), then equation (4.112) becomes a weighted residual statement of the problem (assuming $R_\Gamma = 0$). This equation requires second derivatives of $\hat{\phi}$ to exist in Ω. In light of the discussion in the previous section, it follows that $\hat{\phi}$ (and thus the trial functions) must thus be C^1 continuous. That is, both $\hat{\phi}$ and its first derivative must be continuous in Ω. Experience shows that such requirements are quite restrictive.

In the weak form of equation (4.109), given by equation (4.115), the continuity requirement on $\hat{\phi}$ is *reduced* to C^0, while that on the weighting function is *increased* to C^0. Nonetheless, since the required continuity for both functions is the same, the overall approximation benefits.

4.8 Concluding Remarks

In this chapter, techniques for developing continuous approximate solutions based on the method of weighted residuals were considered. By providing approximations to the primary dependent variables ϕ at all points within the domain Ω and along its boundary Γ, such solutions overcome the fundamental shortcoming of discrete approaches such as the finite difference method. However, continuous approximate solutions require the analyst to choose a set of suitable trial functions, and a set of corresponding weighting functions, for use in the method of weighted residuals. In addition, both of these sets of functions must apply to the *entire* domain. For geometrically simple domains such as lines, rectangles, right prisms, etc., this fact poses no difficulties. However, in the case of domains with curved boundaries, cut-outs, material inhomogeneities, inclusions, etc., the choice of suitable trial solutions becomes substantially more difficult.

4.9 Exercises

4.1

Consider the differential equation

$$\frac{d}{dx}\left[(1+x)\frac{d\phi}{dx}\right] = 0 \quad ; \quad 0 < x < 1$$

with boundary conditions $\phi(0) = 0$ and $\phi(1) = 1$. Using the approximate solution

$$\hat{\phi} = \sum_{m=1}^{2} \alpha_m \sin\frac{m\pi x}{2}$$

determine the general expression for the total residual $R(x) = R_\Omega + R_\Gamma$.

4.2

Consider the approximate solution of the non-linear equation

$$-2u\left(\frac{d^2u}{dx^2}\right) + \left(\frac{du}{dx}\right)^2 = 4 \quad ; \quad 0 < x < 1$$

with boundary conditions $u(0) = 1$ and $u(1) = 0$. Using the approximate solution

$$\hat{u} = \sum_{n=1}^{3} \alpha_n \sin(n\pi x)$$

determine the general expression for the total residual $R(x) = R_\Omega + R_\Gamma$.

4.3

It is required to solve the problem of two-dimensional steady-state heat conduction in a material having unit thermal conductivity. The solution domain is the square region $-1 \leq x \leq 1$, $-1 \leq y \leq 1$. The governing differential equation for steady heat conduction with no heat generation is

$$\frac{\partial^2 T}{\partial x^2} + \frac{\partial^2 T}{\partial y^2} = 0$$

The boundary conditions for the problem are: $T = 0$ on $y = \pm 1$, and $dT/dx = \cos(\pi y/2)$ on $x = \pm 1$. Using the approximate solution

$$\hat{T} = \sum_{m=1}^{5} \alpha_m N_m(x, y)$$

with the trial functions: $N_1 = 1 - y^2$, $N_2 = (1 - y^2)x^2$, $N_3 = (1 - y^2)y^2$, $N_4 = (1 - y^2)x^2y^2$, and $N_5 = (1 - y^2)x^4$, determine the general expression for the total residual $R(x) = R_\Omega + R_\Gamma$.

4.4

Consider the differential equation

$$\frac{d^2u}{dx^2} + u + x^2 = 0 \quad ; \quad 0 < x < 1$$

with boundary conditions $u(0) = u(1) = 0$. The exact solution is

$$u = 2 - 2\cos x - \frac{[1 - 2\cos(1)]}{\sin(1)}\sin x - x^2$$

Using the approximate solution

$$\hat{u} = \sum_{j=1}^{P} \alpha_j N_j = \sum_{j=1}^{P} \alpha_j x^j (1 - x)$$

in conjunction with the collocation, subdomain, least squares and Galerkin methods, compare the approximate solutions with the exact one. Use $P = 1$ and $P = 2$. Discuss your results.

4.5

Consider the uniform rod of unit cross-sectional area shown in Figure 4.1. We are interested in the approximate solution of the steady-state heat conduction equation for a rod of unit length, with $k = 0.01$ and $S = S(x) = x^5$. The following trial solution is proposed:

$$\hat{T} = \sum_{m=1}^{2} \alpha_m N_m = \sum_{m=1}^{2} \alpha_m x(1 - x^m)$$

Please do the following:

1. Determine the exact distribution of temperature $T(x)$ and of heat flux per unit area $q(x)$ for this problem.

2. Define the total residual $R(x)$ associated with the approximate solution.

3. Use the collocation method to determine values for α_1 and α_2. In this case, set the total residual equal to zero at $x = 1/3$ and $x = 2/3$.

4. Use the method of least-squares with a weight of $w_1 = 1.0$ to determine values for α_1 and α_2.

5. Use the method of moments with weights of $w_1 = x$ and $w_2 = x^2$ to determine values for α_1 and α_2.

6. Finally, use the Galerkin method to determine values for α_1 and α_2.

7. On one graph, plot the different approximate solutions $\hat{T}(x)$ and the exact solution $T(x)$ versus distance (x) along the rod. On a second graph, plot the different $\hat{q}(x)$ solutions and the exact solution $q(x)$ versus x.

8. Of the four methods used, which gave the best results? Which was the easiest to use?

4.6

Consider a simply-supported Bernoulli-Euler beam subjected to a uniformly distributed load $\bar{q}$ in the manner shown in Figure 4.11. The beam is prismatic, with a moment of inertia (second moment of area) of I. The material is assumed to be isotropic linear elastic, and is characterized by an elastic modulus E.

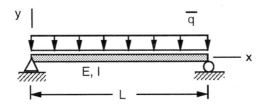

Figure 4.11: Simply-Supported Beam Subjected to Uniformly Distributed Load

Using the approximation

$$\hat{v}(x) = \sum_{i=1}^{P} \alpha_i \sin \frac{i\pi x}{L}$$

with P equal to 1, 2 and 3, obtain approximate solutions using: (1) the *collocation method* (with appropriate number of points); and, (2) The *Galerkin* method. For this problem, the governing differential equation is

$$\frac{d^2}{dx^2}\left(EI\frac{d^2v}{dx^2}\right) - q(x) = 0 \quad ; \quad 0 < x < L$$

where q is positive *upward*. The associated boundary conditions are

$$v = \left(EI\frac{d^2v}{dx^2} \right) = 0 \quad \text{at} \quad x = 0 \text{ and } x = L$$

Graphically compare the approximations for transverse displacement $\hat{v}(x)$, bending moment $\hat{M}(x) = EI\, d^2\hat{v}/dx^2$, and shear force $\hat{V}(x) = d\hat{M}/dx$ with the exact (Bernoulli-Euler beam theory) values. For purposes of specific quantitative comparison (e.g., percent relative error), consider the three points along the beam $x = L/4$, $L/2$ and $3L/4$. Present your results in three separate tables. For all three methods, discuss, in some detail, the convergence of the approximate solution to the exact one.

Chapter 5

Variational Methods

5.1 Introductory Remarks

In the previous chapters, the analysis of physical problems was initiated with investigations of the properties of small differential elements of the continuum. Relationships were established among the mean values of various quantities associated with the infinitesimal elements. Differential or integral equations governing the behavior of the entire domain Ω were then obtained by allowing the dimensions of these elements to approach zero in the limit. Suitable boundary conditions must also be imposed on the boundary Γ.

In many instances, there exists an *equivalent* problem of seeking a function that gives a stationary value to some integral. Problems of this type are called *extremum* or *variational* problems, and their solutions are realized using so-called *variational methods.*

In such methods the approximation for the primary dependent variables is *continuous* over the domain Ω, and has the same general form as for the method of weighted residuals discussed in the previous chapter, viz.,

$$\hat{\phi}(\mathbf{x}) = \psi(\mathbf{x}) + \sum_{m=1}^{P} \alpha_m N_m(\mathbf{x}) \qquad (5.1)$$

In equation (5.1) ψ represents a known function that exactly satisfies the boundary conditions on Γ, the N_m are a linearly independent set of known analytic *trial functions* that satisfy homogeneous boundary conditions on Γ, and the α_m represent unknown constants.[1]

---

[1] Recall that for initial-value problems, ψ and the α_m would be functions of time; the

In formulating a variational problem, a certain class of functions ϕ is selected, and a means for associating a *scalar* value $I(\phi)$ with each function is defined. The function $I(\phi)$ is referred to as a *functional*, that is, a function whose arguments is a class of functions. The functions ϕ must satisfy certain *admissibility conditions*, usually involving the satisfaction of some of the boundary conditions and certain continuity requirements with respect to differentiation. [2] Associated with a particular function ϕ is a single scalar value for $I(\phi)$; as ϕ ranges through the class of neighboring functions, the corresponding values of $I(\phi)$ will, in general, vary. The variational problem then consists of locating those functions ϕ for which $I(\phi)$ remains *stationary* for small variations in ϕ. In some instances, the varied function is not merely stationary but actually attains a local maximum or minimum.

In studying the relationship between equilibrium and variational problems, the answers to the following two questions are typically sought [125]:

1. For a given equilibrium problem, what is the allowable class of functions ϕ and what is the structure of the functional $I(\phi)$ for an equivalent variational (extremum) problem?

2. For a given functional $I(\phi)$ and a given class of functions ϕ, what is the differential equation and associated boundary conditions for the equivalent equilibrium problem?

A complete study of the subject of variational methods is not attempted here. Instead, the allowable class of functions alluded to in the first question is briefly discussed in Section 5.2. The functionals are assumed to be known from variational principles of physics.[3] Consequently, only a few general comments related to functionals shall be made in Section 5.3.

The second question is answered by direct manipulation, using formal procedures of the *calculus of variations*.[4] Since this represents a significant

---

N_m would be functions of time as well as of space.

[2] Recall the discussion of Section 4.6.

[3] In commenting on the term "variational principles," Finlayson and Scriven [160] note that

> $\cdots$ it is no more a principle than any other equally accurate mathematical description of the performance of the same physical system $\cdots$ Indeed, some statements that in fact involve no more than a stationary functional are habitually referred to as 'minimum principles,' Hamilton's principle and the principle of least action being the most famous cases in point.

[4] The name "calculus of variations" was apparently adopted as a result of notations that were introduced by Lagrange around 1760 [84].

undertaking, a detailed consideration of this calculus is not pursued herein.[5]

Instead, in Section 5.4 the ideas associated with variational methods will be introduced in a less rigorous manner, based primarily on the differential calculus. Concerning the nature of problems of the calculus of variations and their relation to the differential calculus, Bliss [84] notes that:

> A fundamental problem of the differential calculus is that of finding in the range of values of an independent variable x one for which a given function $f(x)$ is a maximum or a minimum. The calculus of variations is also a theory of maxima and minima, but for variables and functions which are more complicated than those of the differential calculus.

The main focus of this chapter is the *approximate* solution of physical problems using variational methods. Such solutions entail substituting an approximation of the form of equation (5.1) into an appropriate functional. The functional is then varied with respect to the adjustable parameters α_m, which are then evaluated so as to make the variation vanish in accordance with the variational formulation. Further details pertaining to this so-called *stationary functional method* are given in Section 5.5.

5.2 Admissible Functions

In the analysis of continuous systems using variational methods, certain restrictions must be placed on the primary dependent variables ϕ. A function $\phi(\mathbf{x})$ that meets these restrictions in the domain Ω and on its boundary Γ is said to be an *admissible* function for a particular extremum principle [125].

The admissibility restrictions on ϕ usually consist of suitable continuity or smoothness criteria. For example, in stress-deformation problems analyzed using the principle of minimum potential energy, an admissible displacement function and its derivative must be sufficiently smooth to permit evaluation of the potential energy due to stretching, bending, etc.

An admissible function must also satisfy the *essential* boundary conditions on Γ [125]. For example, in the analysis of bars,[6] the admissible longitudinal displacement function ϕ must satisfy the essential boundary conditions involving prescribed displacements; ϕ does not have to satisfy the natural boundary conditions involving prescribed axial loads.

---

[5] Readers interested in a rigorous treatment of variational calculus are directed to books on the subject, such as those by Courant and Hilbert [123], Bolza [86], Mikhlin [392], Lanczos [302] and Washizu [551], to name a few.

[6] Recall Example 1.2.

5.3 Functionals

The statement of a variational principle implicitly contains the rules for evaluating the functional I for an admissible function ϕ. In many cases $I(\phi)$ is an integral over the domain Ω. In some cases, $I(\phi)$ may, however, be an integral around the boundary Γ, or it may consist of both integrals over Ω and around Γ. If ϕ and its derivatives appear as squares and products (as well as linearly), $I(\phi)$ is said to be a *quadratic* functional [125].

For simplicity, we restrict the discussion to spatially one-dimensional problems. Consider the integral

$$I(\phi) = \int_a^b F\left(x,\,\phi,\,\phi'\right) dx \tag{5.2}$$

where the integrand $F\left(x,\,\phi,\,\phi'\right)$ is a *known* real function of the real arguments x, ϕ and $\phi' \equiv d\phi/dx$. The value of the integral depends upon the choice of $\phi = \phi(x)$, hence the notation $I(\phi)$. For a given ϕ, $I(\phi)$ takes on a *scalar* value; stated mathematically, a functional is an operator that maps ϕ into a scalar $I(\phi)$ [504]. To make $I(\phi)$ meaningful, it is necessary to impose the aforementioned admissibility restrictions on the choice of the argument $\phi(x)$ and on the prescribed function F appearing in the integrand of $I(\phi)$.

Remark

1. Two examples of commonly used variational principles are the *principle of minimum potential energy* and *Hamilton's principle*. The functional I corresponding to the former is the total potential energy associated with a displacement state ϕ. For the latter, I is a time integral of the difference between the kinetic and potential energies, evaluated for a geometrically compatible motion ϕ.

5.3.1 Existence of Functionals

In light of the previous discussion, it is natural to ask what type of physical problems can be represented using a variational equation. This issue has been discussed at length in a paper by Finlayson and Scriven [160]. What follows is a brief summary of the findings from this paper.

If the governing differential equation for a physical problem is *self-adjoint*, there always exists a functional that is stationary. Usually the form of the functional is known and can be found in the literature; if not, it can be constructed by standard procedures or by a general mathematical scheme given in advanced texts on the subject.

For non-self-adjoint problems, a variational formulation does not directly exist. The differential equations corresponding to such problems contain derivatives of odd order. A prime example is the field of fluid mechanics where, for certain types of flows, a pure variational principle is unobtainable. Several approaches for *indirectly* obtaining such a formulation have, however, been proposed [160].

In the case of linear problems that are non-self-adjoint, a variational formulation can be found that involves not only the solution to the problem but also an additional unknown "adjoint" function that is coupled to the former [401]. The resulting variational method leads to an approximation that is equivalent to the method of moments [160]; this variant of the method of weighted residuals has been discussed in Section 4.5.5.

A second alternative for linear systems is to double the order of the differential equation by applying it to the adjoint operator. The variational formulation for this self-adjoint equation has been shown to be equivalent to the method of least squares [392]; this variant of the method of weighted residuals has been discussed in Section 4.5.3.

In the case of non-linear problems that are non-self-adjoint, the situation is more bleak. The class of such problems for which classical variational formulations can be developed is severely limited [160]. In light of this fact, various authors adopted somewhat looser definitions of the term "variational principle." This led to the development of so-called *quasi-variational principles*, *restricted principles* and *inexact principles*. Having surveyed such formulations, Finlayson and Scriven [160] concluded that the formulations lacked the advantages of general variational principles, mainly because the variational integral is not stationary or because no variational integral exists. They further concluded that:

> Apart from self-adjoint, linear systems, which are comparatively rare, there is no practical need for variational formalism. When approximate solutions are in order the applied scientist and engineer are better advised to turn immediately to direct approximation methods for their problems, rather than search for or try to understand quasi-variational formulations and restricted variational principles.

Due to similar conclusions from other researchers, the general consensus appears to be that for non-self-adjoint problems, variational equations are not of great value and should thus be avoided.

5.4   Derivation of Differential Equations

In this section we address the second question raised on page 126, namely: "For a given functional $I(\phi)$ and a given class of functions ϕ, what is the differential equation and associated boundary conditions for the equivalent equilibrium problem?" For simplicity, we again restrict the discussion to spatially one-dimensional problems. This section can be skipped without compromising the understanding of Section 5.5.

Consider the problem of determining the function ϕ that makes the integral of equation (5.2) stationary. In addition, we require that ϕ satisfy all essential boundary conditions on Γ.

In the variational calculus, the interest is in the behavior of $I(\phi)$ when $\phi(x)$ is replaced by a slightly different function $\tilde{\phi}(x)$. This "neighboring" function must be admissible; that is, it must be sufficiently differentiable (in the present case only ϕ' needs to be continuous), and must exactly satisfy at least the essential boundary conditions. The difference between $\tilde{\phi}(x)$ and $\phi(x)$ is called the *variation* of $\phi(x)$, and is denoted by $\delta\phi$; viz.,

$$\delta\phi = \tilde{\phi}(x) - \phi(x) \tag{5.3}$$

It is important to understand that $\delta\phi$ represents a small arbitrary change in $\phi(x)$ from that value which makes $I(\phi)$ stationary; it should not be confused with the differential $d\phi$.

Similarly, the variation in $\phi'(x)$ is denoted by $\delta\phi'$ where

$$\delta\phi' = \tilde{\phi}'(x) - \phi'(x) \tag{5.4}$$

Next consider the behavior of $F(x, \phi, \phi')$ in the neighborhood of the function $\phi(x)$. For fixed x, F depends on $\phi(x)$ and $\phi'(x)$; when these quantities are varied, it follows that F varies.

The difference between $F(x, \tilde{\phi}, \tilde{\phi}')$ and $F(x, \phi, \phi')$ is denoted by ΔF, where

$$\begin{aligned} \Delta F &= F(x, \tilde{\phi}, \tilde{\phi}') - F(x, \phi, \phi') \\ &= F(x, \phi + \delta\phi, \phi' + \delta\phi') - F(x, \phi, \phi') \end{aligned} \tag{5.5}$$

where equations (5.3) and (5.4) have been used.

The function $F(x, \phi + \delta\phi, \phi' + \delta\phi')$ is next expanded in a Taylor series about (x, ϕ, ϕ'), viz.,

$$F(x, \phi + \delta\phi, \phi' + \delta\phi') = F(x, \phi, \phi') + \frac{\partial F}{\partial \phi}\delta\phi + \frac{\partial F}{\partial \phi'}\delta\phi'$$

$$+\frac{1}{2!}\left(\frac{\partial^2 F}{\partial \phi^2}\delta\phi^2 + 2\frac{\partial^2 F}{\partial \phi \partial \phi'}\delta\phi\delta\phi' + \frac{\partial^2 F}{\partial \phi'^2}\delta\phi'^2\right) + \cdots \qquad (5.6)$$

In light of equation (5.5), equation (5.6) becomes

$$\Delta F = \frac{\partial F}{\partial \phi}\delta\phi + \frac{\partial F}{\partial \phi'}\delta\phi'$$

$$+\frac{1}{2!}\left(\frac{\partial^2 F}{\partial \phi^2}\delta\phi^2 + 2\frac{\partial^2 F}{\partial \phi \partial \phi'}\delta\phi\delta\phi' + \frac{\partial^2 F}{\partial \phi'^2}\delta\phi'^2\right) + \cdots \qquad (5.7)$$

Defining the first and second variations of F by

$$\delta F = \frac{\partial F}{\partial \phi}\delta\phi + \frac{\partial F}{\partial \phi'}\delta\phi' \qquad (5.8)$$

$$\delta^2 F = \delta(\delta\phi) = \frac{\partial^2 F}{\partial \phi^2}\delta\phi^2 + 2\frac{\partial^2 F}{\partial \phi \partial \phi'}\delta\phi\delta\phi' + \frac{\partial^2 F}{\partial \phi'^2}\delta\phi'^2 \qquad (5.9)$$

respectively,[7] equation (5.7) becomes

$$\Delta F = \delta F + \frac{1}{2!}\delta^2 F + \cdots \qquad (5.10)$$

The difference between the stationary value of $I(\phi)$ and the value evaluated for $\tilde{\phi}(x)$ is

$$\Delta I = I(x, \tilde{\phi}, \tilde{\phi}') - I(x, \phi, \phi')$$

$$= \int_a^b F(x, \tilde{\phi}, \tilde{\phi}')\, dx - \int_a^b F(x, \phi, \phi')\, dx = \int_a^b \Delta F\, dx \qquad (5.11)$$

Using equation (5.10), equation (5.11) becomes

$$\Delta I = \delta I + \frac{1}{2!}\delta^2 I + \cdots \qquad (5.12)$$

where

----

[7] In general, $\delta^n F = \delta(\delta^{n-1} F)$.

$$\delta I = \int_a^b \delta F\, dx \quad \text{and} \quad \delta^2 I = \int_a^b \delta^2 F\, dx \qquad (5.13)$$

represent the first and second variations,[8] respectively, of $I(\phi)$.

Provided that $\tilde{\phi}$ is "sufficiently close" to $\phi(x)$, the second and higher variations of $I(\phi)$ appearing in equation (5.12) can be shown to be negligible [84]. Thus, the necessary condition for $I(\phi)$ to assume an extremum is that its first variation δI vanish identically, that is, $\delta I = 0$. In light of equation (5.13), it follows that $\delta I = 0$ implies $\delta F = 0$.

Remarks

1. If it is possible to show that $\delta^2 I > 0$ (i.e., positive definite) for all admissible variations, then $\Delta I > 0$ and the extremum is a *minimum*.

2. If $\delta^2 I < 0$ (i.e., negative definite), then the extremum is a *maximum*.

3. Finally, if $\delta^2 I = 0$, the extremum is a *neutral value* or point of inflection.

Noting that $I(\phi)$ will assume an extremum if

$$\delta I = \int_a^b \delta F\, dx = \int_a^b \left(\frac{\partial F}{\partial \phi}\delta\phi + \frac{\partial F}{\partial \phi'}\delta\phi' \right) dx = 0 \qquad (5.14)$$

and that $\delta\phi' = d(\delta\phi)/dx$, we integrate the second term in equation (5.14) by parts. This gives

$$\int_a^b \left(\frac{\partial F}{\partial \phi'}\delta\phi' \right) dx = \left[\frac{\partial F}{\partial \phi'}\delta\phi \right]_a^b - \int_a^b \frac{d}{dx}\left(\frac{\partial F}{\partial \phi'} \right)\delta\phi\, dx \qquad (5.15)$$

Using equation (5.15), equation (5.14) is rewritten as

$$\delta I = \int_a^b \left[\frac{\partial F}{\partial \phi} - \frac{d}{dx}\left(\frac{\partial F}{\partial \phi'} \right) \right] \delta\phi\, dx + \left[\frac{\partial F}{\partial \phi'}\delta\phi \right]_a^b = 0 \qquad (5.16)$$

Since $\delta\phi$ is arbitrary in $a < x < b$, satisfaction of equation (5.16) requires that

---

[8] In general, $\delta^n I = \int_a^b \delta^n F\, dx$.

$$\frac{\partial F}{\partial \phi} - \frac{d}{dx}\left(\frac{\partial F}{\partial \phi'}\right) = 0 \qquad (5.17)$$

This is commonly referred to as the *Euler-Lagrange* equation. It represents a *necessary* but not sufficient condition that $\phi(x)$ must satisfy if it is to yield an extremum for $I(\phi)$ [529].

Finally, note that the rightmost bracketed quantity in equation (5.16) yields the boundary conditions

$$\frac{\partial F}{\partial \phi'} = 0 \quad \text{or} \quad \delta\phi = 0 \qquad (5.18)$$

at $x = a$ and $x = b$. The former is the natural boundary condition and the latter is the essential boundary condition.[9]

Remark

1. If ϕ is a vector of scalars $\{\phi_1 \quad \phi_2 \quad \cdots \quad \phi_P\}^T$, then $\delta F = 0$ represents the stationary points of a function of P variables. Representing $F(\phi + \delta\phi)$ in the neighborhood of ϕ using a Taylor Series, neglecting powers of $\delta\phi$ beyond one, and solving for δF gives

$$\delta F = \frac{\partial F}{\partial \phi_1}\delta\phi_1 + \frac{\partial F}{\partial \phi_2}\delta\phi_2 + \cdots + \frac{\partial F}{\partial \phi_P}\delta\phi_P \qquad (5.19)$$

Requiring that $\delta F = 0$ for arbitrary small $\delta\phi_n$ $(n = 1, 2, \cdots, P)$ implies that

$$\frac{\partial F}{\partial \phi_1} = 0 \quad ; \quad \frac{\partial F}{\partial \phi_2} = 0 \quad ; \quad \cdots \quad ; \quad \frac{\partial F}{\partial \phi_P} = 0 \qquad (5.20)$$

These are, of course, the usual conditions for locating a stationary value for F.

---

[9]Although the words "natural" and "essential" appear to have originated in the context of the calculus of variations [100], as noted in Section 1.3.5, they are now applied to any boundary value problem, irrespective of whether it is in variational or differential equation form.

Example 5.1: Governing Equation for One-Dimensional Heat Flow

Consider the problem of one-dimensional, steady-state heat conduction/convection. The associated domain Ω is shown in Figure 5.1.

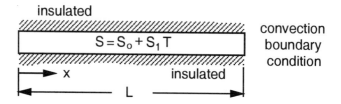

Figure 5.1: One-Dimensional Steady-State Heat Conduction/Convection Problem

The "thermal potential" for the problem is

$$I(T) = \int_0^L \frac{1}{2} k \left(\frac{dT}{dx} \right)^2 dx - \int_0^L \left[S_o T + \frac{1}{2} S_1 (T)^2 \right] dx$$
$$+ \beta \left[\frac{1}{2} \left(T|_{x=L} \right)^2 - T_\infty \left(T|_{x=L} \right) \right] \qquad (5.21)$$

where T is the temperature (the primary dependent variable), S_o and S_1 are parameters quantifying source terms, β is the convection coefficient and T_∞ denotes the ambient temperature. The first variation of $I(T)$ is

$$\delta I = \int_0^L k \left(\frac{dT}{dx} \frac{d\delta T}{dx} \right) dx - \int_0^L [S_o + S_1 T] \, \delta T \, dx$$
$$+ \beta \left[\left(T \, \delta T|_{x=L} \right) - T_\infty \left(\delta T|_{x=L} \right) \right] \qquad (5.22)$$

Integrating the first term in equation (5.22) by parts gives

$$\int_0^L k \left(\frac{dT}{dx} \frac{d\delta T}{dx} \right) dx = \left[k \frac{dT}{dx} \delta T \right]_0^L - \int_0^L \frac{d}{dx} \left(k \frac{dT}{dx} \right) \delta T \, dx \qquad (5.23)$$

Equation (5.22) thus becomes

$$\delta I = \int_0^L \left[\frac{d}{dx} \left(k \frac{dT}{dx} \right) + (S_o + S_1 T) \right] \delta T \, dx$$

$$-\beta \left[(T \, \delta T|_{x=L}) - T_\infty \, (\delta T|_{x=L}) \right] - \left[k \frac{dT}{dx} \delta T \right]_0^L \tag{5.24}$$

The vanishing of the first variation leads to the following Euler-Lagrange equation and second-order governing differential equation:

$$\frac{d}{dx} \left(k \frac{dT}{dx} \right) + (S_o + S_1 T) = 0 \tag{5.25}$$

From the last term in equation (5.24), the boundary conditions at $x = 0$ are

$$\delta T = 0 \quad \text{or} \quad k \frac{dT}{dx} = 0 \tag{5.26}$$

That is, prescribed temperature (*essential* boundary condition) or zero heat flux (*natural* boundary condition), respectively. At $x = L$, the boundary conditions are

$$\delta T = 0 \quad \text{or} \quad -k \frac{dT}{dx} \Big|_{x=L} - \beta(T_L - T_\infty) = 0 \tag{5.27}$$

That is, prescribed temperature (*essential* boundary condition) or prescribed convective heat flux (*natural* boundary condition).

A minimum principle often used in analyzing problems in solid mechanics is the *principle of minimum potential energy*, which states that

Of all the displacement fields that satisfy the prescribed boundary conditions, those which satisfy the equilibrium equations make the potential energy Π an absolute minimum.

A proof that Π actually assumes a minimum value in the case of stable equilibrium is given by Sokolnikoff [504].

The potential energy is the sum of the *internal strain energy* U and the *potential of the external (applied) loads* V_a. More precisely, when an elastic body is loaded, work is done on the body. This work is stored in the body in the form of elastic strain energy, which consists of internal forces and moments acting through internal linear and angular deformations, respectively. As a body is loaded, strain energy is stored; the body thus *gains* potential to do work. The applied loads, however, *lose* potential of doing work. Thus, V_a is equal to the *negative* of the work done by the external loads.

Since $\Pi = U + V_a$, and assuming the loads remain constant, the principle of minimum potential energy requires that $\delta\Pi = \delta U + \delta V_a = 0$. If $\Pi = \Pi(\phi_1, \phi_2, \cdots, \phi_P)$, then $\delta\Pi = 0$ implies that

$$\frac{\partial\Pi}{\partial\phi_1} = 0 \quad , \quad \frac{\partial\Pi}{\partial\phi_2} = 0 \quad , \quad \cdots \quad , \quad \frac{\partial\Pi}{\partial\phi_P} = 0 \qquad (5.28)$$

The principle of minimum potential energy is nothing more than the principle of virtual work applied to elastic structures. Indeed, while the principle of virtual work is valid for both elastic and inelastic bodies subjected to arbitrary loads, the principle of minimum potential energy is applicable only to elastic bodies (linear or non-linear) subjected to forces that are derivable from potential functions [529].

Example 5.2: Governing Equation for a Bernoulli-Euler Beam

To illustrate the application of the principle of minimum potential energy, we derive the equations governing the transverse displacement of a Bernoulli-Euler beam. Such a beam, subjected to a distributed loading $\bar{q}(x)$, to a concentrated end force $\bar{V}$, and to a moment of couple $\bar{M}$ is shown in Figure 5.2.

The total potential energy for the beam is

$$\Pi(v) = \int_0^L \frac{EI}{2}\left(\frac{d^2v}{dx^2}\right)^2 dx - \int_0^L \bar{q}v\, dx - \bar{V}\,v\big|_{x=L} - \bar{M}\,\frac{dv}{dx}\bigg|_{x=L} \qquad (5.29)$$

where E represents the elastic modulus characterizing the material, I denotes the moment of inertia (second moment of area) of the cross-section, and $v(x)$ is the transverse displacement. The first variation of Π is

$$\delta\Pi = \int_0^L EI\left(\frac{d^2v}{dx^2}\frac{d^2\delta v}{dx^2}\right)dx$$

$$- \int_0^L \bar{q}\delta v\, dx - \bar{V}\,\delta v\big|_{x=L} - \bar{M}\,\frac{d\delta v}{dx}\bigg|_{x=L} \qquad (5.30)$$

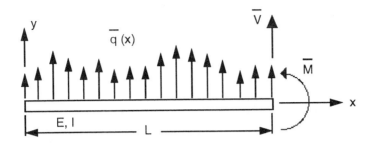

Figure 5.2: Bernoulli-Euler Beam Subjected to Distributed and Concentrated Loadings

where the term $\left(d^2\delta v/dx^2\right)^2$ is assumed to be negligibly small. The first term on the right hand side of equation (5.30) is integrated by parts to give

$$\int_0^L EI\left(\frac{d^2v}{dx^2}\frac{d^2\delta v}{dx^2}\right)dx$$

$$= \left[EI\frac{d^2v}{dx^2}\frac{d\delta v}{dx}\right]_0^L - \int_0^L \frac{d}{dx}\left(EI\frac{d^2v}{dx^2}\right)\frac{d\delta v}{dx}\,dx \qquad (5.31)$$

The second term on the right hand side of equation (5.31) is next integrated by parts. Equation (5.31) thus becomes

$$\int_0^L EI\left(\frac{d^2v}{dx^2}\frac{d^2\delta v}{dx^2}\right)dx = \left[EI\frac{d^2v}{dx^2}\frac{d\delta v}{dx}\right]_0^L$$

$$- \left[\frac{d}{dx}\left(EI\frac{d^2v}{dx^2}\right)\delta v\right]_0^L + \int_0^L \frac{d^2}{dx^2}\left(EI\frac{d^2v}{dx^2}\right)\delta v\,dx \qquad (5.32)$$

Substituting equation (5.32) into equation (5.30), the expression for $\delta\Pi$ becomes

$$\delta\Pi = \int_0^L \left[\frac{d^2}{dx^2}\left(EI\frac{d^2v}{dx^2}\right) - \bar{q}\right]\delta v\,dx + \left[EI\frac{d^2v}{dx^2}\frac{d\delta v}{dx}\right]_0^L$$

$$- \left[\frac{d}{dx}\left(EI\frac{d^2v}{dx^2}\right)\delta v\right]_0^L - \bar{V}\,\delta v\big|_{x=L} - \bar{M}\,\frac{d\delta v}{dx}\bigg|_{x=L} \qquad (5.33)$$

The vanishing of the first variation leads to

$$\int_0^L \left[\frac{d^2}{dx^2} \left(EI \frac{d^2 v}{dx^2} \right) - \bar{q} \right] \delta v \, dx + \left[EI \frac{d^2 v}{dx^2} \frac{d\delta v}{dx} \right]_{x=0}$$
$$+ \left[\left(EI \frac{d^2 v}{dx^2} - \bar{M} \right) \frac{d\delta v}{dx} \right]_{x=L}$$
$$+ \left[\left(-\bar{V} - \frac{d}{dx} \left(EI \frac{d^2 v}{dx^2} \right) \right) \delta v \right]_{x=L}$$
$$- \left[\frac{d}{dx} \left(EI \frac{d^2 v}{dx^2} \right) \delta v \right]_{x=0} = 0 \qquad (5.34)$$

For arbitrary δv, the Euler-Lagrange equation gives the usual fourth-order governing differential equation for the beam, viz.,

$$\frac{d^2}{dx^2} \left(EI \frac{d^2 v}{dx^2} \right) - \bar{q} = 0 \qquad (5.35)$$

From equation (5.34), the boundary conditions at $x = 0$ are

$$\delta v = 0 \quad \text{or} \quad \frac{d}{dx} \left(EI \frac{d^2 v}{dx^2} \right) = 0 \qquad (5.36)$$

That is, prescribed transverse displacement (*essential* boundary condition) or zero prescribed shear force (*natural* boundary condition). In addition, the second set of boundary conditions at $x = 0$ is

$$\frac{d\delta v}{dx} = 0 \quad \text{or} \quad EI \frac{d^2 v}{dx^2} = 0 \qquad (5.37)$$

That is, prescribed slope (*essential* boundary condition) or zero prescribed bending moment (*natural* boundary condition).

At $x = L$ the boundary conditions are

$$\delta v = 0 \quad \text{or} \quad -\frac{d}{dx} \left(EI \frac{d^2 v}{dx^2} \right) = \bar{V} \qquad (5.38)$$

That is, prescribed transverse displacement (*essential* boundary condition) or prescribed shear force (*natural* boundary condition). In addition, the second set of boundary conditions at $x = L$ is

$$\frac{d\delta v}{dx} = 0 \quad \text{or} \quad EI\frac{d^2v}{dx^2} = \bar{M} \tag{5.39}$$

That is, prescribed slope (*essential* boundary condition) or prescribed bending moment (*natural* boundary condition).

If the applied concentrated force $\bar{V}$, the moment of couple $\bar{M}$, or both $\bar{V}$ and $\bar{M}$ are applied at points different from the ends, the situation becomes slightly more complicated. For the case of concentrated forces applied at points different from the ends, $v(x)$ will have discontinuities in its third derivative at these points. The governing differential equation remains unchanged, and must be satisfied in any interval between the applied loads. For the case of concentrated moments applied at points different from the ends, $v(x)$ will have discontinuities in its second derivative at these points. The governing differential equation again remains unchanged, and must be satisfied in any interval between the applied moments.

5.5 Stationary Functional Method

A general approach for generating approximate solutions of physical problems by rendering the functional stationary is now considered. This approach, in the form of a minimization of energy, was first applied by Lord Rayleigh [456] to find the frequencies of vibration (eigenvalues) of complicated systems. Although such a minimization concept was not new, Rayleigh was successful in expanding upon earlier work in the field. The idea of calculating frequencies directly from an energy consideration, without solving differential equations, was subsequently refined and generalized by Ritz [465, 466]. Ritz extended the principle of energy minimization by including the use of multiple independent trial functions. He studied both eigenvalue and equilibrium problems.[10]

The stationary functional method of Rayleigh and Ritz[11] is now widely used not only in studying vibration and elasticity problems, but also in the

---

[10]An overview of Ritz's work in developing approximate solutions for elasticity problems is given by Timoshenko [539].

[11]The name of this method is not universal. According to Crandall [125], the stationary functional method is called the *Ritz method* when it is applied to equilibrium problems and the *Rayleigh-Ritz method* when applied to eigenvalue problems. Finlayson and Scriven [159] acknowledge that basically this is a single method, but distinguish between the Rayleigh-Ritz method when it is applied to minimum or maximum principles and the Ritz method when it is applied to merely stationary principles.

theory of structures, non-linear mechanics, and other branches of physics. In the opinion of Timoshenko [539]:

> Perhaps no other single mathematical tool has led to as much research in the strength of materials and theory of elasticity.

Preliminaries aside, we focus on the description of the stationary functional method.

Let $I(\phi)$ be a functional such that the extremum problem for $I(\phi)$ is equivalent to the equilibrium problem

$$L(\phi) - f = 0 \quad \text{in} \quad \Omega \tag{5.40}$$

$$B(\phi) - r = 0 \quad \text{on} \quad \Gamma \tag{5.41}$$

The stationary functional method treats the extremum problem directly by substituting in the functional the trial solution of equation (5.1) containing a number of adjustable parameters, the α_m. The functional $I(\hat{\phi})$ is then minimized with respect to these parameters by setting $\delta I(\hat{\phi}) = 0$, which implies that

$$\frac{\partial \hat{I}}{\partial \alpha_m} = 0 \tag{5.42}$$

for $m = 1, 2, \cdots, P$.

The resulting set of P simultaneous linear equations is solved for the P unknowns α_m. Mathematical complications often restrict the values of P to the first few integers. An advantage of the stationary functional method is that admissible functions $\hat{\phi}$ for the extremum problem need only satisfy essential boundary conditions. This tends to simplify the selection of the family of trial functions N_m.

The corresponding functional $I(\hat{\phi})$ represents an approximate solution to the extremum problem. That is, it gives $I(\phi)$ a stationary value only for those variations of ϕ that are contained in the expression for $\hat{\phi}$. This solution would not, in general, still be an extremum if more general variations were permitted [125]. Application of the stationary functional method is illustrated in the following examples.

Example 5.3: One-Dimensional Steady-State Heat Conduction

The one-dimensional, steady-state heat conduction problem analyzed in Examples 4.1 to 4.5 is again considered. The solution domain is shown in Figure 5.3.

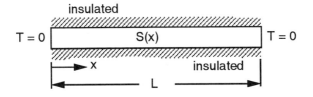

Figure 5.3: Schematic Illustration of an Insulated Rod

The general functional for the problem is given by equation (5.21), only with $S_1 = 0$ and no convection terms. The functional thus reduces to

$$I(T) = \int_0^L \frac{1}{2} k \left(\frac{dT}{dx} \right)^2 dx - \int_0^L S_o T \, dx \qquad (5.43)$$

To be admissible, the approximate temperature must have a smooth first derivative and must satisfy both essential boundary conditions. To facilitate comparison with the weighted residual methods considered in Chapter 4, we choose the same approximation for T, viz.,

$$\hat{T} = \sum_{m=1}^{2} \alpha_m \sin \frac{m \pi x}{L} = \alpha_1 \sin \frac{\pi x}{L} + \alpha_2 \sin \frac{2 \pi x}{L} \qquad (5.44)$$

Substituting equation (5.44) and its first derivative into equation (5.43) gives the expression for the *approximate* functional

$$I(\hat{T}) = \frac{1}{2} \left(\frac{\pi}{L} \right)^2 \int_0^L k \left(\alpha_1 \cos \frac{\pi x}{L} + 2\alpha_2 \cos \frac{2 \pi x}{L} \right)^2 dx$$
$$- \int_0^L S_o x \left(\alpha_1 \sin \frac{\pi x}{L} + \alpha_2 \sin \frac{2 \pi x}{L} \right) dx \qquad (5.45)$$

We next make $I(\hat{T})$ stationary by setting $\dfrac{\partial \hat{I}}{\partial \alpha_1} = \dfrac{\partial \hat{I}}{\partial \alpha_2} = 0$. This gives

$$\frac{\partial I(\hat{T})}{\partial \alpha_1} = \left(\frac{\pi}{L} \right)^2 \int_0^L k \cos \frac{\pi x}{L} \left(\alpha_1 \cos \frac{\pi x}{L} + 2\alpha_2 \cos \frac{2 \pi x}{L} \right) dx$$
$$- \int_0^L S_o x \left(\sin \frac{\pi x}{L} \right) dx = 0 \qquad (5.46)$$

$$\frac{\partial I(\hat{T})}{\partial \alpha_2} = \left(\frac{\pi}{L}\right)^2 \int_0^L 2k \cos \frac{2\pi x}{L} \left(\alpha_1 \cos \frac{\pi x}{L} + 2\alpha_2 \cos \frac{2\pi x}{L}\right) dx$$

$$- \int_0^L S_o x \left(\sin \frac{2\pi x}{L}\right) dx = 0 \tag{5.47}$$

Assuming the thermal conductivity k to be constant and integrating equations (5.46) and (5.47) gives

$$\left\{\begin{array}{c} \alpha_1 \\ \alpha_2 \end{array}\right\} = \frac{S_o L^3}{4k\pi^3} \left\{\begin{array}{c} 8 \\ -1 \end{array}\right\} \tag{5.48}$$

The approximate temperature distribution is thus given by

$$\hat{T}(x) = \frac{S_o L^3}{4k\pi^3} \left(8 \sin \frac{\pi x}{L} - \sin \frac{2\pi x}{L}\right) \tag{5.49}$$

The corresponding approximation to the heat flux distribution is

$$\hat{q}(x) = -k\frac{d\hat{T}}{dx} = -\frac{S_o L^2}{2\pi^2} \left(4 \cos \frac{\pi x}{L} - \cos \frac{2\pi x}{L}\right) \tag{5.50}$$

These results are *identical* to those obtained using the Galerkin method. This phenomenon is explained further in Section 5.6.

Example 5.4: Beam Flexure – Approximation Using the Rayleigh-Ritz Method

Consider a simple (Bernoulli-Euler) cantilever beam subjected to a uniformly distributed load $\bar{q}$ in the manner shown in Figure 5.4. The beam is prismatic, with a moment of inertia (second moment of area) of I. The material is assumed to be isotropic linear elastic, and is characterized by an elastic modulus E.

The exact solutions for the transverse displacement v, the bending moment M and the transverse shear force V distributions are:

$$v(x) = \frac{\bar{q}x^2}{24EI} \left(4Lx - x^2 - 6L^2\right) \tag{5.51}$$

$$M(x) = EI\frac{d^2 v}{dx^2} = -\frac{\bar{q}}{2}(x - L)^2 \tag{5.52}$$

$$V(x) = EI\frac{d^3 v}{dx^3} = \bar{q}(L - x) \tag{5.53}$$

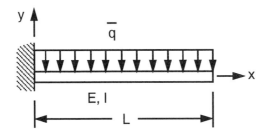

Figure 5.4: Cantilever Beam Subjected to Uniformly Distributed Load

For the beam the elastic strain energy is

$$U = \frac{1}{2} \int_0^L EI \left(\frac{d^2v}{dx^2} \right)^2 dx \tag{5.54}$$

The potential energy due to the applied load is

$$V_a = - \int_0^L q(x)v \, dx = - \int_0^L -\bar{q}v \, dx = \int_0^L \bar{q}v \, dx \tag{5.55}$$

The total potential is thus

$$\Pi = U + V_a = \frac{1}{2} \int_0^L EI \left(\frac{d^2v}{dx^2} \right)^2 dx + \int_0^L \bar{q}v \, dx \tag{5.56}$$

Since the governing differential equation for the beam is fourth-order (recall, equation 5.35), it follows that the essential boundary conditions involve the transverse displacement v and its first derivative (the slope) dv/dx. For the cantilever beam shown in Figure 5.4, the essential boundary conditions occur at $x = 0$, viz.,

$$v\big|_{x=0} = \frac{dv}{dx}\bigg|_{x=0} = 0 \tag{5.57}$$

To be admissible, a trial solution must: (1) Possess smooth derivatives up to degree two appearing in equation (5.56) and, (2) Satisfy the essential boundary conditions of equation (5.57). The following polynomial approximation is thus proposed:

$$\hat{v}(x) = \sum_{i=1}^{P} \alpha_i x^{i-1} \tag{5.58}$$

We first consider the case of a *quadratic* polynomial ($P = 3$). The trial solution is clearly differentiable up to degree two, viz.,

$$\hat{v} = \alpha_1 + \alpha_2 x + \alpha_3 x^2 \tag{5.59}$$

$$\frac{d\hat{v}}{dx} = \alpha_2 + 2\alpha_3 x \tag{5.60}$$

$$\frac{d^2\hat{v}}{dx^2} = 2\alpha_3 \tag{5.61}$$

To satisfy the essential boundary conditions of equation (5.57) it follows that $\alpha_1 = \alpha_2 = 0$. Substituting the resulting approximation for $\hat{v}$ and its second derivative into equation (5.56) gives

$$\hat{\Pi} = \frac{1}{2} \int_0^L EI(2\alpha_3)^2 \, dx + \int_0^L \bar{q}(\alpha_3 x^2) \, dx \tag{5.62}$$

For a stationary system in equilibrium

$$\frac{\partial \hat{\Pi}}{\partial \alpha_3} = 2EI\alpha_3 \int_0^L dx + \bar{q} \int_0^L x^2 \, dx = 0 \tag{5.63}$$

Performing the integrations, evaluating the limits of integration and simplifying gives

$$\alpha_3 = -\frac{\bar{q}L^2}{6EI} \tag{5.64}$$

The approximate transverse displacement, bending moment and shear force distributions for $P = 3$ are thus

$$\hat{v}(x) = -\frac{\bar{q}L^2}{6EI}x^2 \tag{5.65}$$

$$\hat{M}(x) = EI\frac{d^2\hat{v}}{dx^2} = -\frac{\bar{q}}{3} \tag{5.66}$$

$$\hat{V}(x) = EI\frac{d^3\hat{v}}{dx^3} = 0 \tag{5.67}$$

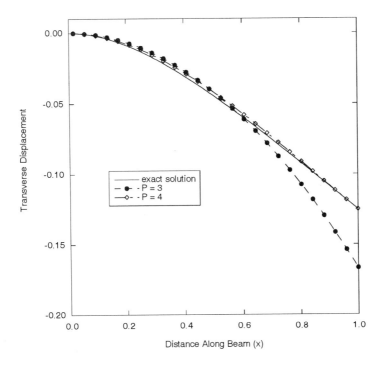

Figure 5.5: Assessment of Rayleigh-Ritz Method: Transverse Displacement

In Figure 5.5, the approximate transverse displacement for $P = 3$ is compared to the exact solution. Since the latter is fourth-order in x, it is not surprising that the quadratic approximation differs from the exact solution, particularly as the free end of the beam is approached.

In Figure 5.6, the approximate bending moment distribution is compared to the exact one. Since the latter is quadratic in x, it is not surprising that the constant $\hat{M}(x)$ distribution is a rather poor approximation. Even more unreasonable is the approximate shear force distribution of zero.

The above results illustrate a general difficulty that arises when moments and forces are derived from approximate expressions for the transverse displacement. Although the approximate displacements may be reasonable (though this was not really the case in this example), the second and third derivatives of these quantities will likely deviate widely from the correct values. Consequently, when the approximate displacements (the primary dependent variables) are used to calculate bending moments and shear forces (the secondary dependent variables), the results must be

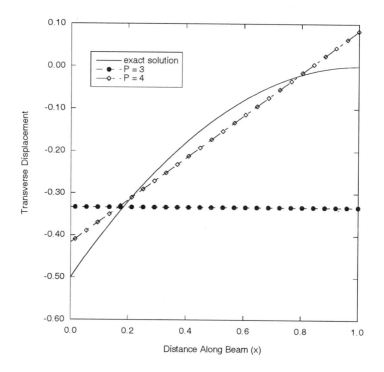

Figure 5.6: Assessment of Rayleigh-Ritz Method: Bending Moment

viewed with caution. This warning is not restricted to displacements and the Rayleigh-Ritz method but, as mentioned in the previous chapters, applies to any primary and secondary dependent variable and to other approximate solution techniques.

In an attempt to improve the approximation, we increase the degree of the polynomial to cubic ($P = 4$). As before, in order to satisfy the essential boundary conditions of equation (5.57), $\alpha_1 = \alpha_2 = 0$. Thus

$$\hat{v} = \alpha_3 x^2 + \alpha_4 x^3 \tag{5.68}$$

$$\frac{d\hat{v}}{dx} = 2\alpha_3 x + 3\alpha_4 x^2 \tag{5.69}$$

$$\frac{d^2\hat{v}}{dx^2} = 2\alpha_3 + 6\alpha_4 x \tag{5.70}$$

The approximate functional thus becomes

$$\hat{\Pi} = \frac{1}{2} \int_0^L EI(2\alpha_3 + 6\alpha_4 x)^2 \, dx + \int_0^L \bar{q}(\alpha_3 x^2 + \alpha_4 x^3) \, dx \qquad (5.71)$$

The requisite two equations required for the determination of α_3 and α_4 are obtained by making the functional stationary, viz.,

$$\frac{\partial \hat{\Pi}}{\partial \alpha_3} = 4EI \int_0^L (\alpha_3 + 3\alpha_4 x) \, dx + \bar{q} \int_0^L x^2 \, dx = 0 \qquad (5.72)$$

$$\frac{\partial \hat{\Pi}}{\partial \alpha_4} = 12EI \int_0^L (\alpha_3 + 3\alpha_4 x) x \, dx + \bar{q} \int_0^L x^3 \, dx = 0 \qquad (5.73)$$

Performing the integrations, evaluating the limits of integration and simplifying gives the following set of equations:

$$\begin{bmatrix} 2 & 3L \\ 1 & 2L \end{bmatrix} \begin{Bmatrix} \alpha_3 \\ \alpha_4 \end{Bmatrix} = -\frac{\bar{q}L^2}{24EI} \begin{Bmatrix} 4 \\ 1 \end{Bmatrix} \qquad (5.74)$$

Solving for α_3 and α_4 gives

$$\begin{Bmatrix} \alpha_3 \\ \alpha_4 \end{Bmatrix} = -\frac{\bar{q}L}{24EI} \begin{Bmatrix} 5L \\ -2 \end{Bmatrix} \qquad (5.75)$$

The approximate transverse displacement, bending moment and shear force distributions for cubic $(P = 4)$ trial solution are thus

$$\hat{v}(x) = -\frac{\bar{q}Lx^2}{24EI}(5L - 2x) \qquad (5.76)$$

$$\hat{M}(x) = EI\frac{d^2\hat{v}}{dx^2} = \frac{\bar{q}L}{12}(6x - 5L) \qquad (5.77)$$

$$\hat{V}(x) = EI\frac{d^3\hat{v}}{dx^3} = \frac{\bar{q}L}{2} \qquad (5.78)$$

In Figure 5.5, the approximate transverse displacement for $P = 4$ is seen to be a marked improvement over the approximation for $P = 3$, and to be rapidly converging to the exact solution. As expected, the convergence of the bending moment, shown in Figure 5.6, is less rapid. Finally, the approximate shear force distribution for $P = 4$, though no longer zero, is constant and thus still differs from the exact linear distribution.

In closing this example, we note that choosing a polynomial of degree four $(P = 5)$ will, of course, give the exact solution.

5.6 Relation to Weighted Residual Method

Using the same trial solution, in Examples 4.4 and 5.3 the solution to the same physical problem was approximated by the Galerkin and stationary functional methods, respectively. The resulting approximate solutions $\hat{T}(x)$ were found to be *identical*. This result is not coincidental, and consequently warrants further discussion.

It is known that where a true variational principle exists (that is, for a self-adjoint problem), Ritz's method based on the stationary functional is equivalent to the Galerkin method [125, 113]. This holds for linear, and certain non-linear systems [160]. In the case of a minimum or maximum principle, the Rayleigh-Ritz and Galerkin methods are known to be equivalent [125, 159].

The equivalence between the Galerkin method and the stationary functional (Ritz and Rayleigh-Ritz) method holds even when the trial functions do not satisfy the natural boundary conditions [159], which they need not do in the stationary functional method.

An important difference between the Galerkin and Rayleigh-Ritz methods is that in the latter some functional — possibly representing an eigenvalue — is being minimized or maximized. The approximate values of the functional thus represent either upper or lower bounds on the exact solution [159, 164]. A more thorough discussion of the advantages of variational methods over weighted residual methods is given by Finlayson and Scriven [160].

5.7 Related Methods

Although not covered in this or the previous chapter, some other methods generating continuous approximations deserve mention.

5.7.1 Kantorovich Method

In the general form of the approximation given by Equation (5.1), the $N(\mathbf{x})$ are *a priori* chosen trial functions and the α_m are unknown constants. Using either a weighted residual or a variational approach, a system of P algebraic equations is developed, the solution of which gives the values of the α_m. From a practical point of view, the main shortcoming of such methods is the strong dependence of the results on the assumed trial functions.

In an effort to eliminate this shortcoming, Kantorovich [277, 278] suggested an approximate solution of the form

$$\hat{\phi}(\mathbf{x}) = \sum_{m=1}^{P} \alpha_m(x_1) N_m(x_1, x_2, \cdots) \qquad (5.79)$$

where the N_m are again known trial functions, but the $\alpha_m(x_1)$ are no longer constants but are unknown functions of one of the independent variables. The functional $I(\hat{\phi})$ now depends upon the P functions of the independent variable x_1. The rendering of $I(\hat{\phi})$ stationary now leads to a set of P ordinary differential equations in the $\alpha_m(x_1)$. This procedure generates a solution that, regardless of the arbitrary nature of the N_m, tends to the exact solution along the x_1 direction.

An extension of the Kantorovich method that involves a rapidly converging iteration process has been proposed by Kerr [288], who showed the final form of the generated solution to be unique.

5.8   Concluding Remarks

In this chapter, techniques for developing continuous approximate solutions based on variational methods were considered. Similar to the solutions based on the method of weighted residuals, variational based approximations are continuous within the domain Ω and along its boundary Γ. As such, they also overcome the fundamental shortcoming of discrete approaches such as the finite difference method.

Since they require the existence of a functional, variational methods can realistically be applied only to physical problems that are self-adjoint. Such a restriction does not apply for approximate solutions based on the method of weighted residuals.

In both the method of weighted residuals and variational methods, the approximate solution $\hat{\phi}(\mathbf{x})$ applies to the *entire* domain Ω. For geometrically simple domains such as lines, rectangles, right prisms, etc., this fact poses no difficulties. However, in the case of domains with curved boundaries, cut-outs, material inhomogeneities, inclusions, etc., the choice of suitable trial solutions becomes substantially more difficult.

An elegant way in which to circumvent such limitations is to break Ω into a finite number of non-overlapping subdomains and to define the trial functions N_m in a piecewise manner. These two actions constitute the basic assumptions underlying the so-called finite element method, the discussion of which commences in the next chapter.

5.9   Exercises

5.1

Consider a bar subjected to a distributed axial load per unit length $f(x)$ in the manner shown in Figure 5.7.

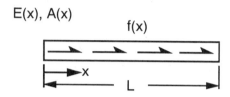

Figure 5.7: Bar Restrained at Both Ends

Given the total potential (functional) in the axial displacement u (the primary dependent variable),

$$\Pi = U + V_a = \int_0^L \frac{EA}{2} \left(\frac{du}{dx}\right)^2 dx - \int_0^L f(x) Au \, du$$

derive the governing differential equation (see Example 1.2).

5.2

Consider the simple bar shown in Figure 5.7 only with both ends restrained (i.e., $u(0) = u(L) = 0$). Use the Rayleigh-Ritz method to determine the displacement at the points $y = L/4$, $L/2$, and $3L/4$, where $L = 3.05$ meters. Compare these values with the exact solution. Also compare the approximate and exact stress distributions.

Assume the elastic modulus $E = 2.07 \times 10^8 \, kPa$, the cross-sectional area $A = 3.20 \times 10^{-3} \, m^2$ and the body force in the *positive* x direction $b_x = 75.0 \, kN/m^3$ (consequently, $f(x) = b_x A$). In the approximation for the axial displacement v, use no more than 3 terms, that is $P \leq 3$.

5.3

Consider a simply-supported Bernoulli-Euler beam subjected to a uniformly distributed load $\bar{q}$ in the manner shown in Figure 5.8. The beam is prismatic, with a moment of inertia (second moment of area) of I. The material is assumed to be isotropic linear elastic, and is characterized by an elastic modulus E.

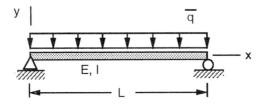

Figure 5.8: Simply-Supported Beam Subjected to Uniformly Distributed Load

Using the approximation

$$\hat{v}(x) = \sum_{i=1}^{P} \alpha_i \sin \frac{i\pi x}{L}$$

with P equal to 1, 2 and 3, obtain approximate solutions using the Rayleigh-Ritz method in conjunction with the above approximation. The total potential (functional) for the beam is

$$\Pi = U + V_a = \int_0^L \frac{EI}{2}\left(\frac{d^2 v}{dx^2}\right)^2 dx - \int_0^L qv\, dx$$

where q is positive *upward*. Graphically compare the approximations for transverse displacement $\hat{v}(x)$, bending moment $\hat{M}(x) = EI\, d^2\hat{v}/dx^2$, and shear force $\hat{V}(x) = d\hat{M}/dx$ with the exact (Bernoulli-Euler beam theory) values. For purposes of specific quantitative comparison (e.g., percent relative error), consider the three points along the beam $x = L/4$, $L/2$ and $3L/4$. Present your results in three separate tables. For all three methods, discuss, in some detail, the convergence of the approximate solution to the exact one.

Chapter 6

Introduction to the Finite Element Method

6.1 Introductory Remarks

In the discussion of continuous approximate solutions in the previous two chapters, the trial functions N_m appearing in the expression

$$\hat{\phi}(\mathbf{x}, t) = \psi(\mathbf{x}, t) + \sum_{m=1}^{P} N_m(\mathbf{x})\alpha_m(t) \tag{6.1}$$

were defined by an expression valid throughout the *entire* domain Ω. Integrals involving the approximating equations, such as the weighted residual statement or a functional, were evaluated over the entire Ω and along its boundary Γ. The result was a set of P simultaneous equations that were solved for the unknown coefficients α_m.

While application of standard weighted residual and variational approaches to problems with simple domains is relatively straightforward, such is not the case for more complicated domains with curved boundaries, cut-outs, material inhomogeneities, inclusions, etc.

In this chapter the finite element method is introduced. The fundamental characteristics that distinguish the method from other approximate procedures are discussed and illustrated in an example problem. The chapter closes with a historical note.

This chapter serves as a transition between the discrete and continuous approximations discussed in the previous three chapters, and a detailed study of the finite element method that begins in the next chapter. As such, the present emphasis is on this transition and not on efficient means for formulating and applying the finite element method.

In this chapter the material is presented without the rigor associated with the mathematical theory underlying the finite element method. Details pertaining to such mathematical aspects are given in books on the subject such as [37, 417, 418].

6.2 The Notion of Nodes

An alternative to the aforementioned classical approximation approach is to identify a finite number of points in Ω where the primary dependent variables will be approximated. These points are called *nodal points*, or simply *nodes*. Equation (6.1) is thus rewritten as

$$\hat{\phi}(\mathbf{x}, t) = \psi(\mathbf{x}, t) + \sum_{m=1}^{P} N_m(\mathbf{x})\hat{\phi}_m(t) \tag{6.2}$$

The nodal values of $\hat{\phi}$, denoted by $\hat{\phi}_m$, will then be solved for in a manner similar to the α_m in equation (6.1). The variation of $\hat{\phi}$ between the nodes is described by the chosen trial or interpolation functions.[1] N_m

Remark

1. The terminology *joints*, which is commonly used in matrix structural analysis of engineering structures, is synonymous with nodes.

2. Equation (6.2) holds for each primary dependent variable being approximated in Ω. The value of P is not necessarily the same for each primary dependent variable.

3. Nodes are essentially identical to grid points in the finite difference method. The primary difference between the two methods is that, whereas equation (6.2) is continuous, the finite difference approximation is discrete. A summary of efforts made to bridge the gap between the finite element and finite difference methods is discussed in [584].

---

[1] Although the designation "trial" and "interpolation" are synonymous, we adopt the latter so as to differentiate between piecewise defined functions and those defined over the entire domain. The interpolation functions are sometimes also referred to as "shape functions," for in applications involving displacement primary dependent variables they describe the deformed shape of the element.

6.3 The Notion of Elements

Instead of dealing with it as a whole, the domain Ω is subdivided into a finite number of non-overlapping *subdomains* or *elements* Ω^e, such that

$$\Omega \approx \sum_{e=1}^{N_e} \Omega^e \tag{6.3}$$

where the integer N_e denotes the total (finite) number of elements. The elements are connected together appropriately at nodes on their boundaries. The collection of elements is commonly referred to as a *mesh*.

The boundary Γ is subdivided in a similar fashion, viz.,

$$\Gamma \approx \sum_{e=1}^{N_e} \Gamma^e \tag{6.4}$$

where Γ^e denotes that portion of Ω^e that lies on the overall boundary Γ. Summations involving Γ^e thus only involve those elements that are adjacent to the boundary.[2]

The approximate nature of equations (6.3) and (6.4) reflects the fact that it is not always possible to *exactly* subdivide Ω into elements. This is illustrated in Figure 6.1, where a quarter-circle is approximated by triangular elements. As the number of elements is increased, the triangles better approximate the curved boundary, and the error associated with the discretization of the area becomes progressively smaller. We shall refer to such error as the *domain discretization error*.

Generalizing the above discussion, it follows that if elements have a relatively simple shape, and if the definition of the interpolation functions over these elements can be made in a repeatable manner, it is possible to easily and accurately deal with complex domains. A fundamental shortcoming of standard continuous approximations will thus be overcome.

In such a "finite element" method,[3] the governing equations describing the continuum are employed in order to arrive at the properties of the respective elements. However, unlike the classical approach for deriving governing differential equations in which the dimensions of infinitesimal material elements approach zero in the limit, the dimensions of the elements remain finite in the analysis.

---

[2]In Chapter 8 this notion of "boundary" points shall be generalized.

[3]The name "finite element method" was coined by Clough [109]. Additional historical notes related to the method are given in Section 6.6.

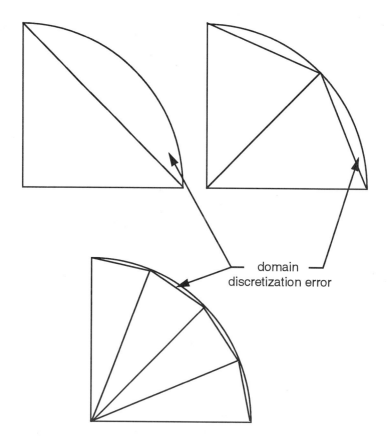

domain
discretization error

Figure 6.1: Schematic Illustration of Discretization Error

6.4   Piecewise Defined Approximations

For its success, the finite element method requires that the interpolation functions ψ and N_m appearing in equation (6.2) be constructed in a *piecewise* manner. In particular, they will be non-zero only over a particular element, and zero throughout the remainder of Ω. As a consequence, the definite integrals evaluated over Ω will be replaced by finite summations of integrals over the respective Ω^e.

The piecewise definition of interpolation functions means that *discontinuities* in the approximating function or in its derivatives will occur. As noted in Section 4.6, some degree of such discontinuity is permissible. Indeed, if certain convergence criteria are satisfied, then, as the number of elements is increased and their dimensions decreased, the behavior of the discrete system will converge to that of the continuous system.

Since they are easy to manipulate mathematically, *polynomial* interpolation functions have traditionally been used for the ψ and N_m. Some of the first general discussions of polynomial approximations for finite elements were given by Felippa [156], Dunne [145] and Silvester [496].

Other functions have, however, also been successfully applied. For example, Krahula and Polhenus [297] used functions formed by adding an appropriate Fourier series to a basic polynomial. Trigonometric interpolation functions have been used [223, 13] for wave propagation analyses because the solutions tend to be decaying sine waves. Examples of other non-polynomial functions are given by Oden [414].

Remarks

1. The finite element method is thus a systematic procedure through which any continuous function is approximated by a discrete model that consists of a set of values of the function at a finite number of nodes in Ω, together with piecewise approximations of the function over a finite number of elements. The local approximation of the function over each element is uniquely defined by equation (6.2) as the product of the discrete values of the function at the finite number of nodes in the element domain Ω^e and the piecewise defined interpolation functions. Assuming triangular elements possessing three nodes and linear interpolation functions, this approximation is shown schematically in Figure 6.2.

2. The approximation of the exact solution by a trial solution with a finite number of terms introduces obviously introduces error into the solution. This error is often referred to as *discretization error*. Since it is at the very heart of the finite element method, discretization error is possibly the most important one [100]. Discretization error is akin to *truncation error*. As discussed in Chapter 2, the latter generally refers to the error involved in using a truncated of finite summation to approximate the sum of an infinite series. Together, discretization and truncation error comprise *approximation errors*. Combined with roundoff error, approximation errors come under the general heading of *algorithmic errors*.

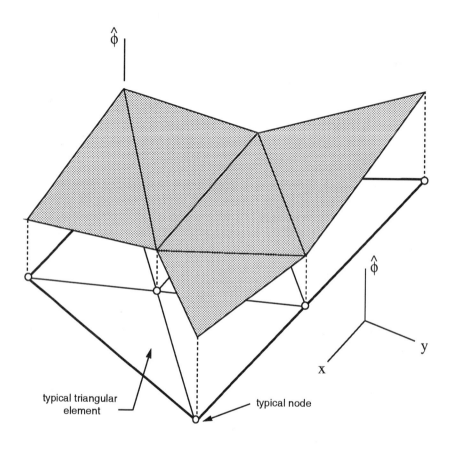

Figure 6.2: Schematic Illustration of Approximate Variation of Primary Dependent Variable Over Two-Dimensional Domain

6.5 Some Specifics

Consider the general equation governing the steady-state response of several one-dimensional physical problems discussed in Chapter 1, viz.,

$$\frac{d}{dx}\left(k\frac{d\phi}{dx}\right) + S = 0 \quad ; \quad 0 < x < L \tag{6.5}$$

where the material parameter k and "source" term S are known constants or possibly functions of x. The associated boundary conditions are

$$\phi = \bar{\phi} \quad \text{on} \quad \Gamma_1 \tag{6.6}$$

and

$$-k\frac{d\phi}{dx} = \bar{q} \quad \text{on} \quad \Gamma_2 \tag{6.7}$$

Multiplying equation (6.5) by an arbitrary weighting function $w(x)$ and integrating over the domain Ω $(0 < x < L)$ gives

$$\int_0^L w\left[\frac{d}{dx}\left(k\frac{d\phi}{dx}\right) + S\right] dx = 0 \tag{6.8}$$

Integrating its first term by parts, equation (6.8) becomes

$$wk\frac{d\phi}{dx}\bigg|_\Gamma - \int_0^L \left(\frac{dw}{dx}k\frac{d\phi}{dx}\right) dx + \int_0^L wS\, dx = 0 \tag{6.9}$$

Recalling from Section 1.3.5 that $\Gamma_1 \cup \Gamma_2 = \Gamma$ and that $\Gamma_1 \cap \Gamma_2 = \emptyset$, the boundary term in equation (6.9) is rewritten as

$$wk\frac{d\phi}{dx}\bigg|_\Gamma = wk\frac{d\phi}{dx}\bigg|_{\Gamma_1} + wk\frac{d\phi}{dx}\bigg|_{\Gamma_2} = wk\frac{d\phi}{dx}\bigg|_{\Gamma_1} - w\bar{q}\big|_{\Gamma_2} \tag{6.10}$$

where equation (6.7) has been used. In Section 4.7 it was shown that on Γ_1 the weighting function $w(x)$ must vanish. Consequently, the first term on the right-hand side of equation (6.10) vanishes. Using this suitably modified expression, equation (6.9) becomes

$$\int_0^L \left(\frac{dw}{dx} k \frac{d\phi}{dx} \right) dx - \int_0^L wS \, dx + w\bar{q}|_{\Gamma_2} = 0 \qquad (6.11)$$

We next approximate the primary dependent variable by equation (6.2), specialized for a one-dimensional domain and steady-state conditions, viz.,

$$\hat{\phi}(x) = \psi(x) + \sum_{m=1}^{P} N_m(x)\hat{\phi}_m \qquad (6.12)$$

The function $\psi(x)$ must satisfy all *essential* boundary conditions on Γ_1. The choice of $\psi(x)$ is clearly *not unique*. For example, assuming $\phi = \bar{\phi}$ at $x = 0$, the following functions would satisfy this essential boundary condition:

$$\psi(x) = \bar{\phi}(1-x)^q \quad ; \quad q \geq 1 \qquad (6.13)$$

$$\psi(x) = \bar{\phi}\left(1 - \frac{x}{L}\right) \qquad (6.14)$$

The error in the approximation will be reduced by taking P in equation (6.12) to be sufficiently large. Stated mathematically,

$$\left| \hat{\phi}(x) - \psi(x) - \sum_{m=1}^{P} N_m(x)\hat{\phi}_m \right| \leq \epsilon \qquad (6.15)$$

for some error tolerance ϵ.

We next consider the weighting function $w(x)$. As previously noted, the only constraint placed on this function is that it vanish on Γ_1. In general, we write

$$w(x) = \sum_{l=1}^{P} G_l(x)\beta_l \qquad (6.16)$$

where the G_l are known interpolation functions and the β_l are arbitrary unknown scalars. Adopting a Galerkin weighted residual statement of the problem, it follows that

$$G_l(x) \equiv N_m(x) \quad ; \quad l, m = 1, 2, \cdots, P \qquad (6.17)$$

To facilitate the development, we write equations (6.12) and (6.16) in matrix form as

$$\hat{\phi} = \psi + \mathbf{N}\hat{\phi}_n \tag{6.18}$$

$$w = \mathbf{N}\beta \tag{6.19}$$

where $\mathbf{N}$ is a $(1 * P)$ row vector of interpolation functions, $\hat{\phi}_n$ is a $(P * 1)$ vector of unknown nodal values of $\hat{\phi}$, and β is a $(P * 1)$ vector of unknown arbitrary coefficients. The derivatives of $\hat{\phi}$ and w with respect to x are

$$\frac{d\hat{\phi}}{dx} = \frac{d\psi}{dx} + \mathbf{B}\hat{\phi}_n \tag{6.20}$$

$$\frac{dw}{dx} = \mathbf{B}\beta \tag{6.21}$$

where $\mathbf{B}$ is a $(1 * P)$ row vector of interpolation function derivatives.

Substituting the above equations for $\hat{\phi}$, w and their derivatives into equation (6.11) gives the Galerkin weighted residual statement of the problem, viz.,

$$G\left(\hat{\phi}, w\right) = \int_0^L \left[\beta^T \mathbf{B}^T k \left(\frac{d\psi}{dx} + \mathbf{B}\hat{\phi}_n\right)\right] dx$$
$$- \int_0^L \beta^T \mathbf{N}^T S \, dx + \beta^T \mathbf{N}^T \bar{q}\Big|_{\Gamma_2} = 0 \tag{6.22}$$

or

$$G\left(\hat{\phi}, w\right) = \beta^T \left[\mathbf{K}\hat{\phi}_n - \mathbf{g} + \mathbf{h}\right] = 0 \tag{6.23}$$

where

$$\mathbf{K} = \int_0^L \mathbf{B}^T k \mathbf{B} \, dx \tag{6.24}$$

$$\mathbf{g} = \int_0^L \mathbf{N}^T S \, dx - \mathbf{N}^T \bar{q} \big|_{\Gamma_2} \tag{6.25}$$

and

$$\mathbf{h} = \int_0^L \left(\mathbf{B}^T k \frac{d\psi}{dx} \right) dx \tag{6.26}$$

For *arbitrary* β, the bracketed expression in equation (6.23) must vanish. We thus solve the "classical" equation

$$\mathbf{K} \hat{\phi}_n = \mathbf{f} \tag{6.27}$$

where $\mathbf{f} = \mathbf{g} - \mathbf{h}$.

Remarks

    1. K shall be referred to as the "property matrix."
    2. g is a vector of equivalent generalized nodal "forces."
    3. h is a vector of nodal "forces" due to specified values of ϕ on Γ_1.

To this point in the development, we have yet to deviate from a standard Galerkin (weak form) weighted residual analysis. We desire to now transform the problem to a finite element analysis.

The first step in such an analysis involves the discretization of the one-dimensional domain $\Omega : [0, L]$ into N_e non-overlapping elements. This is rather easily accomplished by choosing a set of $N_e + 1$ distinct points in Ω having the coordinates $\{x_j ; j = 1, 2, \cdots, (N_e + 1)\}$, with $x_1 = 0$ and $x_{N_e+1} = L$. A typical element domain Ω^e is then defined to be the interval $x_e \leq x \leq x_{e+1}$, where $e = 1, 2, \cdots, N_e$. The length of element e is then $h^{(e)} = x_{e+1} - x_e$. The discretization of the one-dimensional domain is shown in Figure 6.3.

The next step in the finite element analysis involves choosing the number of nodes per element. The simplest choice is to place one node per element, say at the element center. The approximation for $\phi(x)$ will thus be *constant* over each element. However, the overall piecewise constant approximation precludes continuity of $\hat{\phi}(x)$ across element interfaces. More importantly, the derivatives of the corresponding (constant) interpolation functions N_m will have *infinite* values at the element interfaces (see Figure 4.10a).

Next in complexity is the case of two nodes per element. By placing the nodes at the ends of the element, interelement continuity of $\hat{\phi}(x)$ is assured,

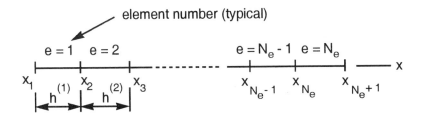

Figure 6.3: Discretization of One-Dimensional Domain

as neighboring elements share a common node. The nodes are numbered sequentially from 1 to P where, for this simple mesh, $P = N_e + 1$. The resulting finite element mesh is shown in Figure 6.4.

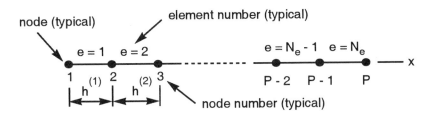

Figure 6.4: One-Dimensional Finite Element Mesh Employing Two Nodes per Element

Remark

1. One-dimensional domains are *always* discretized exactly; the domain discretization error is thus always zero in such cases.

The first step in the analysis involves the selection of suitable piecewise defined interpolation functions N_m. To gain some insight into these functions, arbitrarily assume that the essential boundary condition $\phi = \bar{\phi}$ is specified at node 1. The portion Γ_1 of the overall boundary is thus the point $x = 0$.

Evaluating equation (6.12) at node i, $(i = 2, 3, \cdots, P)$, gives

$$\hat{\phi}(x_i) = 0 + N_1(x_i)\hat{\phi}_1 + N_2(x_i)\hat{\phi}_2 + \cdots + N_P(x_i)\hat{\phi}_P \equiv \hat{\phi}_i \qquad (6.28)$$

It follows that for any node i,

$$N_i(x) = \begin{cases} 1 & \text{if } x = x_i \\ 0 & \text{at all other nodes} \end{cases} \tag{6.29}$$

or, more compactly,

$$N_i(x_j) = \delta_{ij} \tag{6.30}$$

where δ_{ij} is the Kronecker delta.

In light of equation (6.30), and because two nodes have been assigned to each element, we propose a set of *linear* piecewise defined interpolation functions. For the case of an essential boundary condition $\phi = \bar{\phi}$ specified at node 1, we write

$$\psi(x) = \bar{\phi} N_1 = \bar{\phi}\left(\frac{x_2 - x}{x_2 - x_1}\right) = \bar{\phi}\left(\frac{x_2 - x}{h^{(1)}}\right) \tag{6.31}$$

where $x_1 \leq x \leq x_2$. This function is shown in Figure 6.5(a). The derivative of ψ with respect to x is

$$\frac{d\psi}{dx} = -\frac{\bar{\phi}}{h^{(1)}} \tag{6.32}$$

which is obviously a constant.

The interpolation function associated with nodes 2, 3, $P-1$ and P are shown in Figures 6.5(b) to (e), respectively. In general, the interpolation function for node i is defined by

$$N_i(x) = \begin{cases} \dfrac{x - x_{i-1}}{h^{(i-1)}} & , \quad x_{i-1} \leq x \leq x_i \\[3mm] \dfrac{x_{i+1} - x}{h^{(i)}} & , \quad x_i \leq x \leq x_{i+1} \\[3mm] 0 & \quad \text{elsewhere} \end{cases} \tag{6.33}$$

As required by equation (6.30), for $x_{i-1} \leq x \leq x_i$

$$N_i(x_i) = \frac{x_i - x_{i-1}}{h^{(i-1)}} = 1 \tag{6.34}$$

$$N_i(x_{i-1}) = \frac{x_{i-1} - x_{i-1}}{h^{(i-1)}} = 0 \tag{6.35}$$

and for $x_i \leq x \leq x_{i+1}$,

$$N_i(x_i) = \frac{x_{i+1} - x_i}{h^{(i)}} = 1 \tag{6.36}$$

$$N_i(x_{i+1}) = \frac{x_{i+1} - x_{i+1}}{h^{(i-1)}} = 0 \tag{6.37}$$

Remark

1. The piecewise defined interpolation functions are not chosen without guidance. In particular, from equation (6.24) and the discussion of Section 4.6, it is evident that only C^0 continuity will be required of these functions. That is, only the functions need to be continuous at the nodes; the first derivatives of the interpolation functions will be piecewise constant and will exhibit *finite* jumps across element interfaces.

Consider a typical element $e = E$, with node numbers i and j and length $h^{(E)}$. In light of the above discussion, the element interpolation functions are

$$N_i^{(E)}(x) = \begin{cases} \dfrac{x_j - x}{h^{(E)}} & , \quad x_i \leq x \leq x_j \\ \\ 0 & , \quad x < x_i \; ; \; x > x_j \end{cases} \tag{6.38}$$

and

$$N_j^{(E)}(x) = \begin{cases} \dfrac{x - x_i}{h^{(E)}} & , \quad x_i \leq x \leq x_j \\ \\ 0 & , \quad x < x_i \; ; \; x > x_j \end{cases} \tag{6.39}$$

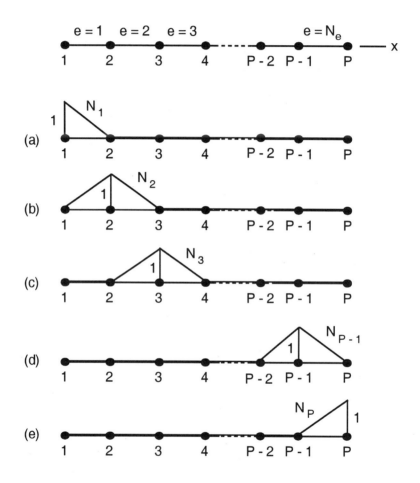

Figure 6.5: Piecewise Defined One-Dimensional Interpolation Functions

A graphical representation of these interpolation functions is given in Figure 6.6. The derivatives of the interpolation functions are constants, viz.,

$$\frac{dN_i^{(E)}}{dx} = -\frac{1}{h^{(E)}} = -\frac{dN_j^{(E)}}{dx} \tag{6.40}$$

Due to the piecewise nature of the interpolation functions, the global problem becomes the sum of N_e element problems, viz.,

$$G\left(\hat{\phi}, w\right) = \sum_{e=1}^{N_e} G^e\left(\hat{\phi}, w\right) \tag{6.41}$$

This will, however, impose certain continuity restrictions on the interpolation functions used.

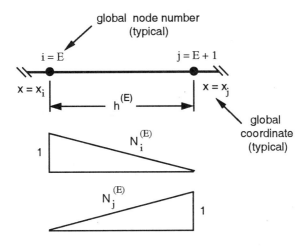

Figure 6.6: Interpolation Functions for Typical Two-Node Element

With the element interpolation functions selected, we next seek the *discrete form* of equation (6.27). The entries in the property matrix, given in general by equation (6.24), now become

$$K_{lm} = \int_0^L k\left(\frac{dN_l}{dx}\frac{dN_m}{dx}\right) dx \quad , \quad l, m = 1, 2, \cdots, P \tag{6.42}$$

Due to the piecewise nature of the interpolation functions, it follows that

$$K_{lm} = \sum_{e=1}^{N_e} K_{lm}^{(e)} \tag{6.43}$$

where, for the typical element $e = E$ with nodes numbers i and j (Figure 6.6):

- If $l \neq i$, $l \neq j$ and $m \neq i$, $m \neq j$, then

$$K_{lm}^{(E)} = 0 \tag{6.44}$$

- If $l = i$ or $l = j$ and $m = i$ or $m = j$ but $l \neq m$, then

$$K_{lm}^{(E)} = k^{(E)} \int_{x_i}^{x_j} \left(\frac{dN_i^{(E)}}{dx} \frac{dN_j^{(E)}}{dx} \right) dx = -\frac{k^{(E)}}{h^{(E)}} = K_{ml}^{(E)} \tag{6.45}$$

where equation (6.40) has been used.

- If $l = i$ or $l = j$ and $m = i$ or $m = j$ and $l = m$, then

$$K_{ll}^{(E)} = K_{mm}^{(E)} = k^{(E)} \int_{x_i}^{x_j} \left(\frac{dN_i^{(E)}}{dx} \frac{dN_i^{(E)}}{dx} \right) dx = \frac{k^{(E)}}{h^{(E)}} \tag{6.46}$$

where equation (6.40) has again been used.

Remark

1. The symmetry of $\mathbf{K}$ results from applying the Galerkin method to a self-adjoint problem.

We next consider the entries in the "forcing" vector in equation (6.27), that is, $\mathbf{f} = \mathbf{g} - \mathbf{h}$. From equation (6.25), the entries in $\mathbf{g}$ now become

$$g_l = \int_0^L N_l S \, dx - \bar{q}|_{\Gamma_2} \tag{6.47}$$

The last term on the right-hand side of equation (6.47) applies only at a node where a natural boundary condition has been specified. The value of $\bar{q}$ is seen to simply be subtracted from the appropriate row of **g**.

As in the case of **K**, the piecewise nature of the interpolation functions implies that

$$g_l = \sum_{e=1}^{N_e} g_l^{(e)} \tag{6.48}$$

where, for the typical element $e = E$ with nodes numbers i and j:

- If $l \neq i$ and $l \neq j$,

$$g_{lm}^{(E)} = 0 \tag{6.49}$$

- If $l = i$,

$$g_l^{(E)} = S^{(E)} \int_{x_i}^{x_j} N_i^{(E)} \, dx = S^{(E)} \int_{x_i}^{x_j} \left(\frac{x_j - x}{h^{(E)}} \right) \, dx = \frac{S^{(E)} h^{(E)}}{2} \tag{6.50}$$

where equation (6.38) has been used.

- If $l = j$,

$$g_l^{(E)} = S^{(E)} \int_{x_i}^{x_j} N_j^{(E)} \, dx = S^{(E)} \int_{x_i}^{x_j} \left(\frac{x - x_i}{h^{(E)}} \right) \, dx = \frac{S^{(E)} h^{(E)}}{2} \tag{6.51}$$

where equation (6.39) has been used.

Remarks

1. In the above equations, no constraints are placed on the element length $h^{(e)}$. Compare this to the finite difference method discussed in Chapter 3, where it is often expedient to use the same grid spacing in a given coordinate direction.

2. Within an element, the material parameter k and "source" term S have been assumed to be *constant*. If the values of these parameters vary within a body, then different, though constant, element values are chosen. For this reason, in the previous equations k and S have been written with the element number as a superscript. This points out a particular strength of the finite element method, namely that material and loading discontinuities can easily be handled. It also implies that element boundaries must coincide with material boundaries. Further discussion of this point is deferred until Chapter 8.

   3. In the above discussion, an absence of nodes possessing essential boundary conditions has tacitly been assumed. Since in one-dimensional problems at least one such boundary condition must be specified in order to have a well-posed problem, this point warrants further discussion. If the essential boundary condition has been specified at a node, contributions to $\mathbf{K}$ and $\mathbf{g}$ for that node are omitted, as the number of unknowns in $\hat{\phi}_n$ is reduced by one. For the case of multiple essential boundary conditions, this logic would be extended accordingly.

   For the node or nodes associated with the essential boundary condition, an appropriate "forcing" term must be added into $\mathbf{f}$ via the vector $\mathbf{h}$, as defined in equation (6.26). In general, $\mathbf{h}$ contains entries associated with each essential boundary condition. In the case of a two node element, these entries apply to the other node in the element in which the essential boundary condition was made. In the present development, $\mathbf{h}$ thus has a single entry, viz.,

$$h_2^{(1)} = k^{(1)} \int_0^L \frac{dN_2^{(1)}}{dx} \left(\frac{d\psi}{dx} \right) dx = k^{(1)} \int_{x_i}^{x_j} \left(\frac{1}{h^{(1)}} \right) \left(-\frac{\bar{\phi}}{h^{(1)}} \right) dx$$

$$= -\frac{k^{(1)} \bar{\phi}}{\left(h^{(1)} \right)^2} \int_{x_i}^{x_j} dx = -\frac{k^{(1)} \bar{\phi}}{h^{(1)}} \tag{6.52}$$

The value of $h_2^{(1)}$ would be subtracted from the entry in $\mathbf{f}$ corresponding to node 2.

Remark

   1. In actual numerical implementations of the finite element method, essential boundary conditions are handled in a more streamlined manner. This shall be illustrated in the following example and more formally discussed in Chapter 8.

   The following example illustrates the application of the finite element method. Many of the seemingly trivial operations appearing in this example are not actually performed in the numerical implementation of the method. This shall be borne out further in Chapters 7 and 8.

Example 6.1. Application of the Finite Element Method

The one-dimensional steady-state problem governed by equation (6.5) is again considered. A domain Ω of unit length is assumed (Figure 6.7a). The material parameter k and "source" term S are taken equal to 0.03 and 1.0, respectively. At both ends of the domain, essential boundary conditions are assumed: at $x = 0$, $\phi = \bar{\phi} = 0.0$ and at $x = 1$, $\phi = \bar{\phi} = 5.0$.

The exact solutions for the primary dependent variable $\phi(x)$ and for the secondary dependent variable $q(x)$ are

$$\phi(x) = \frac{5}{3}(13 - 10x)\,x \tag{6.53}$$

$$q(x) = -k\frac{d\phi}{dx} = -\left(x - \frac{13}{20}\right) \tag{6.54}$$

• **One-Element Approximation**

We first consider the coarsest mesh possible, which consists of a single element. In this case $N_e = 1$ and $P = 2$ (Figure 6.7b). However, since essential boundary conditions are specified at both nodes, these is no need to form the property matrix **K**. Since linear interpolation functions have been assumed, the approximate solution is simply

$$\hat{\phi}(x) = (0.0)N_1 + (5.0)N_2 = 5.0\left(\frac{x - x_1}{h}\right) = 5x \tag{6.55}$$

where $x_1 = 0$ and $h = L = 1$. In Figure 6.8 the approximation given by equation (6.55) is compared to the exact solution. Although $\hat{\phi}(x)$ is obviously exact at each the two nodes, being a linear function, it deviates from the quadratic distribution given by equation (6.53).

In general, the approximate solution for the secondary dependent variable $q(x)$ is given by

$$\hat{q}(x) = -k\frac{d\hat{\phi}}{dx} = -k\left(\frac{d\psi}{dx} + \sum_{m=1}^{P}\frac{dN_m}{dx}\hat{\phi}_m\right) \tag{6.56}$$

For the present single-element mesh, only the $d\psi/dx$ term applies, thus

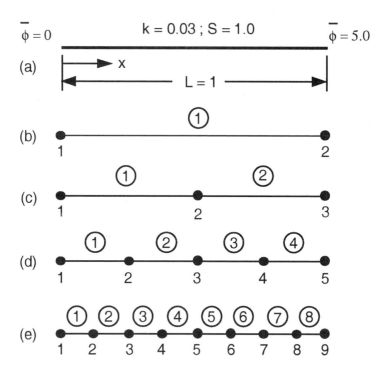

Figure 6.7: One-Dimensional Problem: (a) Domain, (b) One-Element Mesh, (c) Two-Element Mesh, (d) Four-Element Mesh, (e) Eight-Element Mesh

$$\hat{q}(x) = -k^{(1)}\left(\frac{dN_1}{dx}\bar{\phi}_1 + \frac{dN_2}{dx}\bar{\phi}_2\right) = -\frac{k^{(1)}}{h^{(1)}}\left(\hat{\phi}_2\right) = -0.150 \qquad (6.57)$$

In Figure 6.9 the approximation given by equation (6.57) is compared to the exact solution.

Remark

1. When presenting approximate results, care must be taken to accurately depict the order of the approximation. In the present case, within an element $\phi(x)$ is assumed to vary *linearly*; consequently, the flux $q(x)$ is *constant* over an element.

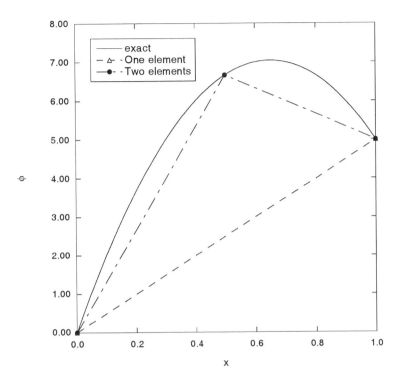

Figure 6.8: Comparison of One- and Two-Element Approximations with Exact Solution for $\phi(x)$

• **Two-Element Approximation**

In an effort to improve upon the woefully inaccurate single element approximation, we *refine* the mesh to two equally-sized elements. Now $N_e = 2$ and $P = 3$ (Figure 6.7c). Since essential boundary conditions are specified at nodes 1 and 3, we need to compute contributions to **K** and **f** only for node 2. The contribution from element 1 is:

$$K_{22}^{(1)} = \frac{k^{(1)}}{h^{(1)}} \tag{6.58}$$

$$g_2^{(1)} = \frac{S^{(1)} h^{(1)}}{2} \tag{6.59}$$

where equations (6.46) and (6.51) have been used.

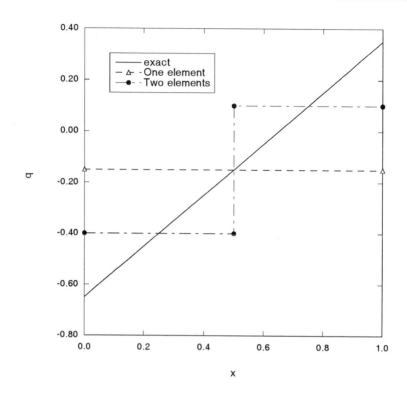

Figure 6.9: Comparison of One- and Two-Element Approximations with Exact Solution for $q(x)$

Similarly, the contribution from element 2 is:

$$K_{11}^{(2)} = \frac{k^{(2)}}{h^{(2)}} \tag{6.60}$$

$$g_1^{(2)} = \frac{S^{(2)} h^{(2)}}{2} \tag{6.61}$$

where equations (6.45) and (6.50) have been used.

Thus, using equations (6.43) and (6.48), we assemble the element contributions:

$$K_{22} = K_{22}^{(1)} + K_{11}^{(2)} = \frac{k^{(1)}}{h^{(1)}} + \frac{k^{(2)}}{h^{(2)}} = \frac{2k}{h} \tag{6.62}$$

and

$$g_2 = g_2^{(1)} + g_1^{(2)} = \frac{S^{(1)}h^{(1)}}{2} + \frac{S^{(2)}h^{(2)}}{2} = Sh \tag{6.63}$$

where $k^{(1)} = k^{(2)} = k$, $S^{(1)} = S^{(2)} = S$, and $h^{(1)} = h^{(2)} = h = 1.0$ for the particular mesh used.

In light of equation (6.52), the contribution to the right-hand side vector of the essential boundary condition at node 1 is

$$h_2^{(1)} = k^{(1)} \int_0^{L/2} \left(\frac{dN_2^{(1)}}{dx} \frac{d\psi^{(1)}}{dx} \right) dx$$

$$= \left(\frac{k^{(1)}}{h^{(1)}} \right) \left(-\frac{\bar{\phi}_{x=0}}{h^{(1)}} \right) \int_0^{L/2} dx = 0 \tag{6.64}$$

Similarly, the contribution of the essential boundary condition at node 3 is

$$h_1^{(2)} = k^{(2)} \int_{L/2}^{L} \left(\frac{dN_1^{(2)}}{dx} \frac{d\psi^{(2)}}{dx} \right) dx = \left(-\frac{k^{(2)}}{h^{(2)}} \right) \left(\frac{\bar{\phi}_{x=1}}{h^{(2)}} \right) \int_{L/2}^{L} dx$$

$$= -\frac{5k}{h} \tag{6.65}$$

The overall set of "global" equations is thus

$$K_{22}\,\hat{\phi}_2 = g_2 - h_2 \tag{6.66}$$

or

$$\frac{2k}{h}\hat{\phi}_2 = Sh - \left(-\frac{5k}{h} \right) \tag{6.67}$$

Substituting values for k, S and $h = 0.5$ and solving gives $\hat{\phi}_2 = 6.667$.

In Figure 6.8 the results of the two-element approximation are compared to the exact solution. Although the accuracy of the finite element approximation has increased substantially over the single element solution, its piecewise linear nature precludes close agreement with the exact (quadratic) distribution of $\phi(x)$.

Using equation (6.56), the approximate secondary dependent variables are computed in the following manner:

$$\hat{q}^{(1)} = -k \left(\frac{dN_1^{(1)}}{dx} \bar{\phi}_{x=0} + \frac{dN_2^{(1)}}{dx} \hat{\phi}_2 \right) = -\frac{k}{h^{(1)}} \left(\hat{\phi}_2 - \bar{\phi}_{x=0} \right) = -0.400$$

$$(6.68)$$

$$\hat{q}^{(2)} = -k \left(\frac{dN_1^{(2)}}{dx} \hat{\phi}_2 + \frac{dN_2^{(2)}}{dx} \bar{\phi}_{x=1} \right) = -\frac{k}{h^{(2)}} \left(\bar{\phi}_{x=1} - \hat{\phi}_2 \right) = 0.100$$

$$(6.69)$$

In Figure 6.9 the approximate distribution of secondary dependent variables is compared to the exact (linear) distribution. Although the accuracy has increased over the single element solution, the fact remains that we are still attempting to approximate a linear distribution with two piecewise constant functions.

Alternate Treatment of Essential Boundary Conditions

In the previous solution, the essential boundary conditions were accounted for via **h**. We now account for these boundary conditions in an *alternate* manner, one that is commonly used in actual computer implementations of the finite element method.

We begin by determining the element contributions without concern for the boundary conditions. For element 1 ($i = 1$, $j = 2$, $x_i = 0.0$ and $x_j = 0.5$), application of equations (6.44) to (6.46) gives the following element property matrix:

$$\mathbf{K}^{(1)} = \frac{k^{(1)}}{h^{(1)}} \begin{bmatrix} 1 & -1 & 0 \\ -1 & 1 & 0 \\ 0 & 0 & 0 \end{bmatrix}$$

$$(6.70)$$

The corresponding element "forcing" vector is determined using equations (6.49) to (6.51), viz.,

$$\mathbf{f}^{(1)} = \frac{S^{(1)} h^{(1)}}{2} \begin{Bmatrix} 1 \\ 1 \\ 0 \end{Bmatrix}$$

$$(6.71)$$

Similarly, for element 2 ($i = 2$, $j = 3$, $x_i = 0.5$ and $x_j = 1.0$):

$$\mathbf{K}^{(2)} = \frac{k^{(2)}}{h^{(2)}} \begin{bmatrix} 0 & 0 & 0 \\ 0 & 1 & -1 \\ 0 & -1 & 1 \end{bmatrix} \tag{6.72}$$

and

$$\mathbf{f}^{(2)} = \frac{S^{(2)} h^{(2)}}{2} \begin{Bmatrix} 0 \\ 1 \\ 1 \end{Bmatrix} \tag{6.73}$$

The "global" arrays $\mathbf{K}$ and $\mathbf{f}$ are next formed by direct summation over all elements in the manner described by equations (6.43) and (6.48), viz.,

$$\mathbf{K} = \sum_{e=1}^{2} \mathbf{K}^{(e)} = \begin{bmatrix} \dfrac{k^{(1)}}{h^{(1)}} & -\dfrac{k^{(1)}}{h^{(1)}} & 0 \\ -\dfrac{k^{(1)}}{h^{(1)}} & \left(\dfrac{k^{(1)}}{h^{(1)}} + \dfrac{k^{(2)}}{h^{(2)}}\right) & -\dfrac{k^{(2)}}{h^{(2)}} \\ 0 & -\dfrac{k^{(2)}}{h^{(2)}} & \dfrac{k^{(2)}}{h^{(2)}} \end{bmatrix} \tag{6.74}$$

$$\mathbf{f} = \sum_{e=1}^{2} \mathbf{f}^{(e)} = \begin{Bmatrix} \dfrac{S^{(1)} h^{(1)}}{2} \\ \dfrac{S^{(1)} h^{(1)}}{2} + \dfrac{S^{(2)} h^{(2)}}{2} \\ \dfrac{S^{(2)} h^{(2)}}{2} \end{Bmatrix} \tag{6.75}$$

Since $k^{(1)} = k^{(2)} = k$, $S^{(1)} = S^{(2)} = S$, and $h^{(1)} = h^{(2)} = h = 0.5$ the above entries in $\mathbf{K}$ and $\mathbf{f}$ could of course be simplified; this was not done above so as to emphasize the respective element contributions.

We next account for the essential boundary conditions. First, the known values of $\bar{\phi}_{x=0} = 0.0$ and $\bar{\phi}_{x=1} = 5.0$ are substituted into the second equation, and all known quantities are transferred to $\mathbf{f}$. Thus,

$$\mathbf{K} = \begin{bmatrix} \dfrac{k}{h^{(1)}} & -\dfrac{k}{h^{(1)}} & 0 \\ 0 & \left(\dfrac{k}{h^{(1)}} + \dfrac{k}{h^{(2)}}\right) & 0 \\ 0 & -\dfrac{k}{h^{(2)}} & \dfrac{k}{h^{(2)}} \end{bmatrix} \tag{6.76}$$

$$
\mathbf{f} = \left\{ \begin{array}{c} \dfrac{Sh^{(1)}}{2} \\[2ex] \dfrac{Sh^{(1)}}{2} + \dfrac{Sh^{(2)}}{2} - \left(-\dfrac{k}{h^{(1)}}\right)(0.0) - \left(-\dfrac{k}{h^{(2)}}(5.0)\right) \\[2ex] \dfrac{Sh^{(2)}}{2} \end{array} \right\} \tag{6.77}
$$

The first and third rows and columns of $\mathbf{K}$ are next zeroed out, and the first and third diagonal entries are set equal to unity. Finally, the first and third rows in $\mathbf{f}$ are replaced by the known values of $\bar{\phi}_{x=0}$ and $\bar{\phi}_{x=1}$. The gives the following arrays:

$$
\mathbf{K} = \begin{bmatrix} 1 & 0 & 0 \\[1ex] 0 & \left(\dfrac{k}{h^{(1)}} + \dfrac{k}{h^{(2)}}\right) & 0 \\[1ex] 0 & 0 & 1 \end{bmatrix} \tag{6.78}
$$

$$
\mathbf{f} = \left\{ \begin{array}{c} 0.0 \\[2ex] \dfrac{Sh^{(1)}}{2} + \dfrac{Sh^{(2)}}{2} - \left(-\dfrac{k}{h^{(2)}}(5.0)\right) \\[2ex] 5.0 \end{array} \right\} \tag{6.79}
$$

Setting $h^{(1)} = h^{(2)} = h = 0.5$, and forming the set of "global" equations $\mathbf{K}\hat{\phi}_n = \mathbf{f}$, the second equation in this set of equations is seen to be *identical* to equation (6.67).

• **Four- and Eight-Element Approximations**

To further investigate the convergence to the finite element solution, the mesh was refined two more times. The resulting four- and eight-element meshes are shown in Figure 6.7(d) and (e), respectively. In Figures 6.10 and 6.11 the associated approximate solutions for $\phi(x)$ and $q(x)$ are compared to exact values. The increase in accuracy of the approximations with mesh refinement is obvious.

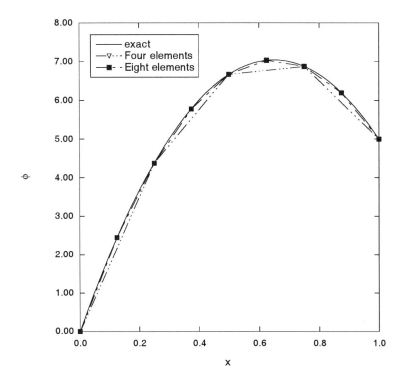

Figure 6.10: Comparison of Four- and Eight-Element Approximations with Exact Solution for $\phi(x)$

Remarks

1. In refining a mesh, it is not necessary to decrease the size of all elements. We shall have more to say about the topic of refinement in Chapters 8 and 13.

2. In Figures 6.8 and 6.10 the finite element solution is seen to be *exact* at the nodes; that is, $\hat{\phi}(x_i) = \phi(x_i)$ for $i = 1, 2, \cdots, P$. This result holds for all Galerkin finite element solutions in one-dimension. A formal proof of this result is given by Hughes [252]. Such exceptional accuracy characteristics are often referred to as *superconvergence* phenomena. They occur only in rather simple problems; in more complicated problems, superconvergence *cannot be guaranteed*.

3. From Figures 6.9 and 6.11 it is evident that the finite element solution for the secondary dependent variable $q(x)$ is less accurate than that for $\phi(x)$. Clearly this is a consequence of approximating a linear function

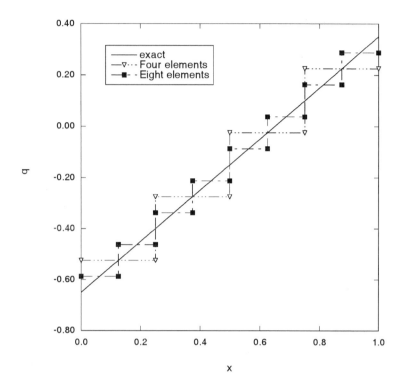

Figure 6.11: Comparison of Four- and Eight-Element Approximations with Exact Solution for $q(x)$

by piecewise constant interpolation functions. However, in the aforementioned figures we see that the solution for $\hat{q}(x)$ is *exact* at one point in each element. Indeed, it can be proven [252] that in general:

- There exists at least one point in each element at which the derivative is exact. If the exact solution is quadratic, the derivative is exact at the midpoint of the element.

- The derivative is second-order accurate at the midpoint of the element. Such special locations are typically referred to as *optimal flux sampling points*. The derivative is said to be *superconvergent* at such points. Further details pertaining to optimal sampling points and superconvergence are given in Appendix F.

6.6 A Historical Note

The finite element method has roots in applied mathematics, in mathematical physics and in engineering. Although knowledge of its exact origins is not crucial to understanding the method, it is none the less instructive to briefly investigate its historical development.

Before proceeding with this discussion, it is timely to point out that the notion of representing solution domains as a collection of non-overlapping subdomains was used by ancient mathematicians to approximate the circumference and area of a circle, as well the as areas and volumes of various geometric shapes [376, 461]. The focus herein is, however, on more recent developments.

6.6.1 Early Contributions of Applied Mathematicians

The mathematical foundation of the finite element method is traced to the work of Galerkin [185] in the method of weighted residuals, and to Rayleigh [456] and Ritz [465, 466] in variational methods. However, as noted in the previous two chapters, the major shortcoming of such methods is the requirement that the trial functions used apply to the *entire* solution domain.

In 1941, the notion of piecewise defined trial functions, which is fundamental to the finite element method, was introduced by Courant [122].[4] Courant noted the equivalence between partial differential equations associated with boundary value problems and the related variational forms, and discussed in detail the Rayleigh-Ritz method. He developed finite difference equations by means of this method, with trial functions that were piecewise linear over triangular subdomains. Using these to approximate the warping function, he studied the classical Saint Venant torsion problem in elasticity. Courant's work represents a significant extension of the Rayleigh-Ritz method, and is notable because of its applicability to problems outside the realm of structural mechanics. Perhaps more importantly, Courant's piecewise application of the Rayleigh-Ritz method involved most of the basic concepts underlying the finite element method.

With the possible exception of an introduction to the theory of splines and advocacy of piecewise defined polynomials for approximation and interpolation presented by Schoenberg [487], Courant's discretization ideas

---

[4]Some earlier contributions of mathematicians, though somewhat more obscure than Courant's work, also deserve mention. The notion of variational principles and approximation by piecewise smooth functions was used (in one dimension) by Leibnitz in 1696 [525]. In two dimensions, Schellbach [484] employed triangulation and piecewise linear functions to solve Plateau's problem of determining the surface of minimum area enclosed by a given closed curve [568, 415].

lay dormant until the mid-1950s. Around this time, several mathematicians [438, 439, 234, 556, 557] employed similar discretization techniques to approximate the upper and lower bounds on eigenvalues associated with boundary value problems in continuum mechanics.

In 1959 Greenstadt [206] outlined a discretization approach that contained many of the fundamental concepts that constitute the mathematical basis for the finite element method. He envisioned the solution domain as being divided into a set of contiguous subdomains or "cells." The unknown primary dependent variable was approximated by a series of functions, each associated with a single cell. After evaluating the appropriate variational principle in each cell, Greenstadt used continuity requirements to suitably relate the equations for all the cells. In this manner the continuous problem was reduced to a discrete one with a finite number of degrees of freedom.

In the early 1960s White [559] and Friedrichs [182], apparently unaware of the development of the finite element method in the engineering community, used regular meshes of triangular subdomains to develop difference equations from variational principles.

6.6.2   Early Contributions of Mathematical Physicists

In the late 1940s, mathematical physicists were also interested in the approximate solution of continuum problems. A notable example is the work of Prager and Synge [441]. The goal of this work was to obtain approximate solutions of elastic boundary value problems with errors that were computable, typically being measured in terms of distance in functional space. Included as part of this work was the development of the so-called "hypercircle" method. In its original form, this method was developed to give a geometric interpretation for the minimum principles of classical elasticity theory. Extensions of the hypercircle method involved the solution of continuum problems using piecewise linear interpolation functions defined over triangulated regions with a Rayleigh-Ritz variational procedure [522, 523]. This was followed by the work of McMahon [383], who used tetrahedral elements and linear trial functions to solve a problem in three-dimensional elastostatics.

6.6.3   Early Contributions of Engineers

Compared to applied mathematicians and mathematical physicists, engineers have approached the discretization of continuous problems in a more physically intuitive manner. In the opinion of Gallagher [187], the origins of the finite element method in the engineering community can be traced back to the pioneering work of Maxwell [380], Castigliano [106], and Mohr [399].

It is from this work that the concepts underlying framework methods of structural analysis emerged. Using such methods, engineers create analogies between actual discrete structural members and portions of the actual continuum. Whenever the number of joints in a structure is finite (e.g., in a truss), such analogies yield accurate accurate results. However, when elastic continua such as plates or shells are encountered, the infinite number of degrees of freedom (joints) associated with such bodies creates difficulties for the analyst.

To overcome such difficulties, a "lattice" analogy for plane stress analysis of elastic bodies was proposed in the early 1900s by Wieghardt [563] and subsequently by Riedel [464]. In this analysis technique the continuum is replaced by a regular pattern of physically separate, one-dimensional elastic bars, interconnected only at their ends (joints). This framework of bars is given the same external outline and boundary constraints as the continuum. All loads are applied at the joints. The bar properties are chosen so that the resulting joint displacements approximate the displacements of the same points in the continuum. Systematic procedures for analyzing complex electrical networks and structural systems were developed by Kron [295]. Hrennikoff [250] presented a more general "framework method" that proved to be a forerunner to the development of general discrete methods of structural analysis. This often cited work motivated the subsequent development of discretization techniques by McHenry [382] and Newmark [407].

The 1950s marked the commercial introduction of high-speed, stored-program digital computers. It was only natural, therefore, that the aforementioned framework analysis procedures were reformulated into a matrix format ideally suited for efficient automatic digital computation. Notable contributions in this area were made by Langefors [303] and Argyris [20, 21, 22]. The latter series of works developed certain generalizations of the linear theory of structures, and presented efficient solution procedures for analyzing discrete structural configurations in forms well-suited to digital computation. It is credited with laying a foundation for further developments in the finite element method.[5] Further details pertaining to computer-related contributions during this important developmental stage of the finite element method can be found in the review article by Argyris and Patton [27].

The framework analysis concepts also represent the basis of matrix structural analysis methodology, the practical usefulness of which was like-

---

[5] The numerical implementation of the finite element method was also profoundly influenced by the introduction of the FORTRAN programming language in 1954. FORTRAN was developed by John Backus and his co-workers as a new compiler for the IBM 704 computer [470].

wise greatly accelerated by the commercial introduction of high-speed digital computers in the 1950s. An excellent discussion of the historical development of the matrix force and stiffness methods, their application in the field of aeroelasticity, and their relation to the development of the finite element method in the engineering community is given by Martin and Carey [376].

In the early 1950s M. J. Turner headed a small group of engineers and academicians at the Boeing Aircraft Company that developed techniques suitable for analyzing the odd-shaped wing panels of high-speed aircraft [376, 110]. In their now classic paper, Turner and his co-workers [545] advanced the framework concept by modeling wing panels as an assemblage of smaller panels (elements) of simple triangular shape, with multiple connection points, under a plane stress idealization. The properties of these triangles were determined from the equations of elasticity. The paper of Turner et al. is also noteworthy in that it was the first to introduce what is now known as the direct stiffness method for determining the properties of finite elements. Overall, this paper represented a conceptual breakthrough in that it provided a more realistic manner in which to model two- and three-dimensional bodies as assemblages of similar, yet geometrically simple two- and three-dimensional elements rather than one-dimensional bars and beams as proposed by Hrennikoff [250]. In a paper that extended the work of Turner et al. [545] on plane stress elasticity, Clough [109] first attached the term "finite element" to the discretization method of analysis used. By the end of the 1960s, this had become the accepted name for the method [218].

The first book devoted specifically to the finite element method appeared in 1967 [587]. In the subsequent years, an increasing number of books on the finite element method were written, a process that continues today.

6.6.4   Synthesis

Conceptually the finite element method began to solidify in the mid-1960s. In this time period, several researchers recognized that the method could be derived from a variational statement of a problem. The ramifications of this finding were that many alternative formulations of the same physical problem were now possible. Besseling [71] presented a complete analogy between the matrix equations of structural analysis and the continuous field equations of elasticity, and recognized that the finite element method was a form of the Rayleigh–Ritz method. Around this same time, Melosh [386] showed the finite element method to be a form of the principle of minimum potential energy, and discussed the possibility of a complementary energy approach. In a paper discussing the upper and lower bounds in matrix structural analysis, Fraeijs de Veubeke [170] also discussed the possibility of defining elements based on the principle of minimum complementary energy. Jones [270] discussed the advantages that could be realized by using the Reissner variational principle, which was shortly thereafter applied by Herrmann [227] to the analysis of plate bending. The above variationally-based approaches led to the subsequent development of so-called "mixed" elements that depend upon assumed displacement and stress fields but possess less stringent continuity requirements for the interpolation functions as compared to standard irreducible elements[6].

The ability to derive finite element equations from variational forms also meant that parameters other than the nodal values of the primary dependent variables could be included in the formulation. Pian [433] presented an element formulation in terms of both assumed displacements and stresses, but with a variational formulation in terms of both the minimum potential and complementary energy principles. He showed how, in the context of elasticity, a certain number of nodeless parameters could be introduced and treated in an otherwise standard manner. This approach led to the subsequent development of so-called "hybrid" elements [434]. Other mixed and hybrid formulations were discussed by Pian and Tong [435].

In the mid-1960s, the finite element method received an even broader interpretation when Zienkiewicz and Cheung [585] showed that it applied to all scalar field problems governed by quasi-harmonic differential forms. This was supported by subsequent application of the method to problems in heat conduction [571], torsion of irregularly shaped cross-sections [228], seepage problems [586], certain fluid flow problems governed by a quasi-linear Poisson equation [573], and three-dimensional field problems [588].

Also in the mid-1960s, efforts to reconcile the purely mathematical and the physically-based engineering approaches to the finite element method were initiated. One of the prime motivations for such an undertaking was the desire to examine the issue of convergence on theoretical (as opposed to simply numerical) grounds [376]. A number of applied mathematicians sought to give the method a firm mathematical foundation. Consequently, a number of studies were performed, aimed at developing suitable basis functions [183, 201], interpolation theorems for (typically Hermite) polynomials [606, 580], estimates of interpolation errors associated with the use of finite-dimensional subspaces [285, 547, 108, 80, 36, 178, 179, 162], and rates of convergence [607, 580, 35, 576]. In the 1970s, the mathematical literature included several monographs [548, 37, 560] and books [515, 417]

---

[6]Further details concerning element interpolation functions and their continuity requirements are given in Chapters 9 and 6, respectively.

devoted to the mathematical foundations of the finite element method. By the early 1980s, all outstanding mathematical issues associated with the method had essentially been resolved [415].

6.6.5   Growth

Motivated by the specific formulations of Turner et al. [545] and Clough [109] for plane stress elements, researchers in the mid-1960s began to establish element formulations for general solids, plates, shells and other structural forms [187].

Once formulations were established for static linear, elastic analyses, they were extended to more complex phenomena such as dynamic response, geometric and material nonlinearities, etc. Examples of early work in these areas are given in an extensive review article by Zienkiewicz [584], and in books by Ames [12] and Oden [414]. The latter was the first book devoted to non-linear finite element analysis of continua.

Beginning in the late sixties and early seventies, the finite element method was applied to fluid mechanics problems such as potential flow, compressible flow and flow through porous media. More complex problems such as flow phenomena with vorticity, solution of the Navier-Stokes equation and flow of non-Newtonian fluids were subsequently studied.

As the finite element method matured, an increasing number of specialized elements were developed. Examples include singular elements for fracture mechanics, "infinite" elements for modeling unbounded domains, boundary-layer elements for viscous flow simulations, etc.

6.6.6   Present State of the Method

The present state of the finite element method is characterized by an ever-increasing number of application areas, the development of new computational strategies and new analysis techniques.

Examples of emerging application areas include: adaptive structures, automotive crash simulations, computational biomechanics, computational probabilistic mechanics, simulation of advanced engineering materials and material forming processes, computational fluid dynamics, simulation of pollutant transport in geomaterials, to name a few.

The development of efficient computational strategies and algorithms is driven by the availability of computer hardware systems such as vector multiprocessors, massively parallel machines and clusters of networked personal computers. Of equal importance is the increased affordability and power of personal computers and workstations.

In the area of analysis techniques, increasing attention is being focused on quality assessment and control of finite element solutions. Major strides have been made towards robust, fully automatic mesh generation.[7] Strategies have also been proposed for adaptive refinement of finite element approximations, with the goal of achieving optimal solutions. Meshless methods are proving to be especially efficient for specific problem areas.[8]

The number of publications related to the finite element method is numerous and ever increasing. Although some general finite element bibliographies were published in the past [561, 413, 335, 411], it is rather pointless, if not impossible, to do so now. A somewhat more rational bibliographic approach involves the compilation of references pertaining to *specific* finite element applications [336] - [369].

6.7 Concluding Remarks

The finite element method has been introduced in this chapter. A finite element solution of a physical problem has the following fundamental characteristics:

1. The governing differential equation is recast in a weak form of the weighted residual statement or, if applicable, as the corresponding variational statement of the problem.

2. A finite number of nodes are identified in the domain Ω. The nodal values of the primary dependent variables represent the unknowns in a finite element analysis.

3. The domain Ω is subdivided into a finite number of non-overlapping elements Ω^e. The elements are connected together at nodes on their boundaries. The discretization of Ω is not necessarily exact.

4. The primary dependent variables are approximated locally over each element by continuous, piecewise defined, finite-dimensional interpolation functions. These functions are uniquely defined in terms of the values of the primary dependent variables (and possibly of the values of their derivatives up to a certain order) at a finite number of nodes in the element.

The first of the above characteristics has been discussed in conjunction with standard weighted residual and variational approximations. It serves to reduce the continuity requirements for the interpolation functions.

---

[7]The subject of mesh generation is discussed further in Chapter 13.

[8]Refinement techniques and meshless methods are discussed further in Chapter 8.

The nodal values of the primary dependent variable are similar to the unknown coefficients α_m appearing in the continuous approximations used in Chapters 4 and 5. However, unlike the α_m, the nodal values of the primary dependent variable have physical significance.

The third characteristic is the means by which the finite element method overcomes the limitations of standard continuous approximations when applied to complex domains.

The final characteristic manifests itself in the ability to formulate equations for individual elements before combining them mathematically to represent the entire problem. This important aspect of the finite element method sets it apart from other approximate solution techniques. For example, comparing the finite element method to the stationary functional method discussed in the previous chapter, we recall that in the latter method the approximation is defined over the entire domain.

Chapter 7

Development of Finite Element Equations

7.1 Introductory Remarks

The basic characteristics of a finite element approximation were introduced in the previous chapter. We now begin to study the finite element method in detail. In this chapter, the general steps followed in developing element equations for a given physical problem are presented. The description of the solution procedure followed in an actual finite element analysis is deferred until Chapter 8.

The development of finite element equations involves the following basic steps[1]:

1. Selection of primary dependent variables

2. Definition of gradient and constitutive relations

3. Identification of the approximate element equations

4. Selection of element interpolation functions

5. Specialization of approximate element equations

Each of these steps is now discussed in detail. The chapter concludes with extended examples that illustrate the formulation of the equations for several different elements.

---

[1]This division into steps is by no means unique; examples of alternate approaches are given in [137, 100, 461].

7.2   Selection of Primary Dependent Variables

As noted in Chapter 6, in finite element analyses the primary dependent variables ϕ are approximated by the sum of products of the nodal values of ϕ and a set of piecewise defined element interpolation functions. Mathematically, this was represented by equation (6.2), viz.,

$$\hat{\phi}(\mathbf{x}, t) = \psi(\mathbf{x}, t) + \sum_{m=1}^{P} N_m(\mathbf{x})\hat{\phi}_m(t) \tag{7.1}$$

In Example 6.1. it was shown that essential boundary conditions could be exactly accounted for without directly involving the function ψ. In addition, because of their piecewise nature, the interpolation functions are known to be non-zero only over a specific element.

As such, equation (7.1) is rewritten in terms of the variation of $\hat{\phi}$ over a specific element. That is, within element e each primary dependent variable is approximated by

$$\hat{\phi}^{(e)}(\mathbf{x}, t) = \sum_{m=1}^{N_{dof}} N_m(\mathbf{x})\hat{\phi}_m^{(e)}(t) \tag{7.2}$$

where m is now a "local" node number in element e, and N_{dof} is the total number of degrees of freedom of $\hat{\phi}^{(e)}$ in the element.[2]

Since the nodal values $\hat{\phi}_m^{(e)}$ constitute the unknowns in a finite element analysis, it follows that prior to developing the element equations themselves we must first clearly identify what primary dependent variables are associated with a specific physical problem.

As shown in the following examples, the choice of primary dependent variables for an element is not necessarily straightforward. Details pertaining to the elements described are not important at this time.

---

[2]The nodal values of the primary dependent variables are also referred to as *degrees of freedom*. A degree of freedom can be defined as an independent (unknown) variable required to represent the effects of all significant actions at a point. For example, consider problems in elasticity. For the case of uniaxial deformation, a point on a body is free to move only axially. This point thus has one degree of freedom. In two-dimensional problems in which bending deformations are ignored, a point is free to move in two independent coordinate directions; the point thus has two degrees of freedom. Finally, if bending deformations are considered, then a rotational degree of freedom must also be included at each point. In the most general (three-dimensional) elasticity analysis involving direct flexural effects, three translational and three rotational degrees of freedom exist at a node.

• Elements for Thermal Analyses

In *irreducible*[3] heat flow formulations, the primary dependent variable is temperature. Each node in the associated element thus has a single degree of freedom. In *mixed* formulations this is supplemented by nodal heat flux degrees of freedom. Both primary dependent variables are not necessarily present at all nodes in the associated element.

• Elements for Stress-Deformation Analyses

In irreducible stress-deformation formulations the primary dependent variables are kinematic quantities. For example, in one- , two- and three-dimensional elements the primary dependent variables are displacements. As such, N_{sdim} displacement components[4] will be present at each node (Figure 7.1).

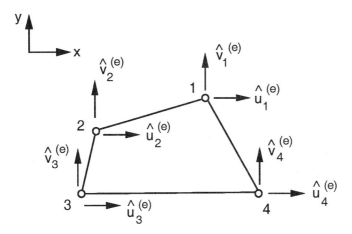

Figure 7.1: Four-Node Quadrilateral Element: Translational Nodal Degrees of Freedom

---

[3]If none of the components of ϕ can be eliminated from the equations $L(\phi) - f = 0$ and $B(\phi) - r = 0$ while still maintaining a well-defined problem, then the formulation is said to be *irreducible*. If this is not the case, the formulation is said to be *mixed*. While the distinction between these two alternate descriptions of a physical problem may appear to be artificial, the approximate solution procedures developed from the two are quite different.

[4]As defined in Chapter 1, N_{sdim} equals the spatial dimension of the problem.

In special "structural" axial-flexural line elements (Figure 7.2), rotational degrees of freedom supplement the nodal displacements.

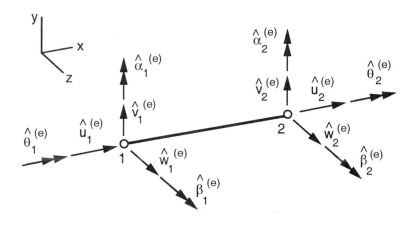

Figure 7.2: Specialized "Space Frame" Element: Translational and Rotational Nodal Degrees of Freedom

In mixed stress-deformation formulations the primary dependent variables are displacements as well as forces, moments or stresses. In Figure 7.3 is shown a mixed quadrilateral "Kirchhoff" plate bending element. The primary dependent variables are transverse displacements $\hat{w}_m^e$ $(m = 1, \cdots, 4)$ and bending moments M_m^e $(m = 5, \cdots, 8)$.

• Elements for Thermo-Mechanical Analyses

In coupled or semi-coupled thermo-mechanical analyses, the primary dependent variables are displacements and temperature. The N_{sdim} nodal displacement degrees of freedom are thus supplemented by nodal temperatures.

• Elements for Geomechanics Applications

For coupled stress-flow problems involving geomaterials, the formulations are typically *mixed*. The primary dependent variables are displacements and pore pressure. The nodal unknowns are thus N_{sdim} displacement components and pore pressure, though the latter are not present at all the

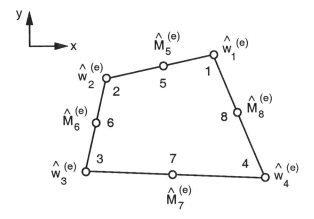

Figure 7.3: Quadrilateral Plate Bending Element: Transverse Displacement and Bending Moment Nodal Degrees of Freedom (after Herrmann [229])

nodes.[5] An example of an element used in the analysis of geomaterials under undrained (zero volume change) conditions is shown in Figure 7.4.

• Elements for Fluid Flow Analyses

In *irreducible* formulations such as groundwater flow or irrotational flow, the primary dependent variable is fluid pressure or hydraulic head. In *mixed* associated with viscous incompressible fluid flow, the primary dependent variables are velocity and pressure. For element stability, the number of pressure degrees of freedom must always be less than the number of velocity degrees of freedom.

• Elements for Analysis of Ice Sheets

For coupled thermo-mechanical modeling of an ice-sheet/ice-shelf flow plane, the primary dependent variables are velocity components, temperature and pressure. As shown in Figure 7.5, the latter degrees of freedom are not present at all nodes.

---

[5]Such a dichotomy is necessary in order to avoid spurious element behavior. This aspect of mixed elements is beyond the scope of this introductory text. Further details can be found in Chapter 4 of [252].

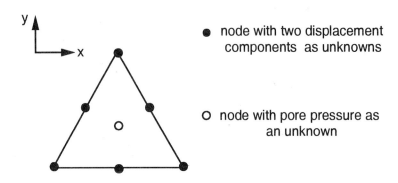

Figure 7.4: Mixed Triangular Element used in Analysis of Geomaterials

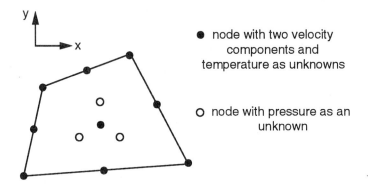

Figure 7.5: Mixed Quadrilateral Element Used in Thermo-Mechanical Modeling of Ice Sheets (after Grace [203])

7.3 Definition of Constitutive Relations

Before proceeding to the actual derivation of element equations, we need to explicitly define certain quantities and relations that enter into these equations. The quantities are the *secondary dependent variables*, which are related to the primary dependent variables through suitable relations involving derivatives of the primary dependent variables.

In addition to the gradient relations, we must also account for the constitution of the material comprising the element domain Ω^e. This is done through the *constitutive relations*. In general, such relations describe the response of a system to an applied loading. The constitutive relations represent one of the most vital parts of a finite element formulation, for unless they are defined to correctly reflect the behavior of the material under consideration, the results of analyses can be rendered all but meaningless. Some examples of simple constitutive relations follow.

Example 7.1. Constitutive Relations for Heat Flow

For one-dimensional variations of temperature, the heat flux is related to the temperature through Fourier's law,[6] viz.,

$$\tilde{q} = \frac{dQ}{dt} = -kA\frac{dT}{dx} \quad \text{or} \quad q = \frac{\tilde{q}}{A} = -k\frac{dT}{dx} \tag{7.3}$$

where Q is the heat, $\tilde{q}$ is the heat flux, q is the heat flux per unit area k is the thermal conductivity, A is the cross-sectional area normal to the direction in which heat flows, and T is the temperature.

For a three-dimensional *anisotropic* continuum, equation (7.3) generalizes to

$$\begin{Bmatrix} q_1 \\ q_2 \\ q_3 \end{Bmatrix} = - \begin{bmatrix} k_{11} & k_{12} & k_{13} \\ k_{12} & k_{22} & k_{23} \\ k_{13} & k_{23} & k_{33} \end{bmatrix} \begin{Bmatrix} \partial T/dx \\ \partial T/dy \\ \partial T/dz \end{Bmatrix} \tag{7.4}$$

or

$$\mathbf{q} = -\mathbf{K}\boldsymbol{\nabla}T \tag{7.5}$$

where the k_{ij} represent thermal conductivity coefficients and ∇ is the gradient operator.

---

[6]See Appendix B.

Example 7.2. Constitutive Relations for Flow of Electric Current[7]

Ohm's law describes a proportionality between the current density vector $\mathbf{j}$ and the gradient of the voltage V [214]. For a one-dimensional domain (e.g., a wire), Ohm's law becomes

$$j = \sigma \frac{dV}{dx} \tag{7.6}$$

where σ represents the conductivity of the material. For conductors such as metals, living tissue, etc., $\sigma > 0$; in insulators such as dry air, vacuum, etc., $\sigma = 0$.

If the electric current i is distributed uniformly across a conductor of cross-sectional area A, the magnitude of the current density is

$$j = \frac{i}{A} \tag{7.7}$$

Noting that an electric current is established if a net charge q passes through any cross-section of the conductor in time t, and assuming that the current varies with time, it follows that

$$i = \frac{dq}{dt} \tag{7.8}$$

Substituting equations (7.7) and (7.8) into equation (7.6) gives

$$\frac{dq}{dt} = -\sigma A \frac{dV}{dx} \tag{7.9}$$

Since positive charge flows in the direction of decreasing V, a minus sign has been introduced into equation (7.9); that is, dq/dt is positive when dV/dt is negative. In closing we note that equation (7.9) is analogous to equation (7.3) for one-dimensional heat flow. In light of the discussion in Section 1.2 of Chapter 1, this should not come as a surprise.

---

[7]Although not covered in detail in this text, the application of the finite element method to problems in electrical engineering is an active area of research. The interested reader is directed to books on the subject such as [498, 92, 430].

Example 7.3. Constitutive Relations for Fluid Flow

The simplest fluid description is the *Newtonian* fluid, wherein the shear stress τ on an interface tangent to the direction of flow is proportional to the velocity gradient or rate of shearing strain γ, viz.,

$$\tau = \mu\dot{\gamma} \tag{7.10}$$

In equation (7.10) the constant of proportionality is the dynamic (absolute) viscosity μ. For a given temperature and pressure the viscosity is independent of pressure and of the shear rate of flow. All gases, and most simple liquids are Newtonian fluids, and hence behave according to equation (7.10).

Plastics, suspensions, pastes, slurries and high polymers are examples of fluids that cannot be considered Newtonian [490]. The essential mechanical difference between plastics and fluids is the presence of a shear stress τ_0 that must be overcome before flow can begin. Thus, for such materials

$$\tau - \tau_0 = \mu\dot{\gamma} \tag{7.11}$$

For most thermoplastic polymers used as matrix materials in conjunction with composite materials, entanglement of the polymer molecules produces (especially at higher deformation rates) a shear rate dependent non-Newtonian shear thinning viscosity. One way in which to describe such a viscosity is with a power-law model [79]

$$\mu(\dot{\gamma}) = m\dot{\gamma}^{n-1} \tag{7.12}$$

where μ is a constant and n represents the shear thinning exponent (for shear thinning fluids, $n < 1$).

A second way in which to describe such a viscosity is with the Carreau viscosity model [103], which is commonly expressed as

$$\mu = \mu_0 \left[1 + (\lambda\dot{\gamma})^2\right]^{(n-1)/2} \tag{7.13}$$

where μ_0 represents the "zero shear rate" viscosity, n is the shear thinning exponent, and λ is a time constant. This three-parameter relation describes the viscosity as Newtonian at low shear rates, followed by a shear thinning power-law behavior at higher rates.

Example 7.4. Constitutive Relations for Elastic Solids

In the case of stress-deformation problems, with infinitesimal displacements as primary dependent variables, the gradient of the displacements is the infinitesimal strain. For example, in a one-dimensional analysis, the axial strain ε_{11} in the x_1-direction is related to the gradient of the displacement u_1 in the x_1-direction as follows:

$$\varepsilon_{11} = \frac{\partial u_1}{\partial x_1} \tag{7.14}$$

Assuming an *anisotropic*, linear elastic material, the stresses σ_{ij} are related to the strains ε_{ij} by the generalized Hooke's law, viz.,

$$\boldsymbol{\sigma} = \mathbf{C}\boldsymbol{\varepsilon} \tag{7.15}$$

Due to the symmetry of both the strain and stresses tensors, only six independent components of each exist. To facilitate numerical implementation, both tensors are represented in vector form. The expanded form of equation (7.15) thus becomes

$$\begin{Bmatrix} \sigma_{11} \\ \sigma_{22} \\ \sigma_{33} \\ \sigma_{12} \\ \sigma_{13} \\ \sigma_{23} \end{Bmatrix} = \begin{bmatrix} c_{11} & c_{12} & c_{13} & c_{14} & c_{15} & c_{16} \\ & c_{22} & c_{23} & c_{24} & c_{25} & c_{26} \\ & & c_{33} & c_{34} & c_{35} & c_{36} \\ & & & c_{44} & c_{45} & c_{46} \\ & \text{symmetric} & & & c_{55} & c_{56} \\ & & & & & c_{66} \end{bmatrix} \begin{Bmatrix} \varepsilon_{11} \\ \varepsilon_{22} \\ \varepsilon_{33} \\ 2\varepsilon_{12} \\ 2\varepsilon_{13} \\ 2\varepsilon_{23} \end{Bmatrix} \tag{7.16}$$

If the material is further assumed to be *isotropic*, the entries in $\mathbf{C}$ become

$$c_{11} = c_{22} = c_{33} = \frac{E(1-\nu)}{(1+\nu)(1-2\nu)} \tag{7.17}$$

$$c_{12} = c_{13} = c_{23} = \frac{E\nu}{(1+\nu)(1-2\nu)} \tag{7.18}$$

$$c_{44} = c_{55} = c_{66} = \frac{E}{2(1+\nu)} \tag{7.19}$$

where E and ν are the elastic modulus and Poisson's ratio, respectively.

Finally, returning to the case of uniaxial loading of a body along the x_1-coordinate direction, the strain is given by equation (7.14). The associated constitutive relation is

$$\sigma_{11} = E\varepsilon_{11} = E\frac{du_1}{dx_1} \qquad (7.20)$$

Example 7.5. Constitutive Relations for Ice Flow

The relation between stress and strain in polycrystalline ice is approximated by Glen's flow law [200] as

$$\dot{\varepsilon} = E\left(\frac{\tau}{B}\right)^n \qquad (7.21)$$

where

$$\dot{\varepsilon} \equiv \sqrt{\frac{1}{2}\dot{\varepsilon}_{ij}\dot{\varepsilon}_{ij}} \quad \text{and} \quad \tau \equiv \sqrt{\frac{1}{2}\sigma_{ij}\sigma_{ij}} \qquad (7.22)$$

are the effective strain rate and effective stress, respectively, and ε_{ij} and σ_{ij} are components of the strain and stress tensors, respectively. In equation (7.22) E is an experimentally derived "enhancement factor," B relates stress and temperature T, and n is a model parameter. For $n = 3$, the following empirical relation between B and T has been proposed [248]:

$$B = 2.207 \exp\left\{\frac{3158}{T} - \frac{0.16612}{(273.39 - T)^{1.17}}\right\} \qquad (7.23)$$

where T has units of degrees Kelvin.

In certain instances, new constitutive relations need to be developed in order to realistically characterize material behavior. The development of robust relations suitable for implementation in finite element analyses consists of the several well-defined steps. Further details concerning this important, and sometimes overlooked, topic can be found in [138, 139].

7.4   Identification of the Element Equations

Once the primary dependent variables have been identified and the constitutive relations selected, attention is turned to deriving the general form of the approximate equations governing the behavior of a typical element. Recall that the feature which sets the finite element method apart from other approximate methods is the ability to formulate solutions for individual elements before combining them to represent the entire domain.

As noted in Section 6.5, the element equations are derived either from a weak form of the weighted residual statement or from a variational statement of the problem, if it exists.[8] Residual and variational formulations are both known as "weak" forms of the governing differential equation for a problem. Recall from Chapter 1 that the governing differential equations themselves comprise the *strong* form.

7.5   Selection of Element Interpolation Functions

When a physical problem is modeled as an assemblage of elements, the *accuracy* of a finite element analysis of the problem depends largely on the number of elements used and on the nature of the assumed interpolation functions. In particular, since the exact solution is typically not known, it is important that the finite element solution converges to the exact one as the number of elements in the model is increased. This in turn requires that the element interpolation functions satisfy certain convergence criteria.

7.5.1   Convergence Criteria

Recalling the discussion of Section 4.6, it is clear that the element interpolation functions cannot be arbitrarily chosen, but must conform to certain continuity requirements. Such requirements were previously presented in conjunction with continuous approximations applicable to the entire domain Ω. Since Ω is now divided into a finite number of non-overlapping elements, the continuity requirements must be supplemented with additional convergence criteria.

The complete set of convergence criteria is mathematically included in a statement of so-called "functional completeness" [605]. We note that some

---

[8]The element equations can also be obtained using the so-called "direct approach." This approach is similar to the direct stiffness method of structural analysis. As such, the approach is limited only to relatively simple elements. A detailed discussion of the direct approach is presented by Gallagher [187].

generalizations of these criteria fall within the "completeness" definition of Mikhlin [393], and have been discussed at length in the literature [290, 542, 18, 269]. Since a full mathematical treatment of the subject is beyond the scope of our discussion, the interested reader should consult these and other references cited herein.

The convergence criteria are applicable to a rather broad class of physical problems, namely those described by governing equations in which the differential operator is linear and self-adjoint. It is assumed the governing differential equation is of order $2s$, where $s = 1, 2, \cdots$.[9]. In addition, the interpolation functions are assumed to be *polynomials*

If the element equations are developed using a weighted residual approach, then it is assumed that the weighted residual equations have been integrated by parts s times. If a Ritz variational approach is used instead, no further manipulations are required. In both cases s is the highest-order derivative of the approximate element solution $\hat{\phi}^{(e)}$ (and thus of the element interpolation functions) appearing in the integral equations identified in step 3.

In order to guarantee that a sequence of finite element solutions converges *monotonically* to the exact solution, the element interpolation functions should satisfy the *compatibility* and *completeness* criteria. The monotonic nature of the convergence means that the accuracy of the analysis, as measured in some norm, increases continuously as the number of elements is increased [59].

Compatibility Criterion

The compatibility criterion consists of two parts.[10] The first part requires that within the element interior Ω^e the interpolation functions should be C^s continuous. This is sometimes also referred to as the *smoothness requirement*.

The second part of the compatibility criterion, sometimes referred to as the *continuity condition*, requires that at interelement boundaries Γ^e the interpolation functions should be C^{s-1} continuous. That is, the interpolation functions should be so chosen that across Γ^e their sth derivatives have, at worst, finite jumps (i.e., the values of the derivatives can be indeterminate at Γ^e). This ensures that all terms appearing in the approximate element equations identified in step 3 will be well-defined. Physically, satisfaction of the compatibility criterion ensures that when elements are assembled together, no "gaps" in $\hat{\phi}$ occur between them.

---

[9] Recall the discussion of Section 1.3.4.

[10] The compatibility criterion is attributed to Melosh [386], though some comments pertaining to the criterion were made earlier by Fraeijs de Veubeke and Sander [173].

Remarks

1. If $s = 1$, the interpolation functions must be C^0 continuous at Γ^e; that is, $\hat{\phi}^{(e)}$ must be continuous at Γ^e. The first derivatives of $\hat{\phi}^{(e)}$ will have, at worst, finite jumps across Γ^e. If, instead, we permit finite discontinuities in the interpolation functions on Γ^e, the first derivatives become delta functions (see Figure 4.10a) and we will be unable to obtain rational results from the element integrals.

2. Interpolation functions satisfying the above requirements for $s = 1$ are of class $C^0(\Omega)$. Elements constructed using such interpolation functions are referred to as C^0-elements [252].

3. If $s = 2$, the compatibility criterion must be strengthened to require C^2 continuity on Ω^e and C^1 continuity at Γ^e. Thus in addition to $\hat{\phi}^{(e)}$, its first derivatives must also be continuous across Γ^e. Interpolation functions satisfying the requirements for $s = 2$ are of class $C^1(\Omega)$ and lead to so-called C^1-elements. Bernoulli-Euler beam elements and Kirchhoff plate bending elements are two examples of C^1-elements.

Completeness Criterion

The completeness criterion ensures that as the number of elements increases, terms in the integral element equations can be approximated as closely as required. The completeness criterion may be stated as follows:

> All uniform states of $\hat{\phi}^{(e)}$ and any of its derivatives up to the order s appearing in the integral equations should have representation in the element when, in the limit, the size of the element shrinks to zero.[11]

To better understand the completeness criterion, imagine that a continuum is discretized by an increasing number of ever smaller elements. As the size of each element becomes very small, the exact solution and its derivatives up to order s approach *constant* values over each Ω^e. To ensure that these constant values are representable, the interpolation functions must contain all constant and monomials up to order s. This argument, commonly attributed to Bazely et al. [60],[12] has been shown to be the key mathematical concept for providing convergence theorems for finite element approximations [515].

Since interpolation functions are commonly developed from polynomials, the completeness criterion can be restated as follows: If a polynomial

---

[11] Although, strictly speaking, this criterion needs to be satisfied in this limit, the imposition of the criterion on elements of finite size leads to improved accuracy [605].

[12] Similar arguments were presented earlier by Irons and Draper [259].

is used for the element interpolation functions, it should be complete[13] at least to degree s; that is, $p \geq s$.

To gain some insight into the completeness criterion, consider a three-dimensional problem (i.e., $N_{sdim} = 3$) for which $s = 1$. In a typical C^0 element used to discretize the continuum, the primary dependent variable is approximated by equation (7.2). The interpolation functions N_m are said to be *complete* if

$$\hat{\phi}_m^{(e)} = \beta_0 + \beta_1 x_m + \beta_2 y_m + \beta_3 z_m \tag{7.24}$$

implies that

$$\hat{\phi}^{(e)} = \beta_0 + \beta_1 x + \beta_2 y + \beta_3 z \tag{7.25}$$

where β_0, β_1, β_2 and β_3 are arbitrary constants and (x_m, y_m, z_m) are the coordinates of node m in element e. Completeness thus requires that the element interpolation functions are capable of exactly representing an arbitrary linear polynomial when the nodal degrees of freedom are assigned values in accordance with it. We will return to this interpretation of the completeness criterion when studying element interpolation functions in greater detail in Chapter 9.

Remarks

1. Satisfaction of the completeness criterion is necessary for convergence to the exact solution.[14] An incomplete element will generally converge to some other limiting function that will be as close to the exact solution as $\hat{\phi}^{(e)}$ will permit [100].

2. Satisfaction of the compatibility criterion is *not* necessary for convergence. The purpose of this criterion is to ensure that discontinuities at the interelement boundaries are not severe enough to introduce errors *in addition to* domain discretization error.

3. If the compatibility criterion is satisfied, along with completeness, then monotonic convergence is ensured; that is, together the two criteria are sufficient for convergence. An element satisfying both criteria is said to be *conforming* or *compatible*.

4. An element that satisfies the completeness criterion but violates the compatibility criterion, yet still converges, is called *non-conforming* or

---

[13] A complete polynomial of degree p in k variables contains $(p+k)!/p!k!$ independent terms; however, the coefficients of the polynomial need not be independent.

[14] A proof of this, restricted to polynomial approximations, was presented by Arates e Oliveira [18]. A generalization of the theorem of Arates e Oliveira was subsequently proved by Oden [414].

incompatible. The convergence of such elements will, in general, not be monotonic [59]. The use of non-conforming elements is quite common, as they often exhibit superior performance relative to conforming elements.[15]

5. When stating the convergence criteria for C^0-elements used in solving elasticity problems, completeness requires representation of all monomials through linear terms. Physically this means that we must have the ability to represent all rigid body translations and rotations, as well as constant strain states. The completeness criterion was first presented in this manner by Melosh [386].

7.5.2   Spatial Isotropy

In addition to satisfying the compatibility and completeness criteria, it is desirable that the approximating polynomial remain unchanged under a linear transformation from one Cartesian coordinate system to another. Polynomials exhibiting this invariance property are said to possess *spatial* or *geometric isotropy*. Spatial isotropy is *not* necessary for convergence, but is a desirable property. Lack of isotropy means that one variable is more accurately represented than another. The performance of elements could thus vary from one point in the mesh to another.

The following types of polynomials ensure geometric isotropy:

1. Complete polynomials have spatial isotropy. It follows that polynomials in one independent variable always possess isotropy.

2. Incomplete polynomials of order s that are "balanced" or "symmetric" with respect to the independent variables have spatial isotropy. Thus, for each term of the form $x^a y^b$, a balanced polynomial in two variables contains a term $x^b y^a$, where a and b are integers. Clearly this property can be imposed on any approximation $\hat{\phi}^{(e)}$.

It follows that a useful guide to selecting terms for spatially isotropic polynomials is the Pascal triangle.

If an element does not possess spatial isotropy, its coefficient matrix $\mathbf{K}^{(e)}$ will depend on the orientation of the local (element) coordinate system used. Although such directional behavior is in general undesirable, some exceptions arise [58]. None the less, in the analysis of two- and

---

[15] Use of the terms "conforming" and "non-conforming" in connection with finite element approximations originated with the paper of Bazely et al. [60], who demonstrated that the compatibility criterion may be too restrictive in certain cases. However, their results indicated that for non-conforming elements convergence may depend upon the orientation of the elements in the finite element mesh.

three-dimensional continua with geometrically invariant behavior, spatial isotropy is a requirement and results in a constraint on the terms in the interpolating polynomial for $\hat{\phi}^{(e)}$. We shall have more to say on this matter in Chapter 9.

Having discussed convergence criteria with a firm mathematical foundation, it is timely to note that the convergence of elements, particularly non-conforming ones, is often ascertained using a more heuristic approach. In particular, specific distributions of the primary dependent variables are imposed on arbitrary "patches" of elements. If compatibility is realized as, in the limit, the size of the elements in the patch is decreased, the element will converge. Further details pertaining to the so-called *patch test* are given in Appendix D.

7.5.3 Rate of Convergence

The rate at which $\hat{\phi}^{(e)}$ converges to ϕ depends on the degree of the polynomial expansions used in representing the former. In this context, the use of *complete* polynomials is most desirable [532].

If the degree of a *complete* polynomial used in the role of an interpolation function is p, $(p \geq s)$, then the error in $\hat{\phi}^{(e)}$ will be of the order $O(h^{p+1})$. In a similar manner, the fluxes given by the sth derivative of $\hat{\phi}^{(e)}$ should converge with an error $O(h^{p+1-s})$. Here h denotes a "representative" element size.

For a linear or bilinear element $(p = 1)$ the convergence rate for $\hat{\phi}^{(e)}$ will thus be $O(h^2)$. As noted in Chapter 3, this means that if h is halved, the error in $\hat{\phi}^{(e)}$ will be reduced by one-quarter. Assuming a second order governing equation $(s = 1)$, the sth derivative of $\hat{\phi}^{(e)}$ should converge with an error $O(h)$. As shown in Appendix F, there are certain locations within an element where the flux is *superconvergent*, that is, it converges at a rate *higher* than $O(h^{p+1-s})$.

7.5.4 Information Regarding Element Nodes

As evident from the discussion of Chapter 7, the interpolation functions are defined in conjunction with the approximate nodal values of the primary dependent variables. The number of nodes assigned to a given element depends upon the type of interpolation function used and upon the kind of nodal variables employed in the formulation. To better understand this, consider a particular primary dependent variable. As noted in the previous discussion, the weak or integral form of the governing differential equations dictates the continuity and completeness requirements imposed on the interpolation functions. These requirements in turn dictate the order of the required polynomial; i.e., the number of unknown coefficients present in the polynomial. This number must equal the total number of degrees of freedom for a given primary dependent variable available in an element. Since these degrees of freedom are used to determine values for the unknown coefficients in the polynomial, this dictates the total number of nodes that must be present in an element. This logic can better be understood by studying the examples presented in Section 7.7 of this chapter.

Remark

1. In concluding this discussion on the selection of the element interpolation functions, we reiterate the following important points related to convergence criteria. If the element interpolation functions are constructed such that $\hat{\phi}^{(e)}$ satisfies the compatibility and completeness criteria, then a sequence of finite element solutions will monotonically converge to the exact solution as $h_{\max}^{(e)} \to 0$ (this assumes that no other errors are introduced into the sequence). Here $h_{\max}^{(e)}$ denotes the size (in one dimension) or representative dimension of the largest element in the mesh. This type of convergence is referred to as *h-convergence* since it is achieved by letting the size of the elements become progressively smaller. We will have more to say about this and other types of convergence in Chapter 8.

7.6   Specialization of Element Equations

With the primary dependent variables, constitutive relations, and element equations identified, and the element interpolation functions selected, the final step in the formulation process involves specialization of the approximate element equations. This involves the substitution of the interpolation functions selected in step 4 into the approximate element equations identified in step 3. Suitable manipulations are then performed, resulting in the *discrete* (matrix) form of the approximate element equations. The general form of these equations is

$$\mathbf{K}^{(e)}\,\hat{\phi}_n^{(e)} = \mathbf{f}^{(e)} \tag{7.26}$$

where $\mathbf{K}^{(e)}$ is the element coefficient matrix, $\hat{\phi}_n^{(e)}$ is the vector of nodal unknowns associated with the element, and $\mathbf{f}^{(e)}$ is the vector of nodal "forces" associated with the element. The discrete element equations are general in nature and apply to all elements of a given type used in mathematically modeling a particular physical problem.

In irreducible thermal analyses $\mathbf{K}^{(e)}$ is the element conductivity matrix, $\hat{\phi}_n^{(e)}$ is the vector of nodal temperatures, and $\mathbf{f}^{(e)}$ is the vector of nodal flux contributions. In irreducible stress-deformation analyses $\mathbf{K}^{(e)}$ is the element stiffness matrix, $\hat{\phi}_n^{(e)}$ is the vector of nodal displacements, and $\mathbf{f}^{(e)}$ is the vector of nodal forces.

7.7 Illustrative Examples

The following examples illustrate the application of the above five-step approach for developing element equations.

Example 7.6. Linear Line Element for Heat Conduction

Equations are derived for a one-dimensional, two-node element for thermal analyses. A two-node element with scalar primary dependent variables represents the simplest possible case for consideration. It thus makes the perfect first example for this chapter.

In particular, we desire to formulate the equations for steady-state heat conduction along an insulated rod arbitrarily oriented in space. The domain Ω is the line shown in Figure 7.6. The x'-coordinate axis is taken coincident with the line.

The governing differential equation for the problem is

$$\frac{d}{dx'}\left(k\frac{dT}{dx'}\right) + S = 0 \quad ; \quad 0 < x' < L \tag{7.27}$$

where T is the temperature, k is the thermal conductivity coefficient (a material parameter), S is a source term, and L is the length of the line along the x'-axis. The associated boundary conditions are

$$T = \bar{T} \quad \text{on} \quad \Gamma_1 \tag{7.28}$$

and

$$-k\frac{dT}{dx'} = \bar{q}' \quad \text{on} \quad \Gamma_2 \tag{7.29}$$

where $\bar{q}'$ is a known heat flux per unit area acting along the x'-axis. For simplicity, convection boundary conditions are not considered herein.

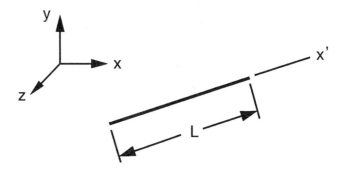

Figure 7.6: Linear Domain in Space

We now begin the five-step procedure for developing the element equations for the problem under consideration.

• **Step 1:** Identification of the Primary Dependent Variable

The primary dependent variable is temperature T. Since temperature is a scalar quantity, it follows that each node in the element will have one degree of freedom associated with it.

• **Step 2:** Definition of the Gradient and Constitutive Relations

The relation between heat flux per unit area along Ω and temperature is assumed to be governed by Fourier's law, viz.,

$$q' = -k\frac{dT}{dx'} \tag{7.30}$$

• **Step 3:** Identification of the Element Equations

Since the present problem constitutes a special case of the mode general one-dimensional problem discussed in Section 6.5, it follows that the weak form of the weighted residual statement of the problem will be given by equation (6.11), viz.,

$$\int_0^L \left(\frac{dw}{dx'}k\frac{d\phi}{dx'}\right) dx' - \int_0^L wS\,dx' + w\bar{q}'|_{\Gamma_2} = 0 \tag{7.31}$$

where $w = w(x')$ is a weighting function that satisfies the condition $w = 0$ on Γ_1.

In light of equation (7.2), the temperature within an element is approximated by

$$\hat{T}^{(e)}(x') = \sum_{m=1}^{N_{dof}} N_m(x')\hat{T}_m^{(e)} \tag{7.32}$$

where N_{dof} is the total number of temperature degrees of freedom in the element, $\hat{T}_m^{(e)}$ are the nodal values of temperature (the unknowns in the problem), and $N_m(x')$ are the element interpolation functions. Since the primary dependent variable (temperature) is a scalar, N_{dof} will equal the number of nodes in the element (N_{en}).

Adopting a Galerkin weighted residual statement of the problem, the weighting functions are written as

$$w(x') = \sum_{l=1}^{N_{dof}} N_l(x')\beta_l^{(e)} \tag{7.33}$$

Writing $\hat{T}^{(e)}$, w and their derivatives with respect to x' in matrix form, and substituting the resulting expressions into equation (7.31) gives the following equation for a typical element:

$$\int_0^L \left[\beta^T \mathbf{B}^T k^{(e)} \mathbf{B}\hat{\mathbf{T}}_n\right] dx - \int_0^L \beta^T \mathbf{N}^T S^{(e)} \, dx + \beta^T \mathbf{N}^T \bar{q}\Big|_{\Gamma_2} = 0 \tag{7.34}$$

where

$$\mathbf{N} = \left\langle N_1 \quad N_2 \quad \cdots \quad N_{N_{dof}} \right\rangle \tag{7.35}$$

$$\mathbf{B} = \left\langle \frac{dN_1}{dx'} \quad \frac{dN_2}{dx'} \quad \cdots \quad \frac{dN_{N_{dof}}}{dx'} \right\rangle \tag{7.36}$$

For *arbitrary* β, it follows that

$$\mathbf{K}^{(e)}\hat{\mathbf{T}}_n^{(e)} = \mathbf{f}^{(e)} \tag{7.37}$$

where

$$\mathbf{K^{(e)}} = \int_{x_i'}^{x_j'} \mathbf{B}^T k^{(e)} \mathbf{B} \, dx' \tag{7.38}$$

$$\mathbf{\hat{T}_n^{(e)}} = \left\{ \hat{T}_1^{(e)} \quad \hat{T}_2^{(e)} \quad \cdots \quad \hat{T}_{N_{dof}}^{(e)} \right\}^T \tag{7.39}$$

and

$$\mathbf{f^{(e)}} = \int_{x_i'}^{x_j'} \mathbf{N}^T S^{(e)} \, dx' - \mathbf{N}^T \bar{q}'\big|_{\Gamma_2} \tag{7.40}$$

In equations (7.38) and (7.40) x_i' and x_j' represent the coordinates of the end points of the element.

• **Step 4:** Selection of Interpolation Functions

We begin this step by observing that the highest order derivative of the interpolation functions appearing in the element equations is one; using the notation of Section 1.3.4, $s = 1$.

The compatibility criterion therefore requires that within Ω^e, the interpolation functions must be C^1 continuous. In addition, across Γ^e, these functions must be C^0 continuous.

Satisfaction of the completeness criterion requires that *constant* states of $\hat{T}^{(e)}$ and its first derivative have representation when, in the limit, the size of the element goes to zero.

Guided by the above observations and by the results of Example 6.1, we propose the following linear polynomial on x':

$$\hat{T}^{(e)}(x') = \alpha_1 + \alpha_2 x' = \chi\alpha \tag{7.41}$$

where the entries in $\chi = \langle 1 \quad x' \rangle$ are called the *basis functions* [370]. The entries in α are unknown *generalized coordinates*, that are a function only of the element geometry.

We desire, however, to express equation (7.41) not in terms of generalized coordinates, but in terms of the nodal degrees of freedom. To achieve this, we evaluate equation (7.41) at each of the element nodes. Since α_1 and α_2 are unknown, we must have two equations to solve for their values. This implies that we must have *two* nodes in the element. Guided by the results of Example 6.1, we place one node at each end of the element, and assign them local (element) node numbers 1 and 2 in the manner shown in Figure 7.7(a). In a multi-element mesh, these nodes will be assigned actual "global" numbers that will, in general, differ from the local ones (Figure 7.7b).

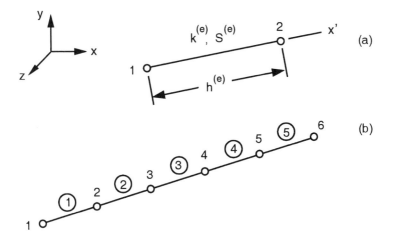

Figure 7.7: Two-Node Heat Conduction Element: (a) Single Element, (b) Mesh Consisting of Five Elements

Evaluating equation (7.41) at the two element nodes and writing the resulting equations in matrix form gives

$$\hat{\mathbf{T}}_n^{(e)} = \left\{ \begin{array}{c} \hat{T}_1^{(e)} \\ \hat{T}_2^{(e)} \end{array} \right\} = \begin{bmatrix} 1 & x_1' \\ 1 & x_2' \end{bmatrix} \left\{ \begin{array}{c} \alpha_1 \\ \alpha_2 \end{array} \right\} = \mathbf{A}\boldsymbol{\alpha} \qquad (7.42)$$

Since $\mathbf{A}$ is obviously non-singular, it follows that

$$\boldsymbol{\alpha} = \mathbf{A}^{-1}\hat{\mathbf{T}}_n^{(e)} = \frac{1}{h^{(e)}} \begin{bmatrix} x_2' & -x_1' \\ -1 & 1 \end{bmatrix} \hat{\mathbf{T}}_n^{(e)} \qquad (7.43)$$

where $h^{(e)} = x_2' - x_1'$ is the length of the element.

From equation (7.43) it is evident that the generalized coordinates α_i are indeed simply functions of the nodal coordinates. Substituting equation (7.43) into equation (7.41) gives

$$\hat{T}^{(e)}(x') = \chi \mathbf{A}^{-1}\hat{\mathbf{T}}_{\mathbf{n}}^{(e)} = \mathbf{N}\hat{\mathbf{T}}_{\mathbf{n}}^{(e)} \tag{7.44}$$

where $\mathbf{N}$ represents the $(1 * N_{dof})$ row vector of interpolation functions. After some straightforward manipulation, the entries in $\mathbf{N}$ are found to be

$$N_1 = \frac{x'_2 - x'}{h^{(e)}} \quad , \quad N_2 = \frac{x' - x'_1}{h^{(e)}} \tag{7.45}$$

These interpolation functions must satisfy the basic requirement for node i given by equation (6.30), viz.,

$$N_i(x'_j) = \delta_{ij} \tag{7.46}$$

where δ_{ij} is the Kronecker delta. It is easily verified that N_1 and N_2 defined in equation (7.45) satisfy this requirement for $i, j = 1, 2$.

The derivatives of the interpolation functions are

$$\frac{dN_1}{dx'} = \frac{-1}{h^{(e)}} = -\frac{dN_2}{dx'} \tag{7.47}$$

• Check of Compatibility Criterion

Checking the compatibility criterion, we note that within Ω^e, the interpolation functions and their first derivatives are clearly continuous. Since the nodes are located at the two ends of an element, and since the interpolation functions N_1 and N_2 are linear in x', it follows that they will likewise be continuous across Γ^e.

• Check of Completeness Criterion

To satisfy the completeness criterion, we must first show that *constant* states of $\hat{T}^{(e)}$ can be represented as $h^{(e)}$ approaches zero in the limit. Assuming $\hat{T}_1^{(e)} = \hat{T}_2^{(e)} = C$ where C is a constant, it follows that

$$\hat{T}^{(e)} = \hat{T}_1^{(e)}N_1 + \hat{T}_2^{(e)}N_2 = C(N_1 + N_2) = C \tag{7.48}$$

This condition will only be satisfied provided that

$$\sum_{m=1}^{2} N_m = 1 \tag{7.49}$$

This important property is known as the *partition of unity* [605]. Checking the interpolation functions given in equation (7.45),

$$N_1 + N_2 = \frac{x'_2 - x'}{h^{(e)}} + \frac{x' - x'_1}{h^{(e)}} = \frac{x'_2 - x'_1}{h^{(e)}} = 1 \qquad (7.50)$$

We must next show that constant states of first derivative of $\hat{T}^{(e)}$ can be represented as $h^{(e)}$ approaches zero in the limit. Making use of equation (7.47), it follows that

$$\frac{d\hat{T}^{(e)}}{dx'} = \sum_{m=1}^{2} \frac{dN_m}{dx'} \hat{T}_m^{(e)} = \frac{1}{h^{(e)}} \left(-\hat{T}_1^{(e)} + \hat{T}_2^{(e)} \right) \qquad (7.51)$$

which is clearly a constant.

Remark

1. For the special case of a constant temperature field given by $\hat{T}_1^{(e)} = \hat{T}_2^{(e)} = C$ where C is again a constant, the temperature flux given by equation (7.51) becomes

$$\frac{d\hat{T}^{(e)}}{dx'} = \frac{1}{h^{(e)}} (-C + C) = 0 \qquad (7.52)$$

which is a physically correct response.

• **Spatial Isotropy**

Finally, we note that polynomials in one spatial variable are always complete. Spatial isotropy is thus guaranteed.

• **Step 5:** Specialization of Approximate Element Equations

Substituting the interpolation function derivatives into equation (7.38), and noting that $x'_i = x'_1$ and $x'_j = x'_2$, gives the following element properties matrix:

$$\mathbf{K}^{(e)} = \int_{x'_1}^{x'_2} \left\{ \begin{array}{c} \dfrac{dN_1}{dx'} \\[2mm] \dfrac{dN_2}{dx'} \end{array} \right\} k^{(e)} \left\langle \begin{array}{cc} \dfrac{dN_1}{dx'} & \dfrac{dN_2}{dx'} \end{array} \right\rangle dx'$$

$$= \frac{k^{(e)}}{\left(h^{(e)}\right)^2} \begin{bmatrix} 1 & -1 \\ -1 & 1 \end{bmatrix} \int_{x'_1}^{x'_2} dx' = \frac{k^{(e)}}{h^{(e)}} \begin{bmatrix} 1 & -1 \\ -1 & 1 \end{bmatrix} \qquad (7.53)$$

Substituting the interpolation functions into equation (7.40), and noting
that $x'_i = x'_1$ and $x'_j = x'_2$, gives the following right-hand side "forcing"
vector:

$$
\begin{aligned}
\mathbf{f}^{(e)} &= \int_{x'_1}^{x'_2} \begin{Bmatrix} N_1 \\ N_2 \end{Bmatrix} S^{(e)}\, dx' - \begin{Bmatrix} N_1 \\ N_2 \end{Bmatrix} \bar{q}' \bigg|_{\Gamma_2} \\
&= \frac{S^{(e)}}{h^{(e)}} \int_{x'_1}^{x'_2} \begin{Bmatrix} x'_2 - x' \\ x' - x'_1 \end{Bmatrix} dx' - \begin{Bmatrix} N_1 \\ N_2 \end{Bmatrix} \bar{q}' \bigg|_{\Gamma_2} \\
&= \frac{S^{(e)} h^{(e)}}{2} \begin{Bmatrix} 1 \\ 1 \end{Bmatrix} - \begin{Bmatrix} \bar{q}'_1|_{\Gamma_2} \\ \bar{q}'_2|_{\Gamma_2} \end{Bmatrix}
\end{aligned}
\tag{7.54}
$$

The element equations are thus

$$
\frac{k^{(e)}}{h^{(e)}} \begin{bmatrix} 1 & -1 \\ -1 & 1 \end{bmatrix} \begin{Bmatrix} \hat{T}_1^{(e)} \\ \hat{T}_2^{(e)} \end{Bmatrix} = \frac{S^{(e)} h^{(e)}}{2} \begin{Bmatrix} 1 \\ 1 \end{Bmatrix} - \begin{Bmatrix} \bar{q}'_1|_{\Gamma_2} \\ \bar{q}'_2|_{\Gamma_2} \end{Bmatrix}
\tag{7.55}
$$

Remarks

1. The term involving $\bar{q}'$ has non-zero entries only if a natural (heat
flux) boundary condition has been specified at the corresponding node. In
Chapter 8 we will discuss a more general way in which to account for both
natural and essential boundary conditions.

2. Although the element equations have been derived with respect to
the x'-coordinate system, the nodal coordinates will typically be known
in global (x, y, z) coordinates. The only practical consequence of this is
the manner in which the element length $h^{(e)}$ is computed. In particular,
denoting by (x_1, y_1, z_1) and (x_2, y_2, z_2) the coordinates of nodes 1 and 2,
respectively, it follows that

$$
h^{(e)} = \sqrt{(x_2 - x_1)^2 + (y_2 - y_1)^2 + (z_2 - z_1)^2}
\tag{7.56}
$$

If the element lies in a coordinate plane, then the calculation of $h^{(e)}$
simplifies. For example, assuming the element lies in the $x - y$ plane, the
length is again computed using equation (7.56), only with $z_1 = z_2$.

3. From the point of view of numerical implementation, the above
remarks show that the *same* subroutines can be used to generate element
equations for both two-dimensional and three-dimensional line elements
with scalar nodal degrees of freedom.

Instead of using *global* x'-coordinates in the derivation of the element equations, it is typically advantageous to instead use *natural* coordinate systems; that is, ones that rely on the element geometry for their definition. Consider the ξ natural coordinate system, which is defined in Appendix C as

$$\xi = \frac{2(x' - x'_c)}{x'_j - x'_i} = \frac{2\left[x' - 0.5(x'_i + x'_j)\right]}{h^{(e)}} \quad ; \quad -1 \leq \xi \leq 1 \tag{7.57}$$

From the chain rule for differentiation, and from the above definition of ξ it follows that

$$\frac{d}{dx'} = \frac{d}{d\xi}\frac{d\xi}{dx'} = \frac{2}{h^{(e)}}\frac{d}{d\xi} \tag{7.58}$$

Also, for integrations,

$$dx' = \frac{h^{(e)}}{2}d\xi \tag{7.59}$$

where $h^{(e)}/2$ represents the determinant of the Jacobian matrix in one dimension (further details pertaining to Jacobian matrices and their determinants are given in Chapter 10).

Solving equation (7.57) for x' and substituting the resulting expression into equation (7.45) gives

$$N_1 = \frac{1-\xi}{2} \quad , \quad N_2 = \frac{1+\xi}{2} \tag{7.60}$$

The derivatives of the interpolation functions with respect to ξ are simply

$$\frac{dN_1}{d\xi} = \frac{-1}{2} = -\frac{dN_2}{d\xi} \tag{7.61}$$

Substituting these derivatives into equation (7.53) gives

$$\mathbf{K}^{(e)} = \int_{-1}^{1} \frac{2}{h^{(e)}} \left\{ \begin{array}{c} \dfrac{dN_1}{d\xi} \\[2mm] \dfrac{dN_2}{d\xi} \end{array} \right\} k^{(e)} \frac{2}{h^{(e)}} \left\langle \begin{array}{cc} \dfrac{dN_1}{d\xi} & \dfrac{dN_2}{d\xi} \end{array} \right\rangle \left(\frac{h^{(e)}}{2}d\xi \right)$$

$$= \frac{k^{(e)}}{2h^{(e)}} \begin{bmatrix} 1 & -1 \\ -1 & 1 \end{bmatrix} \int_{-1}^{1} d\xi' = \frac{k^{(e)}}{h^{(e)}} \begin{bmatrix} 1 & -1 \\ -1 & 1 \end{bmatrix} \tag{7.62}$$

which is *identical* to $\mathbf{K}^{(e)}$ as given by equation (7.53). Substituting next the interpolation functions of equation (7.60) into equation (7.54) gives

$$
\begin{aligned}
\mathbf{f}^{(e)} &= \int_{-1}^{1} \left\{ \begin{matrix} N_1 \\ N_2 \end{matrix} \right\} S^{(e)} \left(\frac{h^{(e)}}{2} d\xi \right) - \left\{ \begin{matrix} N_1 \\ N_2 \end{matrix} \right\} \bar{q}' \Big|_{\Gamma_2} \\
&= \frac{S^{(e)} h^{(e)}}{4} \int_{-1}^{1} \left\{ \begin{matrix} 1-\xi \\ 1+\xi \end{matrix} \right\} d\xi - \left\{ \begin{matrix} N_1 \\ N_2 \end{matrix} \right\} \bar{q}' \Big|_{\Gamma_2} \\
&= \frac{S^{(e)} h^{(e)}}{2} \left\{ \begin{matrix} 1 \\ 1 \end{matrix} \right\} - \left\{ \begin{matrix} \bar{q}_1' |_{\Gamma_2} \\ \bar{q}_2' |_{\Gamma_2} \end{matrix} \right\}
\end{aligned}
\tag{7.63}
$$

which is *identical* to $\mathbf{f}^{(e)}$ as given by equation (7.54).

Example 7.7. Axial Force (Bar) Element

Using similarities among governing differential equations noted in Chapter 1, we derive equations for a one-dimensional, two-node element with vector primary dependent variables. Although seemingly similar to the previous example, the treatment of vector primary dependent variables involves some important differences.

In structural mechanics it is often possible to take advantage of the intrinsic behavior of a particular type of structure to permit an *a priori* simplification of the spatial dependence of stress and deformation in the body. In this manner it becomes possible to define and effectively utilize special theories of beams, cables, membranes, plates, shells, etc. In each of these cases the geometric features of the body, together with the usual pattern of imposed loading, provide an opportunity to simplify the modeling process and to develop a model of lower spatial dimension.

The simplest "structural" element is the so-called *axial force* or *bar* element. The specialization of the general three-dimensional elasticity problem for the case of a bar element is based on the following assumptions related to the actual physical structural member:

- The member is straight. Its cross-sectional dimensions are significantly smaller than its length. If the member is rectangular in cross-section (Figure 7.8a) with width of b, depth of d and length of L, then $b << L$ and $d << L$.

- The member is connected and loaded only at its ends (Figure 7.8b); and,

- The connections are free to rotate (i.e., are not moment resistant).

Consequently, the bar element admits deformations, forces and displacements only along its longitudinal axis. For this reason, such elements are commonly classified as being "one-dimensional."[16]

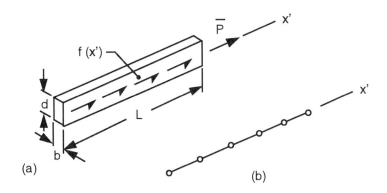

Figure 7.8: (a) Actual Member, (b) Hypothetical Mesh of Two-Node Bar Elements

We desire to develop the equations for a two-node bar element. Rather than do this from "scratch," we make use of the similarities noted in Chapter 1 between governing differential equations for one-dimensional problems.

The governing equation for a one-dimensional axial force member has been derived in Example 1.2, viz.,

$$\frac{d}{dx'}\left(EA\frac{d\delta}{dx'}\right) + f = 0 \quad ; \quad 0 < x' < L \qquad (7.64)$$

where δ is the displacement along the x'-axis, E is the elastic modulus (a material parameter), A is the cross-sectional area of the bar, f is a loading term, and L is the length of the bar along the x'-axis.

---

[16]Since exact solutions are generally available for one-dimensional elements, some consider them to be "trivial examples" of finite elements [605], or not to be finite elements in the true sense [233]. When axial force members constitute the only element type used to model a structure, the structure must, in light of the above assumptions, be a *truss*. Not surprisingly, such elements are discussed in standard texts dealing with matrix structural analysis techniques [33, 300, 325, 375, 381, 384, 448, 473, 477, 555].

Equation (7.64) is similar in form to equation (7.27) for steady-state heat conduction

$$\frac{d}{dx'}\left(k\frac{dT}{dx'}\right) + S = 0 \quad ; \quad 0 < x' < L \tag{7.65}$$

Noting the analogies in equations (7.64) and (7.65) between the primary dependent variables ($\delta \leftrightarrow T$), material and section properties ($EA \leftrightarrow k$), and "source" terms ($f \leftrightarrow S$), we *directly* write the element equations for the one-dimensional elastostatic problem as

$$\frac{E^{(e)}A^{(e)}}{h^{(e)}}\begin{bmatrix} 1 & -1 \\ -1 & 1 \end{bmatrix}\begin{Bmatrix} \hat{\delta}_1^{(e)} \\ \hat{\delta}_2^{(e)} \end{Bmatrix} = \frac{f^{(e)}h^{(e)}}{2}\begin{Bmatrix} 1 \\ 1 \end{Bmatrix} + \begin{Bmatrix} \bar{P}_1'\big|_{\Gamma_2} \\ \bar{P}_2'\big|_{\Gamma_2} \end{Bmatrix} \tag{7.66}$$

where $\bar{P}_1'$ and $\bar{P}_2'$ are specified concentrated nodal forces (natural boundary conditions). The two-node bar element, arbitrarily oriented in space, is shown in Figure 7.9.

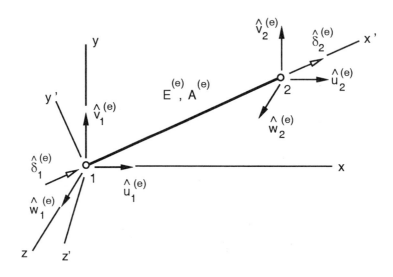

Figure 7.9: Typical Two-Node Bar Element in Space

Remarks

  1. The elastic modulus E, the cross-sectional area A, or both, though constant within a given element, can be different for each and every element.

  2. The forcing term $f(x)$ is typically assumed to consist of two parts: 1) Contributions of the body force $b_{x'}$ (units of FL^{-3}) acting in the x' direction and, 2) Contributions of the applied surface traction $\bar{t}_{x'}$ (units of FL^{-1}) in the same direction. In this case, the right-hand side vector in equation (7.66) becomes

$$\mathbf{f}^{(e)'} = \frac{b_{x'}^{(e)} A^{(e)} h^{(e)}}{2} \begin{Bmatrix} 1 \\ 1 \end{Bmatrix} + \frac{\bar{t}_{x'}^{(e)} h^{(e)}}{2} \begin{Bmatrix} 1 \\ 1 \end{Bmatrix} + \begin{Bmatrix} \bar{P}_1' \big|_{\Gamma_2} \\ \bar{P}_2' \big|_{\Gamma_2} \end{Bmatrix} \qquad (7.67)$$

  3. By assuming the interpolation functions to be linear, one-half of the body force and one-half of the surface traction is equally "lumped" at each of the two element nodes. This illustrates a key point associated with the finite element method. Namely, because the primary dependent variables are solved for only at the nodes, it follows that *any* distributed loading must be "lumped" at the nodes in a consistent manner.

  Having exploited the similarities between the one-dimensional heat conduction and elastostatic problems we note a important difference. Since displacements are *vector* quantities, they require appropriate transformation from the local x'-coordinate system to the global (x, y, z) coordinate system. This was not an issue in the discussion of the one-dimensional heat conduction problem because temperature is a scalar quantity.

  Since some or all of the bar elements in a mesh will have *different orientations* in space, the element equations must be written in terms of the *global* (x, y, z) coordinates. The local (element) and global displacement components are related in the following manner:

$$\hat{\boldsymbol{\delta}}_n^{(e)} = \left\{ \hat{\delta}_1^{(e)} \quad \hat{\delta}_2^{(e)} \right\}^T = \mathbf{R}^{(e)} \hat{\boldsymbol{\phi}}_n^{(e)} \qquad (7.68)$$

where

$$\hat{\boldsymbol{\phi}}_n^{(e)} = \left\{ \hat{u}_1^{(e)} \quad \hat{v}_1^{(e)} \quad \hat{w}_1^{(e)} \quad \hat{u}_2^{(e)} \quad \hat{v}_2^{(e)} \quad \hat{w}_2^{(e)} \right\}^T \qquad (7.69)$$

and

$$\mathbf{R}^{(e)} = \begin{bmatrix} \cos\beta_1^{(e)} & \cos\beta_2^{(e)} & \cos\beta_3^{(e)} & 0 & 0 & 0 \\ 0 & 0 & 0 & \cos\beta_1^{(e)} & \cos\beta_2^{(e)} & \cos\beta_3^{(e)} \end{bmatrix} \quad (7.70)$$

is an orthogonal transformation matrix.

Denoting by (x_1, y_1, z_1) and (x_2, y_2, z_2) the coordinates of nodes 1 and 2, respectively, the direction cosines are determined from

$$\cos\beta_1^{(e)} = \frac{x_2 - x_1}{h^{(e)}} \quad ; \quad \cos\beta_2^{(e)} = \frac{y_2 - y_1}{h^{(e)}} \quad ; \quad \cos\beta_3^{(e)} = \frac{z_2 - z_1}{h^{(e)}} \quad (7.71)$$

where

$$h^{(e)} = \sqrt{(x_2 - x_1)^2 + (y_2 - y_1)^2 + (z_2 - z_1)^2} \quad (7.72)$$

represents the element length.

It can likewise be shown that the vector of axial force components in local (element) coordinates, $\mathbf{f}^{(e)'}$, is related to the vector of nodal loads in global coordinates, $\mathbf{f}^{(e)}$, through the relation

$$\mathbf{f}^{(e)} = \mathbf{R}^{(e)^T} \mathbf{f}^{(e)'} \quad (7.73)$$

The element equations in local coordinates, given by equation (7.66), are written in compact form as

$$\mathbf{K}^{(e)'} \hat{\delta}_n^{(e)} = \mathbf{f}^{(e)'} \quad (7.74)$$

To write these equations in terms of forces and displacements in global coordinates, first substitute equation (7.68) for $\hat{\delta}_n^{(e)}$. This gives

$$\mathbf{K}^{(e)'} \mathbf{R}^{(e)} \hat{\phi}_n^{(e)} = \mathbf{f}^{(e)'} \quad (7.75)$$

Next equation (7.75) is substituted into equation (7.73) to give the desired result, viz.,

$$\mathbf{f}^{(e)} = \mathbf{R}^{(e)^T} \mathbf{K}^{(e)'} \mathbf{R}^{(e)} \hat{\phi}_n^{(e)} = \mathbf{K}^{(e)} \hat{\phi}_n^{(e)} \quad (7.76)$$

where

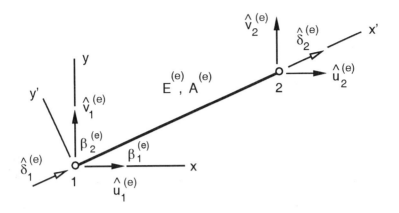

Figure 7.10: Typical Two-Node Planar Bar Element

$$\mathbf{K}^{(e)} = \mathbf{R}^{(e)^T} \mathbf{K}^{(e)'} \mathbf{R}^{(e)} \qquad (7.77)$$

represents the element stiffness matrix with respect to global coordinates.

If, instead, the bar element is arbitrarily oriented in the global $x - y$ plane in the manner shown in Figure 7.10, the relation between local and global displacement degrees of freedom is again given by equation (7.68), only with

$$\hat{\phi}_n^{(e)} = \left\{ \hat{u}_1^{(e)} \quad \hat{v}_1^{(e)} \quad \hat{u}_2^{(e)} \quad \hat{v}_2^{(e)} \right\}^T \qquad (7.78)$$

and

$$\mathbf{R}^{(e)} = \begin{bmatrix} \cos\beta_1^{(e)} & \cos\beta_2^{(e)} & 0 & 0 \\ 0 & 0 & \cos\beta_1^{(e)} & \cos\beta_2^{(e)} \end{bmatrix} \qquad (7.79)$$

Remark

1. In general the size of $\mathbf{R}^{(e)}$ is $(N_{en} * (N_{sdim} * N_{en}))$ and that of $\mathbf{K}^{(e)}$ is $(N_{sdim} * N_{en}) * (N_{sdim} * N_{en})$. The size of the right-hand side vector $\mathbf{f}^{(e)}$ is $(N_{sdim} * N_{en}) * 1$. As before, N_{sdim} equals the spatial dimension of the problem, and N_{en} is the number of nodes contained in the element.

2. From the point of view of numerical implementation, the above remarks show that, unlike elements with scalar primary dependent variables, identical subroutines cannot be used to generate element equations for both two-dimensional and three-dimensional line elements with scalar nodal degrees of freedom. The requisite modifications to accomplish this are, however, trivial.

Line elements, such as those developed in the previous two examples, are often used in conjunction with usual two- or three-dimensional elements (Figure 7.11). In the context of mechanical analyses, one-dimensional elements are used to represent inclusions (e.g., reinforcement) in plane and three-dimensional analyses. For scalar field problems such as those discussed in Chapter 11, one-dimensional elements may be used to represent drains in porous media, or inclusions (e.g., line sources) in a thermal analysis. When one-dimensional elements are used in conjunction with two- and three-dimensional elements, continuity of the interpolation functions across element interfaces must obviously be maintained. Further details pertaining to this topic are given in Chapter 9. We next develop equations for a two-dimensional element.

Example 7.8. Linear Triangular Heat Conduction Element

We desire to formulate the equations for steady state heat conduction analyses involving irreducible linear triangular elements. The triangle is the simplest geometric shape that can be used to discretize a two-dimensional domain Ω. The linear triangle, which is traced to the pioneering work of Courant [122], was one of the first finite elements developed [545, 523, 188].

• **Step 1:** Identification of the Primary Dependent Variable

Since an irreducible formulation shall be used, the primary dependent variable is the temperature T.

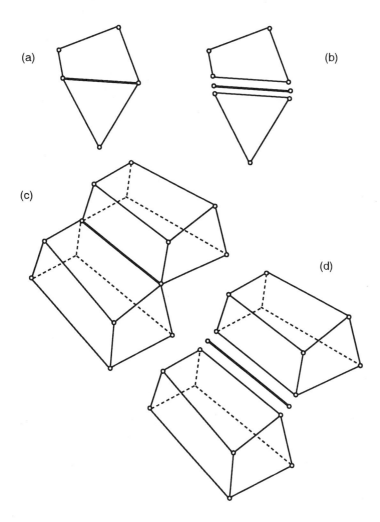

Figure 7.11: Compatible Element Combinations: (a) Linear Two-Dimensional and Line Elements, (b) Exploded View, (c) Linear Three-Dimensional and Line Elements, (d) Exploded View

• **Step 2:** Definition of the Gradient and Constitutive Relations

The relation between heat flux per unit area and the gradient of temperature is assumed to follow Fourier's law. For a two-dimensional problem, Fourier's law generalizes to

$$\left\{\begin{array}{c} q_x \\ q_y \end{array}\right\} = - \begin{bmatrix} k_{11} & k_{12} \\ k_{12} & k_{22} \end{bmatrix} \left\{\begin{array}{c} \partial T/\partial x \\ \partial T/\partial y \end{array}\right\} \tag{7.80}$$

where q_x and q_y represent components of the heat flux (per unit area) and k_{11}, k_{12} and k_{22} are thermal conductivity coefficients.

• **Step 3:** Identification of the Element Equations

Rather than using a weighted residual method as in Example 7.7, we use a variational approach. The element functional is:

$$I^{(e)} = \iint\limits_{\Omega^e} \left[\frac{1}{2} k_{11}^{(e)} \left(\frac{\partial T}{\partial x}\right)^2 + k_{12}^{(e)} \left(\frac{\partial T}{\partial x}\right)\left(\frac{\partial T}{\partial y}\right) + \frac{1}{2} k_{22}^{(e)} \left(\frac{\partial T}{\partial y}\right)^2 \right] d\Omega$$

$$- \iint\limits_{\Omega^e} S^{(e)} \, T \, d\Omega - \int\limits_{\Gamma_2^e} \bar{q} \, T \, d\Gamma \tag{7.81}$$

where $k_{11}^{(e)}$, $k_{12}^{(e)}$ and $k_{22}^{(e)}$ are element thermal conductivity coefficients, $S^{(e)}$ is an element heat source, $\bar{q}$ denotes the heat flux per unit area (natural boundary condition) specified along the part Γ_2 of the element boundary Γ^e, and a unit thickness of the body has been assumed. For simplicity, convection boundary conditions have been omitted from equation (7.81).

The general approximations for temperature and its derivatives are:

$$\hat{T}^{(e)}(x,y) = \sum_{m=1}^{N_{dof}} N_m(x,y) \, \hat{T}_m^{(e)} \tag{7.82}$$

$$\frac{\partial \hat{T}^{(e)}}{\partial x} = \sum_{m=1}^{N_{dof}} \frac{\partial N_m}{\partial x} \, \hat{T}_m^{(e)} \tag{7.83}$$

$$\frac{\partial \hat{T}^{(e)}}{\partial y} = \sum_{m=1}^{N_{dof}} \frac{\partial N_m}{\partial y} \, \hat{T}_m^{(e)} \tag{7.84}$$

Since the primary dependent variable (temperature) is a scalar, the total number of degrees of freedom associated with the element (N_{dof}) will equal the number of nodes in the element (N_{en}).

Substituting these approximations into equation (7.81) gives the following *approximate* functional:

$$
\hat{I}^{(e)} = \iint\limits_{\Omega^e} \left[\frac{1}{2} k_{11}^{(e)} \left(\sum_{m=1}^{N_{dof}} \frac{\partial N_m}{\partial x} \hat{T}_m^{(e)} \right)^2 \right] d\Omega
$$

$$
+ \iint\limits_{\Omega^e} \left[k_{12}^{(e)} \left(\sum_{m=1}^{N_{dof}} \frac{\partial N_m}{\partial x} \hat{T}_m^{(e)} \right) \left(\sum_{m=1}^{N_{dof}} \frac{\partial N_m}{\partial y} \hat{T}_m^{(e)} \right) \right] d\Omega
$$

$$
+ \iint\limits_{\Omega^e} \left[\frac{1}{2} k_{22}^{(e)} \left(\sum_{m=1}^{N_{dof}} \frac{\partial N_m}{\partial y} \hat{T}_m^{(e)} \right)^2 \right] d\Omega
$$

$$
- \iint\limits_{\Omega^e} S^{(e)} \left(\sum_{m=1}^{N_{dof}} N_m \hat{T}_m^{(e)} \right) d\Omega - \int\limits_{\Gamma_2^e} \bar{q} \left(\sum_{m=1}^{N_{dof}} N_m \hat{T}_m^{(e)} \right) d\Gamma \quad (7.85)
$$

- **Step 4:** Selection of Interpolation Functions

The highest order derivative of the primary dependent variable in equation (7.81) is one. The associated interpolation functions must thus be C^0 continuous. Consider a complete first-order polynomial in x and y:

$$
\hat{T}^{(e)}(x,y) = \langle 1 \quad x \quad y \rangle \begin{Bmatrix} \alpha_1 \\ \alpha_2 \\ \alpha_3 \end{Bmatrix} = \chi\alpha \quad (7.86)
$$

where the entries in χ are again the *basis functions*. As in the case of the one-dimensional element considered in Example 7.6, the α_i populating α are unknown *generalized coordinates* that are a function only of the element geometry.

The presence, in equation (7.86), of the three unknown coefficients α_1, α_2 and α_3 requires that three degrees of freedom for $\hat{T}$ be associated with the element. Since temperature is a scalar quantity, one degree of freedom will be associated with each node in the element. The element must thus possess *three* nodes. The simplest, and indeed only practical choice, is the three-node triangular shown in Figure 7.12. In this figure, 1, 2, and 3 denote *local* node numbers – when Ω is discretized into a finite number of elements, these numbers are mapped to appropriate *global* numbers.

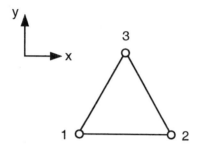

Figure 7.12: Typical 3-Node Triangular Element

Following the procedure used in Example 7.6, we evaluate equation (7.86) at each of the element nodes. This leads to the following set of equations:

$$\mathbf{\hat{T}_n^{(e)}} = \left\{ \begin{matrix} \hat{T}_1^{(e)} \\ \hat{T}_2^{(e)} \\ \hat{T}_3^{(e)} \end{matrix} \right\} = \begin{bmatrix} 1 & x_1 & y_1 \\ 1 & x_2 & y_2 \\ 1 & x_3 & y_3 \end{bmatrix} \left\{ \begin{matrix} \alpha_1 \\ \alpha_2 \\ \alpha_3 \end{matrix} \right\} = \mathbf{A}\boldsymbol{\alpha} \qquad (7.87)$$

For triangular elements, equation (7.87) can always inverted [605]. Thus,

$$\boldsymbol{\alpha} = \mathbf{A}^{-1}\mathbf{\hat{T}_n^{(e)}} \qquad (7.88)$$

where

$$\mathbf{A}^{-1} = \frac{1}{2A^{(e)}} \begin{bmatrix} (x_2 y_3 - x_3 y_2) & (x_3 y_1 - x_1 y_3) & (x_1 y_2 - x_2 y_1) \\ (y_2 - y_3) & (y_3 - y_1) & (y_1 - y_2) \\ (x_3 - x_2) & (x_1 - x_3) & (x_2 - x_1) \end{bmatrix}$$

$$= \frac{1}{2A^{(e)}} \begin{bmatrix} a_1 & a_2 & a_3 \\ b_1 & b_2 & b_3 \\ c_1 & c_2 & c_3 \end{bmatrix} \qquad (7.89)$$

Remarks

1. The determinant of $\mathbf{A}$ is equal to twice the area $A^{(e)}$ of the element.

2. The quantities a_1, a_2, a_3, b_1, b_2, b_3, c_1, c_2 and c_3 appearing in equation (7.89) are *constants*.

From equations (7.88) and (7.89) it is evident that the generalized co-ordinates α_i are indeed simply functions of the nodal coordinates. Substituting equation (7.88) into equation (7.86) gives

$$\hat{T}^{(e)}(x,y) = \chi \mathbf{A}^{-1}\hat{\mathbf{T}}_{\mathbf{n}}^{(e)} = \mathbf{N}\hat{\mathbf{T}}_{\mathbf{n}}^{(e)} \tag{7.90}$$

where $\mathbf{N}$ represents the $(1 * N_{en})$ row vector of interpolation functions. After some straightforward manipulation, the entries in $\mathbf{N}$ are found to be

$$N_i = \frac{1}{A^{(e)}}(a_i + b_i x + c_i y) \quad ; \quad i = 1,2,3 \tag{7.91}$$

where the a_i, b_i and c_i are defined in equation (7.89). Comparing the interpolation functions N_i from equation (7.91) to the natural coordinates L_i for triangular elements presented in Appendix C, we see that these two sets of quantities are *identical*. The N_i are shown in Figure 7.13.

The interpolation functions refer to individual nodes in the element and thus to the individual degrees of freedom. They must satisfy the basic requirement for node i given by equation (6.30), viz.,

$$N_i(x_j, y_j) = \delta_{ij} \tag{7.92}$$

where δ_{ij} is the Kronecker delta, and $i,j = 1,2,3$. We check this requirement for N_1 and leave the verification of N_2 and N_3 as an exercise.

$$N_1(x_1, y_1) = \frac{1}{2A^{(e)}}(a_1 + b_1 x_1 + c_1 y_1) \tag{7.93}$$

$$= \frac{1}{2A^{(e)}}[(x_2 y_3 - x_3 y_2) + (y_2 - y_3)x_1 + (x_3 - x_2)y_1] \tag{7.94}$$

$$= \frac{1}{2A^{(e)}}[x_2 y_3 - x_3 y_2 + x_1 y_2 - x_1 y_3 + x_3 y_1 - x_2 y_1] = 1 \tag{7.95}$$

$$N_1(x_2, y_2) = \frac{1}{2A^{(e)}}(a_1 + b_1 x_2 + c_1 y_2) \tag{7.96}$$

$$= \frac{1}{2A^{(e)}}[(x_2 y_3 - x_3 y_2) + (y_2 - y_3)x_2 + (x_3 - x_2)y_2] \tag{7.97}$$

$$= \frac{1}{2A^{(e)}}[x_2 y_3 - x_3 y_2 + x_2 y_2 - x_2 y_3 + x_3 y_2 - x_2 y_2] = 0 \tag{7.98}$$

$$N_1(x_3, y_3) = \frac{1}{2A^{(e)}} (a_1 + b_1 x_3 + c_1 y_3) \tag{7.99}$$

$$= \frac{1}{2A^{(e)}} [(x_2 y_3 - x_3 y_2) + (y_2 - y_3)x_3 + (x_3 - x_2)y_3] \tag{7.100}$$

$$= \frac{1}{2A^{(e)}} [x_2 y_3 - x_3 y_2 + x_3 y_2 - x_3 y_3 + x_3 y_3 - x_2 y_3] = 0 \tag{7.101}$$

• Check of Compatibility Criterion

We next must check whether the interpolation functions satisfy the criteria for monotonic convergence. The first of these is the *compatibility* criterion. As evident from equation (7.81), the highest order derivative of T appearing in the element equations is one (i.e., $s = 1$). Consequently, with the element domain Ω^e, the first derivatives of $\hat{T}^{(e)}$ must be continuous. This in turn requires that the first derivatives of the interpolation functions must be continuous within Ω^e. In light of equations (7.83), (7.84) and (7.91), it follows that

$$\frac{\partial \hat{T}^{(e)}}{\partial x} = \sum_{m=1}^{3} \frac{\partial N_m}{\partial x} \hat{T}_m^{(e)} = \frac{1}{2A^{(e)}} \sum_{m=1}^{3} b_m \hat{T}_m^{(e)} \tag{7.102}$$

$$\frac{\partial \hat{T}^{(e)}}{\partial y} = \sum_{m=1}^{3} \frac{\partial N_m}{\partial y} \hat{T}_m^{(e)} = \frac{1}{2A^{(e)}} \sum_{m=1}^{3} c_m \hat{T}_m^{(e)} \tag{7.103}$$

From these equations, and from the definitions of the b_m and c_m, it is evident that within Ω^e the derivatives of the interpolation functions will indeed be continuous.

The compatibility criterion also requires that the primary dependent variable $\hat{T}^{(e)}$ be continuous across the element interfaces Γ^e. To be continuous, $\hat{T}^{(e)}$, and thus the associated interpolation functions (recall equation 7.82), must thus be *linear* along Γ^e. Stated another way, since the nodal unknowns must be continuous, and since two points uniquely define a straight line, it follows that if the interpolation functions are shown to be linear along an element interface, then no "gaps" in $\hat{T}^{(e)}$ can occur there. Since the check for linearity of the interpolation functions along each of the three element edges is straightforward, it is not pursued further in this example.

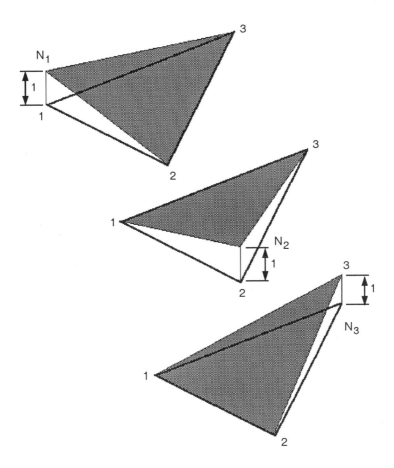

Figure 7.13: Interpolation Functions Associated with a Linear Triangle

• Check of Completeness Criterion

The second criterion to be satisfied by the element interpolation functions is *completeness*. Since in the present problem $s = 1$, the completeness criterion requires that constant states of $\hat{T}^{(e)}$ and its first derivatives be continuous as the size of the element goes to zero.

A state of constant temperature corresponds to the "rigid body mode" in stress-deformation problems. Assume that $\hat{T}_1^{(e)} = \hat{T}_2^{(e)} = \hat{T}_3^{(e)} = C$, where C is some constant. Consequently, we must thus show that the partition of unity property is satisfied, viz.,

$$\hat{T}^{(e)} = \sum_{m=1}^{3} C N_m = C N_1 + C N_2 + C N_3 = C \Rightarrow \sum_{m=1}^{3} N_m = 1 \quad (7.104)$$

From equation (7.91) we have

$$\sum_{m=1}^{3} N_m = \frac{1}{2A^{(e)}} \left[\sum_{m=1}^{3} a_m + \left(\sum_{m=1}^{3} b_m \right) x + \left(\sum_{m=1}^{3} c_m \right) y \right] \quad (7.105)$$

Examining the respective terms in this equation, we find the following results:

$$\sum_{m=1}^{3} a_m = (x_2 y_3 - x_3 y_2) + (x_3 y_1 - x_1 y_3) + (x_1 y_2 - x_2 y_1)$$

$$= 2A^{(e)} \quad (7.106)$$

$$\sum_{m=1}^{3} b_m = (y_2 - y_3) + (y_3 - y_1) + (y_1 - y_2) = 0 \quad (7.107)$$

$$\sum_{m=1}^{3} c_m = (x_3 - x_2) + (x_1 - x_3) + (x_2 - x_1) = 0 \quad (7.108)$$

Equation (7.104) is thus satisfied. Next we must check that constant states of first derivatives of $\hat{T}^{(e)}$ can be represented by the element interpolation functions. Since $A^{(e)}$, $\hat{T}_m^{(e)}$, b_m and c_m are constant in equations (7.102) and (7.103), it follows that constant states of $\partial T / \partial x$ and $\partial T / \partial y$ will indeed be realized.

Remark

1. For the special case where all the nodal temperatures are equal to the same value C (recall equation 7.104), equations (7.102) and (7.103) give

$$\frac{\partial \hat{T}^{(e)}}{\partial x} = C \sum_{m=1}^{3} \frac{\partial N_m}{\partial x} = \frac{C}{2A^{(e)}} \sum_{m=1}^{3} b_m = 0 \qquad (7.109)$$

$$\frac{\partial \hat{T}^{(e)}}{\partial y} = C \sum_{m=1}^{3} \frac{\partial N_m}{\partial y} = \frac{C}{2A^{(e)}} \sum_{m=1}^{3} c_m = 0 \qquad (7.110)$$

which is the physically consistent result.

Finally, since the interpolation functions are linear in x and in y, we must also show that if $\hat{T}_m^{(e)} = C_1 x_m$ for $m = 1, 2, 3$, where C_1 is some constant, then $\hat{T}^{(e)} = C_1 x_m$. Likewise, we must show that if $\hat{T}_m^{(e)} = C_2 y_m$ where C_2 is some constant, then $\hat{T}^{(e)} = C_2 x_m$. Since the manipulations are rather straightforward, this check is left as an exercise. The completeness criterion is thus satisfied.

The element interpolation functions of equation (7.91) thus satisfy all of the necessary criteria to ensure monotonic convergence to the exact solution.

- **Spatial Isotropy**

The last issue to be addressed related to the interpolation functions is spatial isotropy. Since the two-dimensional polynomial of equation (7.86) is *complete*, spatial isotropy is guaranteed.

- **Step 5:** Specialization of Approximate Element Equations

The functional of equation (7.85) is specialized by appropriately substituting the interpolation functions of equation (7.91) and their derivatives from equations (7.102) and (7.103). To facilitate this process, we write the approximation of nodal temperatures and their derivatives in matrix form as

$$\hat{T}^{(e)}(x, y) = \langle N_1 \quad N_2 \quad N_3 \rangle \begin{Bmatrix} \hat{T}_1^{(e)} \\ \hat{T}_2^{(e)} \\ \hat{T}_3^{(e)} \end{Bmatrix} = \mathbf{N}\hat{\mathbf{T}}_\mathbf{n} \qquad (7.111)$$

$$\left\{\begin{array}{c} \dfrac{\partial \hat{T}^{(e)}}{\partial x} \\[2ex] \dfrac{\partial \hat{T}^{(e)}}{\partial y} \end{array}\right\} = \dfrac{1}{2A^{(e)}} \begin{bmatrix} b_1 & b_2 & b_3 \\ c_1 & c_2 & c_3 \end{bmatrix} \left\{\begin{array}{c} \hat{T}_1^{(e)} \\ \hat{T}_2^{(e)} \\ \hat{T}_3^{(e)} \end{array}\right\} = \begin{bmatrix} \mathbf{B_1} \\ \mathbf{B_2} \end{bmatrix} \hat{\mathbf{T}}_n^{(e)} = \mathbf{B}\hat{\mathbf{T}}_n^{(e)} \quad (7.112)$$

where $\mathbf{B_1}$ and $\mathbf{B_2}$ represent $(1 * 3)$ sub-matrices. Substitution of equations (7.111) and (7.112) into equation (7.85) gives

$$\hat{I}^{(e)} = \iint\limits_{\Omega^e} \left[\frac{k_{11}^{(e)}}{2} \hat{\mathbf{T}}_n^{(e)T} \mathbf{B_1}^T \mathbf{B_1} \hat{\mathbf{T}}_n^{(e)} \right] dx\, dy$$

$$+ \iint\limits_{\Omega^e} \left[k_{12}^{(e)} \hat{\mathbf{T}}_n^{(e)T} \mathbf{B_1}^T \mathbf{B_2} \hat{\mathbf{T}}_n^{(e)} \right] dx\, dy$$

$$+ \iint\limits_{\Omega^e} \left[\frac{k_{22}^{(e)}}{2} \hat{\mathbf{T}}_n^{(e)T} \mathbf{B_2}^T \mathbf{B_2} \hat{\mathbf{T}}_n^{(e)} \right] dx\, dy - \iint\limits_{\Omega^e} S^{(e)} \mathbf{N}\hat{\mathbf{T}}_n \, dx\, dy \,(7.113)$$

where due account has been made for conformability of matrix products. In equation (7.113) the surface flux term has been omitted; it will be discussed separately following the derivation of the element equations.

After some preliminary manipulations, the approximate functional is first made stationary with respect to $\hat{T}_1^{(e)}$:

$$\frac{\partial \hat{I}^{(e)}}{\partial \hat{T}_1^{(e)}} = \iint\limits_{\Omega^e} \left\{ \frac{k_{11}^{(e)}}{2\left(A^{(e)}\right)^2} \left[(b_1)^2 \hat{T}_1^{(e)} + b_1 b_2 \hat{T}_2^{(e)} + b_1 b_3 \hat{T}_3^{(e)} \right] \right.$$

$$+ \frac{k_{12}^{(e)}}{4\left(A^{(e)}\right)^2} \left[2b_1 c_1 \hat{T}_1^{(e)} + (b_1 c_2 + b_2 c_1)\hat{T}_2^{(e)} + (b_1 c_3 + b_3 c_1)\hat{T}_3^{(e)} \right]$$

$$\left. + \frac{k_{22}^{(e)}}{2\left(A^{(e)}\right)^2} \left[(c_1)^2 \hat{T}_1^{(e)} + c_1 c_2 \hat{T}_2^{(e)} + c_1 c_3 \hat{T}_3^{(e)} \right] \right\} dx\, dy$$

$$- S^{(e)} \iint\limits_{\Omega^e} \frac{1}{2A^{(e)}} \left[a_1 + b_1 x + c_1 y \right] dx\, dy = 0 \qquad (7.114)$$

A closer look at the last integral in equation (7.113) reveals that it assumes a very simple form, viz.,

$$S^{(e)} \iint\limits_{\Omega^e} \frac{1}{2A^{(e)}} \left[a_1 + b_1 x + c_1 y\right] dx\, dy = S^{(e)} \iint\limits_{\Omega^e} N_1\, dx\, dy$$

$$= S^{(e)} \left(\frac{A^{(e)}}{3}\right) \qquad (7.115)$$

where $S^{(e)}$ has been assumed *constant* over the element, and equation (F.34) from Appendix F has been used.

Similarly, the approximate functional is made stationary with respect to $\hat{T}_2^{(e)}$ and $\hat{T}_3^{(e)}$, viz.,

$$\frac{\partial \hat{I}^{(e)}}{\partial \hat{T}_2^{(e)}} = \iint\limits_{\Omega^e} \left\{ \frac{k_{11}^{(e)}}{2\left(A^{(e)}\right)^2} \left[b_1 b_2 \hat{T}_1^{(e)} + (b_2)^2 \hat{T}_2^{(e)} + b_2 b_3 \hat{T}_3^{(e)}\right] \right.$$

$$+ \frac{k_{12}^{(e)}}{4\left(A^{(e)}\right)^2} \left[(b_1 c_2 + b_2 c_1)\hat{T}_1^{(e)} + 2b_2 c_2 \hat{T}_2^{(e)} + (b_2 c_3 + b_3 c_2)\hat{T}_3^{(e)}\right]$$

$$\left. + \frac{k_{22}^{(e)}}{2\left(A^{(e)}\right)^2} \left[c_1 c_2 \hat{T}_1^{(e)} + (c_2)^2 \hat{T}_2^{(e)} + c_2 c_3 \hat{T}_3^{(e)}\right] \right\} dx\, dy$$

$$- \frac{S^{(e)} A^{(e)}}{3} = 0 \qquad (7.116)$$

$$\frac{\partial \hat{I}^{(e)}}{\partial \hat{T}_3^{(e)}} = \iint\limits_{\Omega^e} \left\{ \frac{k_{11}^{(e)}}{2\left(A^{(e)}\right)^2} \left[b_1 b_3 \hat{T}_1^{(e)} + b_2 b_3 \hat{T}_2^{(e)} + (b_3)^2 \hat{T}_3^{(e)}\right] \right.$$

$$+ \frac{k_{12}^{(e)}}{4\left(A^{(e)}\right)^2} \left[(b_1 c_3 + b_3 c_1)\hat{T}_1^{(e)} + (b_2 c_3 + b_3 c_2)\hat{T}_2^{(e)} + 2b_2 c_2 \hat{T}_3^{(e)}\right]$$

$$\left. + \frac{k_{22}^{(e)}}{2\left(A^{(e)}\right)^2} \left[c_1 c_3 \hat{T}_1^{(e)} + c_2 c_3 \hat{T}_2^{(e)} + (c_3)^2 \hat{T}_3^{(e)}\right] \right\} dx\, dy$$

$$- \frac{S^{(e)} A^{(e)}}{3} = 0 \qquad (7.117)$$

Since the thermal conductivity coefficients $k_{11}^{(e)}$, $k_{12}^{(e)}$ and $k_{22}^{(e)}$ are assumed to be constant over an element, and noting that the b_i, c_i are constants, the integrations in equations (7.114), (7.116) and (7.117) are trivial. The element equations, in matrix form, are thus

$$\mathbf{K}^{(e)}\mathbf{T_n^{(e)}} = \mathbf{f}^{(e)} \tag{7.118}$$

where

$$K_{11}^{(e)} = \frac{1}{4A^{(e)}}\left[k_{11}^{(e)}(b_1)^2 + 2k_{12}^{(e)}b_1c_1 + k_{22}^{(e)}(c_2)^2\right] \tag{7.119}$$

$$K_{12}^{(e)} = K_{21}^{(e)} = \frac{1}{4A^{(e)}}\left[k_{11}^{(e)}b_1b_2 + k_{12}^{(e)}(b_1c_2 + b_2c_1) + k_{22}^{(e)}c_1c_2\right] \tag{7.120}$$

$$K_{13}^{(e)} = K_{31}^{(e)} = \frac{1}{4A^{(e)}}\left[k_{11}^{(e)}b_1b_3 + k_{12}^{(e)}(b_1c_3 + b_3c_1) + k_{22}^{(e)}c_1c_3\right] \tag{7.121}$$

$$K_{22}^{(e)} = \frac{1}{4A^{(e)}}\left[k_{11}^{(e)}(b_2)^2 + k_{12}^{(e)}b_2c_2 + k_{22}^{(e)}(c_2)^2\right] \tag{7.122}$$

$$K_{23}^{(e)} = K_{32}^{(e)} = \frac{1}{4A^{(e)}}\left[k_{11}^{(e)}b_2b_3 + k_{12}^{(e)}(b_2c_3 + b_3c_2) + k_{22}^{(e)}c_2c_3\right] \tag{7.123}$$

$$K_{33}^{(e)} = \frac{1}{4A^{(e)}}\left[k_{11}^{(e)}(b_3)^2 + k_{12}^{(e)}b_3c_3 + k_{22}^{(e)}(c_3)^2\right] \tag{7.124}$$

$$\mathbf{T_n^{(e)}} = \left\{\hat{T}_1^{(e)} \quad \hat{T}_2^{(e)} \quad \hat{T}_3^{(e)}\right\}^T \tag{7.125}$$

$$\mathbf{f^{(e)}} = \frac{S^{(e)}A^{(e)}}{3}\{1 \quad 1 \quad 1\}^T \tag{7.126}$$

• Computation of Equivalent Nodal Fluxes

In the development of the approximate element equations, the distributed edge flux term

$$\int_{\Gamma_2^e} \bar{q}\left(\sum_{m=1}^{N_{dof}} N_m\,\hat{T}_m^{(e)}\right)d\Gamma$$

was omitted. This term is now considered in some detail.

In the finite element analysis of thermal problems with temperature as the primary dependent variable, the only permissible form of loading is by the specification of concentrated nodal heat fluxes. Consequently, distributed fluxes $\bar{q}$ must be reduced to *equivalent* nodal fluxes.[17]

For two-dimensional elements such as the linear triangle, the boundary Γ^e consists of the element edges. Since a flux can be specified along *any* element edge, the contribution of each specified flux to the functional must appropriately be summed.

Consider the loaded edge of a linear triangular element such as that shown in Figure 7.14a). By its definition, a distributed flux $\bar{q}$ specified along Γ^e acts *normal* to the edge. The distributed flux need not be constant, but can vary along the edge. The magnitude of the flux is given by the nodal values of $\bar{q}$ along the element edge. For linear elements such as the three-node triangle, the magnitude of the distribution is given by the values $\bar{q}_i$ and $\bar{q}_j$ associated with nodes i and j, respectively, as in Figure 7.14(a).

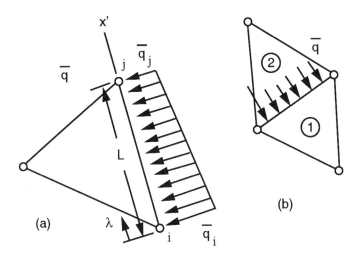

Figure 7.14: Distribution of Heat Flux per Unit Length Along: (a) Single Element Edge, (b) Interface Between Two Elements

The sign of a distributed heat flux depends on the order in which node numbers along the edge are specified, and on the signs of the nodal values $\bar{q}_i$

---

[17]Due to the fact that a unit thickness of the element has been assumed, the units of $\bar{q}$ will be heat per length.

and $\bar{q}_j$. In defining the connectivity[18] of an element, the node numbers are typically specified sequentially in a counterclockwise order. A distributed flux is assumed to be *positive* if, in conjunction with such a numbering sequence, the nodal values of heat flux are positive and the flux is directed *into* the element.

Referring to Figure 7.14(a), if the nodal flux values are specified in a counterclockwise manner, and if $\bar{q}_i$ and $\bar{q}_j$ are positive, then the flux will be directed into the element in the manner shown. Physically, heat is being added along this element edge.

The above definitions are necessary in order to avoid confusion when distributed fluxes are specified along element interfaces. For example, assuming a counterclockwise specification of node numbers along the loaded edge of element 1 in Figure 7.14(b), the flux distribution will be *positive*. If the same distribution is instead assumed to be associated with element 2, then it will be *negative*.

We begin the determination of equivalent nodal fluxes by defining the λ natural coordinate system along the loaded element edge of length L shown in Figure 7.14(a). In light of the discussion in Appendix C, we have

$$\lambda \equiv \frac{x' - x_i'}{L} \quad ; \quad 0 \le \lambda \le 1 \tag{7.127}$$

Since the variation of $\hat{T}^{(e)}$ along an element edge is linear, we write

$$\hat{T}^{(e)} = (1 - \lambda)\hat{T}_1^{(e)} + \lambda\hat{T}_2^{(e)} \tag{7.128}$$

The distributed edge flux term thus becomes

$$\int_{\Gamma_2^e} \bar{q} \left(\sum_{m=1}^{N_{dof}} N_m \hat{T}_m^{(e)} \right) d\Gamma = \int_0^1 \bar{q}(\lambda) \left[(1 - \lambda)\hat{T}_1^{(e)} + \lambda\hat{T}_2^{(e)} \right] (L\, d\lambda) \tag{7.129}$$

Differentiating equation (7.129) with respect to $\hat{T}_1^{(e)}$ and $\hat{T}_2^{(e)}$ gives the following equivalent fluxes for nodes i and j:

$$g_i = L \int_0^1 \bar{q}(\lambda)(1 - \lambda)\, d\lambda \tag{7.130}$$

---

[18]This topic will be discussed further in Chapter 10.

$$g_j = L \int_0^1 \bar{q}(\lambda) \lambda \, d\lambda \qquad (7.131)$$

The most general heat flux distribution for a linear (two-node) element edge is *trapezoidal*, viz.,

$$\bar{q}(\lambda) = (1 - \lambda)\bar{q}_i + \lambda \bar{q}_j \qquad (7.132)$$

Substituting this expression for $\bar{q}(\lambda)$ into equations (7.130) and (7.131) and integrating gives

$$g_i = L \left(\frac{1}{3}\bar{q}_i + \frac{1}{6}\bar{q}_j \right) \quad , \quad g_j = L \left(\frac{1}{6}\bar{q}_i + \frac{1}{3}\bar{q}_j \right) \qquad (7.133)$$

As a check, we note that the total area under the flux distribution is $\frac{1}{2}(\bar{q}_i + \bar{q}_j) L$, which is exactly equal to the sum of g_i and g_j.

When $\bar{q}(\lambda) = \bar{q}_0$ where $\bar{q}_0$ is a constant, then $\bar{q}_i = \bar{q}_j = \bar{q}_0$. Substituting these values into equations (7.130) and (7.131) and integrating gives

$$g_i = g_j = \frac{\bar{q}_0 L}{2} \qquad (7.134)$$

As expected, one-half of the distributed flux is thus "lumped" to each node along the loaded element edge.

If we next consider the *triangular* flux distribution $\bar{q}(\lambda) = \lambda \bar{q}_0$, then $\bar{q}_i = 0$ and $\bar{q}_j = \bar{q}_0$. Substituting these values into equations (7.130) and (7.131) and integrating gives

$$g_i = \frac{\bar{q}_0 L}{6} \quad , \quad g_j = \frac{\bar{q}_0 L}{3} \qquad (7.135)$$

which is perhaps *not* what would intuitively be expected, since even though $\bar{q}_i = 0$, $g_i \neq 0$. This result is, none the less, consistent with the finite element formulation. As a check, we note that

$$g_i + g_j = \frac{\bar{q}_0 L}{2} \qquad (7.136)$$

which is the total area under the flux distribution. The "intuitive" assumption of $g_i = 0$ would thus lead to erroneous results.

Remarks

    1. The above discussion of prescribed heat flux holds for *any* element edge where the distribution of the scalar primary dependent variable is *linear*. Thus, although the equivalent nodal fluxes g_i and g_j were derived for a linear triangle, they likewise apply to a four-node rectangular element.

    2. A geometric error, akin to the domain discretization error, may be introduced when a distributed flux is applied to a curved boundary. If the boundary is discretized using straight-sided elements, the flux, which is applied normal to the boundary, will be oriented incorrectly to some extent. A solution to this problem is to use curved-sided elements (i.e., quadratic and higher order) [100].

    3. As will be shown in Chapter 10, the integrations associated with the development of the element equations must, in general, be performed *numerically*. This has the potential of introducing *numerical integration* or *quadrature* error into the solution. Further discussion of this source of error is deferred until Section 8.3.2 of the next chapter.

7.8 Concluding Remarks

In concluding this chapter, the importance of formulating an element that correctly models the physical problem is perhaps best summed up by Bathe [58], who notes that:

> It is clear that the finite element solution will solve only the selected mathematical model and that this model will be reflected in the predicted response. We cannot expect any more information in the prediction of physical phenomena than the information contained in the mathematical model. Hence the choice of an appropriate mathematical model is crucial and completely determines the insight into the actual physical problem that we can obtain by the analysis.

7.9   Exercises

7.1

Consider a string[19] of length L which, when unloaded, is oriented in a straight line parallel to the (longitudinal) x-axis (Figure 7.15).

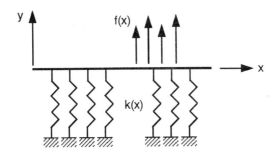

Figure 7.15: Schematic Illustration of "String" on Elastic Foundation

The string is stretched under a tension $T(x)$, which may vary along the length (for example, if the string were hanging vertically in a gravitational field). The only loads on the string are applied transversely to the longitudinal x-axis and result in displacements that are also transverse to this axis. The string rests on an elastic (Winkler) foundation with modulus $k(x)$.

Assuming "small" displacements,[20] the equation governing the transverse displacement of the string is:

$$\frac{d}{dx}\left(T\frac{du}{dx}\right) - ku + f = 0 \quad ; \quad 0 < x < 1 \tag{7.137}$$

where $T(x)$ is the tension in the string (units: F), $u(x)$ is the transverse displacement from the unloaded (straight) configuration (units: L), $k(x)$

---

[19]By "string" is meant any body with one characteristic cross-sectional dimension greater than the other two, for which the extensional stiffness is much greater than the flexural stiffness. As such, cables, flexible conduits, etc. can often be considered to be "strings."

[20]This restriction has the following implications. First, it assumed that $T(x)$ remains the same in both loaded and unloaded configurations. Second, it assumes that the slope θ of any section of the string relative to the x-axis is small enough that $\sin\theta \approx \tan\theta$, or, equivalently, that $\cos\theta \approx 1$. Thirdly, it assumes that bending stresses generated by the change in curvature are small relative to the axial stress (i.e., $T(x)$ divided by the cross-sectional area of the string).

is the material parameter (elastic modulus) characterizing the foundation (units: FL^{-2}), and $f(x)$ is a known applied load, distributed along the string (units: FL^{-1}).

Given the above information, develop the equations for the *simplest* one-dimensional element suitable for analysis of strings on elastic foundations. The primary dependent variable is identified as being the transverse displacement $u(x)$. The constitutive relation is simply $F(x) = -T(x)\,du/dx$, where $F(x)$ is the transverse component of the string tension. The element functional associated with the problem is:

$$I^{(e)} = \frac{1}{2}\int_0^{h^e}\left[T(x)\left(\frac{du}{dx}\right)^2 + k(x)u^2\right]dx - \int_0^{h^e} f(x)\,u\,dx \qquad (7.138)$$

Proceed in the following manner:

1. Based on the element functional, discuss the requisite continuity requirements for the approximation of the primary dependent variable.

2. Using the ξ natural coordinate system $(-1 \le \xi \le 1)$, derive suitable element interpolation functions for the primary dependent variable. Verify that these functions satisfy the completeness criterion.

3. Substituting the element interpolation functions into the approximate element functional, explicitly derive the element equations. In doing this assume that $T(x)$, $k(x)$ and $f(x)$ are *constant* over a given element.

$$\mathbf{K}^{(e)}\hat{\boldsymbol{\phi}}_n^{(e)} = \mathbf{f}^{(e)} \qquad (7.139)$$

7.2

Consider the quadratic one-dimensional element shown in Figure 7.16. Given the functional associated with the approximate solution

$$\hat{F} = \int_{x_1}^{x_3}\frac{1}{2}k\left(\frac{d\hat{T}}{dx}\right)^2 dx - \int_{x_1}^{x_3} S\hat{T}\,dx$$

where $\hat{T}$ is the approximate temperature, k is the thermal conductivity (to be assumed constant within an element), and S is the heat source term (likewise to be assumed constant within an element). Using the following

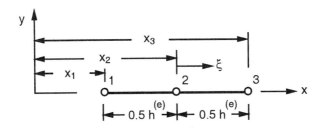

Figure 7.16: Quadratic Heat Conduction Line Element

interpolation functions written in terms of natural coordinates (see Appendix C):

$$N_1 = \frac{\xi}{2}(\xi - 1) \quad ; \quad N_2 = 1 - \xi^2 \quad ; \quad N_3 = \frac{\xi}{2}(\xi + 1)$$

in conjunction with the above expression for $\hat{F}$ (suitably modified to apply to ξ-space), use a variational approach to determine the entries in: (a) the element conductivity matrix $\mathbf{K}^{(e)}$, and (b) the element "loading" vector $\mathbf{f}^{(e)}$.

Express your answers in terms of k, $h^{(e)}$ and S. To save yourself time and computational effort, minimize the functional and then perform the integrations (the use of symbolic algebra software is strongly encouraged). Also, whenever possible, make use of symmetry.

7.3

Although it is typically assumed constant within an element, the cross-sectional area can easily be varied. Consider, for example, the standard linear two-node element used in one-dimensional stress-deformation analyses (Figure 7.17).

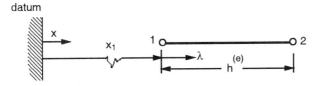

Figure 7.17: Linear Line Element for Stress-Deformation Analyses

From Appendix C we recall the λ natural coordinate system, defined by

$$\lambda = \frac{x - x_1}{h^{(e)}} \quad ; \quad 0 \le \lambda \le 1$$

If the cross-sectional area varies linearly with λ as

$$A = (1 - \lambda)A_1 + \lambda A_2$$

where A_1 and A_2 represent the values of the cross-sectional area at nodes 1 and 2, respectively, using a *variational based approach*, derive the stiffness matrix $\mathbf{K}^{(e)}$ for this element. What check do you have on your result?

7.4

Consider the quadratic three-node element shown in Figure 7.16, only applied to stress-deformation analyses. (a) First assume a *linear* distribution of surface traction T_x described by

$$T_x = 0.5(1 - \xi)T_{x1} + \xi T_{x3}$$

where T_{x1} and T_{x3} represent the constant values of T_x at nodes 1 and 3, respectively. Derive the contribution to the element load vector $\mathbf{f}^{(e)}$ due to T_x.

(b) Next assume the following *quadratic* distribution of surface traction:

$$T_x = N_1 T_{x1} + N_2 T_{x2} + N_3 T_{x3}$$

where T_{x1}, T_{x2} and T_{x3} represent the constant values of T_x at nodes 1, 2 and 3, respectively. Repeat the derivation of part (a) above.

7.5

Generalize the approximate element equations derived for the uniaxial elastostatic problem to include the effects of initial strains ε^0. That is, strains that are independent of stressing (e.g., thermal strains). Carry out your work in the following manner:

1) Noting the elastic strain is equal to

$$\varepsilon_{x'} = \frac{d\delta}{dx'} - \varepsilon^0$$

suitably modify the equations of Example 7.7.

2) Then, using the interpolation functions determined in Example 7.7, derive the explicit expressions for the approximate element equations.

Chapter 8

Steps in Performing Finite Element Analyses

8.1 Introductory Remarks

In the previous chapter the basic steps followed in developing finite element equations for a specific physical problem were described. Attention is now turned to performing actual finite element analyses that employ the equations developed in the previous chapter.[1] This involves the following tasks:

1. Discretization of the domain

2. Assembly of element equations to form the global equations

3. Application of nodal specifications

4. Solution of the global equations

5. Calculation of the secondary dependent variables

6. Postprocessing of the results

7. Interpretation of the results

Each of these tasks is now discussed in detail. The chapter culminates with some simple examples that illustrate these tasks.

---

[1] In actual finite element analyses the steps described herein are, of course, performed on a digital computer. Further details regarding programming aspects associated with the finite element method are given in Chapter 13.

8.2   Discretization of the Domain

The goal of the first task is to discretize the domain Ω as accurately as possible. This is realized by breaking up Ω into a finite number of non-overlapping *elements*. This process is commonly called *meshing*; the collection of elements is referred to as a *mesh*. A high-quality mesh is requisite to convergent and accurate finite element solutions.

8.2.1   Domain Discretization Error

As noted in Chapter 6, the meshing of two- and three-dimensional domains has the potential of introducing *domain discretization error* into the analysis. This geometric error arises when the elements along the boundary do not exactly fit the geometry of the actual body.

This is schematically shown in Figure 8.1, where a half-circle is approximated by two, four and eight straight-sided triangular elements. It is evident that the as the number of elements increases, the domain discretization error decreases.

If the boundary of a general solution domain is polygonal (i.e., all sides are straight), then any type of element can be used to match it exactly. Similarly, if the boundary is defined by higher-order polynomials (quadratic, cubic, etc.), then corresponding higher-order elements can likewise match the boundary exactly. Further details pertaining to the use of such elements to approximate boundaries are given in Chapter 10.

Domain discretization errors arise when non-polynomial boundaries are approximated by polynomial elements. Depending on the types of elements used and on the placement of the associated nodes, this error manifests itself in small portions of the domain not being covered by an element (Figure 8.1), or extra portions of the mesh "spilling over" the boundaries of the domain. In either case, since the outer boundary of the mesh defines the extent of the approximate solution domain, this domain differs from the actual one and some amount of additional error is introduced into the solution.

The subject of domain discretization error is particularly well summarized by Burnett [100], who notes that the maximum radial error between a circular arc (the most common non-polynomial curve) and the side of a quadratic element with nodes lying on the arc has been shown to be about 1 percent for a 90 degree arc and less than 0.1 percent for a 45 degree arc [222]. If a linear element is used, the corresponding values become 29 and 7.6 percent, respectively. Thus, for any three consecutive nodes lying along the arc, a significant reduction in error is realized by fitting a single quadratic element, rather than two linear ones.

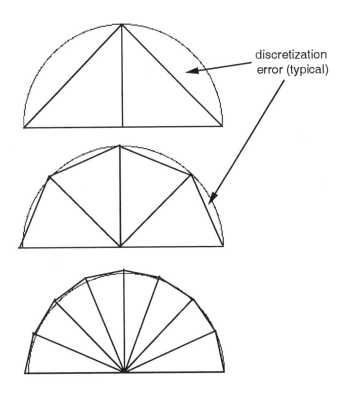

Figure 8.1: Schematic Illustration of Domain Discretization Error

When analyzing second-order problems, the use of straight-sided elements to approximate curved boundaries sometimes results in convergence to an incorrect solution in the vicinity of the boundary; for fourth-order problems, the solution may not converge at all [34, 454, 35, 515, 608, 610]. According to this phenomenon, typically referred to as the *Babuška paradox* [34, 35], within a few elements of the boundary, the normal components of flux may have exceptionally high error. This has been demonstrated both theoretically [515, 610] and numerically [454, 608], though other numerical experiments [100] have failed to produce such an error. Finally, on the upside, curved-sided isoparametric elements (i.e., quadratics or higher order) do not exhibit these convergence problems, and are thus recommended whenever a solution is sought in the vicinity of a curved boundary [298, 100].

8.2.2   Common Element Types

In theory, a variety of element shapes could be used in a given mesh. However, in practice only a *few* different shapes are typically employed. Some examples of commonly used one-, two- and three-dimensional elements are given below. The treatment is not meant to be exhaustive, but rather introductory. A more detailed discussion of these elements, including methods for determining their interpolation functions, is deferred until Chapter 9.

Remark

1. The designations "one-dimensional", "two-dimensional" and "three-dimensional" refer only to the spatial extent of the element domain; they have nothing to do with the number of nodes per element, unknowns per node, etc.

One-dimensional elements

In the most general case, a one-dimensional element domain is a line in space. The discretization of one-dimensional domains is quite straightforward and introduces *no domain discretization error.* Element interfaces are vertex nodes. As evident from Examples 7.6 and 7.7, one-dimensional (line) elements are quite simple to construct. A sampling of such elements is shown in Figure 8.2.

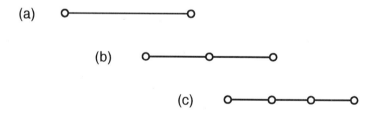

Figure 8.2: Family of One-Dimensional Elements : (a) Linear (simplex), (b) Quadratic, (c) Cubic Members.

When using one-dimensional elements in conjunction with two- or three-dimensional elements, care must be exercised to use elements that are compatible with the polynomial approximation along the element edges where they are applied (recall Figure 7.11).

Two-dimensional elements

Two-dimensional elements are used to discretize surfaces. Element interfaces consist of edges.

The simplest polygonal shape that will span a two-dimensional domain is the *triangle*. The first three members of the triangular family of elements are shown in Figure 8.3. Large assemblies of triangular elements can represent any arbitrary two-dimensional domain, with minimal discrepancies occurring along curved boundaries.

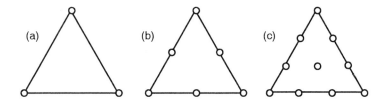

Figure 8.3: Triangular Family of Elements: (a) Linear (simplex), (b) Quadratic, (c) Cubic Members.

Quadrilaterals constitute the second type of element commonly used to discretize surfaces. The first three members of the so-called "serendipity" family of elements are shown in Figure 8.4.

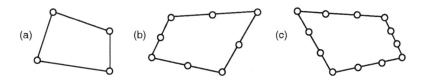

Figure 8.4: "Serendipity" Family of Quadrilateral Elements : (a) Linear, (b) Quadratic, (c) Cubic Members.

Three-dimensional elements

Three-dimensional elements are described in a manner analogous to two-dimensional ones, only with greater complexity. Element interfaces now consist of element faces. The simplest polygonal shape that will span a three-dimensional domain is the *tetrahedron*. The first two members of the tetrahedral family of elements are shown in Figure 8.5.

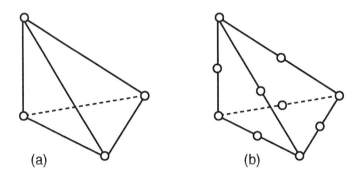

(a)   (b)

Figure 8.5: Tetrahedral Family of Elements : (a) Linear (simplex), (b) Quadratic Members.

Hexahedra constitute the second type of element commonly used to discretize volumes. The first three members of the "serendipity" family of hexahedral elements are shown in Figure 8.6.

A third element type used to discretize volumes are *pentahedra*. The first two members of the "serendipity" family of pentahedral elements are shown in Figure 8.7. Pentahedral elements serve as "fillers" between hexahedral elements.

8.2.3   Element Characteristics

Every element in a mesh is characterized by the following information:

- Its *shape* (e.g., a line, triangle, quadrilateral, tetrahedron, hexahedron, etc.).

- The number of nodes in the element and the global node numbers assigned to these nodes.

- The *type* of nodes present in the element. Nodes are either *interior* or *exterior* to the element (Figure 8.8). Interior nodes are those nodes not shared by neighboring elements. Exterior nodes are either corner nodes located at the vertices, or are those located along an element edge.

- The type of nodal degrees of freedom (recall step 1 in Chapter 7).

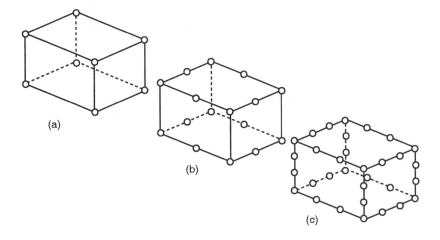

Figure 8.6: "Serendipity" Family of Hexahedral Elements : (a) Linear, (b) Quadratic, (c) Cubic Members.

- The order of interpolation functions used to describe: a) the element geometry and, b) the approximation of the nodal values of the primary dependent variables.[2]

- Parameters describing the material constitution of the continuum being discretized by the element (recall step 2 in Chapter 7).

8.2.4 Placement of Elements

The accuracy of finite element analyses is largely dependent upon the type of element used and upon the distribution of these elements. In discretizing a domain, the distribution and location of elements is based largely on experience of the analyst. Usually one starts with a simple mesh and subsequently refines the mesh to a finer one. None the less, certain guidelines concerning the placement of elements exist, namely:

1. Element boundaries should be placed along all exterior and interior boundaries of the body being discretized, in a manner that minimizes the domain discretization error.

---

[2]Further details pertaining to the description of the element geometry and element interpolation functions are given in Chapters 10 and 9, respectively.

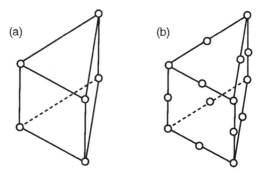

Figure 8.7: "Serendipity" Family of Pentahedral Elements : (a) Linear, (b) Quadratic Members.

2. Since material and geometric parameters are typically assumed to be constant within an element, element interfaces must coincide with material interfaces.

3. Nodes should be placed at any location in Ω at which a value of the primary dependent variable(s) is desired.

4. Elements should not be overly distorted.[3]

Even in simple meshes, it is common practice to place the bulk of the elements in locations where the gradient of the primary dependent variables would be expected (or is known) to be high. For example, in stress-deformation problems a large number of elements are placed in the vicinity of high stress concentrations. These can be caused by geometric discontinuities (Figure 8.9), or by loads applied over small portions of the continuum (e.g., applied concentrated forces or regions of high shear stress).

8.2.5   Element Shapes

It is well known that elements perform best if their shapes are "compact" and "regular." [116] Elements tend to "stiffen" and lose accuracy as their aspect ratios increase, as their vertex angles become markedly different from one another, as sides become curved, or if edge nodes (Figure 8.8) are non-uniformly spaced. In Section 10.6 we shall have more to say about acceptable (and unacceptable) element shapes.

---

[3] A quantification of element distortion and the associated consequences, are discussed in Chapter 10.

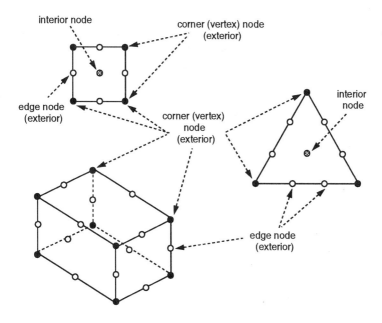

Figure 8.8: Node Types.

Different elements have different sensitivities to shape distortion. Consequently, few general guidelines apply to all elements. None the less, for each element:

- In quadrilateral elements, the aspect ratio should be kept near unity.

- The vertex angles in quadrilateral elements should be near 90 degrees. In triangular elements these angles should be near 60 degrees.

- The edges in higher-order elements should be straight.

- Edge nodes should be uniformly distributed.

These guidelines are not without exceptions. A detailed discussion of these exceptions is, however, postponed until Chapter 13.

Fortunately, poorly shaped elements may produce only *locally* poor results [116]. Provided that the surrounding elements have acceptable shapes, any spurious gradients attributed to the poorly shaped elements tend to die out rather than propagate.

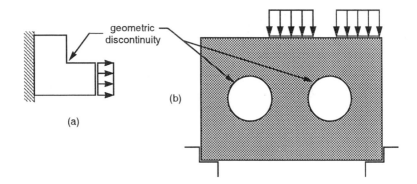

Figure 8.9: Examples of Geometric Discontinuities: (a) "Corner" Problem, (b) Bridge Cross-Section with Circular Holes

8.2.6 Mesh Generation

It is timely to note that the generation of high quality meshes is a very active area of research. Driving this research is the realization that possibly the most important task to automate in the entire finite element simulation process is the generation of high quality meshes.

Mesh generation is complicated by the realization that a mesh must satisfy almost contradictory requirements: it must conform to the shape of the domain Ω; it must be composed of elements that are the "right" size and shape, avoiding excessive distortion (e.g., overly small or overly large interior angles); and, over relatively short distances it will typically have to grade from large elements (coarse discretization) to small elements (fine discretization).

Numerous algorithms for the construction of two-, 2.5-[4] and three-dimensional meshes have been developed. The choice of the specific algorithm is strongly related to the geometric characteristics of the solution domain under consideration. We defer further discussion of mesh generation until Chapter 13.

8.2.7 Meshless Methods

The difficulties associated with generating high quality meshes, particularly in three dimensions, have fueled a growing interest in so-called *meshless* or *element free* methods. The essential feature underlying such methods is

---

[4]2.5-dimensional meshes refer to surfaces in 3-space.

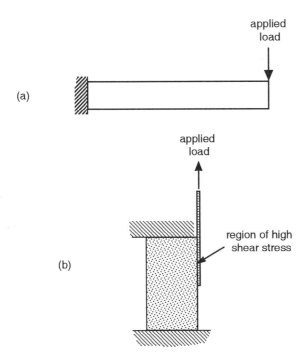

Figure 8.10: Examples of Stress Concentrations: (a) In Vicinity of Applied Load, (b) Along Material Interface

the complete description of the discrete model by nodes that describe the domain Ω. The discrete equations are formulated so that the interaction between the nodes is transparent to the analyst [64].

To date, a number of meshless methods have been proposed to date. These include the smooth particle hydrodynamics (SPH) method [400, 453, 87], the diffuse element method (DEM) [405], the reproducing kernal particle method (RKPM) [320, 321, 322, 323], the element free Galerkin (EFG) method [68, 332], the moving least-squares reproducing kernal method (MLSRK) [313, 324], the partition of unity finite element method (PUFEM) [385, 47], the h-p clouds method [142], the reproducing kernal hierarchical partition of unity method [314, 315], the finite point method [420], the local boundary integral equation (LBIE) method [582], the meshless local Petrov-Galerkin (MLPG) method [31], and the method of finite spheres (MFS) [134].

Meshless methods in which requisite interpolations and integrations can be performed without the need of a "background" mesh are commonly referred to as being "truly meshless." The finite point method, the MLPG method, and the MFS are examples of truly meshless methods. In contrast, SPH, the RKPM, DEM and EFG method are considered "pseudo-meshless" since they use a background mesh for numerical integration and sometimes even for imposing the essential boundary conditions. Further details pertaining to "pseudo-meshless" techniques can be found in the survey article by Belytschko et al. [69]. A more general overview of meshless methods is available in Chapter 16 of [605].

Although considerable progress has been made in the development of meshless methods, most currently available techniques are still computationally less efficient that the standard finite element method. The primary reason for this is the complicated (non-polynomial) interpolation functions employed and the associated numerical integration, which is very difficult to perform efficiently [134]. None the less, meshless methods are proving to be efficient for specific problem areas such as fracture (e.g., progressive crack growth), damage mechanics and localization, contact problems with moving interfaces, and multi-resolution analyses.

8.2.8   Computer Implementation Issues

Discretization of the solution domain requires the following three sets of arrays to store the requisite data:

1. First are floating point arrays containing the global coordinates of all nodes in the mesh. For example, the following arrays are used in the APES computer program [274] to store the nodal coordinates:

    `x_coords, y_coords, z_coords`

    The size of each of these arrays is *max_nodes*, which equals the maximum number of nodes contained in the mesh. Since contiguous node numbering is assumed, *max_nodes* also equals the largest node number in the mesh.

2. Second is a connectivity array that stores the global node numbers associated with each element in the mesh. Although two-dimensional arrays are often proposed for this task [252], this approach becomes somewhat inefficient when elements possessing different numbers of nodes are used in the same computer program. A more efficient approach is to use a one-dimensional array with a second "pointer" array. For example, in the APES computer program [274] the following two integer arrays are used:

```
nodes(max_elm_nodes), node_loc(max_elements+1)
```

Here max_elements is the total number of elements contained in the mesh, and max_elm_nodes is computed in the following manner:

$$\text{max\_elm\_nodes} = 3 * \text{max\_T3} + 4 * \text{max\_Q4} + \cdots \qquad (8.1)$$

where *max_T3* is the maximum number of 3-node triangles in the mesh, *max_Q4* is the maximum number of 4-node quadrilaterals in the mesh, etc. For each element in the mesh, the global node numbers are assigned to the *nodes* array in the following manner:

```
DO m=1,nen
   nodes(node_loc(num_elements) + m - 1) = nodes_read(m)
END DO

node_loc(num_elements+1) = node_loc(num_elements) + nen
```

where *nen* is the total number of nodes in the element, and *nodes_read* is a temporary array containing the global node numbers read from a line of input data for the element.

3. Finally is an array that contains a link to the material idealization that corresponds to each element. In the simplest case, this would be an integer array containing the material number for each element. For example

```
mat_num (max_elements)
```

For each material number, an appropriate array would then contain the parameters associated with the particular material idealization being used. It is timely to note that material information does not necessarily have to be referenced by integers. This can also be accomplished using descriptive quoted string [409], a practice that equally applies to the description of groups of nodes, elements and nodal specifications.

8.3   Assembly of Element Equations

The outcome of Chapter 7 was a set of element equations having the following general form:

$$\mathbf{K}^{(e)} \hat{\phi}_n^{(e)} = \mathbf{f}^{(e)} \qquad (8.2)$$

where $\mathbf{K}^{(e)}$ is the element coefficient matrix, $\hat{\phi}_n^{(e)}$ is the vector of nodal unknowns associated with the element, and $\mathbf{f}^{(e)}$ is the vector of nodal "forces" associated with the element. The discrete element equations are general in nature and apply to all elements of a given type used in mathematically modeling a particular physical problem.

Once the equations are determined for an element, they must next be suitably incorporated into the "global" equations applicable to the entire domain. This so-called *assembly process* is performed for all the elements in the mathematical model. It essentially consists of mapping element (local) degrees of freedom to global degrees of freedom.

The assembly process is based on the premise of *compatibility* or *continuity*. That is, since in accordance with the compatibility criterion, the element interpolation functions must exhibit C^0 continuity across element interfaces. The values of primary dependent variables, and possibly their derivatives, associated with two adjacent elements must be the same at node points along the common interface.

The result of the assembly process is a set of *global* equations

$$\mathbf{K} \hat{\phi}_n = \mathbf{f} \qquad (8.3)$$

where $\mathbf{K}$ is the global property matrix, $\hat{\phi}_n$ is the vector of all nodal unknowns associated with the mathematical model, and $\mathbf{f}$ is the vector of all nodal "forcing" parameters associated with the mathematical model. The assembly process is best illustrated through the following examples.

Example 8.1: Assembly Procedure Applied to Elements with Scalar Unknowns

Consider the arrangement of two-node, one-dimensional (line) elements shown in Figure 8.11(a). The element numbers are circled; the arrows indicate the order in which the local node numbers have been specified in defining the respective elements.

Assuming scalar nodal degrees of freedom, and in the absence of concentrated nodal fluxes, the element equations are given by

$$\frac{k^{(e)}}{h^{(e)}} \begin{bmatrix} 1 & -1 \\ -1 & 1 \end{bmatrix} \begin{Bmatrix} \hat{\phi}_1^{(e)} \\ \hat{\phi}_2^{(e)} \end{Bmatrix} = \frac{S^{(e)} h^{(e)}}{2} \tag{8.4}$$

where $k^{(e)}$, $h^{(e)}$ and $S^{(e)}$ represent the material parameter, the length and the "source" term associated with a typical element e possessing two nodes. The quantities $\hat{\phi}_1^{(e)}$ and $\hat{\phi}_2^{(e)}$ are nodal element unknowns. The numbers 1 and 2 appearing as subscripts denote *local* degrees of freedom.

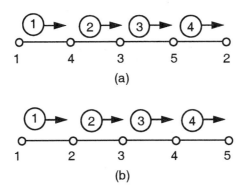

(a)

(b)

Figure 8.11: Mesh of One-Dimensional Elements: (a) Arbitrary Numbering, (b) Contiguous Numbering

The key to the assembly process is the mapping of these local degrees of freedom to the global degrees of freedom. This is shown schematically in Figure 8.12 for the case of element 1. The mapping for this element is simply:

local 1 → global 1   ;   local 2 → global 4

It follows that the element contribution

$$K_{11}^{(1)} = \frac{k^{(1)}}{h^{(1)}} \tag{8.5}$$

must be *summed* into the $(1,1)$ location in the global $\mathbf{K}$ matrix. Likewise,

$$K_{12}^{(1)} = -\frac{k^{(1)}}{h^{(1)}} \rightarrow K_{14} \tag{8.6}$$

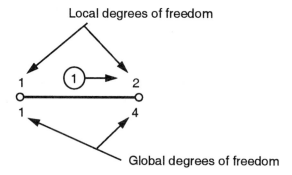

Figure 8.12: Local and Global Node Numbers for Element 1

$$K_{21}^{(1)} = -\frac{k^{(1)}}{h^{(1)}} \to K_{41} \tag{8.7}$$

$$K_{22}^{(1)} = \frac{k^{(1)}}{h^{(1)}} \to K_{44} \tag{8.8}$$

The contribution of the element "forcing" vector $\mathbf{f}^{(1)}$ is assembled in a similar manner, viz.,

$$f_{1}^{(1)} = \frac{S^{(1)}h^{(1)}}{2} \to f_{1} \tag{8.9}$$

$$f_{2}^{(1)} = \frac{S^{(1)}h^{(1)}}{2} \to f_{4} \tag{8.10}$$

Proceeding in a similar manner leads to the following global $\mathbf{K}$ matrix:

$$\mathbf{K} = \begin{bmatrix} K_{11}^{(1)} & 0 & 0 & K_{12}^{(1)} & 0 \\ 0 & K_{22}^{(4)} & 0 & 0 & K_{21}^{(4)} \\ 0 & 0 & K_{22}^{(2)} + K_{11}^{(3)} & K_{21}^{(2)} & K_{12}^{(3)} \\ K_{21}^{(1)} & 0 & K_{12}^{(2)} & K_{22}^{(1)} + K_{11}^{(2)} & 0 \\ 0 & K_{12}^{(4)} & K_{21}^{(3)} & 0 & K_{22}^{(3)} + K_{11}^{(4)} \end{bmatrix} \tag{8.11}$$

Note that as expected, global degree of freedom 4 has contributions from elements 1 and 2; global degree of freedom 3 has contributions from elements 2 and 3; and global degree of freedom 5 has contributions from elements 3 and 4.

The associated global right hand side vector is

$$
\mathbf{f} = \left\{ \begin{array}{c} f_1^{(1)} \\ f_2^{(4)} \\ f_2^{(2)} + f_1^{(3)} \\ f_2^{(1)} + f_1^{(2)} \\ f_2^{(3)} + f_1^{(4)} \end{array} \right\} \tag{8.12}
$$

The global $\mathbf{K}$ matrix is seen to be *sparse* and *banded*. If the $\mathbf{K}^{(e)}$ are symmetric, then $\mathbf{K}$ will likewise be symmetric. In this case the bandedness is quantified by the *half-bandwidth* n_{bw} of the matrix, where

$$
n_{bw} = (dif_{max} + 1) * node_{dof} \tag{8.13}
$$

where dif_{max} equals the maximum difference (over all the elements in the mesh) between any two node numbers associated with the *same* element, and $node_{dof}$ is the number of unknowns per node ($node_{dof} = 1$ for scalar analyses). For this mesh it easily verified that $dif_{max} = 3$ and thus $n_{bw} = 4$.

If the mesh is next renumbered in the manner shown in Figure 8.11(b), the assembly process leads to the following global arrays:

$$
\mathbf{K} = \begin{bmatrix} K_{11}^{(1)} & K_{12}^{(1)} & 0 & 0 & 0 \\ K_{21}^{(1)} & K_{22}^{(1)} + K_{11}^{(2)} & K_{12}^{(2)} & 0 & 0 \\ 0 & K_{21}^{(2)} & K_{22}^{(2)} + K_{11}^{(3)} & K_{12}^{(3)} & 0 \\ 0 & 0 & K_{21}^{(3)} & K_{22}^{(3)} + K_{11}^{(4)} & K_{12}^{(4)} \\ 0 & 0 & 0 & K_{21}^{(4)} & K_{22}^{(4)} \end{bmatrix} \tag{8.14}
$$

$$
\mathbf{f} = \left\{ \begin{array}{c} f_1^{(1)} \\ f_2^{(1)} + f_1^{(2)} \\ f_2^{(2)} + f_1^{(3)} \\ f_2^{(3)} + f_1^{(4)} \\ f_2^{(4)} \end{array} \right\} \tag{8.15}
$$

For this mesh $dif_{max} = 1$ and $n_{bw} = 2$.

Example 8.2: Assembly Procedure Applied to Elements with Vector Unknowns

We next investigate the assembly procedure as applied to elements possessing vector nodal degrees of freedom. In particular, we consider the assembly of linear, three-node triangular elements in conjunction with the mesh shown in Figure 8.13.

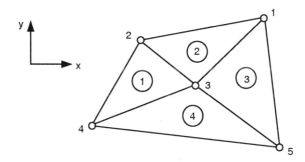

Figure 8.13: Hypothetical Mesh of Linear Triangular Elements

Each node m $(m = 1, 2, 3)$ in a typical element e has two degrees of freedom associated with it, namely a displacement component $\hat{u}_m^{(e)}$ parallel to the global x-direction and a displacement component $\hat{v}_m^{(e)}$ parallel to the global y-direction. Each element thus has a total of *six* degrees of freedom; these are shown in Figure 8.14.

The equations for a typical element have the following general form:

$$
\begin{bmatrix}
K_{11}^{(e)} & K_{12}^{(e)} & K_{13}^{(e)} & K_{14}^{(e)} & K_{15}^{(e)} & K_{16}^{(e)} \\
K_{12}^{(e)} & K_{22}^{(e)} & K_{23}^{(e)} & K_{24}^{(e)} & K_{25}^{(e)} & K_{26}^{(e)} \\
K_{13}^{(e)} & K_{23}^{(e)} & K_{33}^{(e)} & K_{34}^{(e)} & K_{35}^{(e)} & K_{36}^{(e)} \\
K_{14}^{(e)} & K_{24}^{(e)} & K_{34}^{(e)} & K_{44}^{(e)} & K_{45}^{(e)} & K_{46}^{(e)} \\
K_{15}^{(e)} & K_{25}^{(e)} & K_{35}^{(e)} & K_{45}^{(e)} & K_{55}^{(e)} & K_{56}^{(e)} \\
K_{16}^{(e)} & K_{26}^{(e)} & K_{36}^{(e)} & K_{46}^{(e)} & K_{56}^{(e)} & K_{66}^{(e)}
\end{bmatrix}
\begin{Bmatrix}
\hat{u}_1^{(e)} \\
\hat{v}_1^{(e)} \\
\hat{u}_2^{(e)} \\
\hat{v}_2^{(e)} \\
\hat{u}_3^{(e)} \\
\hat{v}_3^{(e)}
\end{Bmatrix}
=
\begin{Bmatrix}
f_1^{(e)} \\
g_1^{(e)} \\
f_2^{(e)} \\
g_2^{(e)} \\
f_3^{(e)} \\
g_3^{(e)}
\end{Bmatrix}
\tag{8.16}
$$

where the element stiffness matrix is symmetric. The quantities $f_m^{(e)}$ and $g_m^{(e)}$ are the x and y components of forces at node m.

The global vector of nodal unknowns corresponding to this problem will thus contain the following entries

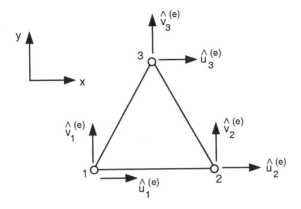

Figure 8.14: Typical Linear Triangular Element

$$\hat{\phi}_n = \left\{ \hat{u}_1 \quad \hat{v}_1 \quad \cdots \quad \hat{u}_5 \quad \hat{v}_5 \right\}^T \tag{8.17}$$

where the numbers appearing as subscripts are now global node numbers. In general, for two-dimensional elements with nodal displacements as unknowns, degrees of freedom $2i - 1$ and $2i$ are present at global node i. For three-dimensional elements the corresponding degrees of freedom are $3i - 2$, $3i - 1$ and $3i$.

Consider element 1. In describing the element connectivity, the global node numbers are specified in the order: 2 - 4 - 3. The mapping for this element is thus

$$\begin{array}{ll} \text{local } 1 \to \text{global } 3 & ; \quad \text{local } 2 \to \text{global } 4 \\ \text{local } 3 \to \text{global } 7 & ; \quad \text{local } 4 \to \text{global } 8 \\ \text{local } 5 \to \text{global } 5 & ; \quad \text{local } 6 \to \text{global } 6 \end{array}$$

The contributions of element 1 to the global stiffness matrix $\mathbf{K}$ are thus:

$$K_{11}^{(1)} \to K_{33} \quad ; \quad K_{12}^{(1)} \to K_{34} \quad ; \quad K_{13}^{(1)} \to K_{37}$$

$$K_{14}^{(1)} \to K_{38} \quad ; \quad K_{15}^{(1)} \to K_{35} \quad ; \quad K_{16}^{(1)} \to K_{36}$$

$$K_{22}^{(1)} \to K_{44} \quad ; \quad K_{23}^{(1)} \to K_{47} \quad ; \quad K_{24}^{(1)} \to K_{48}$$

$$K_{25}^{(1)} \to K_{45} \quad ; \quad K_{26}^{(1)} \to K_{46}$$

$$K_{33}^{(1)} \to K_{77} \quad ; \quad K_{34}^{(1)} \to K_{78} \quad ; \quad K_{35}^{(1)} \to K_{75} \quad ; \quad K_{36}^{(1)} \to K_{76}$$

$$K_{44}^{(1)} \to K_{88} \quad ; \quad K_{45}^{(1)} \to K_{85} \quad ; \quad K_{46}^{(1)} \to K_{86}$$

$$K_{55}^{(1)} \to K_{55} \quad ; \quad K_{56}^{(1)} \to K_{56} \quad ; \quad K_{66}^{(1)} \to K_{66}$$

The assembly of the remaining elements is left as an exercise.

8.3.1   Insight Into Node and Element Numbering

From the results of the previous examples, we draw the following conclusions:

1. The assembly of element matrices $\mathbf{K}^{(e)}$, which themselves are full, produces a global $\mathbf{K}$ matrix that is *banded* and *sparse*. If the $\mathbf{K}^{(e)}$ are *symmetric*, then $\mathbf{K}$ will likewise be symmetric.

2. The manner in which the nodes in a mesh are numbered globally has a direct impact on the bandwidth and sparseness of $\mathbf{K}$.

8.3.2   Errors Associated with Formation and Assembly of Element Arrays

The primary source of error in the formation and assembly of element arrays is the chopping and rounding of numbers associated with their finite precision representation.[5] [59] Also, if the element arrays are not formed in a local or natural coordinate system, then subtraction of large numbers of similar magnitude will cause loss of significant digits. This type of error is sometimes referred to as *cancellation* error [376], and is commonly lumped under the general heading of truncation error.

As mentioned in Appendix F, the numerical quadrature schemes used in generating the element arrays (e.g., Newton-Cotes, Gauss-Legendre, Lobatto, etc.) compute discrete approximations to the continuous spatial

---

[5] Recall the discussion of Chapter 2.

integration. As such, a *numerical integration* or *quadrature* error is introduced into this portion of the analysis. In discussions of errors, quadrature error is sometimes lumped under the heading of truncation error [512].

To this is added (in some manner) roundoff error. Investigations of quadrature error indicate that this error can be expected to grow approximately in proportion to the inverse of the number n of quadrature points in a given direction [133]. The roundoff error, on the other hand, grows at worst as the first power of n. Thus as n increases, roundoff error does likewise while discretization error decreases. With regard to roundoff error, quadrature formulas employing both positive and negative weights are less desirable [133].

In general, if, in the computation of element arrays, efficient routines are employed, roundoff error should be minor. The use of interpolation functions avoids numerical inversion of non-orthogonal transformation matrices that would otherwise be required (and would be a further source of roundoff error) [376].

8.4 Nodal Specifications

Since it has not been modified to account for constraints imposed at boundary and possibly interior nodes, the system property matrix $\mathbf{K}$ is singular at this stage of the analysis. In order to specialize the problem, the nodal constraints must next be considered.

In finite element analyses, constraints on the *nodal* values of the primary dependent variables or their derivatives are required in order to account for, or approximate, the actual constraints imposed on the body being analyzed. It is only through such constraints that a problem can be uniquely posed. Since the values of the primary dependent variables, or their derivatives, can be specified at *any node* in the mesh, the previous notion of "boundary conditions" must be generalized to *nodal specifications*.

Along the total boundary Γ of a body, at nodes along element boundaries, or at any node within an element, two types of nodal specifications are typically possible[6]:

- Known values of the primary dependent variables are specified along the portion of Γ denoted by Γ_1. These are referred to as essential or Dirichlet specifications.

---

[6]Convection boundaries in heat conduction analyses and concentrated springs represent examples of a third (mixed) type of nodal specification. Although such specifications are easily accounted for in finite element analyses, for brevity their discussion is deferred to later chapters.

- Known values of the gradients of the primary dependent variables are specified along the portion of Γ denoted by Γ_2. These are referred to as natural or Neumann specifications.

With regard to a *specific* primary dependent variable at a *specific* node in a body, *either* an essential constraint or a natural constraint, but not both can be specified. Stated mathematically, $\Gamma_1 \cup \Gamma_2 = \Gamma$ and $\Gamma_1 \cap \Gamma_2 = \emptyset$.

For the subsequent discussion, assume element and system coefficient matrices that are symmetric. As such, in the process of imposing nodal specifications, we do not want to destroy this symmetry. For imposing nodal specifications, two general approaches are available: *elimination* and *penalty*. Each of these approaches can be applied at either the global or element level. These approaches are now each discussed in some detail.

8.4.1   Elimination Approach at the Global Level

Consider a hypothetical body possessing three degrees of freedom. The global equations are thus

$$
\begin{bmatrix} K_{11} & K_{12} & K_{13} \\ K_{21} & K_{22} & K_{23} \\ K_{31} & K_{32} & K_{33} \end{bmatrix} \begin{Bmatrix} \hat{\phi}_1 \\ \hat{\phi}_2 \\ \hat{\phi}_3 \end{Bmatrix} = \begin{Bmatrix} f_1 \\ f_2 \\ f_3 \end{Bmatrix} \tag{8.18}
$$

where symmetry of $\mathbf{K}$ is assumed.

If a known nodal "load" (for problems in heat conduction the "load" would be the heat flux; in fluid flow problems it would be the velocity of flow; etc.) is specified in degree of freedom m, then this "load" is simply *summed* into the mth row of the global right-hand side vector $\mathbf{f}$.

If, on the other hand, a known value of the primary dependent variable (e.g., displacement, temperature, hydraulic head, etc.) is specified at a node, the following procedure is used. For the present discussion, assume that $\hat{\phi}_2 = \bar{\phi}$ (i.e., $m = 2$), where $\bar{\phi}$ is a known value of the primary dependent variable. Referring to equation (8.18), we subtract $\bar{\phi}$ times the second column of $\mathbf{K}$ from both sides of the equation, to give

$$
\begin{bmatrix} K_{11} & 0 & K_{13} \\ K_{21} & 0 & K_{23} \\ K_{31} & 0 & K_{33} \end{bmatrix} \begin{Bmatrix} \hat{\phi}_1 \\ \hat{\phi}_2 \\ \hat{\phi}_3 \end{Bmatrix} = \begin{Bmatrix} f_1 - K_{12}\bar{\phi} \\ f_2 - K_{22}\bar{\phi} \\ f_3 - K_{32}\bar{\phi} \end{Bmatrix} \tag{8.19}
$$

Since the second (mth) equation is not necessary for determining $\hat{\phi}_1$ and $\hat{\phi}_2$, we restore symmetry by zeroing the second row

$$\begin{bmatrix} K_{11} & 0 & K_{13} \\ 0 & 0 & 0 \\ K_{31} & 0 & K_{33} \end{bmatrix} \begin{Bmatrix} \hat{\phi}_1 \\ \hat{\phi}_2 \\ \hat{\phi}_3 \end{Bmatrix} = \begin{Bmatrix} f_1 - K_{12}\bar{\phi} \\ 0 \\ f_3 - K_{32}\bar{\phi} \end{Bmatrix} \tag{8.20}$$

Finally, we force the second (mth) equation to return the trivial result of $\hat{\phi}_2 = \bar{\phi}$. This involves setting the mth diagonal entry in $\mathbf{K}$ equal to unity (1.0) and the mth right hand side entry equal to $\bar{\phi}$. The resulting equations are thus

$$\begin{bmatrix} K_{11} & 0 & K_{13} \\ 0 & 1.0 & 0 \\ K_{31} & 0 & K_{33} \end{bmatrix} \begin{Bmatrix} \hat{\phi}_1 \\ \hat{\phi}_2 \\ \hat{\phi}_3 \end{Bmatrix} = \begin{Bmatrix} f_1 - K_{12}\bar{\phi} \\ \bar{\phi} \\ f_3 - K_{32}\bar{\phi} \end{Bmatrix} \tag{8.21}$$

When this set of equations is solved by an actual algorithm, the desired results shall obviously be obtained. To complete the discussion, a segment of "pseudo code" for the elimination approach at the system level is provided.

```
do i=1,ndof
   if (f_bar is specified in degree of freedom i)  then
      f(i) = f(i) + f_bar
   elseif (phi_bar is specified in degree of freedom i)  then
      for (j=1,2, . . . ,i-1,i+1, . . . ,ndof)
         f(j) = f(j) - K(j,i)*phi_bar
      zero out row i
      zero out column i
      set f(i) = phi_bar
      set K(i,i) = 1.0
   endif
end do
```

Here *ndof* is the total number of degrees of freedom in the global equations.

8.4.2 Elimination Approach at the Element Level

In this approach the nodal specifications are applied to the element arrays *immediately prior* to assembly. Trivial equations associated with degrees of freedom in which primary dependent variables are specified are suitably "marked" — in the solution of the system equations these trivial equations are skipped during the reduction process. The following "pseudo code" segment, which assumes one degree of freedom per node, better illustrates this approach.

```
do i=1,nen
    if(specification is associated with node i)  then
        if(f_bar is specified in d.o.f. i)  then
            if(node "i" not considered in other element) then
                f_e(i) = f_e(i) + f_bar
                flag node ''i'' as having been considered
            endif
        elseif(phi_bar is specified in d.o.f. i)  then
            if(node i not considered in other element) then
            set f_e(i) = phi_bar
            flag node i as having been considered
        else
            set f_e(i) = 0.0
        endif
            for (j=1,2, . . , i-1, i+1, . . , nen)
                f_e(j) = f_e(j) - k_e(ji)*phi_bar
            zero row i in [ k_e ]
            zero column i in [ k_e ]
            set k_e(i,i) = 0.0
        endif
    endif
end do
```

Here *nen* is the total number of nodes in the element.

8.4.3   Penalty Approach at the Global Level

Consider the same three degree of freedom system corresponding to equation (8.18).

If a known nodal "load" is specified in degree of freedom m, then this "load" is simply *summed* into the mth row of the system load vector **f**.

If a known value of the primary dependent variable is specified at a node, the following procedure is used. Once again we assume that $\hat{\phi}_2 = \bar{\phi}$ (i.e., $m = 2$), where $\bar{\phi}$ is a known value of the primary dependent variable. Referring to equation (8.18), we replace K_{22} by the product GK_{22}, where

G is a large positive penalty number.[7] We next replace f_2 by the product $GK_{22}\bar{\phi}$. The resulting system equations are thus

$$\begin{bmatrix} K_{11} & K_{12} & K_{13} \\ K_{21} & GK_{22} & K_{23} \\ K_{31} & K_{32} & K_{33} \end{bmatrix} \begin{Bmatrix} \hat{\phi}_1 \\ \hat{\phi}_2 \\ \hat{\phi}_3 \end{Bmatrix} = \begin{Bmatrix} f_1 \\ GK_{22}\bar{\phi} \\ f_3 \end{Bmatrix} \qquad (8.22)$$

This has the same effect as placing a "spring" of stiffness GK_{22} in degree of freedom 2. To better understand the above "penalty" approach, consider the second (mth) equation from the above system (note that the remaining equations, and the symmetry of the system equations, are unaffected)

$$K_{21}\hat{\phi}_1 + GK_{22}\hat{\phi}_2 + K_{23}\hat{\phi}_3 = GK_{22}\bar{\phi} \qquad (8.23)$$

Since the magnitude of G has been chosen to be large, and since the elements of $\mathbf{K}$ are typically of the similar order, it follows that the term GK_{22} will dominate the left hand side of the equation (8.23), effectively reducing it to

$$\hat{\phi}_1 + G\hat{\phi}_2 + \hat{\phi}_3 \approx G\bar{\phi} \qquad (8.24)$$

which implies that $G\hat{\phi}_2 \approx G\bar{\phi} \quad \Rightarrow \quad \hat{\phi}_2 \approx \bar{\phi}$.

Although the nodal constraint is enforced in an approximate manner, the results of the analysis will be *identical* to those obtained using the "elimination" procedure described above. This is attributed to the fact that finite precision arithmetic is used in digital computers. It is important to again emphasize that the remaining equations are unaltered, as is the symmetry of $\mathbf{K}$.

To complete the present discussion, a "pseudo code" segment for the "penalty" approach at the system level is provided.

```
do i=1,ndof
    if (f_bar is specified in degree of freedom i)  then
        f(i) = f(i) + f_bar
    elseif (phi_bar is specified in degree of freedom i)  then
```

---

[7]Concerning penalty numbers, Hughes [252] states that: "In our experience with words of length 60-64 bits, we have found that the range $10^7 \le G \le 10^9$ is effective." In the Fortran 90 computer program APES [274], a penalty number of 10^{20} has been found to be effective in conjunction with double precision floating point numbers.

```
      set K(i,i) = G*K(i,i)
      set f(i) = G*K(i,i)*phi_bar
   endif
end do
```

8.4.4   Penalty Approach at the Element Level

The "penalty" approach can likewise be implemented at the element level.
The requisite logic is actually simpler than in the case of the "elimination"
approach. The following segment of Fortran 90 code, taken from the APES
computer program [274], better illustrates the penalty approach at the
element level:

```
                         !   get down to business
DO n=1,ndof
! .............................................. !
!       concentrated nodal force is summed in    !
! .............................................. !
   IF(iflag(n) == 0)  THEN
      IF(.NOT. used)  erhs(nrow(n))  &
                      = erhs(nrow(n)) + value(n)

! .............................. !
!  nodal displacement is specified  !
! .............................. ... !
   ELSEIF(iflag(n) == 1)  THEN
      IF(used)  THEN
         erhs(nrow(n)) = zero
         IF(reform)  estiff(nrow(n),nrow(n)) = zero
      ELSE
         temp = estiff(nrow(n),nrow(n))*penalty
         erhs(nrow(n)) = (value(n) - soloc(n))*temp
         IF(reform)  estiff(nrow(n),nrow(n)) = temp
      ENDIF
   ELSE

! .............................................. !
!   concentrated boundary "spring" is specified   !
! .............................................. !
      IF(.NOT. used .AND. reform)  &
```

```
        estiff(nrow(n),nrow(n))  &
            = estiff(nrow(n),nrow(n)) + value(n)
  ENDIF
  END DO
```

Here *ndof* degrees of freedom per element are assumed, and the logic is used for each node in each element.

Insight Into the Penalty Approach

Consider the simple elastostatic bar loaded by a concentrated force P in the manner shown in Figure 8.15. The bar is discretized by a single element. As such, the system equations, which are identical to the element equations, are

$$\frac{EA}{h} \begin{bmatrix} 1 & -1 \\ -1 & 1 \end{bmatrix} \begin{Bmatrix} \hat{v}_1 \\ \hat{v}_2 \end{Bmatrix} = \begin{Bmatrix} f_1 \\ f_2 \end{Bmatrix} \tag{8.25}$$

where the superscript (e) has, for brevity, been omitted from E, A and h.

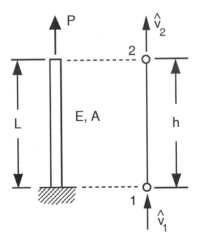

Figure 8.15: Hypothetical Single Element Example

Using the penalty approach to enforce the essential specification at node 1, equation (8.25) becomes

$$\frac{EA}{h}\begin{bmatrix} G & -1 \\ -1 & 1 \end{bmatrix}\begin{Bmatrix} \hat{v}_1 \\ \hat{v}_2 \end{Bmatrix} = \begin{Bmatrix} \frac{EA}{h}G(1) * (0) \\ P \end{Bmatrix} \tag{8.26}$$

where G is the penalty number. From the first row of equation (8.26):

$$\frac{EA}{h}(G\hat{v}_1 - \hat{v}_2) = 0 \quad \Rightarrow \quad \hat{v}_2 = G\hat{v}_1 \tag{8.27}$$

From the second row of equation (8.26):

$$\frac{EA}{h}(-\hat{v}_1 + \hat{v}_2) = P \tag{8.28}$$

Substituting equation (8.27) for $\hat{v}_2$ gives

$$\hat{v}_1 = \frac{1}{(G-1)}\frac{Ph}{EA} \tag{8.29}$$

and thus

$$\hat{v}_2 = G\hat{v}_1 = \frac{G}{(G-1)}\frac{Ph}{EA} \tag{8.30}$$

The next question to ask is what would the value of $\hat{v}_2$ be if $\hat{v}_1$ was *identically* zero, say as a result of the "elimination" procedure for enforcing nodal specifications. In such a case, the modified global equations would be

$$\begin{bmatrix} 1 & 0 \\ 0 & EA/h \end{bmatrix}\begin{Bmatrix} \hat{v}_1 \\ \hat{v}_2 \end{Bmatrix} = \begin{Bmatrix} 0 \\ P \end{Bmatrix} \tag{8.31}$$

implying that

$$\hat{v}_1 = 0 \quad \text{and} \quad \hat{v}_2 = \frac{Ph}{EA} \tag{8.32}$$

The difference between equations (8.30) and (8.32) is seen to be the factor $G/(G-1)$. Since G is a very large number, it follows that this factor is very close to unity. This factor also shows that G can never equal one — returning to equation (8.25) we see that for $G = 1$ the system matrix would be singular.

8.4.5 Computer Implementation Issues

The computer implementation of nodal specifications requires the following information:

1. An integer array that is used to indicate whether a constraint has been specified at a given node. For the elimination and penalty approaches applied at the *element* level, the same array can be used to "mark" nodes that have been considered in previous elements. In this manner, the same specification is not applied to a given node more than once.

2. An integer array or arrays that store the *type* of nodal specification (i.e., essential or natural) that is made at a given node. For scalar field problems (e.g. heat transfer), only a single array is typically required. Vector field problems (e.g., elastostatics) require two or more arrays.

3. A floating point array or arrays that store the *value* of the quantity being specified. For scalar field problems, only a single array is typically required. Vector field problems require two or more arrays.

Since it requires no modification to the element or global equations, the default specifications at a given node are typically a specified value of the secondary dependent variable equal to zero.

Example 8.3: Nodal Specifications for One-Dimensional Elastostatics

Consider the strut modeled shown in Figure 8.16. The domain is discretized using linear (two-node) elements such as those developed in Example 7.7. Nodal displacements are thus the primary dependent variables. For the entire strut, the elastic modulus is equal to E. The quantities A_1 and A_2 are cross-sectional areas. The body force (self-weight) of the strut is equal to b_y, and acts in the negative y-coordinate direction.

Assembling the element arrays gives the following global equations:

$$
E \begin{bmatrix} \dfrac{A_2}{r} & -\dfrac{A_2}{r} & 0 & 0 \\[2mm] -\dfrac{A_2}{r} & \dfrac{A_2}{r}+\dfrac{A_1}{s} & -\dfrac{A_1}{s} & 0 \\[2mm] 0 & -\dfrac{A_1}{s} & \dfrac{A_1}{s}+\dfrac{A_1}{t} & -\dfrac{A_1}{t} \\[2mm] 0 & 0 & -\dfrac{A_1}{t} & \dfrac{A_1}{t} \end{bmatrix} \begin{Bmatrix} \hat{v}_1 \\ \hat{v}_2 \\ \hat{v}_3 \\ \hat{v}_4 \end{Bmatrix}
$$

$$
= -\frac{b_y}{2} \begin{Bmatrix} A_2 r \\ A_2 r + A_1 s \\ A_1 s + A_1 t \\ A_1 t \end{Bmatrix} \tag{8.33}
$$

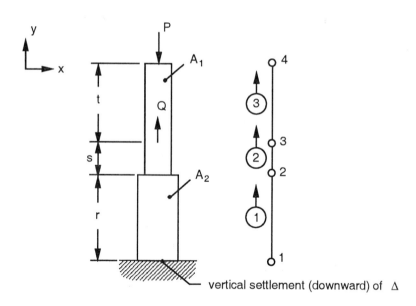

vertical settlement (downward) of Δ

Figure 8.16: Simple Mesh Consisting of One-Dimensional Bar Elements

Using an "elimination" approach on the global level, the above equations are modified to account for the nodal specifications shown. Note that because the known settlement Δ and the applied force P act in the negative y-coordinate direction they must have a negative sign associated with them.

$$
\begin{bmatrix}
1 & 0 & 0 & 0 \\
0 & \dfrac{A_2 E}{r} + \dfrac{A_1 E}{s} & -\dfrac{A_1 E}{s} & 0 \\
0 & -\dfrac{A_1 E}{s} & \dfrac{A_1 E}{s} + \dfrac{A_1 E}{t} & -\dfrac{A_1 E}{t} \\
0 & 0 & -\dfrac{A_1 E}{t} & \dfrac{A_1 E}{t}
\end{bmatrix}
\begin{Bmatrix}
\hat{v}_1 \\ \hat{v}_2 \\ \hat{v}_3 \\ \hat{v}_4
\end{Bmatrix}
$$

$$
= \begin{Bmatrix}
-\Delta \\[4pt]
-\dfrac{b_y}{2}(A_2 r + A_1 s) + \dfrac{A_2 E}{r}\Delta \\[4pt]
-\dfrac{b_y}{2}(A_1 s + A_1 t) + Q \\[4pt]
-\dfrac{b_y}{2}(A_1 t) - P
\end{Bmatrix}
\tag{8.34}
$$

The above exercise is next repeated, only using the "penalty" approach on the global level. The resulting system equations are

$$
E \begin{bmatrix}
\dfrac{A_2}{r}G & -\dfrac{A_2}{r} & 0 & 0 \\
-\dfrac{A_2}{r} & \dfrac{A_2}{r} + \dfrac{A_1}{s} & -\dfrac{A_1}{s} & 0 \\
0 & -\dfrac{A_1}{s} & \dfrac{A_1}{s} + \dfrac{A_1}{t} & -\dfrac{A_1}{t} \\
0 & 0 & -\dfrac{A_1}{t} & \dfrac{A_1}{t}
\end{bmatrix}
\begin{Bmatrix}
\hat{v}_1 \\ \hat{v}_2 \\ \hat{v}_3 \\ \hat{v}_4
\end{Bmatrix}
$$

$$
= \begin{Bmatrix}
\dfrac{A_2 E}{r}G(-\Delta) \\[4pt]
-\dfrac{b_y}{2}(A_2 r + A_1 s) \\[4pt]
-\dfrac{b_y}{2}(A_1 s + A_1 t) + Q \\[4pt]
-\dfrac{b_y}{2}(A_1 t) - P
\end{Bmatrix}
\tag{8.35}
$$

where G represents a large penalty number. Note that in the coefficient matrix only the $(1, 1)$ entry has been modified.

8.5   Solution of Global Equations

The matrix *global* equations, generated from the element equations through the assembly process, have the following general form:

$$\mathbf{K}\hat{\phi}_n = \mathbf{f} \tag{8.36}$$

where the global property matrix $\mathbf{K}$ is a square matrix with known coefficients, $\hat{\phi}_n$ is the vector of all nodal unknowns associated with the mathematical model, and $\mathbf{f}$ is the vector of all nodal "forcing" parameters associated with the mathematical model.

Through the nodal specifications, equation (8.36) has been specialized for a particular problem. In order to obtain the nodal values (approximations) of the primary dependent variables $\hat{\phi}_n$, theses equations must next be solved. The subsequent discussion is restricted to *linear* systems of equations that are associated with *steady-state* analyses.

In theory, equation (8.36) can be solved using *any* numerical technique suitable for simultaneous systems of linear equations. In practice, however, the solution of the global equations represents a significant portion of the overall computational effort associated with finite element analyses. Consequently, an equation solver must be selected judiciously.

Equation solvers best suited for typical finite element analyses are those that take advantage of the *symmetry* (if applicable), *sparseness* and *bandedness* of $\mathbf{K}$. In addition, the solvers should be robust, general in scope (e.g., possibly handle multiple right hand side vectors), efficient, automated, scalable (on computers with multiple processors), and should give predictable results. Further details pertaining to equation solvers are given in Appendix E.

8.5.1   Mesh Renumbering Schemes

One of the important observations emanating from the discussion of the assembly process was the dependence of the bandwidth and sparseness of $\mathbf{K}$ on the manner in which the nodes in the mesh were numbered globally. Having briefly discussed direct and iterative equation solvers, it is evident that, in the interest of efficiency, storage and computational effort associated with the solution of the global equations should be *minimized*.

One way in which to achieve this is to number the nodes (or elements) in as optimal a manner as possible. To this end, many algorithms have been developed that automatically renumber a mesh in order to maximize efficiency. These algorithms are *solver-specific*. Thus no single strategy is best for all objectives or all programs. For example, in the case of banded solvers, the algorithms renumber the mesh to minimize the bandwidth (or, for symmetric systems, the half-bandwidth) of **K**. For profile solvers, the nodes are renumbered in a manner that reduces the profile (and may leave the bandwidth unchanged). Finally, for frontal solvers, the wave front is minimized by renumbering the elements.

The algorithms for nodal and element renumbering both require a mesh with initial node and element numbers assigned. This starting mesh is typically created using a generation scheme such as those alluded to above in Section 8.2.6. Renumbering algorithms do not guarantee an optimum numbering, or even an improvement over the original numbering, though such an improvement is usually realized.

In order not to detract from the goals of this chapter, we defer further discussion of mesh renumbering schemes to Chapter 13. We conclude with the following basic guidelines for numbering a mesh so as to minimize the bandwidth of **K** [100]:

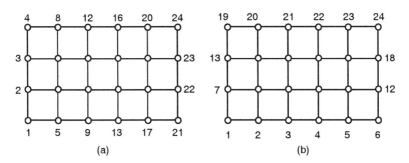

Figure 8.17: Two-Dimensional Mesh Numbered Across (a) Shortest and (b) Longest Direction

1. Nodes should be numbered along paths that contain the least number of nodes *from one boundary to another*. The numbering should proceed from one element to an adjacent element; that is, without skipping over elements.

2. For a mesh in which the spacing between nodes is *comparable* in all directions, the nodes should be numbered across the *shortest* dimension of the domain. The node numbering scheme shown in Figure 8.17(a) is thus preferred over that shown in Figure 8.17(b).

3. If the internodal spacings are *not* comparable, then numbering across the shortest dimension might give a larger bandwidth. It is thus the

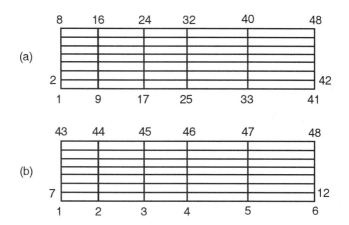

Figure 8.18: Two-Dimensional Mesh Numbered Across (a) Shortest and (b) Longest Direction

number of degrees of freedom from one side of a domain to the other that should be the determining factor, not the dimension of Ω. The node numbering scheme shown in Figure 8.18(b) is thus preferred over that shown in Figure 8.18(a).

8.6   Secondary Dependent Variables

Typically the nodal values of the primary dependent variables are used to calculate values of the secondary dependent variables. This rather straightforward procedure is illustrated in the following examples.

Example 8.4: Stress-Deformation Analyses

In stress-deformation finite element analyses involving irreducible formulations, the primary dependent variables are components of displacement, and the secondary dependent variables are strains and stresses.

For the case of the one-dimensional elastostatic element developed in Example 7.7, the primary dependent variable is a single component of displacement $\hat{\delta}^{(e)}$. For the two-node element developed, this displacement is approximated by

$$\hat{\delta}^{(e)} = N_1 \hat{\delta}_1^{(e)} + N_2 \hat{\delta}_2^{(e)} = 0.5(1 - \xi)\hat{\delta}_1^{(e)} + 0.5(1 + \xi)\hat{\delta}_2^{(e)} \tag{8.37}$$

The approximate strains are thus

$$\hat{\varepsilon}_{x'}^{(e)} = \frac{d\hat{\delta}^{(e)}}{dx'} = \frac{d\hat{\delta}^{(e)}}{d\xi}\frac{d\xi}{dx'} = \frac{2}{h^{(e)}}\frac{d\hat{\delta}^{(e)}}{d\xi} = \frac{2}{h^{(e)}}\left[-0.5\,\hat{\delta}_1^{(e)} + 0.5\,\hat{\delta}_2^{(e)}\right]$$

$$= \frac{1}{h^{(e)}}\begin{bmatrix} -1 & 1 \end{bmatrix}\begin{Bmatrix} \hat{\delta}_1^{(e)} \\ \hat{\delta}_2^{(e)} \end{Bmatrix} = \mathbf{B}\hat{\delta}_n^{(e)} \tag{8.38}$$

where infinitesimal displacements and displacement gradients have been assumed. The approximate stress are related to the strain through Hooke's law, viz.,

$$\hat{\sigma}_{x'}^{(e)} = E\hat{\varepsilon}_{x'}^{(e)} \tag{8.39}$$

This constitutive relation is written in the more general form as

$$\hat{\sigma}_{x'}^{(e)} = \mathbf{CB}\hat{\delta}_n^{(e)} \tag{8.40}$$

where for one-dimensional response $\mathbf{C} = E$, the elastic modulus. For two- and three-dimensional elastostatic analyses, the form of the equations for computing secondary quantities is exactly the same as for the one-dimensional case; only the number of stress and strain components and thus the size of the arrays is increased. Further details pertaining to the computation of secondary dependent variables associated with multi-axial stress-deformation analyses are given in Chapter 12.

Example 8.5: Heat Conduction Analyses

In heat conduction finite element analyses involving irreducible formulations, the primary dependent variable is the temperature T. The secondary dependent variables are components of the heat conduction vector. For the case of the two-dimensional heat conduction element developed in Example 7.8, the secondary dependent variables are related to the temperature through Fourier's Law, viz.,

$$\begin{Bmatrix} q_x \\ q_y \end{Bmatrix} = -\begin{bmatrix} k_{11} & k_{12} \\ k_{12} & k_{22} \end{bmatrix}\begin{Bmatrix} \partial T/\partial x \\ \partial T/\partial y \end{Bmatrix} \tag{8.41}$$

where q_x and q_y represent components of the heat flux (per unit area) vector, and k_{11}, k_{12} and k_{22} are thermal conductivity coefficients. The general approximations for temperature and its derivatives are given by equations (7.82) to (7.84), viz.,

$$\hat{T}^{(e)}(x,y) = \sum_{m=1}^{nen} N_m(x,y)\,\hat{T}_m^{(e)} \tag{8.42}$$

$$\frac{\partial \hat{T}^{(e)}}{\partial x} = \sum_{m=1}^{nen} \frac{\partial N_m}{\partial x}\,\hat{T}_m^{(e)} \tag{8.43}$$

$$\frac{\partial \hat{T}^{(e)}}{\partial y} = \sum_{m=1}^{nen} \frac{\partial N_m}{\partial y}\,\hat{T}_m^{(e)} \tag{8.44}$$

where nen is the number of nodes in an element. For the specific case of a linear, three-node triangular element, approximate derivatives become

$$\frac{\partial \hat{T}^{(e)}}{\partial x} = \sum_{m=1}^{3} \frac{\partial N_m}{\partial x}\,\hat{T}_m^{(e)} = \frac{1}{2A^{(e)}} \sum_{m=1}^{3} b_m\hat{T}_m^{(e)} \tag{8.45}$$

$$\frac{\partial \hat{T}^{(e)}}{\partial y} = \sum_{m=1}^{3} \frac{\partial N_m}{\partial y}\,\hat{T}_m^{(e)} = \frac{1}{2A^{(e)}} \sum_{m=1}^{3} c_m\hat{T}_m^{(e)} \tag{8.46}$$

where the b_i and c_i are defined in equation (7.89) of Example 7.8. The components of the approximate heat flux vector are thus computed from

$$\left\{ \begin{matrix} \hat{q}_x^{(e)} \\ \hat{q}_y^{(e)} \end{matrix} \right\} = - \begin{bmatrix} k_{11} & k_{12} \\ k_{12} & k_{22} \end{bmatrix} \frac{1}{A^{(e)}} \begin{bmatrix} b_1 & b_2 & b_3 \\ c_1 & c_2 & c_3 \end{bmatrix} \left\{ \begin{matrix} \hat{T}_1^{(e)} \\ \hat{T}_2^{(e)} \\ \hat{T}_3^{(e)} \end{matrix} \right\} = -\mathbf{CB}\hat{\mathbf{T}}_n^{(e)} \tag{8.47}$$

This equation has the same general form as equation (8.40) of the previous example. This is true for a large number of physical problems.

Remark

1. The primary source of error associated with the calculation of secondary dependent variables is *cancellation* error [376]. One remedy for such errors is the use of double precision floating point arithmetic for all calculations. Other approaches to minimize error involve the averaging of secondary dependent variables among neighboring elements, and the mapping of these variables to the nodes, or both.

8.7 Postprocessing of Results

This task involves the presentation of data generated in the finite element analysis in a form easily understood by the analyst. Typically this entails one or more of the following methods of postprocessing:

1. On-screen computer graphics.

2. Plotting.

3. Hardcopy computer printout.

Computer graphics are used to view the finite element mesh, contour plots of the primary and secondary dependent variables and so on. Because of the typically voluminous nature of the latter method of presenting results, hardcopy computer printouts are usually the least desirable (and typically least readable) form of presenting results.

Examples of some the above postprocessing methods are given in select example problems throughout the remainder of this text.

8.8 Interpretation of Results

This final, and sometimes hardest step involves a check of accuracy and possible refinement of the solution.

8.8.1 Validity of Finite Element Analyses

The finite element method is a numerical (approximate) method that is used to analyze a mathematical model which itself is an approximation to some physical problem. As such, in using any finite element program the following questions must be answered:

1. Is the mathematical model of the physical problem correct?

2. Is the mesh used fine enough?

3. Are there any errors in the program itself? The specific problem may not have been considered previously using the particular program — "bugs" may thus have been discovered.

4. Is roundoff error a problem? Even if the coding and the theory are correct, as noted in Chapter 2 roundoff errors can lead to incorrect results. A potential way to check on the effect of roundoff error is to re-analyze a particular problem using double the precision.

8.8.2   Refinement of the Finite Element Solution

Ever since their introduction in conjunction with piecewise continuous approximate solutions in Chapter 6, interpolation functions that exhibit monotonic convergence to the exact solution have been sought. Such functions ensure that the approximate solution converges to the correct one as the finite element solution is refined.

Traditionally, the accuracy of the finite element solution has been assessed by *mesh refinement*. Rigorous mathematical proofs of convergence assume that the process of mesh refinement occurs in a regular fashion defined by the following conditions:

1. The elements must be made smaller in such a way that every point of the solution domain can always be within an element (regardless of how small the element may be).

2. All previous meshes must be contained in the refined mesh.

3. The form of the interpolation functions must remain unchanged during the process of mesh refinement.

When elements with straight boundaries are used to model solution domains with curved boundaries, the first two conditions are not satisfied and rigorous mathematical proofs of convergence may not be obtainable. Thus, if within engineering accuracy, the result of the finite element analysis is the same after sub-division as before, and if the mathematical model has been verified for correctness, the solution is more likely to be a good one. Indeed, during the early years of development of the finite element method, discretization errors were controlled by uniform or nearly uniform mesh refinement.

Practically speaking, in all but the simplest problems it is uneconomical to refine the *entire* mesh and we must instead consider the broader issue of *refinement of the finite element solution,* a special case of which is mesh refinement. From their rather heuristic beginnings, solution refinement techniques have evolved significantly.

As noted in Section 6.6, a rigorous mathematical treatment of the finite element method began in the mid-1960s. As an outgrowth of this effort, error estimation techniques were extensively investigated in the 1970s. During this period, *adaptive* mesh refinement procedures designed to reduce the discretization errors with improved efficiency received a great deal of attention [39, 40]. The general idea behind adaptive procedures is to analyze an initial mesh, estimate the resulting errors, modify the mesh in regions where the error exceeds some preset tolerance, and repeat the analysis.

Various procedures exist for refining finite element solutions. These are commonly placed into three general categories: *h-* , *p-* and *hp*-refinement.

h-Refinement

Discretization errors can be reduced by judiciously changing the size, number or location of the elements in the mesh. This approach is commonly referred to as the *h-version* of solution refinement. The symbol h represents the size of the finite elements. Convergence occurs when the size of the largest element ($h_{max}^{(e)}$) is progressively reduced (hence the name *h-version*).

The *h*-refinement category is commonly divided into the following subclasses of methods [605]:

1. The first subclass involves a complete mesh *regeneration* or *remeshing*. Based on the results of an analysis, new element sizes are predicted for the entire domain and a totally new mesh is generated. Consequently, a refinement (addition of nodes) and derefinement (removal of nodes) are simultaneously allowed. Adaptive mesh generation is a natural choice for problems in which element distortions become excessive.

2. The second subclass involves *element subdivision* or *enrichment*. Here, elements with excessive error are divided into smaller ones, keeping the original element boundaries intact. Mesh derefinement is possible, but its implementation involves somewhat complex data structures that may reduce the efficiency of the approach [581, 605].

3. The final subclass keeps the total number of nodes the same, but adjusts their position to obtain an optimal approximation while maintaining a boundary conforming mesh [18, 455]. This technique is typically referred to as the *r-version* of refinement. Although the element connectivity does not change, the mesh will become somewhat distorted. Practically speaking, *r*-refinement is limited to rather small mesh distortions and thus does not enjoy widespread popularity [605].

p-Refinement

A solution can also be modified by using the original element layout but increasing the order of the interpolating polynomial used in the approximation. The polynomial degree of elements is usually denoted by the symbol p. Convergence occurs when the lowest polynomial degree (p_{min}) is progressively increased. This technique is thus commonly referred to as the *p-version* of refinement.

The p-version grew in popularity in the 1970s, as numerical experiments indicated that discretization errors could be reduced by increasing the degree of the polynomial interpolation functions [524]. The theoretical basis of the p-version was established by Babuška et al. [41]. Since then, significant progress has been made in the area, including error estimate theorems, programming implementation and engineering applications [44].

The p-refinement category is commonly divided into the following subclasses of methods [605]: 1) Those in which the interpolating polynomial degree is increased uniformly throughout the whole domain, and 2) Those in which the interpolating polynomial degree is increased locally using hierarchical refinement.

Since the accuracy of finite element analyses with a particular mesh and p-distribution cannot be known a priori, a posteriori error estimators have been developed in order to estimate the error in the energy norm and to determine the contribution of each element to this error [599, 5, 45].

Armed with such error estimators, the p-version of the finite element method lends itself particularly well to an *adaptive* solution strategy [605]. The basic idea behind such a strategy is as follows: The analyst selects an acceptable level of error for the finite element analysis (say 5 percent). The solution domain is discretized in a particular manner and an analysis performed. Using suitable posteriori error estimators, the error associated with each element is computed and compared to the acceptable level selected by the analyst. For those elements in which the error exceeds this level, the order of the interpolating polynomial is increased and the analysis repeated. This process continues until all the error in all elements is less that the level set by the analyst. The advantages of a p-version adaptive solution are [597]: 1) A faster rate of convergence than the h-version, 2) The possibility of using mixed-order interpolation, 3) The utilization of previous solutions and computations, 4) An accurate starting value for an iterative solution procedure, and 5) Improved equation conditioning.

Remarks

1. Since it uses smooth functions to generate approximate solutions, the p-version of the finite element method is similar to the Rayleigh-Ritz

method discussed in Chapter 5. The difference between the two methods is that in the *p*-version the solution domain is divided into elements with polynomial approximations that are piecewise smooth only over individual subdomains. Conversely, in the Rayleigh-Ritz method, the entire solution domain is approximated by smooth functions.

2. Practical application of *p*-refinement requires the use of so-called *hierarchical elements* [596]. The degree of interpolating polynomial associated with such elements is increased by adding higher-order terms that do not modify the lower-order ones. To achieve this, the *hierarchic interpolation functions* cannot all be associated with nodal unknowns. This is counter to the "standard" interpolation functions used in the previous chapters, and discussed in detail in Chapter 9. Consequently, hierarchical elements incorporate some of the advantages of the more classical continuous approximation techniques such as the method of weighted residuals and the stationary functional method. A detailed discussion of hierarchical elements and their associated interpolation functions is beyond the scope of this introductory text; for further details, the interested reader is referred to [370, 605].

3. The *p*-version was developed primarily for solid mechanics applications; in heat transfer and fluid flow applications the *p*-version is considered to be impractical [218]. A closely related development, primarily for applications in fluid mechanics, is the *spectral element method* [372, 44].

hp-Refinement

A third general approach to improving the accuracy of a finite element solution combines the *h*- and *p*-versions of refinement. This approach is usually referred to as the *hp*-version of refinement. The first paper describing *hp*-refinement appeared in 1981 [42] and has been followed by a rather substantial amount of work in this area [208, 209, 43, 600, 136, 419, 450, 46].

The *hp*-version of the finite element method is often the most optimum approach for obtaining a given level of accuracy. It is also the most complex to implement numerically, as the architecture of the application program differs from the *h*-version. The *hp*-version offers various possibilities that the *h*-version does not offer [44], namely: 1) A posteriori assessment of errors, 2) New ways in which to model plates and shells, and 3) Inherent parallelization.

A more detailed discussion of *hp*-refinement is beyond the scope of this introductory text. The interested reader is directed to the references cited above and to Chapter 15 in [605].

8.9   Illustrative Examples

Example 8.6: One-dimensional Steady State Heat Conduction

Consider a hypothetical rod 28 centimeters long, having a conductivity of $k = 0.10$ cal/$^\circ$C-cm-sec and a cross-sectional area of 1.0 cm^2. Over the middle portion of the rod, a heat source S is uniformly distributed (Figure 8.19).

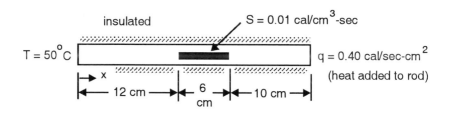

Figure 8.19: Hypothetical Insulated Rod

Since the problem will be solved by hand, the domain (the rod) is discretized in the coarsest possible manner (Figure 8.20). The presence of the heat source over the middle portion of the rod necessitates the use of a three element mesh, as the source is assumed constant over a given element. The primary dependent variable is temperature.

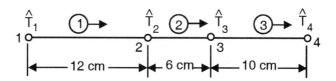

Figure 8.20: Finite Element Mesh for Insulated Rod

The element equations are next generated.

• For Element 1:

$$\frac{k^{(1)}}{h^{(1)}} = \frac{0.10}{12.0} = 0.0083 \quad ; \quad \frac{S^{(1)}h^{(1)}}{2} = 0.0 \tag{8.48}$$

The arrays associated with element 1 are thus

$$0.0083 \begin{bmatrix} 1 & -1 \\ -1 & 1 \end{bmatrix} \begin{Bmatrix} \hat{T}_1^{(1)} \\ \hat{T}_2^{(1)} \end{Bmatrix} = \begin{Bmatrix} 0.0 \\ 0.0 \end{Bmatrix} \tag{8.49}$$

• For Element 2:

$$\frac{k^{(2)}}{h^{(2)}} = \frac{0.10}{6.0} = 0.0167 \quad ; \quad \frac{S^{(2)} h^{(2)}}{2} = \frac{(0.010)(6.0)}{2} = 0.030 \tag{8.50}$$

Thus,

$$0.0167 \begin{bmatrix} 1 & -1 \\ -1 & 1 \end{bmatrix} \begin{Bmatrix} \hat{T}_1^{(2)} \\ \hat{T}_2^{(2)} \end{Bmatrix} = 0.030 \begin{Bmatrix} 1.0 \\ 1.0 \end{Bmatrix} \tag{8.51}$$

• For Element 3:

$$\frac{k^{(3)}}{h^{(3)}} = \frac{0.10}{10.0} = 0.010 \quad ; \quad \frac{S^{(3)} h^{(3)}}{2} = 0.0 \tag{8.52}$$

The arrays associated with element 3 are thus

$$0.010 \begin{bmatrix} 1 & -1 \\ -1 & 1 \end{bmatrix} \begin{Bmatrix} \hat{T}_1^{(3)} \\ \hat{T}_2^{(3)} \end{Bmatrix} = \begin{Bmatrix} 0.0 \\ 0.0 \end{Bmatrix} \tag{8.53}$$

The element arrays are next assembled to give the global arrays. The requisite mapping of local (element) degrees of freedom to global degrees of freedom is as follows:

• For Element 1:

local 1 → global 1 ; local 2 → global 2

• For Element 2:

local 1 → global 2 ; local 2 → global 3

• For Element 3:

local 1 → global 3 ; local 2 → global 4

The global arrays are thus

$$\begin{bmatrix} 0.0083 & -0.0083 & 0.0 & 0.0 \\ -0.0083 & 0.0083 + 0.0167 & -0.0167 & 0.0 \\ 0.0 & -0.0167 & 0.0167 + 0.010 & -0.010 \\ 0.0 & 0.0 & -0.010 & 0.010 \end{bmatrix} \begin{Bmatrix} \hat{T}_1 \\ \hat{T}_2 \\ \hat{T}_3 \\ \hat{T}_4 \end{Bmatrix}$$

$$\begin{Bmatrix} 0.0 \\ 0.0 + 0.030 \\ 0.030 + 0.0 \\ 0.0 \end{Bmatrix} \tag{8.54}$$

Using the elimination approach at the global level, the nodal specifications are next specified. The modified global arrays are thus

$$\begin{bmatrix} 1.0 & 0.0 & 0.0 & 0.0 \\ 0.0 & 0.0083 + 0.0167 & -0.0167 & 0.0 \\ 0.0 & -0.0167 & 0.0167 + 0.010 & -0.010 \\ 0.0 & 0.0 & -0.010 & 0.010 \end{bmatrix} \begin{Bmatrix} \hat{T}_1 \\ \hat{T}_2 \\ \hat{T}_3 \\ \hat{T}_4 \end{Bmatrix}$$

$$\begin{Bmatrix} 50.0 \\ 0.030 - (-0.0083)(50) \\ 0.030 \\ 0.40 \end{Bmatrix} \tag{8.55}$$

Solution of the equations gives the following nodal temperatures:

$$\hat{\mathbf{T}}_\mathbf{n} = \begin{Bmatrix} 50.0 \\ 105.4 \\ 131.2 \\ 171.2 \end{Bmatrix} \, {}^\circ\mathrm{C} \tag{8.56}$$

The values of the secondary dependent variables (the heat fluxes) are next computed. For a typical two-node element (e) such as those used above, with local node numbers i and j, we have

$$\hat{q}^{(e)} = -k^{(e)} \frac{d\hat{T}}{dx} = -k^{(e)} \sum_{m=1}^{2} \hat{T}_m^{(e)} \frac{dN_m}{dx} = -\frac{k^{(e)}}{h^{(e)}} \{-1 \quad 1\} \begin{Bmatrix} \hat{T}_1^{(e)} \\ \hat{T}_2^{(e)} \end{Bmatrix} \tag{8.57}$$

Thus,

- For Element 1:

$$\hat{q}^{(1)} = \frac{0.10}{12.0}(50.0 - 105.4) = -0.462 \text{ cal/s-cm}^2 \qquad (8.58)$$

- For Element 2:

$$\hat{q}^{(2)} = \frac{0.10}{6.0}(105.4 - 131.2) = -0.430 \text{ cal/s-cm}^2 \qquad (8.59)$$

- For Element 3:

$$\hat{q}^{(3)} = \frac{0.10}{10.0}(131.2 - 171.2) = -0.400 \text{ cal/s-cm}^2 \qquad (8.60)$$

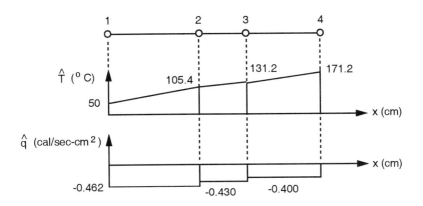

Figure 8.21: Plot of Finite Element Results

When presenting the results graphically, care must be taken to correctly depict the approximation used. For example, in the present case the temperature is assumed to vary *linearly* within an element. The plot of approximate temperature must thus be *linear* within each element. As a consequence of the assumption on temperature, the heat flux (per unit area) is assumed to be *constant* within an element, thus resulting in finite jumps in flux across element boundaries. The corresponding plot must be consistent with this piecewise constant distribution. The plots for both the approximate temperature and heat flux for the insulated rod are shown in Figure 8.21.

Example 8.7: One-dimensional Stress-Deformation Analysis

A hypothetical 4 foot long strut is loaded at two locations, and is rigidly connected at both of its ends in the manner shown in Figure 8.22(a). The cross-sectional area of the strut is constant and equal to 10 square inches. Both materials are idealized as isotropic linear elastic, and the self-weight of the strut is assumed negligible. The strain, stress and force distribution in the strut shall be computed using the SS1D computer program [275].

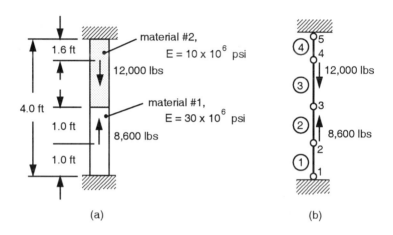

Figure 8.22: Simple Strut Composed of Two Materials (a) Physical Model, (b) Mathematical Model

In creating the mathematical model, refer to the discussion of Section 8.2 concerning the placement of elements. First note that nodes must be placed at either end of the solution domain (i.e., the strut). Also, since concentrated loads are applied at two locations along the strut, nodes must likewise be placed there. Finally, since element interfaces must coincide with material boundaries, a node must also be placed at the middle of the strut. The simplest finite element mesh that can be used to analyze the strut is thus shown in Figure 8.22(b), where units of pounds and inches are used.

Since the self-weight (body force) of the strut has been assumed to be negligible, and since no surface traction and/temperature changes are applied, it follows that no distributed load data needs to be supplied. The nodal specifications at either end of the strut are zero displacement (i.e., a homogeneous, essential boundary condition). The results of the SS1D analysis are shown below.

```
OUTPUT  DATA:
------------
   node          nodal            nodal
  number       coordinate      displacement
 ------        ----------      ------------
     1         0.000E+00        1.300E-15
     2         1.200E+01        1.300E-05
     3         2.400E+01       -3.180E-04
     4         2.880E+01       -7.152E-04
     5         4.800E+01       -7.152E-14


         element
 number   center      strain        stress        force
 ------   ------      ------        ------        -----
    1    6.000E+00   1.083E-06     3.250E+01     3.250E+02
    2    1.800E+01  -2.758E-05    -8.275E+02    -8.275E+03
    3    2.640E+01  -8.275E-05    -8.275E+02    -8.275E+03
    4    3.840E+01   3.725E-05     3.725E+02     3.725E+03
```

Remarks

1. The displacements at nodes 1 and 5 are not identically zero (as specified). This is because a "penalty" approach is used in the SS1D computer program to enforce nodal specifications.

2. The strain, stress and force distributions are constant over an element, and must be correctly depicted if results are presented graphically.

3. The end reactions, which follow from the force distribution, are shown in Figure 8.23. Vertical force equilibrium is verified by inspection.

Instead of performing a "strength of materials" solution, we investigate the accuracy of the finite element solution by uniformly refining the mesh. Consequently, the element sizes are all halved, and the previous meshes is contained in the refined mesh.

The results of the SS1D analysis of the 8 element mesh are shown below. By comparing nodal displacements at common the common locations of 12.0, 24.0 and 28.8 inches, it is evident that results of the refined analysis are identical to those of the first analysis. This likewise holds for the approximate strains, stresses and forces. In light of the fact that that no distributed loads were applied to the strut, these results are indeed reasonable.

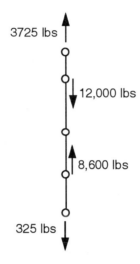

Figure 8.23: Reactions and Applied Loads Acting on Strut

```
OUTPUT   DATA:
------------
     node            nodal              nodal
    number         coordinate         displacement
    ------         ----------         ------------
        1          0.000E+00           6.500E-16
        2          6.000E+00           6.500E-06
        3          1.200E+01           1.300E-05
        4          1.800E+01          -1.525E-04
        5          2.400E+01          -3.180E-04
        6          2.640E+01          -5.166E-04
        7          2.880E+01          -7.152E-04
        8          3.840E+01          -3.576E-04
        9          4.800E+01          -3.576E-14

        element
number   center       strain        stress         force
------   ------       ------        ------         -----
     1  3.000E+00    1.083E-06     3.250E+01      3.250E+02
     2  9.000E+00    1.083E-06     3.250E+01      3.250E+02
     3  1.500E+01   -2.758E-05    -8.275E+02     -8.275E+03
     4  2.100E+01   -2.758E-05    -8.275E+02     -8.275E+03
```

5	2.520E+01	-8.275E-05	-8.275E+02	-8.275E+03
6	2.760E+01	-8.275E-05	-8.275E+02	-8.275E+03
7	3.360E+01	3.725E-05	3.725E+02	3.725E+03
8	4.320E+01	3.725E-05	3.725E+02	3.725E+03

Example 8.8: One-Dimensional Steady-State Heat Conduction Revisited

The insulated rod, first analyzed in Example 3.2 and then subsequently throughout Chapter 4, is re-analyzed using the SS1D program [275]. For completeness, the physical model is shown in Figure 8.24.

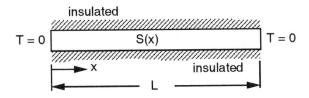

Figure 8.24: Schematic Illustration of an Insulated Rod

For this problem the exact temperature and heat flux per unit area solutions are

$$T = \frac{S_o}{6k} \left(L^2 - x^2 \right) x \qquad (8.61)$$

$$q = \frac{S_o}{6} \left(3x^2 - L^2 \right) \qquad (8.62)$$

As in earlier analyses of this problem, let $S_o = k = L = 1.0$. The development of the present mathematical model is complicated by the fact that in a given element the heat source is assumed *constant*. The actual source term, in the other hand, varies with x. As such, the analyst is faced with the decision of which (constant) heat source value to use for a given element. For the present analysis, the element source term is computed using the x-coordinate associated with the element center. Thus, for element e, $S^{(e)} = \bar{x}^{(e)}$, where $\bar{x}^{(e)}$ represents the x-coordinate of the center of the element.

To better assess the ramifications of this assumption, a mesh refinement study involving 2, 4, and 8 element meshes, was undertaken. In each mesh, the elements were of *equal* size.

The results are graphically summarized in Figures 8.25 and 8.26. It is evident that the approximate temperature distribution is converging to the exact solution. It is also evident that the convergence of the approximate heat flux distribution converges more slowly than the temperature approximation.

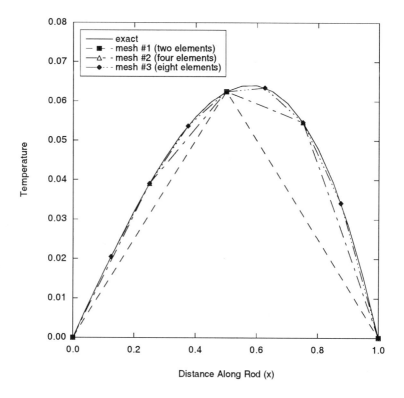

Figure 8.25: Approximate and Exact Temperature Distributions

From Figures 8.26 it is interesting to note that at the midpoint of each element, the approximate flux is equal to the exact value. This result was first observed in the context of linear elastic analyses of bars by Barlow [53], who first noted that points of optimal accuracy existed within such elements. The midpoints of linear elements are thus sometimes referred to as *Barlow stress points*. Further details pertaining to the theory of locating points for optimal accuracy in computing element fluxes are given in Appendix F and in references such as [242, 252, 605].

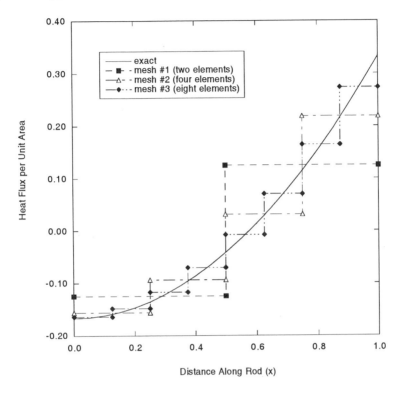

Figure 8.26: Approximate and Exact Heat Flux Distributions

8.10   Concluding Remarks

The basic tasks associated with finite element analyses have been discussed in this chapter. These include the following:

1. Discretization of the domain

2. Assembly of element equations to form the global equations

3. Application of nodal specifications

4. Solution of the global equations

5. Calculation of the secondary dependent variables

6. Postprocessing of the results

7. Interpretation of the results

The next two chapters are devoted to detailed discussions of two important topics related to finite element analyses, namely: (1) the derivation of element interpolation functions, and (2) the mapping of elements to allow a more accurate discretization of the domain.

These are followed by two chapters that describe the application of the finite element method to scalar field problems and ones in linear elastostatics. The final chapter focuses on complete finite element programs and on related issues such as the role of finite element analysis in engineering design, mesh generation, modeling techniques and the associated errors.

8.11 Exercises

8.1
Consider the circular domain of hypothetical one-dimensional elements shown in Figure 8.27(a). Assume one degree of freedom per node. How should the nodes be numbered in order to minimize the half-bandwidth of the system coefficient matrix **K**? What is the half-bandwidth of **K**?

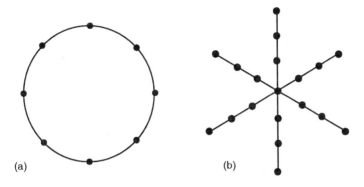

Figure 8.27: Hypothetical (a) Circular Mesh, (b) "Star" Shaped Mesh

8.2
Consider the "star" shaped domain [100] of hypothetical one-dimensional elements shown in Figure 8.27(b). Assume one degree of freedom per node. How should the nodes be numbered in order to minimize the half-bandwidth of the system coefficient matrix **K**? What is the half-bandwidth of **K**?

8.3
The mesh shown in Figure 8.28 is to be used in a combined thermal-mechanical analysis, where the effects of temperature on material strength are modeled. The element numbers associated with the mesh are shown. At each node in the mesh there will be three unknowns; i.e., the temperature and two components of displacement.

1. Assign node numbers to the elements in such a way as to minimize the computational cost associated with solution of the resulting system equations.

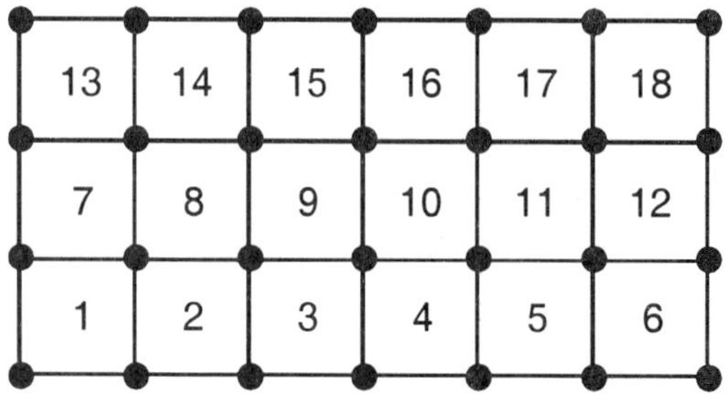

Figure 8.28: Hypothetical Rectangular Mesh

2. Compute the half-bandwidth for the node numbering scheme selected in step 1.

8.4

Consider the insulated rod of unit cross-sectional area shown in Figure 8.29. Using meshes of 4, 8 and 16 elements (not necessarily equally spaced) in conjunction with a computer program containing one-dimensional elements, study the convergence of solutions for the temperature and heat flux distributions. Present your results in the form of two plots: (1) temperature vs. distance along the rod; and, (2) heat flux vs. distance. In both cases, also plot the exact solution. Discuss your findings. Note that this study of convergence involves mesh refinement, not re-meshing.

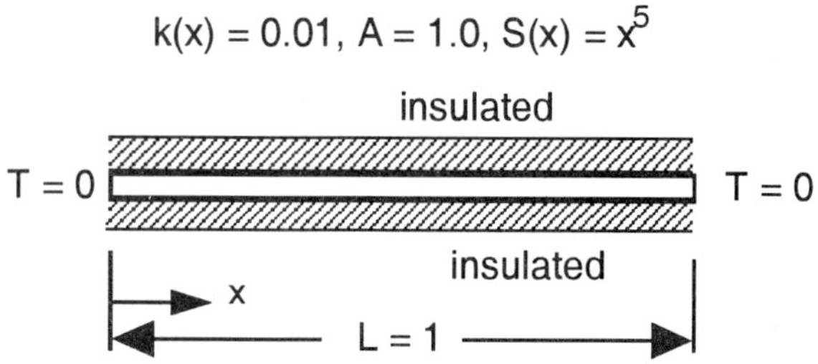

$$k(x) = 0.01, \ A = 1.0, \ S(x) = x^5$$

Figure 8.29: Insulated Rod of Unit Cross-Sectional Area

8.5

Using a computer program with one-dimensional heat conduction elements, determine the temperature distribution through the wall shown. The wall, which has the dimensions shown, can be considered to be of infinite extent in the other two directions. Be sure to use consistent units.

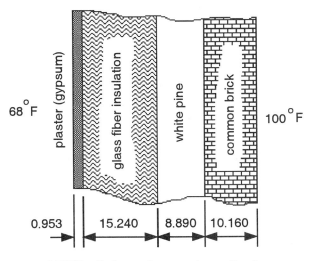

NOTE: all dimensions are in centimeters

Figure 8.30: Hypothetical Wall

The thermal conductivities, in units of $W/m - {}^oK$, for the materials comprising the wall are: plaster (gypsum): 0.50, glass fiber insulation: 0.035, white pine: 0.11, and common brick: 0.70.

8.6

Consider the non-prismatic bar shown in Figure 8.31. Using a *hand solution* with the coarsest possible mesh, determine the nodal displacements, and the element strains, stresses and forces. When applying the nodal specifications, use the elimination approach at the global level. The self-weight of the bar is negligible. The cross-sectional areas of the two portions of the bar are equal to 17.2 and 5.8 square inches. The elastic modulus is equal to 30×10^6 pounds per square inch (psi). NOTE: 1 kip = 1000 pounds.

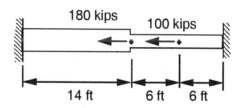

Figure 8.31: Non-prismatic Bar

8.7

A stainless steel rod consisting of two different cross-sectional areas is loaded by a surface traction equal to 50 pounds per inch, and by a concentrated force of 6000 pounds in the manner shown in Figure 8.32.

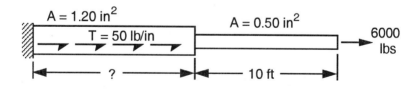

Figure 8.32: Non-prismatic Bar

Using a computer program with one-dimensional bar elements, determine the length of the thicker portion of the rod such that the displacement of the right end is approximately 0.072 inches.

8.8

An aluminum rod, attached to rigid walls in the manner shown in Figure 8.33, is subjected to a distributed traction force $T(x)$. Assuming that $T_o = 1000$, and using grids of 2, 4, 8 and 16 elements in conjunction with a computer program with one-dimensional bar elements, study the convergence of solutions for stresses in the rod. Discuss your results.

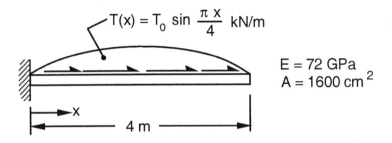

Figure 8.33: Bar with Varying Traction

8.9

The bar shown in Figure 8.34 is subjected to a concentrated force of 20 kN applied 40 centimeters from the left end of the bar. The cross-sectional area of the bar is equal to $4.0\,\text{cm}^2$, and the material is characterized by an elastic modulus equal to $20,000\,\text{kN/cm}^2$.

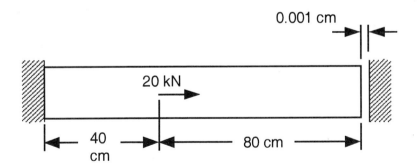

Figure 8.34: Bar with Initial Gap

8.10

The hypothetical non-prismatic, circular column shown in Figure 8.35 is adequately braced so as to prevent lateral buckling. The self-weight of the column is negligible; the material is elastic with a modulus of 5.80×10^6 pounds per square inch (psi). Using a computer program with one-dimensional bar elements, please do the following:

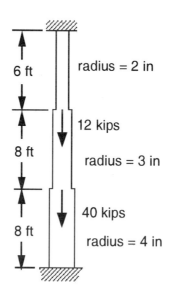

Figure 8.35: Non-prismatic Column

1. For the applied loads shown, determine the force and stress distributions in the column.

2. For the same applied loading, assume that the lower support settles downward by 1 inch. Compare the force and stress distributions for this case with those from part (a).

8.11

Using meshes of 4, 8 and 16 equally spaced elements in conjunction with a computer program with one-dimensional bar elements, study the convergence of solutions for displacements and stresses at $y = 16$ and at $y = 48$ centimeters in the non-prismatic circular column shown in Figure 8.36. Present your results in the form of two plots (note that interpolation may, in certain cases, be required): (1) displacement vs. number of elements; and, (2) longitudinal stress vs. number of elements. Discuss your findings.

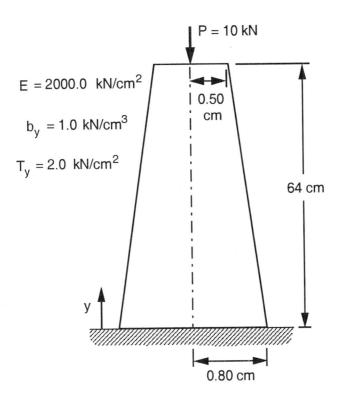

Figure 8.36: Non-prismatic Column

NOTE: assume that the body force b_y acts in the *negative* y-direction.

Chapter 9

Element Interpolation Functions

9.1 Introductory Remarks

All finite element formulations make use of element interpolation functions.[1] While such functions have been introduced and used in the previous chapters, details regarding their derivation have not been presented in a formal manner. Such details are described in this chapter.

As mentioned in Chapters 6 and 7, element interpolation functions cannot be chosen arbitrarily, but must satisfy certain requirements. The most basic among these is the requirement that at some node i within the element, the corresponding interpolation function N_i must equal to unity; at all others nodes N_i must be zero. Stated mathematically,

$$N_i(x_j, y_j, z_j) = \delta_{ij} \tag{9.1}$$

where δ_{ij} is the Kronecker delta.

In addition, in order to ensure monotonic convergence to the exact solution as the number of elements is increased, the interpolation functions must satisfy the *compatibility* and *completeness* criteria. As part of the latter criterion, recall that as the element size goes to zero, uniform states of $\hat{\phi}$ must be approximated as closely as required. Such a "rigid body"

---

[1] As noted in Chapter 6, the interpolation functions are also referred to as "shape functions," for in solid mechanics applications they describe the deformed shape of the element.

mode is realized only if the interpolation functions for a given element sum to one.

In the subsequent chapter it will be shown that geometrically simple "parent" domains Ω^p can uniquely be mapped into distorted element geometries that more accurately discretize the domains being analyzed. If the interpolation functions used in the approximation of the primary dependent variables via

$$\hat{\phi}^{(e)} = \sum_{m=1}^{N_{dof}} N_m \hat{\phi}_m^{(e)} \tag{9.2}$$

are such that continuity of $\phi_m^{(e)}$ is preserved in the parent coordinates, then continuity requirements will be satisfied in the distorted elements [605]. It follows that in the present discussion, the element interpolation functions need to be developed only for the parent domains.

In addition, although in earlier chapters the interpolation functions have been derived using global coordinates, herein the functions are developed using natural coordinates. Such an approach significantly facilitates element mapping.

Perhaps the most intuitive, though not necessarily the simplest or most general, approach for deriving element interpolation functions involves the direct manipulation of basis functions and solution for so-called generalized coordinates. This approach, which was used in Examples 7.6 to 7.8, is illustrated in the following example.

Example 9.1. Interpolation Functions for Linear Rectangular Elements

Consider the four-node rectangular element shown in Figure 9.1. We derive interpolation functions for this element in terms of natural coordinates.

The element node points are numbered in ascending order corresponding to the counterclockwise direction. The global x and y coordinates of the element nodes are denoted by $x_i^{(e)}$ and $y_i^{(e)}$, respectively, where $i = 1, 2, 3, 4$.

The natural coordinates ξ and η are related to the global x and y coordinates through the following relations[2]:

$$\xi = \frac{x^{(e)} - x_c}{a} \quad ; \quad \eta = \frac{y^{(e)} - y_c}{b} \tag{9.3}$$

---

[2] Further details pertaining to this coordinate transformation are given in Appendix C.

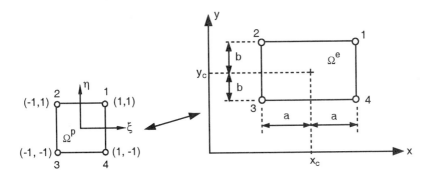

Figure 9.1: Typical Linear Rectangular Element and its Parent Domain

where

$$x_c = \frac{x_1^{(e)} + x_2^{(e)}}{2} = \frac{x_3^{(e)} + x_4^{(e)}}{2} \tag{9.4}$$

and

$$y_c = \frac{y_1^{(e)} + y_4^{(e)}}{2} = \frac{y_2^{(e)} + y_3^{(e)}}{2} \tag{9.5}$$

The result is that the actual nodal coordinates are mapped to a (ξ, η) natural coordinate system associated with the parent domain Ω^p shown in Figure 9.1.

A *bilinear* approximation of the primary dependent variable ϕ is assumed; this approximation is written in terms of natural coordinates. Since four degrees of freedom are available for the approximation, we write

$$\hat{\phi}^{(e)} = \alpha_1 + \alpha_2\xi + \alpha_3\eta + \alpha_4\xi\eta = \chi\alpha \tag{9.6}$$

where the entries in $\chi = \langle 1 \quad \xi \quad \eta \quad \xi\eta \rangle$ are called the *basis functions* [370], and the α_i populating α are unknown *generalized coordinates* that are a function only of the element geometry. Although this approximation is linear along the sides of the element, due to the presence of the $\xi\eta$ term, within the element it is bilinear. Note that equation (9.6) constitutes an *incomplete* second-order polynomial in two-dimensions (the terms involving ξ^2 and η^2 are missing). The symmetry of the polynomial with respect to the Pascal triangle (Figure 9.2) none the less ensures spatial isotropy.[3]

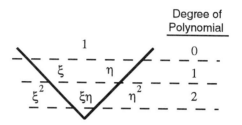

Figure 9.2: Pascal Triangle for Linear Rectangular Element

Evaluating equation (9.6) at each of the nodes gives

$$
\left\{
\begin{array}{l}
\hat{\phi}_1^{(e)} = \hat{\phi}(\xi_1, \eta_1) \\
\hat{\phi}_2^{(e)} = \hat{\phi}(\xi_2, \eta_2) \\
\hat{\phi}_3^{(e)} = \hat{\phi}(\xi_3, \eta_3) \\
\hat{\phi}_4^{(e)} = \hat{\phi}(\xi_4, \eta_4)
\end{array}
\right\}
=
\begin{bmatrix}
1 & 1 & 1 & 1 \\
1 & -1 & 1 & -1 \\
1 & -1 & -1 & 1 \\
1 & 1 & -1 & -1
\end{bmatrix}
\left\{
\begin{array}{l}
\alpha_1 \\
\alpha_2 \\
\alpha_3 \\
\alpha_4
\end{array}
\right\}
\tag{9.7}
$$

or

$$
\hat{\phi}_n^{(e)} = \mathbf{A}\boldsymbol{\alpha}
\tag{9.8}
$$

Provided that $\mathbf{A}$ is non-singular, we write

$$
\boldsymbol{\alpha} = \mathbf{A}^{-1}\hat{\phi}_n^{(e)} = \frac{1}{4}
\begin{bmatrix}
1 & 1 & 1 & 1 \\
1 & -1 & -1 & 1 \\
1 & 1 & -1 & -1 \\
1 & -1 & 1 & -1
\end{bmatrix}
\hat{\phi}_n^{(e)}
\tag{9.9}
$$

Finally, substituting equation (9.9) into equation (9.6) gives the desired expression, viz.,

$$
\hat{\phi}^{(e)} = \boldsymbol{\chi}\mathbf{A}^{-1}\hat{\phi}_n^{(e)} = \mathbf{N}\hat{\phi}_n^{(e)}
\tag{9.10}
$$

---

[3] Recall the discussion of Section 7.5.2.

Table 9.1: Values of ξ_i and η_i for Bilinear 4-Node Elements

i	ξ_i	η_i
1	1	1
2	-1	1
3	-1	-1
4	1	-1

Expansion and simplification of equation (9.10) leads to

$$\hat{\phi}^{(e)} = \sum_{i=1}^{4} N_i \hat{\phi}_i^{(e)} = \sum_{i=1}^{4} \frac{1}{4}(1 + \xi_i\xi)(1 + \eta_i\eta)\hat{\phi}_i^{(e)} \qquad (9.11)$$

The constants ξ_i and η_i appearing in equation (9.11) are defined in Table 9.1.

Having derived the element interpolation functions, it is easily verified that they satisfy the basic requirement represented by equation (9.1). In addition, these functions satisfy the partition of unity property, viz.,

$$\sum_{i=1}^{4} N_i = \frac{1}{4}(1 + \xi)(1 + \eta) + \frac{1}{4}(1 - \xi)(1 + \eta)$$

$$+ \frac{1}{4}(1 - \xi)(1 - \eta) + \frac{1}{4}(1 + \xi)(1 - \eta)$$

$$= \frac{1}{2}(1 + \eta) + \frac{1}{2}(1 - \eta) = 1 \qquad (9.12)$$

In light of the above example, the general approach using basis functions and matrix inversion to compute generalized coordinates for the determination of element interpolation functions is now summarized. The polynomial approximation for a primary dependent variable is written in terms of natural coordinates as

$$\hat{\phi}^{(e)} = \alpha_1 + \alpha_2\xi + \alpha_3\eta + \alpha_4\zeta + \cdots = \chi\alpha \qquad (9.13)$$

where χ is the $(1 * N_{dof})$ row vector of basis functions and α is the $(N_{dof} * 1)$ column vector of generalized coordinates.

Evaluating equation (9.13) at each of the N_{dof} nodes in the element at which $\hat{\phi}^{(e)}$ is an unknown gives

$$\hat{\phi}_n^{(e)} = \mathbf{A}\boldsymbol{\alpha} \tag{9.14}$$

where $\mathbf{A}$ is an $(N_{dof} * N_{dof})$ matrix of constants. Provided $\mathbf{A}$ is non-singular, equation (9.14) is inverted. The resulting expression for $\boldsymbol{\alpha}$ is then substituted into equation (9.13) to give

$$\hat{\phi}^{(e)} = \boldsymbol{\chi}\boldsymbol{\alpha}\mathbf{A}^{-1}\hat{\phi}_n^{(e)} = \mathbf{N}\hat{\phi}_n^{(e)} \tag{9.15}$$

where $\mathbf{N}$ is the $(1 * N_{dof})$ row vector of interpolation functions.

Remarks

1. For some types of elements $\mathbf{A}^{-1}$ may not exist for all orientations of the element in the global coordinate system [605].

2. For large values of N_{dof} the analytic determination of $\mathbf{A}^{-1}$ may require a substantial computational effort. This effort is, however, lessened by the availability of software capable of carrying out symbolic arithmetic operations.

3. In developing element interpolation functions using generalized coordinates, it is not always an easy task to satisfy spatial isotropy [59]. The difficulties of obtaining spatially isotropic generalized coordinate formulations, for which $\mathbf{A}^{-1}$ is well-defined, have enhanced the development of formulations in which interpolation functions are used in defining the element geometry. Further details pertaining to such "parametric mappings" are given in Chapter 10.

It is thus desirable to develop a procedure by which the interpolation functions can be written down *directly*, thus avoiding the potential pitfalls and excessive computational effort associated with the aforementioned approach. Such a direct approach is particularly useful when higher-order elements are required.

Higher-order elements maintain the interelement continuity of lower order elements, but employ a higher-order approximation (e.g., more terms in the polynomial) for $\hat{\phi}^{(e)}$. Relatively small numbers of higher-order elements are typically capable of more accurate representations than the linear ones discussed in the previous chapters. Furthermore, as discussed in Chapter 10, the boundary edges and surfaces of such elements can be curved, thus allowing for more accurate representations of element domains. As compared to the basic linear elements, higher-order elements are, however, more expensive to formulate; as a result, the cost-effectiveness of various elements represents an area of on-going dispute. Since the optimal choice of element type is very often problem-dependent, it follows that no single element is exclusively preferred.

The development of interpolation functions for some "families" of C^0 elements, including higher-order ones, are now discussed.

9.2 Lagrangian Elements

Elements comprising the so-called Lagrangian family are systematically derived with the aid of one-dimensional Lagrange polynomials.[4] In the present discussion, we denote a typical Lagrange polynomial by

$$
\Lambda_i^m(\xi) = \frac{\prod\limits_{q=1,\,q\neq i}^{m+1} (\xi - \xi_q)}{\prod\limits_{q=1,\,q\neq i}^{m+1} (\xi_i - \xi_q)}
$$

$$
= \frac{(\xi - \xi_1)\,\cdots\,(\xi - \xi_{i-1})(\xi - \xi_{i+1})\,\cdots\,(\xi - \xi_{m+1})}{(\xi_i - \xi_1)\,\cdots\,(\xi_i - \xi_{i-1})(\xi_i - \xi_{i+1})\,\cdots\,(\xi_i - \xi_{m+1})} \tag{9.16}
$$

where m denotes the order of the polynomial, i represents the local (i.e., element) node numbers, and ξ is a natural coordinate. From equation (9.16) it is evident that Lagrange polynomials satisfy the basic requirement of equation (9.1) for interpolation functions; that is, $\Lambda_i^m(\xi_j) = \delta_{ij}$.

9.2.1 One-Dimensional Lagrangian Elements

For a general one-dimensional Lagrangian element containing N_{en} nodes, the interpolation function associated with node i will be the Lagrange polynomial of degree $(N_{en} - 1)$ that takes on the value of one at node i and the value of zero at the remaining nodes. This is written as

$$
N_i = \Lambda_i^{(N_{en}-1)} \quad ; \quad i = 1, 2, \cdots, N_{en} \tag{9.17}
$$

Although the present development is limited to the one-dimensional parent domain $\Omega^p : \xi \in [-1, 1]$, as will be shown in Chapter 10, this

---

[4]Although named in honor of the French mathematician Joseph-Louis Lagrange (1736-1813), the real credit for discovering the interpolation theorem underlying these polynomials should, according to Boole [88], be assigned to Euler, who had obtained a closely analogous expression long before Lagrange.

domain is easily mapped to actual element geometries. As such, the interpolation functions derived herein apply to any one-dimensional element possessing the requisite number of nodes.

For example, consider the linear one-dimensional element shown in Figure 9.3(a). For this element, $N_{en} = 2$. It follows that in deriving the element interpolation functions, a first-order Lagrange polynomial should be used.

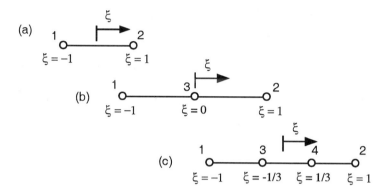

Figure 9.3: Parent Domains for One-Dimensional Lagrangian Elements

Thus for $i = 1$,

$$N_1 = \Lambda_1^1(\xi) = \frac{\xi - \xi_2}{\xi_1 - \xi_2} = \frac{\xi - 1}{-1 - 1} = \frac{1}{2}(1 - \xi) \qquad (9.18)$$

and for $i = 2$,

$$N_2 = \Lambda_2^1(\xi) = \frac{\xi - \xi_1}{\xi_2 - \xi_1} = \frac{\xi - (-1)}{1 - (-1)} = \frac{1}{2}(1 + \xi) \qquad (9.19)$$

Next consider the quadratic element shown in Figure 9.3(b). For this element $N_{en} = 3$, and the interpolation functions are derived using the following second-order Lagrange polynomials:

$$N_1 = \Lambda_1^2(\xi) = \frac{(\xi - \xi_2)(\xi - \xi_3)}{(\xi_1 - \xi_2)(\xi_1 - \xi_3)} = \frac{(\xi - 0)(\xi - 1)}{(-1 - 0)(-1 - 1)} = \frac{1}{2}\xi(\xi - 1) \quad (9.20)$$

$$N_2 = \Lambda_2^2(\xi) = \frac{(\xi - \xi_1)(\xi - \xi_2)}{(\xi_2 - \xi_1)(\xi_2 - \xi_2)} = \frac{(\xi + 1)(\xi - 0)}{(1 + 1)(1 - 0)} = \frac{1}{2}\xi(\xi + 1) \qquad (9.21)$$

$$N_3 = \Lambda_3^2(\xi) = \frac{(\xi - \xi_1)(\xi - \xi_3)}{(\xi_3 - \xi_1)(\xi_3 - \xi_3)} = \frac{(\xi + 1)(\xi - 1)}{(0 + 1)(0 - 1)} = (1 - \xi^2) \qquad (9.22)$$

These interpolation functions can also be derived in an alternate manner. Beginning with the linear element, we add node 3. The associated interpolation function must be a parabola with value of one at $\xi = 0$ and zero at $\xi = \pm 1$. These requirements are easily satisfied by the function $N_3 = 1 - \xi^2$. The interpolation functions associated with the linear element are, however, not zero at the newly added node 3; indeed, $N_1^{(N_{en}=2)}(0) = N_2^{(N_{en}=2)}(0) = 0.50$. To remedy this situation, subtract one-half of N_3 from the linear interpolation functions, giving

$$N_1 = N_1^{(N_{en}=2)} - \frac{1}{2}(1 - \xi^2) = \frac{1}{2}(1 - \xi) - \frac{1}{2}(1 - \xi^2)$$

$$= \frac{1}{2}\xi(\xi - 1) \qquad (9.23)$$

$$N_2 = N_2^{(N_{en}=2)} - \frac{1}{2}(1 - \xi^2) = \frac{1}{2}(1 + \xi) - \frac{1}{2}(1 - \xi^2)$$

$$= \frac{1}{2}\xi(\xi + 1) \qquad (9.24)$$

which are precisely the expressions obtained in equations (9.20) and (9.21).

Finally, the interpolation functions associated with the cubic one-dimensional element shown in Figure 9.3(c) are:

$$N_1 = \frac{1}{16}(1 - \xi)(9\xi^2 - 1) \qquad (9.25)$$

$$N_2 = \frac{1}{16}(1 + \xi)(9\xi^2 - 1) \qquad (9.26)$$

$$N_3 = \frac{27}{16}\left(\frac{1}{3} - \xi\right)(1 - \xi^2) \qquad (9.27)$$

$$N_4 = \frac{27}{16}\left(\frac{1}{3} + \xi\right)(1 - \xi^2) \qquad (9.28)$$

Their derivation is left as an exercise.

9.2.2   Two-Dimensional Lagrangian Elements

In general, two-dimensional Lagrangian elements are quadrilateral in shape. The rectangular version[5] of the linear Lagrangian element was studied in Example 9.1.

The first three members of the two-dimensional Lagrangian element family, along with their respective Pascal triangles, are shown in Figure 9.4. Although the present development is limited to the two-dimensional parent domain (a biunit square) Ω^p : $\xi, \eta \in [-1, 1]$, as shown in Chapter 10, this domain is easily mapped to actual *quadrilateral* element geometries, possibly with curved edges. As such, the interpolation functions derived herein apply to any two-dimensional element possessing the requisite number of nodes.

The interpolation functions for such elements are derived by forming products of two one-dimensional Lagrange polynomials in the following manner:

$$N_i(\xi, \eta) = \Lambda_j^{(N_{en\,\xi}-1)}(\xi) * \Lambda_k^{(N_{en\,\eta}-1)}(\eta) \qquad (9.29)$$

where i denotes the node number in the parent element domain Ω^p, j denotes the node number for a one-dimensional interpolation function parallel to the ξ-axis, k is the node number for a one-dimensional interpolation function parallel to the η-axis. The quantities $N_{en\xi}$ and $N_{en\eta}$ are the number of element nodes in the ξ- and η-directions, respectively. The relation between i, j and k is element specific.

We now present some details pertaining to the development of interpolation functions for two-dimensional Lagrangian elements.

Bilinear Lagrangian Element

The schematic relationship between the bilinear Lagrangian element and the two linear one-dimensional elements used in determining its interpolation functions is shown in Figure 9.5.

---

[5]The linear rectangular element was first proposed by Argyris [20] and used by Gallagher et al. [188] and Melosh [386]. The linear quadrilateral element is attributed to Taig [526]. This element was also used by Argyris et al. [29], by Zienkiewicz et al. [589], and by Ergatoudis et al. [151]. The latter group of investigators was one of the first to use the quadratic member of the Lagrangian family.

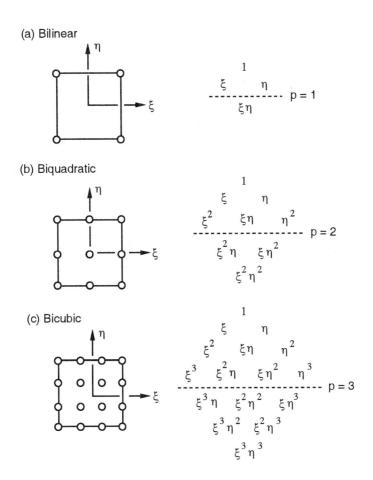

Figure 9.4: Two-Dimensional Lagrangian Elements: Parent Domains and Corresponding Pascal Triangles

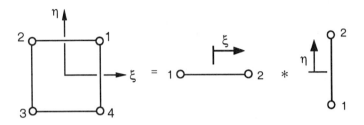

Figure 9.5: Parent Element Domain for Bilinear Lagrangian Element

Since $N_{en\xi} = N_{en\eta} = 2$, the element interpolation functions are determined from the following expression:

$$N_i(\xi, \eta) = \Lambda_j^{(1)}(\xi) * \Lambda_k^{(1)}(\eta) \tag{9.30}$$

where the relationship between the indices i (the node number in the parent element), j (the index associated with the ξ-axis in a one-dimensional element), and k (the index associated with the η-axis in a one-dimensional element) is given in Table 9.2.

The element interpolation functions are thus

$$N_1 = \Lambda_2^1(\xi) * \Lambda_2^1(\eta) = \frac{(\xi - \xi_1)}{(\xi_2 - \xi_1)} \frac{(\eta - \eta_1)}{(\eta_2 - \eta_1)} = \frac{(\xi + 1)}{(1 + 1)} \frac{(\eta + 1)}{(1 + 1)}$$

$$= \frac{1}{4}(1 + \xi)(1 + \eta) \tag{9.31}$$

$$N_2 = \Lambda_1^1(\xi) * \Lambda_2^1(\eta) = \frac{(\xi - \xi_2)}{(\xi_1 - \xi_2)} \frac{(\eta - \eta_1)}{(\eta_2 - \eta_1)} = \frac{(\xi - 1)}{(-1 - 1)} \frac{(\eta + 1)}{(1 + 1)}$$

$$= \frac{1}{4}(1 - \xi)(1 + \eta) \tag{9.32}$$

$$N_3 = \Lambda_1^1(\xi) * \Lambda_1^1(\eta) = \frac{(\xi - \xi_2)}{(\xi_1 - \xi_2)} \frac{(\eta - \eta_2)}{(\eta_1 - \eta_2)} = \frac{(\xi - 1)}{(-1 - 1)} \frac{(\eta - 1)}{(-1 - 1)}$$

$$= \frac{1}{4}(1 - \xi)(1 - \eta) \tag{9.33}$$

Table 9.2: Relation Between Indices for Bilinear Lagrangian Elements

i	j	k
1	2	2
2	1	2
3	1	1
4	2	1

$$N_4 = \Lambda_2^1(\xi) * \Lambda_1^1(\eta) = \frac{(\xi - \xi_1)}{(\xi_2 - \xi_1)} \frac{(\eta - \eta_2)}{(\eta_1 - \eta_2)} = \frac{(\xi + 1)}{(1 + 1)} \frac{(\eta - 1)}{(-1 - 1)}$$
$$= \frac{1}{4}(1 + \xi)(1 - \eta) \qquad (9.34)$$

These interpolation functions, which are identical to those derived using the "generalized coordinate" approach illustrated in Example 9.1, are again summarized as in equation (9.11), viz.,

$$N_i = \frac{1}{4}(1 + \xi_i \xi)(1 + \eta_i \eta) \qquad (9.35)$$

with the constants ξ_i and η_i defined in Table 9.1.

As noted in Example 9.1, these interpolation functions satisfy the basic requirement described by equation (9.1), as well as the partition of unity property.

Finally, the derivatives[6] of the interpolation functions with respect to the natural coordinates follow from equation (9.35), viz.,

$$\frac{\partial N_i}{\partial \xi} = \frac{1}{4}\xi_i(1 + \eta_i \eta) \quad ; \quad \frac{\partial N_i}{\partial \eta} = \frac{1}{4}\eta_i(1 + \xi_i \xi) \qquad (9.36)$$

Biquadratic Lagrangian Element

We next consider the biquadratic member of the Lagrangian element family. The schematic relationship between this element and two quadratic one-dimensional elements used in determining its interpolation functions is shown in Figure 9.6.

---

[6]The reason for computing these derivatives will become evident in Chapter 10.

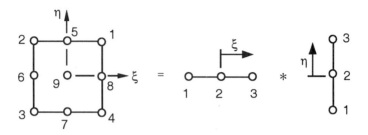

Figure 9.6: Parent Element Domain for Biquadratic Lagrangian Element

The interpolation functions are determined from the following general expression:

$$N_i(\xi, \eta) = \Lambda_j^2(\xi) * \Lambda_k^2(\eta) \tag{9.37}$$

where the relationship between the indices i, j, and k is given in Table 9.3. Using equation (9.37), the interpolation function associated with node 1 is determined in the following manner:

$$
\begin{aligned}
N_1 = \Lambda_3^2(\xi) * \Lambda_3^2(\eta) &= \frac{(\xi - \xi_1)(\xi - \xi_2)}{(\xi_3 - \xi_1)(\xi_3 - \xi_2)} \frac{(\eta - \eta_1)(\eta - \eta_2)}{(\eta_3 - \eta_1)(\eta_3 - \eta_2)} \\
&= \frac{(\xi + 1)(\xi - 0)}{(1 + 1)(1 - 0)} \frac{(\eta + 1)(\eta - 0)}{(1 + 1)(1 - 0)} = \frac{1}{4}\xi\eta(1 + \xi)(1 + \eta) \tag{9.38}
\end{aligned}
$$

The derivation of the remaining interpolation functions is left as an exercise.

The element interpolation functions and their derivatives with respect to the natural coordinates are summarized below in compact form. The constants ξ_i and η_i are defined in Table 9.3.

- For $i = 1$ to 4:

$$N_i = \frac{1}{4}(\xi_i\xi)(\eta_i\eta)(1 + \xi_i\xi)(1 + \eta_i\eta) \tag{9.39}$$

$$\frac{\partial N_i}{\partial \xi} = \frac{1}{4}\xi_i(1 + 2\xi_i\xi)(\eta_i\eta)(1 + \eta_i\eta) \tag{9.40}$$

$$\frac{\partial N_i}{\partial \eta} = \frac{1}{4}(\xi_i\xi)\eta_i(1 + \xi_i\xi)(1 + 2\eta_i\eta) \tag{9.41}$$

Table 9.3: Indices and ξ_i and η_i Values for Biquadratic Lagrangian Elements

i	j	k	ξ_i	η_i
1	3	3	1	1
2	1	3	-1	1
3	1	1	-1	-1
4	3	1	1	-1
5	2	3	0	1
6	1	2	-1	0
7	2	1	0	-1
8	3	2	1	0
9	2	2	0	0

- For $i = 5$ and 7:

$$N_i = \frac{1}{2}(\eta_i\eta)(1 - \xi^2)(1 + \eta_i\eta) \tag{9.42}$$

$$\frac{\partial N_i}{\partial \xi} = -\xi(\eta_i\eta)(1 + \eta_i\eta) \tag{9.43}$$

$$\frac{\partial N_i}{\partial \eta} = \frac{1}{2}(1 - \xi^2)\eta_i(1 + 2\eta_i\eta) \tag{9.44}$$

- For $i = 6$ and 8:

$$N_i = \frac{1}{2}(\xi_i\xi)(1 + \xi_i\xi)(1 - \eta^2) \tag{9.45}$$

$$\frac{\partial N_i}{\partial \xi} = \frac{1}{2}\xi_i(1 + 2\xi_i\xi)(1 - \eta^2) \tag{9.46}$$

$$\frac{\partial N_i}{\partial \eta} = -(\xi_i\xi)\eta(1 + \xi_i\xi) \tag{9.47}$$

- For $i = 9$:

$$N_9 = (1 - \xi^2)(1 - \eta^2) \tag{9.48}$$

$$\frac{\partial N_9}{\partial \xi} = 2\xi(\eta^2 - 1) \tag{9.49}$$

$$\frac{\partial N_9}{\partial \eta} = 2\eta(\xi^2 - 1) \tag{9.50}$$

9.2.3 Three-Dimensional Lagrangian Elements

In general, three-dimensional Lagrangian elements are hexahedral in shape. The first two members of the three-dimensional Lagrangian element family are shown in Figure 9.7. Although the present development is limited to the three-dimensional parent domain (a biunit cube) Ω^p : $\xi, \eta, \zeta \in [-1, 1]$, as shown in Chapter 10, this domain is easily mapped to actual *hexahedral* element geometries, possibly with curved faces. As such, the interpolation functions derived herein apply to any three-dimensional element possessing the requisite number of nodes.

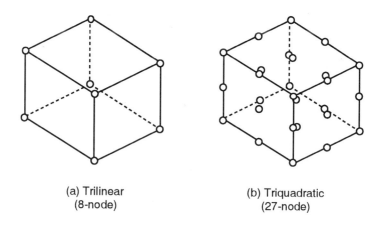

(a) Trilinear   (b) Triquadratic
(8-node)   (27-node)

Figure 9.7: Parent Domains for Three-Dimensional Lagrangian Elements

The interpolation functions for such elements are derived by forming products of *three* one-dimensional Lagrange polynomials. That is,

$$N_i(\xi, \eta, \zeta) = \Lambda_j^{(N_{en\xi}-1)}(\xi) * \Lambda_k^{(N_{en\eta}-1)}(\eta) * \Lambda_l^{(N_{en\zeta}-1)}(\zeta) \qquad (9.51)$$

where i denotes the node number in the parent element domain Ω^p, j denotes the node number for a one-dimensional interpolation function parallel to the ξ-axis, k denotes the node number for a one-dimensional interpolation function parallel to the η-axis, l denotes the node number for a one-dimensional interpolation function parallel to the ζ-axis, and $N_{en\xi}$, $N_{en\eta}$, and $N_{en\zeta}$ denote the number of element nodes in the ξ-, η-, ζ-directions, respectively. The relation between i, j, k and l is element specific.

Trilinear Lagrangian Element

The trilinear Lagrangian element,[7] commonly referred to as the "8-node brick," is shown in Figure 9.8. The interpolation functions for this are derived from the following general expression:

$$N_i(\xi, \eta, \zeta) = \Lambda_j^1(\xi) * \Lambda_k^1(\eta) * \Lambda_l^1(\zeta) \tag{9.52}$$

where the relationship between the indices i, j, k, and l is given in Table 9.4. The basis functions associated with this element are: 1, ξ, η, ζ, $\xi\eta$, $\xi\zeta$, $\eta\zeta$, and $\xi\eta\zeta$. The interpolating polynomial is complete to degree $p = 1$.

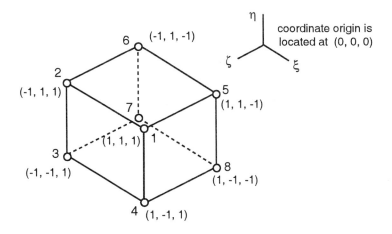

Figure 9.8: Parent Element Domain for Trilinear Lagrangian Element

Using equation (9.52), the interpolation function associated with node 1 is determined in the following manner:

$$N_1 = \Lambda_2^1(\xi) * \Lambda_2^1(\eta) * \Lambda_2^1(\zeta) = \frac{(\xi + 1)}{(1 + 1)} \frac{(\eta + 1)}{(1 + 1)} \frac{(\zeta + 1)}{(1 + 1)}$$

$$= \frac{1}{8}(1 + \xi)(1 + \eta)(1 + \zeta) \tag{9.53}$$

The derivation of the remaining interpolation functions is left as an exercise.

---

[7]The analytically integrated, right prism version of this element was first presented by Taig [526] and subsequently generalized to orthotropic materials by Melosh [386]. The hexahedral version of the element is attributed to Irons [254], and was discussed by Zienkiewicz et al. [589].

Table 9.4: Indices and ξ_i, η_i and ζ_i Values for Trilinear Lagrangian Elements

i	j	k	l	ξ_i	η_i	ζ_i
1	2	2	2	1	1	1
2	1	2	2	-1	1	1
3	1	1	2	-1	-1	1
4	2	1	2	1	-1	1
5	2	2	1	1	1	-1
6	1	2	1	-1	1	-1
7	1	1	1	-1	-1	-1
8	2	1	1	1	-1	-1

The element interpolation functions and their derivatives with respect to the natural coordinates are summarized below in compact form. The constants ξ_i, η_i and ζ_i are defined in Table 9.4.

$$N_i = \frac{1}{8}(1 + \xi_i\xi)(1 + \eta_i\eta)(1 + \zeta_i\zeta) \tag{9.54}$$

$$\frac{\partial N_i}{\partial \xi} = \frac{1}{8}\xi_i(1 + \eta_i\eta)(1 + \zeta_i\zeta) \tag{9.55}$$

$$\frac{\partial N_i}{\partial \eta} = \frac{1}{8}\eta_i(1 + \xi_i\xi)(1 + \zeta_i\zeta) \tag{9.56}$$

$$\frac{\partial N_i}{\partial \zeta} = \frac{1}{8}\zeta_i(1 + \xi_i\xi)(1 + \eta_i\eta) \tag{9.57}$$

Triquadratic Lagrangian Element

The basis functions associated with this element are: 1, ξ, η, ζ, ξ^2, η^2, ζ^2, $\xi\eta$, $\xi\zeta$, $\eta\zeta$, $\xi^2\eta$, $\xi^2\zeta$, $\xi\eta^2$, $\xi\zeta^2$, $\eta^2\zeta$, $\eta\zeta^2$, $\xi\eta\zeta$, $\xi^2\eta\zeta$, $\xi\eta^2\zeta$, $\xi\eta\zeta^2$, $\xi^2\eta^2$, $\xi^2\zeta^2$, $\eta^2\zeta^2$, $\xi^2\eta^2\zeta$, $\xi^2\eta\zeta^2$, $\xi\eta^2\zeta^2$, and $\xi^2\eta^2\zeta^2$. The interpolating polynomial is complete to degree $p = 2$. Due to the presence of the interior nodes, the implementation and use of quadratic and higher order three-dimensional Lagrangian elements is typically deemed to be inefficient [370, 605].

9.2.4   Lagrangian Elements: A Summary

The previous examples illustrate the relative ease with which interpolation functions for Lagrangian elements are derived. It is evident that, for the interpolation functions considered, interelement continuity of the primary dependent variables $\hat{\phi}^{(e)}$ is assured. This is because the number of parameters associated with any element side or surface is sufficient to uniquely define the interpolating polynomial there. This in turn insures C^0 continuity of $\hat{\phi}^{(e)}$. These advantages notwithstanding, Lagrangian elements have two fundamental disadvantages.

First, they become computationally less efficient as the order of the element increases. To better understand this, note that the interpolation functions for two-dimensional Lagrangian elements are obtained by forming the product of complete Lagrange polynomials of degree $(N_{en} - 1)$ in both ξ and η via equation (9.29); for three-dimensional elements this is extended to products of complete polynomials of degree $(N_{en} - 1)$ in ξ, η and ζ via equation (9.51). The number of terms in the polynomial resulting from this product is in excess of the number required to produce a *complete* polynomial of degree $(N_{en} - 1)$ in ξ and η or in ξ, η and ζ[8]. Due to the "symmetry" of terms present in the polynomial, spatial isotropy is, however, maintained.

For example, a biquadratic Lagrangian element requires an interpolating polynomial with nine terms, which is three more than required for a complete second degree polynomial in ξ and η (Figure 9.4). The three additional terms include two terms of degree three ($\xi^2\eta$ and $\xi\eta^2$) and one term of degree four ($\xi^2\eta^2$). In other words, the element interpolating polynomial is a complete quadratic but an incomplete cubic and quartic.

In general, a two-dimensional Lagrangian element complete to degree p contains $(p + 1)^2$ terms, of which only $(p + 1)(p + 2)/2$ of the terms are needed for completeness. The remaining $p(p + 1)/2$ "parasitic" terms typically do not improve the rate of convergence.

The second disadvantage of Lagrangian elements is the presence of interior nodes in the quadratic and higher order members of the family. Interior nodes potentially complicate the meshing process in that their location requires additional computation in order to minimize the mapping distortion of curved-sided elements. Since only edge and vertex nodes are common to adjacent elements, it is possible to eliminate the interior nodes from the element *prior* to assembly by so-called *nodal condensation*. This computational process is discussed further in Section 9.7.

The shortcomings of Lagrangian elements can be overcome by developing elements with interpolation functions that are capable of producing more closely only the terms present in a *complete* polynomial of appropriate degree. This is realized by so-called "serendipity" elements.

---

[8] Recall that the rate of convergence of an element depends on the degree of the highest *complete* polynomial contained in the element interpolation functions.

9.3   Serendipity Elements

In serendipity[9] elements the nodes are arranged to lie, as much as possible, only along the element boundaries. In general, the interpolation functions associated with such elements are constructed from a series of terms that are products of pth degree Lagrange interpolating polynomials in one variable with linear Lagrange interpolating polynomials in the other variable [532]. On the boundaries of serendipity elements the form of the approximation $\hat{\phi}$ is identical to that produced by the Lagrangian family of elements. Interelement C^0 continuity of the approximation is thus maintained.

Originally interpolation functions for serendipity elements were derived by inspection. The progression to higher order elements was difficult and required some ingenuity. With time, systematic approaches for generating interpolation functions for serendipity elements were, however, developed [590, 532, 609, 50].

9.3.1   Two-Dimensional Serendipity Elements

Like their Lagrangian counterparts, two-dimensional serendipity elements are, in general, quadrilaterals.[10]

Linear Two-Dimensional Serendipity Element

The linear serendipity element is *identical* to the bilinear Lagrangian element. The interpolating polynomial is complete to degree $p = 1$ (Figure 9.4a). These are the only two identical two-dimensional elements in the Lagrangian and serendipity families.

Quadratic Two-Dimensional Serendipity Element

The quadratic serendipity element has eight nodes (Figure 9.9). Four terms must thus be added to the interpolating polynomial corresponding to the linear element. These include two missing quadratic terms (ξ^2 and η^2), and two cubic terms ($\xi^2\eta$ and $\xi\eta^2$) that are symmetric in ξ and η in order to preserve spatial isotropy. The interpolating polynomial is complete to degree $p = 2$.

---

[9]The word serendipity means "an apparent aptitude for making fortunate discoveries accidently." It was coined (c. 1754) by Horace Walpole after *The Three Princes of Serendip* (i.e., Sri Lanka), a Pers fairy tail in which the princes make such discoveries.

[10]Linear, quadratic and cubic serendipity elements were first described by Ergatoudis et al. [151]; a more formal presentation of these elements was given in [589].

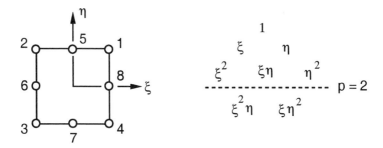

Figure 9.9: Parent Domain and Pascal Triangle for Quadratic Serendipity Element

The interpolation functions and their derivatives with respect to the natural coordinates are summarized below in compact form.

- For $i = 1$ to 4:

$$N_i = \frac{1}{4}(1 + \xi_i\xi)(1 + \eta_i\eta)(\xi_i\xi + \eta_i\eta - 1) \qquad (9.58)$$

$$\frac{\partial N_i}{\partial \xi} = \frac{1}{4}\xi_i(1 + \eta_i\eta)(2\xi_i\xi + \eta_i\eta) \qquad (9.59)$$

$$\frac{\partial N_i}{\partial \eta} = \frac{1}{4}(1 + \xi_i\xi)\eta_i(\xi_i\xi + 2\eta_i\eta) \qquad (9.60)$$

- For $i = 5$ and 7:

$$N_i = \frac{1}{2}(1 - \xi^2)(1 + \eta_i\eta) \qquad (9.61)$$

$$\frac{\partial N_i}{\partial \xi} = -\xi(1 + \eta_i\eta) \qquad (9.62)$$

$$\frac{\partial N_i}{\partial \eta} = \frac{1}{2}\eta_i(1 - \xi^2) \qquad (9.63)$$

• For $i = 6$ and 8:

$$N_i = \frac{1}{2}(1 + \xi_i\xi)(1 - \eta^2) \tag{9.64}$$

$$\frac{\partial N_i}{\partial \xi} = \frac{1}{2}\xi_i(1 - \eta^2) \tag{9.65}$$

$$\frac{\partial N_i}{\partial \eta} = -(1 + \xi_i\xi)\eta \tag{9.66}$$

The constants ξ_i and η_i appearing in the above equations are defined in the first eight rows of Table 9.3.

Cubic Two-Dimensional Serendipity Element

To progress to the cubic serendipity element from the quadratic one requires the addition of four nodes, one per side (Figure 9.10). The corresponding four new terms in the interpolating polynomial include two missing cubic terms (ξ^3 and η^3), and two quartic terms ($\xi^3\eta$ and $\xi\eta^3$) that are symmetric in ξ and η. The interpolating polynomial is complete to degree $p = 3$.

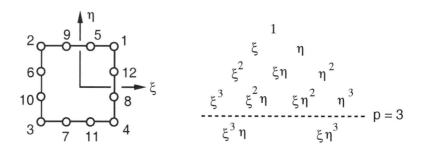

Figure 9.10: Parent Domain and Pascal Triangle for Cubic Serendipity Element

The interpolation functions and their derivatives with respect to natural coordinates are summarized below in compact form.

- For $i = 1$ to 4:

$$N_i = \frac{1}{32}(1 + \xi_i\xi)(1 + \eta_i\eta)\left[9(\xi^2 + \eta^2) - 10\right] \qquad (9.67)$$

$$\frac{\partial N_i}{\partial \xi} = \frac{1}{32}(1 + \eta_i\eta)\left[\xi_i(9\eta^2 - 10) + 9\xi(2 + 3\xi_i\xi)\right] \qquad (9.68)$$

$$\frac{\partial N_i}{\partial \eta} = \frac{1}{32}(1 + \xi_i\xi)\left[\eta_i(9\xi^2 - 10) + 9\eta(2 + 3\eta_i\eta)\right] \qquad (9.69)$$

- For $i = 6, 8, 10$ and 12:

$$N_i = \frac{9}{32}(1 + \xi_i\xi)(1 - \eta^2)(1 + 9\eta_i\eta) \qquad (9.70)$$

$$\frac{\partial N_i}{\partial \xi} = \frac{9}{32}\xi_i(1 - \eta^2)(1 + 9\eta_i\eta) \qquad (9.71)$$

$$\frac{\partial N_i}{\partial \eta} = \frac{9}{32}(1 + \xi_i\xi)\left[9\eta_i - \eta(2 + 27\eta_i\eta)\right] \qquad (9.72)$$

- For $i = 5, 7, 9$ and 11:

$$N_i = \frac{9}{32}(1 + \eta_i\eta)(1 - \xi^2)(1 + 9\xi_i\xi) \qquad (9.73)$$

$$\frac{\partial N_i}{\partial \xi} = \frac{9}{32}(1 + \eta_i\eta)[9\xi_i - \xi(2 + 27\xi_i\xi)] \qquad (9.74)$$

$$\frac{\partial N_i}{\partial \eta} = \frac{9}{32}\eta_i(1 - \xi^2)(1 + 9\xi_i\xi) \qquad (9.75)$$

The constants ξ_i and η_i appearing in the above equations are defined in Table 9.5.

Quartic Two-Dimensional Serendipity Element

The progression from the cubic serendipity element to the quartic one would seemingly involve the addition of four nodes, one on each side of the element. However, this cannot be done while still maintaining symmetry in ξ and η, thus precluding spatial isotropy. Consequently, the quartic element includes a fifth additional node, located at the element center in the manner shown in Figure 9.11.

Table 9.5: Values of ξ_i and η_i for Cubic Two-Dimensional Serendipity Elements

i	ξ_i	η_i	i	ξ_i	η_i
1	1	1	7	-1/3	-1
2	-1	1	8	1	-1/3
3	-1	-1	9	-1/3	1
4	1	-1	10	-1	-1/3
5	1/3	1	11	1/3	-1
6	-1	1/3	12	1	1/3

Associated with this interior node is the interpolation function

$$N_{17} = (1 - \xi^2)(1 - \eta^2) \tag{9.76}$$

which introduces the $\xi^2\eta^2$ term into the interpolating polynomial. We note that, as required, N_{17} is zero at the other sixteen nodes along the element's sides. Typically the equations corresponding to the interior node are eliminated at the element level by static condensation. This computational process is discussed further in Section 9.7. In summary, the five additional terms associated with the quartic serendipity element are thus ξ^4, $\xi^2\eta^2$, η^4, $\xi^4\eta$ and $\xi\eta^4$. The interpolating polynomial is complete to degree $p = 4$.

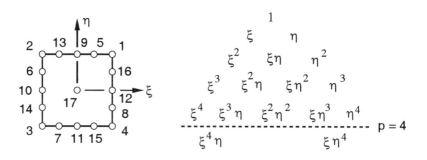

Figure 9.11: Parent Domain and Pascal Triangle for Quartic Serendipity Element

9.3.2   Three-Dimensional Serendipity Elements

Similar to their counterparts in the Lagrangian family, three-dimensional serendipity elements are, in general, hexahedrals.[11] The first three members of the three-dimensional serendipity element family are shown in Figure 9.12.

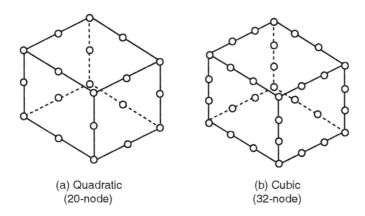

(a) Quadratic   (b) Cubic
(20-node)   (32-node)

Figure 9.12: Parent Domains for Three-Dimensional Serendipity Elements

Linear Three-Dimensional Serendipity Element

The linear serendipity element is *identical* to the trilinear Lagrangian element shown in Figure 9.7(a). These are the only two identical three-dimensional elements in the Lagrangian and serendipity families.

Quadratic Three-Dimensional Serendipity Element

Beginning with the quadratic element (Figure 9.12a), the interpolation functions for the three-dimensional serendipity family differ from those for Lagrangian elements. The quadratic serendipity element has twenty nodes, which are numbered in the manner shown in Figure 9.13.

The basis functions associated with this element are: 1, ξ, η, ζ, ξ^2, η^2, ζ^2, $\xi\eta$, $\xi\zeta$, $\eta\zeta$, $\xi^2\eta$, $\xi^2\zeta$, $\xi\eta^2$, $\xi\zeta^2$, $\eta^2\zeta$, $\eta\zeta^2$, $\xi\eta\zeta$, $\xi^2\eta\zeta$, $\xi\eta^2\zeta$, $\xi\eta\zeta^2$. The

---

[11] Early use of such serendipity elements is attributed to Ergatoudis et al. [151, 152] and to Zienkiewicz et al. [590].

highest order *complete* polynomial used is two. The associated interpolation functions and their derivatives with respect to natural coordinates are summarized below.

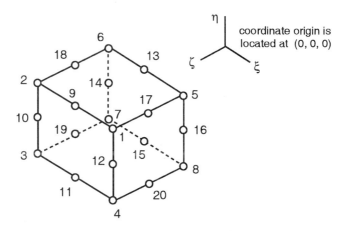

Figure 9.13: Parent Domain for Quadratic Three-Dimensional Serendipity Element

- For $i = 1$ to 8 (i.e, vertex nodes):

$$N_i = \frac{1}{8}(1 + \xi_i\xi)(1 + \eta_i\eta)(1 + \zeta_i\zeta)(\xi_i\xi + \eta_i\eta + \zeta_i\zeta - 2) \qquad (9.77)$$

$$\frac{\partial N_i}{\partial \xi} = \frac{1}{8}\xi_i(1 + \eta_i\eta)(1 + \zeta_i\zeta)(2\xi_i\xi + \eta_i\eta + \zeta_i\zeta - 1) \qquad (9.78)$$

$$\frac{\partial N_i}{\partial \eta} = \frac{1}{8}\eta_i(1 + \xi_i\xi)(1 + \zeta_i\zeta)(\xi_i\xi + 2\eta_i\eta + \zeta_i\zeta - 1) \qquad (9.79)$$

$$\frac{\partial N_i}{\partial \zeta} = \frac{1}{8}\zeta_i(1 + \xi_i\xi)(1 + \eta_i\eta)(\xi_i\xi + \eta_i\eta + 2\zeta_i\zeta - 1) \qquad (9.80)$$

- For $i = 9, 11, 13$ and 15 (midside nodes with $\xi_i = 0$, $\eta_i = \pm 1$, $\zeta_i = \pm 1$):

$$N_i = \frac{1}{4}(1 - \xi^2)(1 + \eta_i\eta)(1 + \zeta_i\zeta) \qquad (9.81)$$

$$\frac{\partial N_i}{\partial \xi} = -\frac{1}{2}\xi(1 + \eta_i\eta)(1 + \zeta_i\zeta) \tag{9.82}$$

$$\frac{\partial N_i}{\partial \eta} = \frac{1}{4}(1 - \xi^2)\eta_i(1 + \zeta_i\zeta) \tag{9.83}$$

$$\frac{\partial N_i}{\partial \zeta} = \frac{1}{4}(1 - \xi^2)(1 + \eta_i\eta)\zeta_i \tag{9.84}$$

• For $i = 10, 12, 14$ and 16 (midside nodes with $\xi_i = \pm 1$, $\eta_i = 0$, $\zeta_i = \pm 1$):

$$N_i = \frac{1}{4}(1 + \xi_i\xi)(1 - \eta^2)(1 + \zeta_i\zeta) \tag{9.85}$$

$$\frac{\partial N_i}{\partial \xi} = \frac{1}{4}\xi_i(1 - \eta^2)(1 + \zeta_i\zeta) \tag{9.86}$$

$$\frac{\partial N_i}{\partial \eta} = -\frac{1}{2}\eta(1 + \xi_i\xi)(1 + \zeta_i\zeta) \tag{9.87}$$

$$\frac{\partial N_i}{\partial \zeta} = \frac{1}{4}(1 + \xi_i\xi)(1 - \eta^2)\zeta_i \tag{9.88}$$

• For $i = 17$ to 20 (midside nodes with $\xi_i = \pm 1$, $\eta_i = \pm 1$, $\zeta_i = 0$):

$$N_i = \frac{1}{4}(1 + \xi_i\xi)(1 + \eta_i\eta)(1 - \zeta^2) \tag{9.89}$$

$$\frac{\partial N_i}{\partial \xi} = \frac{1}{4}\xi_i(1 + \eta_i\eta)(1 - \zeta^2) \tag{9.90}$$

$$\frac{\partial N_i}{\partial \eta} = \frac{1}{4}(1 + \xi_i\xi)\eta_i(1 - \zeta^2) \tag{9.91}$$

$$\frac{\partial N_i}{\partial \zeta} = -\frac{1}{2}(1 + \xi_i\xi)(1 + \eta_i\eta)\zeta \tag{9.92}$$

The constants ξ_i, η_i, and ζ_i appearing in the above equations are defined in Table 9.6.

Table 9.6:   Values of ξ_i, η_i, and ζ_i for Quadratic Three-Dimensional Serendipity Elements

i	ξ_i	η_i	ζ_i	i	ξ_i	η_i	ζ_i
1	1	1	1	11	0	-1	1
2	-1	1	1	12	1	0	1
3	-1	-1	1	13	0	1	-1
4	1	-1	1	14	-1	0	-1
5	1	1	-1	15	0	-1	-1
6	-1	1	-1	16	1	0	-1
7	-1	-1	-1	17	1	1	0
8	1	-1	-1	18	-1	1	0
9	0	1	1	19	-1	-1	0
10	-1	0	1	20	1	-1	0

Cubic Three-Dimensional Serendipity Element

The cubic three-dimensional serendipity element shown in Figure 9.12(b) has thirty-two nodes (i.e., four nodes per edge). Use of this element is rather limited [370].

9.4   Triangular and Tetrahedral Elements

Members of the triangular and tetrahedral element families always employ *complete* interpolating polynomials. As such, the nodal locations in such elements are arranged in correspondence to the locations of terms in a two- or three-dimensional Pascal triangles. Although the present development is limited to a two- and three-dimensional parent domains, as shown in Chapter 10, such domains are easily mapped to the actual distorted element geometries. As such, the interpolation functions derived herein apply to any triangular and tetrahedral element possessing the requisite number of nodes.

The number of nodes, N_{en}, contained in a triangular or tetrahedral element is equal to

$$N_{en} = \frac{(p+k)!}{p!\,k!} \tag{9.93}$$

where p equals the order of the element, and k is equals the number of variables (for triangles $k = 2$; for tetrahedra, $k = 3$).

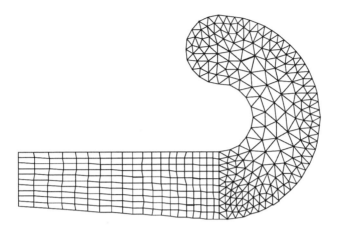

Figure 9.14: Curved Boundary Approximated using Triangular Elements (after Bickerton [74]).

As evident from Figure 9.14, arbitrarily shaped triangular elements can accurately approximate curved boundaries. For this reason, many different types of mesh generation schemes involving triangular and tetrahedral elements have been developed [194]. Further discussion of these schemes is deferred until Chapter 13, where an overview of mesh generation is presented.

Interestingly enough, the usefulness of triangles and tetrahedra is not universally acknowledged. For example, Hughes [252] states that:

It is often stated that triangular and tetrahedral elements are responsible for the geometric flexibility of the finite element method. This is perhaps somewhat of an exaggeration as triangular and tetrahedral shapes are often not needed in practice. Most regions are conveniently discretized by arbitrary quadrilateral and brick-shaped elements ⋯ . It is also often stated that triangles and tetrahedra enable modeling of particularly intricate geometries and that these shapes facilitate transition from coarsely meshed zones of a grid to finely meshed zones. This is, of course, true, but quadrilaterals and bricks are capable of doing the same thing at least to some degree ⋯ .

The points raised by Hughes have merit. Indeed, as shown in Figure 9.15, triangular domains can easily be meshed by quadrilateral elements. Thus, any triangular element could, in theory, be replaced by a suitable number of quadrilaterals.

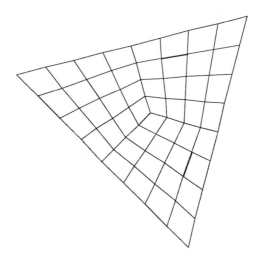

Figure 9.15: Triangular Domain Discretized Using Quadrilateral Elements.

In addition, as evident from Figure 9.16, transitions in meshes consisting of quadrilaterals can indeed be realized without the need for triangular elements. The subject of transition elements will be discussed further in Section 9.6.

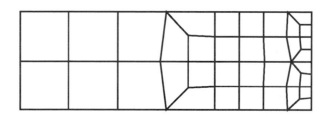

Figure 9.16: Mesh Transition Using Only Quadrilateral Elements.

9.4.1 Element "Degeneration"

Certain triangles and tetrahedra can be formed by degenerating (collapsing) quadrilateral and hexahedral elements, respectively. For example, a simple way in which to generate a linear triangle is by degenerating a four-node quadrilateral. This is achieved by assigning exactly the same coordinates to two consecutive nodes. The process is illustrated in Figure 9.17.

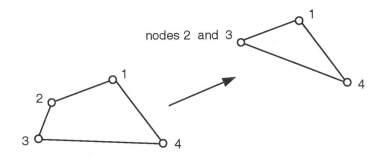

Figure 9.17: Degeneration of Four-Node Quadrilateral to a 3-Node Triangular Element.

Similarly, six-node (quadratic) triangle can be formed by collapsing two nodes in the manner shown in Figure 9.18. Further details pertaining to this and the previous element degeneration procedure are given in [58] and [252].

In the case of three-dimensional elements, linear (4-node) tetrahedra can formed by degenerating an eight-node Lagrangian/serendipity hexahedral. Beginning with this element (Figure 9.8), node 1 is first collapsed onto node 2, and node 4 onto node 3. This gives the triangular prism shown in Figure 9.19(a). Finally, nodes 1, 2 and 5 are collapsed onto node 6. The result is the linear tetrahedral element shown in Figure 9.19(b). Further details pertaining to this procedure are given in [408, 58].

The degeneration of quadrilateral and hexahedral elements to produce triangles and tetrahedra, respectively, is not without limitations. First, a number of useful triangular and tetrahedral elements exist for which there are no quadrilateral or hexahedral counterparts. Secondly, if extensive use of triangles and/or tetrahedra is anticipated, then the degeneration procedure proves inefficient [252], and the associated element interpolation functions should be derived *directly*.

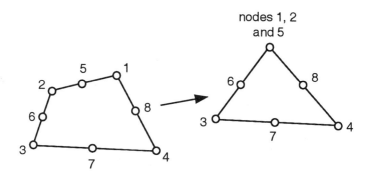

Figure 9.18: Degeneration of Eight-Node Quadrilateral to a 6-Node Triangular Element.

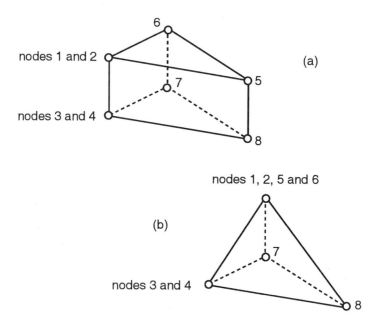

Figure 9.19: Linear Tetrahedral Formed by Degenerating a Hexahedral Element.

9.4.2 Triangular Elements

The triangular family of elements is shown in Figure 9.20.

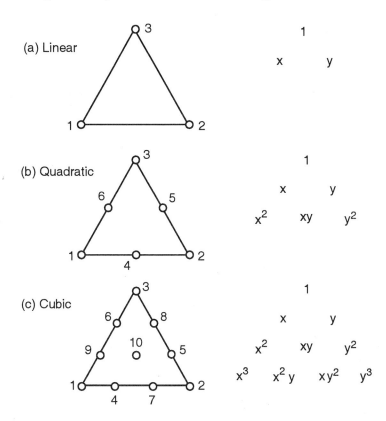

Figure 9.20: Triangular Elements and Their Associated Pascal Triangles.

The interpolation functions for triangular elements can be derived in a manner quite similar to that used in defining Lagrangian elements.[12] The approach presented herein was advanced in the infancy of the finite element method [28, 496, 532]. For purposes of determining the order of the one-dimensional Lagrange interpolating polynomials, the nodes in a triangular element (not necessarily equally spaced) are assigned the indices (I, J, K), where $I + J + K = p$. Each vertex node has one index equal to

---

[12] If the interpolation functions are instead determined by explicitly solving for generalized coordinates in the manner described in Section 9.1, triangular elements have the distinction of always having non-singular χ matrices in equation (9.13) [605].

p, and the other two indices equal to zero. A typical element thus has the configuration shown in Figure 9.21(a).

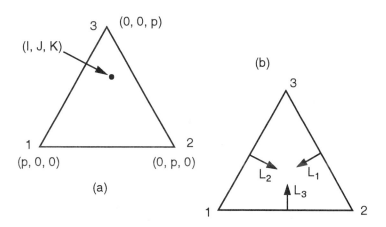

Figure 9.21: (a) Notation Used to Determine Order of Lagrange Polynomials, and (b) Natural Coordinate System Used for Triangular Elements.

It is important to point out that the use of the above notation is restricted only to determining the *order* of the one-dimensional Lagrange polynomials for use in developing the element interpolation functions; this notation should not be confused with the (L_1, L_2, L_3) natural ("area") coordinate system introduced in Appendix C and shown in Figure 9.21(b).

In a triangular element, the interpolation function for node i, with indices (I, J, K), is obtained from the following expression:

$$N_i(L_1,\ L_2,\ L_3) = \Lambda_I^I(L_1)\Lambda_J^J(L_2)\Lambda_K^K(L_3) \qquad (9.94)$$

where a typical interpolating polynomial is given by

$$\Lambda_I^I(L_1) = \frac{(L_1 - L_{1_0})(L_1 - L_{1_1}) \cdots}{(L_{1_I} - L_{1_0})(L_{1_I} - L_{1_1}) \cdots} \qquad (9.95)$$

with

$$\Lambda_0^0(L_1) = \Lambda_0^0(L_2) = \Lambda_0^0(L_3) = 1 \qquad (9.96)$$

The nodal coordinates are then [532]

$$(L_1)_I = \frac{I}{p}; \quad (L_2)_J = \frac{J}{p}; \quad (L_3)_K = \frac{K}{p} \qquad (9.97)$$

Recall that the three natural coordinates are not independent but must satisfy the relation

$$L_1 + L_2 + L_3 = 1 \qquad (9.98)$$

The procedure for determining interpolation functions is illustrated by considering the first two members of the triangular family of elements.

Linear Triangular Element

The linear triangular element $(p = 1)$ is shown in Figure 9.20(a). Since it is the simplest polygonal shape that will span a two-dimensional domain, it is commonly referred to as the two-dimensional "simplex" element.[13]
The interpolation function N_1 $(I = 1, J = K = 0)$ is determined in the following manner:

$$N_1 = \Lambda_1^1(L_1)\Lambda_0^0(L_2)\Lambda_0^0(L_3)\Lambda_0^0(L_4) = \Lambda_1^1(L_1)(1)(1)(1) = \frac{L_1 - 0}{1 - 0} = L_1 \qquad (9.99)$$

Similarly,

$$N_2 = \Lambda_0^0(L_1)\Lambda_1^1(L_2)\Lambda_0^0(L_3) = (1)\Lambda_1^1(L_2)(1) = \frac{L_2 - 0}{1 - 0} = L_2 \qquad (9.100)$$

$$N_3 = \Lambda_0^0(L_1)\Lambda_0^0(L_2)\Lambda_1^1(L_3) = (1)(1)\Lambda_1^1(L_3) = \frac{L_3 - 0}{1 - 0} = L_3 \qquad (9.101)$$

The derivatives of N_1, N_2 and N_3 with respect to the natural coordinates are next determined. For $i, j = 1, 2$ we have

$$\frac{\partial N_i}{\partial L_j} = \delta_{ij} \qquad (9.102)$$

---

[13]The two-dimensional simplex was one of the first finite elements conceived and used [122, 545, 523, 188]. Properties of the simplex approximation have been rather widely discussed; an extensive list of pertinent references is given by Oden [414].

In light of equation (9.98) it follows that

$$\frac{\partial N_3}{\partial L_1} = \frac{\partial N_3}{\partial L_3}\frac{\partial L_3}{\partial L_1} = (1)(-1) = -1 \tag{9.103}$$

$$\frac{\partial N_3}{\partial L_2} = \frac{\partial N_3}{\partial L_3}\frac{\partial L_3}{\partial L_2} = -1 \tag{9.104}$$

Quadratic Triangular Element

The quadratic triangle[14] ($p = 2$) is shown in Figure 9.20(b). For this element N_1 ($I = 2$, $J = K = 0$) is determined in the following manner:

$$N_1 = \Lambda_2^2(L_1)\Lambda_0^0(L_2)\Lambda_0^0(L_3) = \frac{L_1 - 0}{1.0 - 0.0}\frac{L_1 - 0.5}{1.0 - 0.5}(1)(1)$$
$$= L_1(2L_1 - 1) \tag{9.105}$$

The derivatives of N_1 with respect to the natural coordinates are

$$\frac{\partial N_1}{\partial L_1} = 4L_1 - 1 \quad ; \quad \frac{\partial N_1}{\partial L_2} = 0 \tag{9.106}$$

Similarly, for the remaining vertex nodes:

$$N_2 = \Lambda_0^0(L_1)\Lambda_2^2(L_2)\Lambda_0^0(L_3) = (1)\frac{L_2 - 0}{1.0 - 0.0}\frac{L_2 - 0.5}{1.0 - 0.5}(1)$$
$$= L_2(2L_2 - 1) \tag{9.107}$$

$$\frac{\partial N_2}{\partial L_1} = 0 \quad ; \quad \frac{\partial N_2}{\partial L_2} = 4L_2 - 1 \tag{9.108}$$

$$N_3 = \Lambda_0^0(L_1)\Lambda_0^0(L_2)\Lambda_2^2(L_3) = (1)(1)\frac{L_3 - 0}{1.0 - 0.0}\frac{L_3 - 0.5}{1.0 - 0.5}$$
$$= L_3(2L_3 - 1) \tag{9.109}$$

---

[14]This element was first derived by Veubeke [171]. The first published results, in the context of plane stress analysis, are attributed to Argyris [23, 24]. The formulation of the element was later "justified" by Zlámal [606].

$$\frac{\partial N_3}{\partial L_1} = \frac{\partial N_3}{\partial L_3}\frac{\partial L_3}{\partial L_1} = (4L_3 - 1)(-1) = 4(L_1 + L_2) - 3 \qquad (9.110)$$

$$\frac{\partial N_3}{\partial L_1} = \frac{\partial N_3}{\partial L_3}\frac{\partial L_3}{\partial L_1} = (4L_3 - 1)(-1) = 4(L_1 + L_2) - 3 \qquad (9.111)$$

Finally, for the midside nodes:

$$N_4 = \Lambda_1^1(L_1)\Lambda_1^1(L_2)\Lambda_0^0(L_3) = \frac{(L_1 - 0.0)}{(0.5 - 0.0)}\frac{(L_2 - 0.0)}{(0.5 - 0.0)}(1) = 4L_1L_2 \qquad (9.112)$$

$$\frac{\partial N_4}{\partial L_1} = 4L_2 \quad ; \quad \frac{\partial N_4}{\partial L_2} = 4L_1 \qquad (9.113)$$

$$N_5 = \Lambda_0^0(L_1)\Lambda_1^1(L_2)\Lambda_1^1(L_3) = (1)\frac{(L_2 - 0.0)}{(0.5 - 0.0)}\frac{(L_3 - 0.0)}{(0.5 - 0.0)} = 4L_2L_3 \qquad (9.114)$$

$$\frac{\partial N_5}{\partial L_1} = -4L_2 \quad ; \quad \frac{\partial N_5}{\partial L_2} = 4(1 - L_1 - 2L_2) \qquad (9.115)$$

$$N_6 = \Lambda_1^1(L_1)\Lambda_0^0(L_2)\Lambda_1^1(L_3) = \frac{(L_1 - 0.0)}{(0.5 - 0.0)}(1)\frac{(L_3 - 0.0)}{(0.5 - 0.0)} = 4L_1L_3 \qquad (9.116)$$

$$\frac{\partial N_6}{\partial L_1} = 4(1 - 2L_1 - L_2) \quad ; \quad \frac{\partial N_6}{\partial L_2} = -4L_1 \qquad (9.117)$$

9.4.3 Tetrahedral Elements

The three-dimensional counterpart of the triangle is the tetrahedron. The first two members of the tetrahedral family of elements are shown in Figure 9.22.

As with triangles, *complete* interpolating polynomials, in this case in three coordinates, are used for each member of the tetrahedral element family. Each face of a tetrahedral element contains the same number of

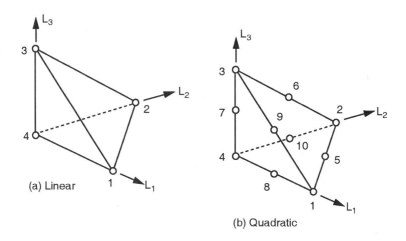

Figure 9.22: Tetrahedral Elements

nodes as found in the corresponding triangular element of the same order. As such, in the plane of each tetrahedral face, the order of the two-
dimensional polynomial is the same as for the corresponding triangular
element.

In light of the relation between triangular and tetrahedral elements, it
is not surprising that the determination of interpolation functions for the
two classes of elements is quite similar. For purposes of determining the
order of the one-dimensional Lagrange interpolating polynomial associated
with an element of order M, the nodes in the element are assigned the
indices (I, J, K, L), where $I + J + K + L = M$. The interpolation function
for node i is then determined from the following expression

$$N_i(L_1,\ L_2,\ L_3,\ L_4) = \Lambda_I^I(L_1)\Lambda_J^J(L_2)\Lambda_K^K(L_3)\Lambda_L^L(L_4) \qquad (9.118)$$

where equation (9.95) applies, and

$$\Lambda_0^0(L_1) = \Lambda_0^0(L_2) = \Lambda_0^0(L_3) = \Lambda_0^0(L_4) = 1 \qquad (9.119)$$

The quantities $L_1,\ L_2,\ L_3$ and L_4 represent the natural ("volume")
coordinates introduced in Appendix C for tetrahedral elements, where

$$L_1 + L_2 + L_3 + L_4 = 1 \qquad (9.120)$$

The results of the above procedure, as applied to the first two members of the tetrahedral family of elements, is presented below.

Linear Tetrahedral Element

The linear tetrahedral element ($p = 1$) shown in Figure 9.22(a) is the three-dimensional "simplex."[15] The interpolation function N_1 ($I = 1$, $J = K = L = 0$) is determined in the following manner:

$$N_1 = \Lambda_1^1(L_1)\Lambda_0^0(L_2)\Lambda_0^0(L_3)\Lambda_0^0(L_4) = \frac{L_1 - 0}{1 - 0} = L_1 \qquad (9.121)$$

Similarly, $N_2 = L_2$, $N_3 = L_3$, and $N_4 = L_4$.

The derivatives of N_1, N_2, N_3 and N_4 with respect to the natural coordinates are next determined. For $i, j = 1, 2, 3$ we have

$$\frac{\partial N_i}{\partial L_j} = \delta_{ij} \qquad (9.122)$$

In light of equation (9.120) it can be easily verified that

$$\frac{\partial N_3}{\partial L_1} = \frac{\partial N_3}{\partial L_2} = \frac{\partial N_3}{\partial L_3} = -1 \qquad (9.123)$$

Quadratic Tetrahedral Element

The quadratic ($p = 2$) tetrahedral element[16] is shown in Figure 9.22(b). The element interpolation functions and their derivatives with respect to natural coordinates are summarized in Tables 9.7 and 9.8, respectively.

---

[15] Examples of early use of the linear tetrahedral element include the work of Gallagher et al. [188], Melosh [387] and Argyris [23].

[16] The first applications of this element appear to have been by Argyris [25, 26].

Table 9.7: Interpolation Functions for Quadratic Tetrahedral Elements

Node i	Indices (I, J, K, L)	N_i
1	(2,0,0,0)	$(2L_1 - 1)L_1$
2	(0,2,0,0)	$(2L_2 - 1)L_2$
3	(0,0,2,0)	$(2L_3 - 1)L_3$
4	(0,0,0,2)	$(2L_4 - 1)L_4$
5	(1,1,0,0)	$4L_1L_2$
6	(0,1,1,0)	$4L_2L_3$
7	(0,0,1,1)	$4L_3L_4$
8	(1,0,0,1)	$4L_1L_4$
9	(1,0,1,0)	$4L_1L_3$
10	(0,1,0,1)	$4L_2L_4$

Table 9.8: Interpolation Function Derivatives for Quadratic Tetrahedral Elements

i	$\partial N_i / \partial L_1$	$\partial N_i / \partial L_2$	$\partial N_i / \partial L_3$
1	$4L_1 - 1$	0	0
2	0	$4L_2 - 1$	0
3	0	0	$4L_3 - 1$
4	$4(L_1 + L_2 + L_3) - 3$	$4(L_1 + L_2 + L_3) - 3$	$4(L_1 + L_2 + L_3) - 3$
5	$4L_2$	$4L_1$	0
6	0	$4L_3$	$4L_2$
7	$-4L_3$	$-4L_3$	$4(1 - L_1 - L_2 - 2L_3)$
8	$4(1 - 2L_1 - L_2 - L_3)$	$-4L_1$	$-4L_1$
9	$4L_3$	0	$4L_1$
10	$-4L_2$	$4(1 - L_1 - 2L_2 - L_3)$	$-4L_2$

9.5 Triangular Prism Elements

While not as general as tetrahedra or hexahedra, elements based on triangular prisms have also proven useful in three-dimensional finite element analyses [370, 605]. Although the present development is limited to the three-dimensional parent domains shown in Figure 9.23, this domain is easily mapped to general *pentahedral* element geometries, possibly with curved faces.

Interpolation functions for triangular prisms are obtained by multiplying the basis functions for triangular elements by appropriate functions of ζ. Triangular prisms can be constructed using either variants of the Lagrange polynomial product approach or the serendipity type approach. Elements of the latter type tend to be more economical, since they do not contain interior nodes. Two examples are given below.

Six-Node Triangular Prism

The first member of either the Lagrangian or serendipity family is the six-node element shown in Figure 9.23(a). The basis functions associated with this element are: 1, ξ, η, ζ, $\xi\zeta$, $\eta\zeta$. The interpolating polynomial is complete to degree $p = 1$.

The interpolation functions for the six-node triangular prism are:

$$N_i = \frac{1}{2}\left(1 - \xi - \eta\right)\left(1 + \zeta_i\zeta\right) \quad ; \quad i = 1, 4 \tag{9.124}$$

$$N_i = \frac{1}{2}\xi\left(1 + \zeta_i\zeta\right) \quad ; \quad i = 2, 5 \tag{9.125}$$

$$N_i = \frac{1}{2}\eta\left(1 + \zeta_i\zeta\right) \quad ; \quad i = 3, 6 \tag{9.126}$$

Fifteen-Node Triangular Prism

The quadratic serendipity element is shown in Figure 9.23(b). The basis functions associated with this element are: 1, ξ, η, ζ, ξ^2, η^2, ζ^2, $\xi\eta$, $\xi\zeta$, $\eta\zeta$, $\xi^2\zeta$, $\xi\eta\zeta$, $\xi\zeta^2$, $\eta^2\zeta$, $\eta\zeta^2$. The interpolating polynomial is complete to degree $p = 2$.

The interpolation functions for the fifteen-node triangular prism are summarized below.

$$N_i = \frac{1}{2}\left(1 - \xi - \eta\right)\left(1 + \zeta_i\zeta\right)\left(\zeta_i\zeta - 2\xi - 2\eta\right) \quad ; \quad i = 1, 4 \qquad (9.127)$$

$$N_i = \frac{1}{2}\xi\left(1 + \zeta_i\zeta\right)\left(\zeta_i\zeta + 2\xi - 2\right) \quad ; \quad i = 2, 5 \qquad (9.128)$$

$$N_i = \frac{1}{2}\eta\left(1 + \zeta_i\zeta\right)\left(\zeta_i\zeta + 2\xi - 2\right) \quad ; \quad i = 3, 6 \qquad (9.129)$$

$$N_i = 2\xi\left(1 - \xi - \eta\right)\left(1 + \zeta_i\zeta\right) \quad ; \quad i = 7, 13 \qquad (9.130)$$

$$N_i = 2\xi\eta\left(1 + \zeta_i\zeta\right) \quad ; \quad i = 8, 14 \qquad (9.131)$$

$$N_i = 2\eta\left(1 - \xi - \eta\right)\left(1 + \zeta_i\zeta\right) \quad ; \quad i = 9, 15 \qquad (9.132)$$

$$N_{10} = \left(1 - \xi - \eta\right)\left(1 + \zeta_i\zeta\right) \qquad (9.133)$$

$$N_{11} = \xi\left(1 + \zeta_i\zeta\right) \qquad (9.134)$$

$$N_{12} = \eta\left(1 + \zeta_i\zeta\right) \qquad (9.135)$$

Further details pertaining to triangular prisms are given in [370, 605].

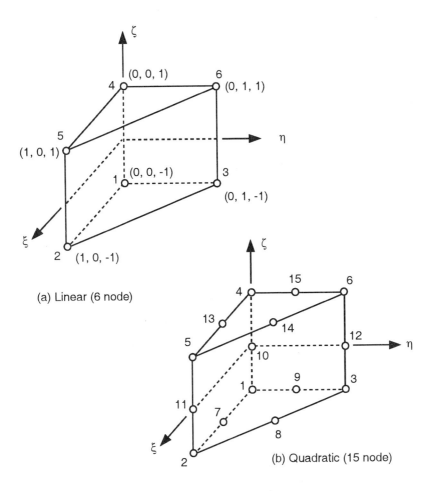

Figure 9.23: Triangular Prism Family of Elements

9.6   Transition Elements

As mentioned in Chapter 8, finite element meshes are rarely uniform. Instead, higher element densities are typically employed in portions of the solution domain where the gradients of the primary dependent variables are expected or known to be high. Away from such locations, a less dense mesh is often warranted. The connection between the fine and coarse mesh regions is realized through a suitable transition region. Some approaches to realizing such regions are presented and discussed in this section.

(a)

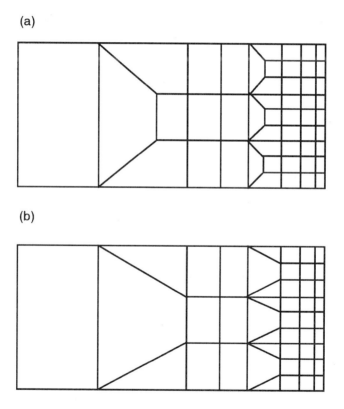

(b)

Figure 9.24: Graded Mesh Consisting of: (a) Only Quadrilateral Elements, (b) A Mixture of Quadrilateral and Triangular Elements

If the mesh is generated using an automatic or semi-automatic preprocessor, then the transition in element density is typically realized using the same element type as is used to create the surrounding mesh. For example, the hypothetical transition from nine elements to a single element shown in Figure 9.24(a) is realized using only quadrilateral elements. As shown in Figure 9.24(b), essentially the same mesh transition can likewise be realized by using a mixture of quadrilateral and triangular elements.

Transitions in element density can also be realized by using special "transition elements." Such elements are typically available in general purpose finite element analysis programs, especially programs that have been around for a while. Some examples of special transition elements are presented below.

Mesh transitions for quadratic quadrilateral elements can be realized using special five- and six-node elements (Figure 9.25). The interpolation functions for these transition elements are determined in an exercise.

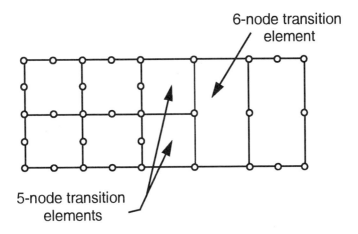

Figure 9.25: Application of Five- and Six-Node Quadrilateral Transition Elements

The transition from biquadratic to bilinear quadrilateral elements shown in Figure 9.26 also requires the five-node element. Using a different five-node element, mesh transitions for four-node quadrilateral elements can be realized (Figure 9.27). The interpolation functions for this transition element are determined in an exercise.

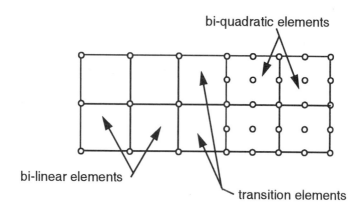

Figure 9.26: Transition from Biquadratic to Bilinear Quadrilateral Elements

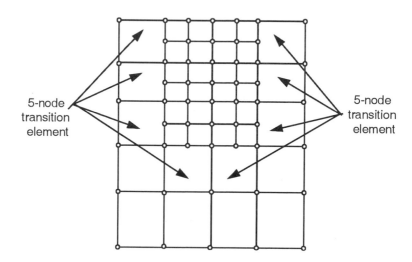

Figure 9.27: Use of Five-Node Transition Element for Bilinear Quadrilaterals

9.7 Nodal Condensation

In the discussion of the Lagrangian family of elements, mention was made of the process of *nodal* or *static condensation*. This process involves the removal, at the element level, of degrees of freedom associated with interior nodes. Some details pertaining to nodal condensation are presented in this section.

The static condensation method was initially used to eliminate internal degrees of freedom in a quadrilateral element constructed from four triangles [569]. Noting that the method was more general than the application at the element level, Wilson [570] showed that it could likewise be used to reduce the number of degrees of freedom associated with the global equations.

To better understand the process of nodal condensation, consider the hypothetical four node triangular element shown in Figure 9.28. Each element degree of freedom is approximated in the usual fashion, viz.,

$$\hat{\phi}^{(e)} = N_1 \hat{\phi}_1^{(e)} + N_2 \hat{\phi}_2^{(e)} + N_3 \hat{\phi}_3^{(e)} + \mu \hat{\phi}_4^{(e)} \qquad (9.136)$$

where μ represents an "element unknown" function. In the case of Lagrangian elements with interior nodes, μ would be the interpolation function associated with an interior node. Whereas nodes 1, 2 and 3 contribute to interelement continuity, node 4 is associated only with the particular element. Thus, prior to forming the global equations, the interior degrees of freedom are often "condensed" out of the element equations. For the four node triangle under consideration, this is realized in the following manner.

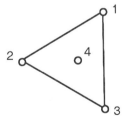

Figure 9.28: Hypothetical Triangular Element

Assuming one degree of freedom per node, the element equations are qualitatively represented as

$$\begin{bmatrix} k_{11}^e & k_{12}^e & k_{13}^e & k_{14}^e \\ k_{21}^e & k_{22}^e & k_{23}^e & k_{24}^e \\ k_{31}^e & k_{32}^e & k_{33}^e & k_{34}^e \\ k_{41}^e & k_{42}^e & k_{43}^e & k_{44}^e \end{bmatrix} \begin{Bmatrix} \hat{\phi}_1^{(e)} \\ \hat{\phi}_2^{(e)} \\ \hat{\phi}_3^{(e)} \\ \hat{\phi}_4^{(e)} \end{Bmatrix} = \begin{Bmatrix} f_1^e \\ f_2^e \\ f_3^e \\ f_4^e \end{Bmatrix} \tag{9.137}$$

While still on the element level, we proceed to eliminate $\hat{\phi}_4^{(e)}$. Solving the fourth equation for this nodal unknown gives

$$\hat{\phi}_4^{(e)} = \frac{1}{k_{44}^e} \left(f_4^e - k_{41}^e \hat{\phi}_1^{(e)} - k_{42}^e \hat{\phi}_2^{(e)} - k_{43}^e \hat{\phi}_3^{(e)} \right) \tag{9.138}$$

Substituting equation (9.138) into the first three equations (9.137) gives, after straightforward manipulation

$$\begin{bmatrix} k_{11}^{e*} & k_{12}^{e*} & k_{13}^{e*} \\ k_{21}^{e*} & k_{22}^{e*} & k_{23}^{e*} \\ k_{31}^{e*} & k_{32}^{e*} & k_{33}^{e*} \end{bmatrix} \begin{Bmatrix} \hat{\phi}_1^{(e)} \\ \hat{\phi}_2^{(e)} \\ \hat{\phi}_3^{(e)} \end{Bmatrix} = \begin{Bmatrix} f_1^{e*} \\ f_2^{e*} \\ f_3^{e*} \end{Bmatrix} \tag{9.139}$$

where

$$k_{ij}^{e*} = \left(k_{ij}^e - \frac{k_{i4}^e k_{4j}^e}{k_{44}^e} \right) \tag{9.140}$$

and

$$f_i^{*e} = f_i^e - \frac{k_{i4}^e}{k_{44}^e} f_4^e \tag{9.141}$$

where $i, j = 1, 2, 3$. Using the standard assembly procedure discussed in Chapter 8, the modified element equations (9.139) are then summed into the global arrays. Once the nodal unknowns $\hat{\phi}_1^{(e)}$, $\hat{\phi}_2^{(e)}$ and $\hat{\phi}_3^{(e)}$ have been determined, $\hat{\phi}_4^{(e)}$ can be computed from equation (9.138).

To generalize the above process, assume that n_p degrees of freedom are to be "condensed" from a total of $(n_d + n_p)$ element degrees of freedom. Assume that the element arrays are partitioned as

$$\begin{bmatrix} \mathbf{K}_{11}^e & \mathbf{K}_{12}^e \\ \mathbf{K}_{21}^e & \mathbf{K}_{22}^e \end{bmatrix} \begin{Bmatrix} \hat{\phi}_1^{(e)} \\ \hat{\phi}_2^{(e)} \end{Bmatrix} = \begin{Bmatrix} \mathbf{f}_1^e \\ \mathbf{f}_2^e \end{Bmatrix} \tag{9.142}$$

where $\mathbf{K}_{11}^e$, $\mathbf{K}_{12}^e$, $\mathbf{K}_{21}^e$, $\mathbf{K}_{22}^e$ have the sizes $(n_d * n_d)$, $(n_d * n_p)$, $(n_p * n_d)$, and $(n_p * n_p)$, respectively. It follows that the vectors $\phi_1^{(e)}$ and $\phi_2^{(e)}$ have the sizes $(n_d * 1)$ and $(n_p * 1)$, respectively, with the latter vector containing all the degrees of freedom to be "condensed."

Using the second matrix equation in (9.142), we solve for $\hat{\phi}_2^{(e)}$. Although in actual applications this solution would be obtained using Gauss elimination, symbolically we write

$$\hat{\phi}_2^{(e)} = \left(\mathbf{K}_{22}^e\right)^{-1} \left(\mathbf{f}_2^e - \mathbf{K}_{21}^e \hat{\phi}_1^{(e)}\right) \qquad (9.143)$$

Finally, equation (9.143) is substituted into the first matrix equation in (9.142) to obtain the following "condensed" set of equations

$$\left(\mathbf{K}_{11}^e - \mathbf{K}_{12}^e \left(\mathbf{K}_{22}^e\right)^{-1} \mathbf{K}_{21}^e\right) \hat{\phi}_1^{(e)} = \mathbf{f}_1^e - \mathbf{K}_{12}^e \left(\mathbf{K}_{22}^e\right)^{-1} \mathbf{f}_2^e \qquad (9.144)$$

Physically, the term $\mathbf{K}_{12}^e \left(\mathbf{K}_{22}^e\right)^{-1} \mathbf{K}_{21}^e$ represents the modification of $\mathbf{K}^e$ due to the "release" of the degrees of freedom $\hat{\phi}_2^{(e)}$. The right hand side term $\mathbf{K}_{12}^e \left(\mathbf{K}_{22}^e\right)^{-1} \mathbf{f}_2^e$ represents the "force" carried over from $\hat{\phi}_2^{(e)}$ to the $\hat{\phi}_1^e$ degrees of freedom [569].

In light of the above discussion, it is evident that the process of nodal condensation reduces the number of degrees of freedom associated with the global equations as well as the size of the element arrays. However, if nodal condensation is carried out without taking advantage of identical elements, the total computational effort associated with condensation and the solution of the reduced set of global equations will essentially be *the same* as solving the global equations assembled from the *uncondensed* element arrays [570].

Remarks

1. Readers familiar with matrix structural analysis will note the similarity between nodal condensation and the topic of force/moment releases [329].

2. When applied to groups of elements, nodal condensation is more commonly referred to as *substructuring* [447].

9.8   Concluding Remarks

In this chapter, procedures for directly writing interpolation functions for several classes of elements have been presented. To facilitate the understanding of such procedures, we note the following:

- A systematic approach for deriving the interpolation functions for variable-node quadrilateral elements is described in [58, 252, 605].

- A systematic approach for deriving the interpolation functions for hexahedral elements possessing eight to twenty nodes is described in [58].

- A systematic approach for deriving the interpolation functions for triangular elements possessing three to six nodes is described in [58].

- A systematic approach for deriving the interpolation functions for tetrahedral elements possessing four to ten nodes is described in [58].

For other element families whose interpolation functions cannot be written down directly, the associated functions have been listed in this chapter for reference.

Remark

1. The aforementioned systematic approaches for deriving the interpolation functions for elements with variable numbers of nodes are *not* the most efficient way in which to implement interpolation function routines on a computer. For an element with a given number of nodes, it is typically much better to explicitly list the respective interpolation functions in an IF – ELSE IF – END IF construction.

9.9 Exercises

9.1
Using suitable Lagrange polynomials, derive the interpolation functions associated with the cubic one-dimensional element shown in Figure 9.3c.

9.2
Using equation (9.37), explicitly derive the interpolation functions for the biquadratic Lagrange element. The parent domain for this element is shown in Figure 9.6.

9.3
Using equation (9.52), explicitly derive the interpolation functions for the trilinear Lagrange element. The parent domain for this element is shown in Figure 9.8.

9.4
The grading of finite element meshes involving biquadratic continuum elements is often facilitated by the use of five- and six-noded transition elements such as those shown in Figure 9.25. Determine the interpolation functions associated with the five- and six-node transition elements. Is the completeness criterion satisfied for these elements? To facilitate your work, adopt the local node numbering shown in Figure 9.29.

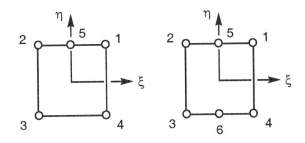

Figure 9.29: Five- and Six-Node Transition Elements

Note that the five-node element is *quadratic* along edge 1-5-2, and is *linear* along the remaining edges. Along its edge 1-5-2, the six-node element is *piecewise linear*; along edge 4-6-3 it is *quadratic*.

9.5

Using the approach presented in Section 9.2.2, determine the interpolation functions associated with a six-node Lagrange element. To facilitate your work, adopt the local node numbering of the six-node element shown in Figure 9.29. Note that along the edges 1-5-2 and 4-6-3 this element is *quadratic*. Along the remaining two edges, it is *linear*. Does the element satisfy stipulations related to the completeness criterion?

9.6

It is desired to develop a transition element that will facilitate abrupt changes in mesh refinement between domains consisting of bilinear quadrilateral elements such as those shown in Figure 9.27. The transition element is to be designed to maintain continuity across all element interfaces. An interpolation function associated with node 5 that accomplishes this is given by

$$N_5 = \begin{cases} 0.5(1+\eta)(1+\xi) & \text{if } \xi \leq 0 \\ 0.5(1+\eta)(1-\xi) & \text{if } \xi > 0 \end{cases}$$

Starting with the usual bilinear interpolation functions, determine the appropriate modifications required for the above transition element and derive the complete set of interpolation functions. To facilitate your work, adopt the local node numbering for the five-node element shown in Figure 9.29. Does the element satisfy the completeness criterion?

9.7

Consider the quadratic (3-node) one-dimensional line element shown in Figure 9.30.

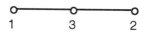

Figure 9.30: Hypothetical Quadratic Line Element

Assuming one degree of freedom per node, the element equations are given by

$$\begin{bmatrix} 6.0 & -3.0 & -2.0 \\ -3.0 & 4.0 & -1.0 \\ -2.0 & -1.0 & 2.0 \end{bmatrix} \begin{Bmatrix} \hat{\phi}_1^{(e)} \\ \hat{\phi}_2^{(e)} \\ \hat{\phi}_3^{(e)} \end{Bmatrix} = \begin{Bmatrix} 1.0 \\ 3.0 \\ 5.0 \end{Bmatrix} \tag{9.145}$$

Using the process of "nodal condensation," reduce the element to one with two nodes (i.e., only nodes 1 and 2); determine the associated element arrays.

9.8

Using the approach presented in Section 9.4.2, derive the interpolation functions for a 10-node cubic triangular element such as that shown in Figure 9.31(a). As an example, consider node 10. The associated interpolation function is

$$N_{10} = \Lambda_1^1(L_1)\Lambda_1^1(L_2)\Lambda_1^1(L_3) = \frac{L_1 - 0}{\frac{1}{3} - 0}\frac{L_2 - 0}{\frac{1}{3} - 0}\frac{L_3 - 0}{\frac{1}{3} - 0} = 27L_1L_2L_3$$

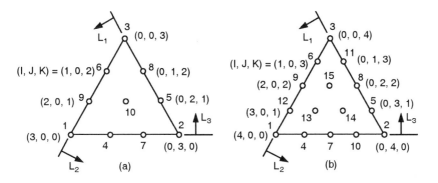

Figure 9.31: Triangular Elements: (a) Cubic and (b) Quartic

9.9

Using the approach presented in Section 9.4.2, derive the interpolation functions for a 15-node quartic triangular element such as that shown in Figure 9.31(b). Such cubic strain elements were used by Sloan and Randolph [500] in their critical examination and extension of the work of Nagtegaal et al. [402] on numerical simulation of collapse loads.

9.10

Using the method described in Section 9.4.3, derive the interpolation functions for a 20-node cubic tetrahedral element such as that shown in Figure 9.32.

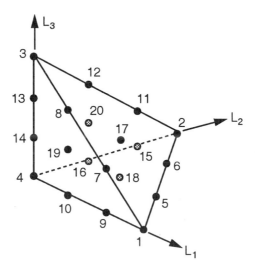

Figure 9.32: Cubic Tetrahedral Element

Chapter 10

Element Mapping

10.1 Introductory Remarks

The elements studied to this point have possessed simple shapes such as straight-sided equilateral triangles, rectangles, straight-sided tetrahedra, and rectangular prisms. When analyzing problems involving curved boundaries, large numbers of such elements must be used in order to minimize the domain discretization error discussed in Section 8.2.

To overcome this shortcoming requires elements possessing more general, distorted shapes. Thus, curved-sided triangles are used in lieu of straight-sided elements. Instead of rectangles, the more general case of quadrilateral elements, possibly with curved sides, are adopted. In three-dimensions, rectangular prisms are generalized to hexahedra, possibly with curved faces; tetrahedra possessing curved faces can likewise be used.

The distortion of basic element shapes is realized by mapping simple one-, two-, or three-dimensional "parent" domains Ω^p into new, distorted configurations that constitute the element domains Ω^e. For example, in two dimensions Ω^p for quadrilateral elements is a biunit square in (ξ, η) natural coordinates as discussed in Appendix C. Through suitable manipulations, Ω^p is mapped into a new, curvilinear set of Cartesian coordinates (x, y) (Figure 10.1).

A similar mapping, though employing (L_1, L_2, L_3) natural coordinates, is used for triangles (Figure 10.2).

In three dimensions, Ω^p for rectangular prisms is a triunit cube in (ξ, η, ζ) natural coordinates. Through suitable manipulations, Ω^p is mapped into a new, curvilinear set of Cartesian coordinates (x, y, z). For tetrahedra, Ω^p is described in (L_1, L_2, L_3, L_4) natural coordinates.

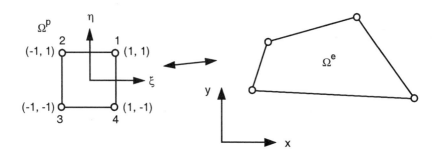

Figure 10.1: Mapping of a Quadrilateral Element

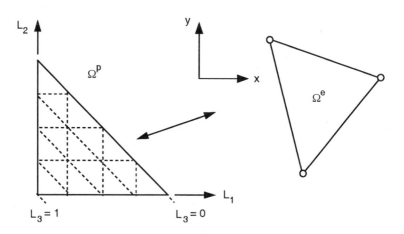

Figure 10.2: Mapping of a Triangular Element

The development of element equations is greatly simplified by carrying out the requisite differentiations and integrations over Ω^p. Once developed, these equations are transformed back to global coordinates.

10.2 General Aspects of Mapping

The mapping of Ω^p to distorted Ω^e and the development of element equations over Ω^p is predicated upon the ability to transform or map natural coordinates to global ones and vice versa. In a general sense mapping implies a function that assigns to a point in space a point in space [575]. A mapping is typically described in terms of some functional relationship between the two coordinate systems.

The general mapping from natural (ξ, η, ζ) coordinates to Cartesian (x, y, z) coordinates is mathematically expressed by[1]

$$x = f_1(\xi, \eta, \zeta) \tag{10.1}$$
$$y = f_2(\xi, \eta, \zeta) \tag{10.2}$$
$$z = f_3(\xi, \eta, \zeta) \tag{10.3}$$

Naturally, we wish a point with natural coordinates (ξ, η, ζ) to have a unique set of global coordinates (x, y, z). Thus, we require that throughout the element domain Ω^e, the coordinates (x, y, z) be *single-valued* functions of the (ξ, η, ζ). Moreover, we wish the point with coordinates (x, y, z) to have a unique set of natural coordinates (ξ, η, ζ). Thus, we also require that equations (10.1) to (10.3) be solvable for (ξ, η, ζ) as *single-valued* functions of the (x, y, z), viz.,

$$\xi = f_4(x, y, z) \tag{10.4}$$
$$\eta = f_5(x, y, z) \tag{10.5}$$
$$\zeta = f_6(x, y, z) \tag{10.6}$$

In advanced calculus, it is shown [98] that if the first partial derivatives of the coordinate functions in equations (10.1) to (10.3) are continuous and

---

[1]For tetrahedral elements, the natural coordinates (ξ, η, ζ) appearing in the subsequent development must appropriately be replaced by the volume coordinates (L_1, L_2, L_3, L_4), where $L_4 = 1 - L_1 - L_2 - L_3$. For triangular elements, the natural coordinates (ξ, η) must be replaced by the area coordinates (L_1, L_2, L_3), where $L_3 = 1 - L_1 - L_2$.

if, throughout Ω^e, the *Jacobian* determinant[2] is *non-zero*, then around any interior point of Ω^e there exists a neighborhood in which equations (10.1) to (10.3) have a single-valued inverse described by equations (10.4) to (10.6) [575]. Naturally, since the inverse mapping equations (10.4) to (10.6) are to be uniquely solvable for (ξ, η, ζ), the Jacobian determinant associated with the *inverse* mapping must likewise be non-zero throughout Ω^e.

In three dimensions, the Jacobian matrix has the following form:

$$
\mathbf{J} = \begin{bmatrix}
\dfrac{\partial x}{\partial \xi} & \dfrac{\partial y}{\partial \xi} & \dfrac{\partial z}{\partial \xi} \\[2mm]
\dfrac{\partial x}{\partial \eta} & \dfrac{\partial y}{\partial \eta} & \dfrac{\partial z}{\partial \eta} \\[2mm]
\dfrac{\partial x}{\partial \zeta} & \dfrac{\partial y}{\partial \zeta} & \dfrac{\partial z}{\partial \zeta}
\end{bmatrix}
\tag{10.7}
$$

Once the functions f_1, f_2 and f_3 are selected in equations (10.1) to (10.3), and coordinates are chosen for every element such that these map into contiguous spaces, then interpolation functions written in the local element (ξ, η, ζ) space can be used to represent the function variation over the element in the global (x, y, z) space. This must, of course, be achieved without upsetting the requirement of interelement continuity.

In addition, the coordinate change must not distort the element excessively, or else the Jacobian determinant may vanish within the region of integration. Excessive distortion will also destroy the accuracy built into the element [515]. Polynomials in the new variables do not correspond to polynomials in the old. To preserve the finite element approximation requires that the coordinate changes should be smooth. Further discussion of element distortions is deferred until Section 10.6.

10.3   Treatment of Derivatives

The element equations contain the derivatives of the interpolation functions with respect to the global (x, y, z) coordinates. However, these interpolation functions are typically written in the (ξ, η, ζ) natural coordinate system. Derivatives in these two coordinate systems are related through the chain rule of differentiation, viz.,

---

[2]Named for the German mathematician Karl Gustav Jacob Jacobi (1804-1851), who introduced it in 1841. The *Jacobian matrix* or *Jacobian operator* $\mathbf{J}$ is frequently referred to simply as the "Jacobian" of the mapping or transformation.

$$\frac{\partial}{\partial x} = \frac{\partial}{\partial \xi}\frac{\partial \xi}{\partial x} + \frac{\partial}{\partial \eta}\frac{\partial \eta}{\partial x} + \frac{\partial}{\partial \zeta}\frac{\partial \zeta}{\partial x} \tag{10.8}$$

$$\frac{\partial}{\partial y} = \frac{\partial}{\partial \xi}\frac{\partial \xi}{\partial y} + \frac{\partial}{\partial \eta}\frac{\partial \eta}{\partial y} + \frac{\partial}{\partial \zeta}\frac{\partial \zeta}{\partial y} \tag{10.9}$$

$$\frac{\partial}{\partial z} = \frac{\partial}{\partial \xi}\frac{\partial \xi}{\partial z} + \frac{\partial}{\partial \eta}\frac{\partial \eta}{\partial z} + \frac{\partial}{\partial \zeta}\frac{\partial \zeta}{\partial z} \tag{10.10}$$

However, the evaluation of the partial derivatives $\partial/\partial x$, $\partial/\partial y$ and $\partial/\partial z$ requires the calculation of $\partial \xi/\partial x$, $\partial \eta/\partial x$, $\partial \zeta/\partial x$, etc., which means that the inverse relationships given by equations 10.4 to 10.6 need to be evaluated. Since these relationships are, in general, difficult to establish explicitly, the necessary derivatives are instead computed in the following manner. Using the chain rule, we write suitable expressions for $\partial/\partial \xi$, $\partial/\partial \eta$ and $\partial/\partial \zeta$, viz.,

$$\frac{\partial}{\partial \xi} = \frac{\partial}{\partial x}\frac{\partial x}{\partial \xi} + \frac{\partial}{\partial y}\frac{\partial y}{\partial \xi} + \frac{\partial}{\partial z}\frac{\partial z}{\partial \xi} \tag{10.11}$$

$$\frac{\partial}{\partial \eta} = \frac{\partial}{\partial x}\frac{\partial x}{\partial \eta} + \frac{\partial}{\partial y}\frac{\partial y}{\partial \eta} + \frac{\partial}{\partial z}\frac{\partial z}{\partial \eta} \tag{10.12}$$

$$\frac{\partial}{\partial \zeta} = \frac{\partial}{\partial x}\frac{\partial x}{\partial \zeta} + \frac{\partial}{\partial y}\frac{\partial y}{\partial \zeta} + \frac{\partial}{\partial z}\frac{\partial z}{\partial \zeta} \tag{10.13}$$

Written in matrix form, equations (10.11) to (10.13) become

$$\left\{ \begin{array}{c} \dfrac{\partial}{\partial \xi} \\[2mm] \dfrac{\partial}{\partial \eta} \\[2mm] \dfrac{\partial}{\partial \zeta} \end{array} \right\} = \mathbf{J} \left\{ \begin{array}{c} \dfrac{\partial}{\partial x} \\[2mm] \dfrac{\partial}{\partial y} \\[2mm] \dfrac{\partial}{\partial z} \end{array} \right\} \tag{10.14}$$

where **J** represents the Jacobian matrix defined in equation (10.7). Provided that **J** is non-singular, the desired derivatives with respect to x, y and to z are obtained by inverting equation (10.14) to give

$$\left\{ \begin{array}{c} \dfrac{\partial}{\partial x} \\[2ex] \dfrac{\partial}{\partial y} \\[2ex] \dfrac{\partial}{\partial z} \end{array} \right\} = \mathbf{J}^{-1} \left\{ \begin{array}{c} \dfrac{\partial}{\partial \xi} \\[2ex] \dfrac{\partial}{\partial \eta} \\[2ex] \dfrac{\partial}{\partial \zeta} \end{array} \right\} \tag{10.15}$$

As stated in the previous section, $\mathbf{J}^{-1}$ will exist provided that a unique one-to-one correspondence exists between the (ξ, η, ζ) and (x, y, z) coordinate systems. That is, to each value of ξ, η, and ζ there corresponds only one value of x, y and z, respectively. This one-to-one mapping of coordinates will be realized provided that, within the element or on its boundary, the *sign* of the determinant of **J** remains *unchanged* [605].

To gain some insight into equations (10.14) and (10.15), consider spatially one-dimensional problem. For this case the Jacobian matrix given in equation (10.7) is simply

$$\mathbf{J} = \frac{\partial x}{\partial \xi} \tag{10.16}$$

Equations equations (10.14) and (10.15) thus simplify to

$$\frac{\partial}{\partial \xi} = \frac{\partial x}{\partial \xi} \frac{\partial}{\partial x} \quad \text{and} \quad \frac{\partial}{\partial x} = \frac{\partial \xi}{\partial x} \frac{\partial}{\partial \xi} \tag{10.17}$$

These equations represent the familiar chain rule for differentiation in one-dimension.

10.4   Treatment of Integrals

When elements have curved boundaries, the integrals associated with the element matrices are most easily evaluated using suitable parent domains Ω^p in natural coordinate space. There is thus no need to deal with equations for the curved boundaries.

This evaluation of integrals is, however, not without its limitations. For example, as the order of the element increases, so does the complexity of terms involved in the integration. Furthermore, the mapping used to distort the shape of Ω^p involves inversion of the Jacobian matrix. The complexity of the associated integrals makes their exact evaluation impractical. In such cases the integration must be approximated using *numerical integration*, which is also called *quadrature*.

The numerical approximation of an integral has the general form

$$\int_a^b f(x)\,dx \approx \sum_{i=1}^n f(x_i) w_i \tag{10.18}$$

where $f(x)$ is assumed to be smooth and integrable and $[a, b]$ is a closed interval.

For finite element applications the most efficient approach approach to numerical integration typically involves the use of Gaussian quadrature. The idea behind Gaussian quadrature is to choose the quadrature points x_i and the weighting coefficients w_i appearing in equation (10.18) in an *optimal* manner. By using polynomials orthogonal to the polynomial being integrated, Gaussian quadrature formulae achieve a degree of precision of $2n - 1$. For finite element applications, the *Legendre* polynomials represent the most commonly used set of orthogonal polynomials. Further details pertaining to Gaussian quadrature, including values for x_i and w_i associated with Gaussian-Legendre quadrature, are given in Appendix F.

The following examples illustrate some simple coordinate mappings and the calculation of the Jacobian determinant.

Example 10.1. Mapping Between Cylindrical and Rectangular Coordinates

Consider the well-known relation between cylindrical (r, θ) and rectangular (x, y) Cartesian coordinates:

$$x = r \cos\theta \quad ; \quad y = r \sin\theta \tag{10.19}$$

This relationship represents a mapping by which points in (r, θ) space are mapped into (x, y) space. This is shown in Figure 10.3.

The Jacobian matrix is thus given by

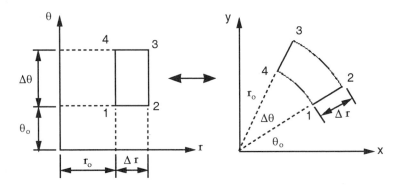

Figure 10.3: Mapping Between Cylindrical and Rectangular Coordinates

$$\mathbf{J} = \begin{bmatrix} \dfrac{\partial x}{\partial r} & \dfrac{\partial y}{\partial r} \\[2ex] \dfrac{\partial x}{\partial \theta} & \dfrac{\partial y}{\partial \theta} \end{bmatrix} = \begin{bmatrix} \cos\theta & \sin\theta \\[1ex] -r\sin\theta & r\cos\theta \end{bmatrix} \tag{10.20}$$

It follows that the inverse of the Jacobian is

$$\mathbf{J}^{-1} = \begin{bmatrix} \cos\theta & -\dfrac{\sin\theta}{r} \\[2ex] \sin\theta & \dfrac{\cos\theta}{r} \end{bmatrix} \tag{10.21}$$

The derivatives are thus related in the following manner:

$$\left\{ \begin{array}{c} \dfrac{\partial}{\partial x} \\[2ex] \dfrac{\partial}{\partial y} \end{array} \right\} = \begin{bmatrix} \cos\theta & -\dfrac{\sin\theta}{r} \\[2ex] \sin\theta & \dfrac{\cos\theta}{r} \end{bmatrix} \left\{ \begin{array}{c} \dfrac{\partial}{\partial r} \\[2ex] \dfrac{\partial}{\partial \theta} \end{array} \right\} \tag{10.22}$$

The Jacobian determinant follows from equation (10.20), viz.,

$$\det \mathbf{J} = \det \begin{bmatrix} \cos\theta & \sin\theta \\ -r\sin\theta & r\cos\theta \end{bmatrix} = r(\cos\theta)^2 + r(\sin\theta)^2 = r \qquad (10.23)$$

Thus, for the purpose of evaluating integrals,

$$dx\,dy = r\,dr\,d\theta \qquad (10.24)$$

Example 10.2. Mapping Associated with a Rectangular Element

Consider the bilinear rectangular element shown in Figure 10.4.

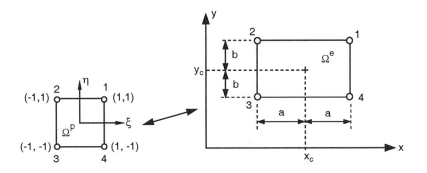

Figure 10.4: Mapping Associated with Typical Rectangular Element

The mapping from global (x, y) coordinates to the parent domain in (ξ, η) space is given by the relations developed in Appendix C, viz.,

$$\xi = \frac{x - x_c}{a} \quad ; \quad \eta = \frac{y - y_c}{b} \qquad (10.25)$$

The Jacobian and its inverse are thus

$$
\mathbf{J} = \begin{bmatrix} \dfrac{\partial x}{\partial \xi} & \dfrac{\partial y}{\partial \xi} \\[2ex] \dfrac{\partial x}{\partial \eta} & \dfrac{\partial y}{\partial \eta} \end{bmatrix} = \begin{bmatrix} a & 0 \\ 0 & b \end{bmatrix} \quad ; \quad \mathbf{J}^{-1} = \begin{bmatrix} \dfrac{1}{a} & 0 \\[2ex] 0 & \dfrac{1}{b} \end{bmatrix} \tag{10.26}
$$

The Jacobian determinant is thus equal to ab, which is one-quarter of the element area. It can easily be shown that this result holds for the more general case where the element is a parallelogram [100].

The relation between derivatives with respect to global and natural coordinates is thus

$$
\left\{ \begin{array}{c} \dfrac{\partial}{\partial x} \\[2ex] \dfrac{\partial}{\partial y} \end{array} \right\} = \mathbf{J}^{-1} \left\{ \begin{array}{c} \dfrac{\partial}{\partial \xi} \\[2ex] \dfrac{\partial}{\partial \eta} \end{array} \right\} = \begin{bmatrix} \dfrac{1}{a} & 0 \\[2ex] 0 & \dfrac{1}{b} \end{bmatrix} \left\{ \begin{array}{c} \dfrac{\partial}{\partial \xi} \\[2ex] \dfrac{\partial}{\partial \eta} \end{array} \right\} \tag{10.27}
$$

The extension of the above example to a three-dimensional rectangular prism element is straightforward.

10.5 Parametric Mapping

Having discussed the treatment of derivatives and integrals associated with element mapping, it remains to explicitly define the mapping described analytically by equations (10.1) to (10.3). A very common and convenient form of element mapping is of the *parametric* kind.[3] In parametric mapping the relationship between the natural coordinates (ξ, η, ζ) and the global coordinates (x, y, z) is written employing the *same* type of element interpolation functions as used to approximate the primary dependent variables over a given element, viz.,

$$
\hat{\phi}^{(e)} = \sum_{m=1}^{N_{dof}} \hat{\phi}_m^{(e)} N_m(\xi, \eta, \zeta) \tag{10.28}
$$

---

[3] The first use of parametric mapping for establishing curvilinear coordinates in the context of finite element analysis is attributed to Taig [526], who considered only linear quadrilateral elements. Irons [254, 255] subsequently generalized the concept to two- and three-dimensional elements. Quite independently, Coons [118] applied the concept to the generation of curved surfaces of engineering interest.

where, for the case of irreducible elements, N_{dof} is equal to the total number of nodes (N_{en}) in the element.

If $G_m(\xi, \eta, \zeta)$ represents a suitable polynomial interpolation function for the element geometry, then for each element the general mapping relations of equations (10.1) to (10.3) are written as

$$x = \sum_{m=1}^{N_{pt}} x_m^{(e)} G_m(\xi, \eta, \zeta) \qquad (10.29)$$

$$y = \sum_{m=1}^{N_{pt}} y_m^{(e)} G_m(\xi, \eta, \zeta) \qquad (10.30)$$

$$z = \sum_{m=1}^{N_{pt}} z_m^{(e)} G_m(\xi, \eta, \zeta) \qquad (10.31)$$

where $(x_m^{(e)}, y_m^{(e)}, z_m^{(e)})$ are the global (x, y, z) coordinates of the point into which point m in (ξ, η, ζ) coordinates shall be mapped, and N_{pt} equals the number of points in the element that are used in defining its geometry.

Remarks

1. In general, the local (element) node numbers 1 to N_{pt} will be mapped to some different global numbers that are assigned to the entire mesh.

2. N_{pt} does not necessarily equal to N_{en}, which is the number of nodes in an element that are used in the approximation of primary dependent variables via equation (10.28). This point is discussed further below.

The derivatives of x, y and z with respect to ξ, η and ζ are required for the Jacobian matrix of equation (10.7). For a parametric mapping, the general expressions for the derivatives of x with respect to the natural coordinates are

$$\frac{\partial x}{\partial \xi} = \sum_{m=1}^{N_{pt}} x_m^{(e)} \frac{\partial G_m}{\partial \xi} \qquad (10.32)$$

$$\frac{\partial x}{\partial \eta} = \sum_{m=1}^{N_{pt}} x_m^{(e)} \frac{\partial G_m}{\partial \eta} \qquad (10.33)$$

$$\frac{\partial x}{\partial \zeta} = \sum_{m=1}^{N_{pt}} x_m^{(e)} \frac{\partial G_m}{\partial \zeta} \tag{10.34}$$

The general expressions for the derivatives of y and z with respect to the natural coordinates are computed in a similar fashion.

In order that conforming elements in (ξ, η, ζ) coordinates be conforming in (x, y, z) coordinates, the following *global* geometric continuity condition on the mapping must be satisfied: If the element equations contain sth derivatives of the primary dependent variables, then the mapping must be of class C^{s-1} between elements [515].

For the case of $s = 1$, arising from second-order differential equations, C^0 interelement geometric continuity is required of the interpolation functions $G_m(\xi, \eta, \zeta)$. This ensures that a point common to two adjacent elements will not split into two separate points when the elements are mapped into distorted shapes.

The $G_m(\xi, \eta, \zeta)$ must also be chosen so that the condition

$$x(\xi_i, \eta_i, \zeta_i) = x_i^{(e)} \tag{10.35}$$

$$y(\xi_i, \eta_i, \zeta_i) = y_i^{(e)} \tag{10.36}$$

$$z(\xi_i, \eta_i, \zeta_i) = z_i^{(e)} \tag{10.37}$$

is satisfied. This imposes the following restriction of the geometric interpolation functions:

$$G_m(\xi_i, \eta_i, \zeta_i) = \delta_{mi} \tag{10.38}$$

where δ_{mi} is the Kronecker delta.

In general, the $G_m(\xi, \eta, \zeta)$ are not necessarily the same as the interpolation functions $N_m(\xi, \eta, \zeta)$ associated with the approximation of the primary dependent variables via equation (10.28). This raises the question of whether this approximation will be continuous once elements are distorted. It turns out that provided the N_m are such that continuity is preserved in the parent domain, then continuity requirements will likewise be satisfied in distorted configurations [605].

In light of the above observations regarding the G_m and N_m, the following three types of parametric mappings are possible:

- If *fewer* nodes are used to define the element geometry than are used to approximate the primary dependent variables, then $N_{pt} < N_{en}$ and the interpolation functions G_m are different from the functions N_m. Such an element is said to be *subparametric*.

- If the number of nodes used to define the element geometry is *greater* than the number of nodes used to approximate the primary dependent variables, then $N_{pt} > N_{en}$ and the interpolation functions G_m are again different from the functions N_m. Such an element is said to be *superparametric*.[4]

- Finally, if the *same* nodes define the element geometry and the interpolation of the primary dependent variables, then $N_{pt} = N_{en}$ and the functions G_m and N_m are *identical*. In this case, the element is said to be *isoparametric*.

10.5.1 Isoparametric Elements

Since their early development [526, 254, 255, 151, 152, 260, 589, 4], isoparametric elements have proven to be quite effective in most practical finite element analyses. Consequently, of the three possible types of parametric mappings, the isoparametric case is the most widely used. We thus investigate further some details pertaining to isoparametric elements.

Since the interpolation functions G_m and N_m are *identical* for isoparametric elements, it follows that all of the functions developed in Chapter 9 are used to describe the parametric mapping between natural and global coordinates. Thus shall be illustrated in the subsequent examples.

Example 10.3. Isoparametric Mapping for One-Dimensional Elements

Consider the linear, isoparametric one-dimensional element shown in Figure 10.5. The ξ natural coordinate system, first introduced in Appendix C, is again employed. Since the element is isoparametric, it follows that

$$x' = \sum_{m=1}^{2} x_m^{(e)} N_m(\xi) \qquad (10.39)$$

---

[4]Such elements have successfully been applied to the analysis of shells [4, 210].

The associated interpolation functions are given by equations (9.18) and (9.19), viz.,

$$N_1 = \frac{1}{2}(1 - \xi) \quad ; \quad N_2 = \frac{1}{2}(1 + \xi) \tag{10.40}$$

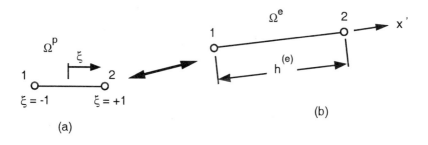

(a)

(b)

Figure 10.5: Linear One-Dimensional Element: (a) Parent Domain, (b) Actual Element

The Jacobian matrix, defined in equation (10.7), reduces to

$$\mathbf{J} = \frac{dx'}{d\xi} = \sum_{i=1}^{2} x'^{(e)}_i \frac{dN_i}{d\xi} = \frac{1}{2}\left(-x'^{(e)}_1 + x'^{(e)}_2\right) = \frac{h^{(e)}}{2} \tag{10.41}$$

The determinant of $\mathbf{J}$ is thus equal to one-half the element length, and remains *constant* throughout the element. For the purpose of computing derivatives, note that

$$\frac{d}{dx'} = \mathbf{J}^{-1}\frac{d}{d\xi} = \frac{2}{h^{(e)}}\frac{d}{d\xi} \tag{10.42}$$

Finally, for the evaluation of integrals,

$$dx' = \det \mathbf{J}\, d\xi = \frac{h^{(e)}}{2}d\xi \tag{10.43}$$

Next consider a quadratic, isoparametric one-dimensional element such as that shown in Figure 10.6. The coordinate mapping is now given by

$$x' = \sum_{m=1}^{3} x_m^{(e)} N_m(\xi) \tag{10.44}$$

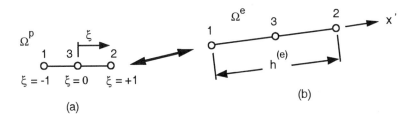

(a)

(b)

Figure 10.6: Quadratic One-Dimensional Element: (a) Parent Domain, (b) Actual Element

The associated interpolation functions are given by equations (9.20) to (9.22), viz.,

$$N_1 = \frac{1}{2}\xi(\xi - 1) \quad ; \quad N_2 = \frac{1}{2}\xi(\xi + 1) \quad ; \quad N_3 = 1 - \xi^2 \tag{10.45}$$

The Jacobian determinant is now *linear* in ξ, viz.,

$$\mathbf{J} = \frac{dx'}{d\xi} = \sum_{i=1}^{3} x'^{(e)}_i \frac{dN_i}{d\xi}$$

$$= x'^{(e)}_1 \left(\xi - \frac{1}{2}\right) + x'^{(e)}_2 \left(\xi + \frac{1}{2}\right) - 2x'^{(e)}_3 \xi$$

$$= \left(x'^{(e)}_1 + x'^{(e)}_2 - 2x'^{(e)}_3\right) \xi + \frac{1}{2}\left(x'^{(e)}_2 - x'^{(e)}_1\right) \tag{10.46}$$

If node 3 is located midway between the end nodes; that is,

$$x'^{(e)}_3 = \frac{1}{2}\left(x'^{(e)}_2 + x'^{(e)}_1\right) \tag{10.47}$$

the Jacobian determinant will be equal to $h^{(e)}/2$. Letting

$$x'^{(e)}_3 = x_1^{(e)} + a\left(x'^{(e)}_2 - x'^{(e)}_1\right) \quad ; \quad 0 < a < 1 \tag{10.48}$$

it is instructive to investigate the effect of other locations of node 3 on the Jacobian determinant. Assuming a *unit* element length, the results are summarized in Figure 10.7. From this figure it is evident that for $a \leq 0.25$ and $a \geq 0.75$, the Jacobian determinant will be zero or will change sign. Consequently, the isoparametric mapping will not be one-to-one.

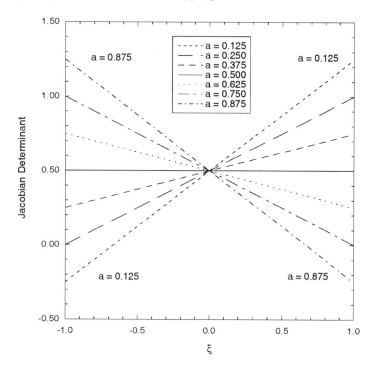

Figure 10.7: Effect of Node 3 Location on Value of Jacobian Determinant

The above findings underscore an important conclusion concerning the location of non-vertex nodes within an element. Namely, for certain elements there exist locations for which the coordinate mapping ceases to be one-to-one.

Example 10.4. : Isoparametric Mapping of a Four-Node Quadrilateral Element

The isoparametric mapping of the parent domain (biunit square) into a four node quadrilateral element is considered. The mapping is shown schematically in Figure 10.8. The node numbers associated with Ω^e are local numbers; they will be mapped to global numbers that, in general, differ from the local ones.

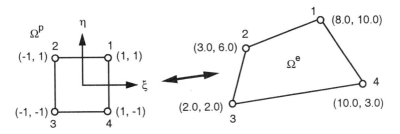

Figure 10.8: Linear Quadrilateral Element: (a) Parent Domain, (b) Actual Element

Since $N_{pt} = N_{en} = 4$, the nodal coordinates map according to the relations

$$x = \sum_{m=1}^{4} x_m^{(e)} G_m(\xi, \eta) = \sum_{m=1}^{4} x_m^{(e)} N_m(\xi, \eta) \tag{10.49}$$

$$y = \sum_{m=1}^{4} y_m^{(e)} G_m(\xi, \eta) = \sum_{m=1}^{4} y_m^{(e)} N_m(\xi, \eta) \tag{10.50}$$

where the interpolation functions are represented by equation (9.35) of Chapter 9, viz.,

$$N_i = \frac{1}{4}(1 + \xi_i \xi)(1 + \eta_i \eta) \tag{10.51}$$

The constants ξ_i and η_i are defined in Table 10.1.

Expanding equations (10.49) and (10.50), and substituting for the nodal coordinates and interpolation functions gives

$$x = 2.00(1 + \xi)(1 + \eta) + 0.75(1 - \xi)(1 + \eta) + 0.50(1 - \xi)(1 - \eta)$$
$$+2.50(1 + \xi)(1 - \eta) \tag{10.52}$$

$$y = 2.50(1 + \xi)(1 + \eta) + 1.50(1 - \xi)(1 + \eta) + 0.50(1 - \xi)(1 - \eta)$$
$$+0.75(1 + \xi)(1 - \eta) \tag{10.53}$$

Table 10.1: Values of ξ_i and η_i for Bilinear 4-Node Elements

i	ξ_i	η_i
1	1	1
2	-1	1
3	-1	-1
4	1	-1

As a check, consider the mapping of node number 2. Substituting the values $\xi = -1$ and $\eta = 1$ into the above equations gives the desired result of $x = 3.0$ and $y = 6.0$.

Note also that at the element center ($\xi = \eta = 0.0$),

$$x(0,0) = \frac{1}{4}\left(x_1^{(e)} + x_2^{(e)} + x_3^{(e)} + x_4^{(e)}\right)$$

$$= \frac{1}{4}\left(8.0 + 3.0 + 2.0 + 10.0\right) = 5.750 \tag{10.54}$$

$$y(0,0) = \frac{1}{4}\left(y_1^{(e)} + y_2^{(e)} + y_3^{(e)} + y_4^{(e)}\right)$$

$$= \frac{1}{4}\left(10.0 + 6.0 + 2.0 + 3.0\right) = 5.250 \tag{10.55}$$

These values simply represent the *average* of the nodal x and y coordinates, respectively. Finally, consider the mapping of a straight line in $\xi - \eta$ space into $x - y$ space. For the specific case of $\xi = 0.50$, equations (10.52) and (10.53) become

$$x = 3.000(1 + \eta) + 0.375(1 + \eta) + 0.250(1 - \eta) + 3.750(1 - \eta)$$

$$= 7.375 - 0.625\eta \tag{10.56}$$

$$y = 3.750(1 + \eta) + 0.750(1 + \eta) + 0.250(1 - \eta) + 1.125(1 - \eta)$$

$$= 5.875 + 3.125\eta \tag{10.57}$$

For any value of η, these equations give the coordinates of points along a straight line in $x - y$ space.

Convergence Criteria

As discussed in Chapter 7, an element will exhibit monotonic convergence provided its interpolation functions satisfy the compatibility and completeness criteria. It can be shown that the former criterion is easily satisfied for isoparametric elements [252]. To investigate the latter criterion, consider the specific case of a irreducible three-dimensional C^0 isoparametric element. The highest order derivative of $\hat{\phi}^{(e)}$ appearing in the element equations is one. Thus, according to the completeness criterion, uniform states of $\hat{\phi}^{(e)}$ and its first derivatives must be represented as the size of the element goes to zero. For this to be possible, the following expression for $\hat{\phi}^{(e)}$, defined in global coordinates, must be contained in the isoparametric formulation for each of the nodal primary dependent variables:

$$\hat{\phi}^{(e)} = \alpha_1 + \alpha_2 x + \alpha_3 y + \alpha_4 z \tag{10.58}$$

where α_1, α_2, α_3 and α_4 are constants. The nodal values of $\hat{\phi}^{(e)}$ corresponding to the field described by equation (10.58) are

$$\hat{\phi}_i^{(e)} = \alpha_1 + \alpha_2 x_i^{(e)} + \alpha_3 y_i^{(e)} + \alpha_4 z_i^{(e)} \tag{10.59}$$

for $i = 1, 2, \cdots, \ldots N_{en}$. Consider first the derivatives of $\hat{\phi}^{(e)}$ with respect to global coordinates. From equation (10.58) it follows that

$$\frac{\partial \hat{\phi}^{(e)}}{\partial x} = \alpha_2 \quad ; \quad \frac{\partial \hat{\phi}^{(e)}}{\partial y} = \alpha_3 \quad ; \quad \frac{\partial \hat{\phi}^{(e)}}{\partial z} = \alpha_4 \tag{10.60}$$

which are clearly constant and independent of element size.

The test for completeness next requires that constant values of $\hat{\phi}^{(e)}$ in equation (10.58) are indeed obtained within the element when the element nodal values of $\hat{\phi}^{(e)}$ are given by equation (10.59). Next substituting equation (10.59) into equation (10.28) gives

$$\hat{\phi}^{(e)} = \sum_{i=1}^{N_{en}} N_i \hat{\phi}_i^{(e)}$$

$$= \alpha_1 \sum_{i=1}^{N_{en}} N_i + \alpha_2 \sum_{i=1}^{N_{en}} N_i x_i^{(e)} + \alpha_3 \sum_{i=1}^{N_{en}} N_i y_i^{(e)} + \alpha_4 \sum_{i=1}^{N_{en}} N_i z_i^{(e)} \tag{10.61}$$

However, since the element is isoparametric and $N_{pt} = N_{en}$,

$$x = \sum_{i=1}^{N_{en}} N_i x_i^{(e)}, \quad y = \sum_{i=1}^{N_{en}} N_i y_i^{(e)}, \quad z = \sum_{i=1}^{N_{en}} N_i z_i^{(e)} \qquad (10.62)$$

In light of equation (10.62) , equation (10.61) reduces to

$$\hat{\phi}^{(e)} = \alpha_1 \sum_{i=1}^{N_{en}} N_i + \alpha_2 x + \alpha_3 y + \alpha_4 z \qquad (10.63)$$

The value of $\hat{\phi}^{(e)}$ in equation (10.63) will be identical to that in equation (10.58) only if, at any point within the element

$$\sum_{i=1}^{N_{en}} N_i = 1 \qquad (10.64)$$

Note that at any node, equation (10.64) is automatically satisfied because of the fundamental requirement that $N_i(x_j, y_j, z_j) = \delta_{ij}$, where δ_{ij} denotes the Kronecker delta. In summary, for isoparametric C^0 elements, satisfaction of equation (10.64) *automatically* satisfies the completeness criterion.

Remark

1. The relative ease with which the convergence criteria are satisfied is an important benefit of isoparametric elements.

10.5.2 Evaluation of Element Arrays

Some details pertaining to the development of element equations in natural coordinates are presented. Emphasis is placed on expressions related to the isoparametric mapping of elements. We defer discussion of computer implementation issues to Section 10.7.

Example 10.5. : Linear Triangular Elements

Consider a typical three node isoparametric triangular element such as that shown in Figure 10.2. From Appendix C recall that for triangular elements, the natural coordinates L_1, L_2 and L_3 are defined by

$$\left\{\begin{array}{c} L_1 \\ L_2 \\ L_3 \end{array}\right\} = \frac{1}{2A^{(e)}} \begin{bmatrix} a_1 & a_2 & a_3 \\ b_1 & b_2 & b_3 \\ c_1 & c_2 & c_3 \end{bmatrix} \left\{\begin{array}{c} 1 \\ x \\ y \end{array}\right\} \tag{10.65}$$

where $A^{(e)}$ is the area of the element, and

$$a_1 = x_2 y_3 - x_3 y_2 \quad , \quad b_1 = y_2 - y_3 \quad , \quad c_1 = x_3 - x_2 \quad (10.66)$$

$$a_2 = x_3 y_1 - x_1 y_3 \quad , \quad b_2 = y_3 - y_1 \quad , \quad c_2 = x_1 - x_3 \quad (10.67)$$

$$a_3 = x_1 y_2 - x_2 y_1 \quad , \quad b_3 = y_1 - y_2 \quad , \quad c_3 = x_2 - x_1 \quad (10.68)$$

As shown in Section 9.4.2, the interpolation functions for the linear triangle are identical to the natural coordinates; i.e., $N_i = L_i$; $i = 1, 2, 3$. Consequently, the derivatives of the interpolation functions with respect to the coordinate directions are obtained in a straightforward manner, viz.,

$$\frac{\partial N_i}{\partial x} = \frac{b_i}{2A^{(e)}} \quad ; \quad \frac{\partial N_i}{\partial y} = \frac{c_i}{2A^{(e)}} \tag{10.69}$$

Since an *isoparametric* mapping is assumed, it follows that

$$x = \sum_{i=1}^{3} x_i^{(e)} N_i \quad ; \quad y = \sum_{i=1}^{3} y_i^{(e)} N_i \tag{10.70}$$

where $(x_i^{(e)}, y_i^{(e)})$ represent the nodal coordinates. Noting that $L_3 = 1 - L_1 - L_2$ and expanding the above equations gives

$$x = L_1 x_1^{(e)} + L_2 x_2^{(e)} + (1 - L_1 - L_2) x_3^{(e)} \tag{10.71}$$

$$y = L_1 y_1^{(e)} + L_2 y_2^{(e)} + (1 - L_1 - L_2) y_3^{(e)} \tag{10.72}$$

The two-dimensional Jacobian matrix is written as

$$\mathbf{J} = \begin{bmatrix} \dfrac{\partial x}{\partial L_1} & \dfrac{\partial y}{\partial L_1} \\[2mm] \dfrac{\partial x}{\partial L_2} & \dfrac{\partial y}{\partial L_2} \end{bmatrix} = \begin{bmatrix} (x_1^{(e)} - x_3^{(e)}) & (y_1^{(e)} - y_3^{(e)}) \\[2mm] (x_2^{(e)} - x_3^{(e)}) & (y_2^{(e)} - y_3^{(e)}) \end{bmatrix} \tag{10.73}$$

The Jacobian matrix thus remains *unchanged* throughout the element. Computing the determinant of $\mathbf{J}$,

$$\det \mathbf{J} = (x_1^{(e)} - x_3^{(e)})(y_2^{(e)} - y_3^{(e)}) - (x_2^{(e)} - x_3^{(e)})(y_1^{(e)} - y_3^{(e)})$$

$$= \begin{vmatrix} 1 & 1 & 1 \\ x_1 & x_2 & x_3 \\ y_1 & y_2 & y_3 \end{vmatrix} = (a_1 + a_2 + a_3) = 2A^{(e)} \qquad (10.74)$$

we see that it is equal to twice the area of the element and is likewise obviously constant throughout the element.

Example 10.6. : Isoparametric Quadrilateral Elements

The geometry of a general quadrilateral element is defined by appropriately mapping the parent domain in (ξ, η) space onto the global (x, y) space. Since the element is isoparametric, this mapping is given by

$$x = \sum_{m=1}^{N_{en}} x_m^{(e)} N_m(\xi, \eta) \qquad (10.75)$$

$$y = \sum_{m=1}^{N_{en}} y_m^{(e)} N_m(\xi, \eta) \qquad (10.76)$$

The derivatives of interpolation functions with respect to global coordinates are computed using equation (10.15), suitably specialized for two dimensions, viz.,

$$\begin{Bmatrix} \dfrac{\partial N_i}{\partial x} \\[2ex] \dfrac{\partial N_i}{\partial y} \end{Bmatrix} = \mathbf{J}^{-1} \begin{Bmatrix} \dfrac{\partial N_i}{\partial \xi} \\[2ex] \dfrac{\partial N_i}{\partial \eta} \end{Bmatrix} \qquad (10.77)$$

where $\mathbf{J}$ represents the two-dimensional Jacobian matrix. Specializing equation (10.7) for two dimensions, equation (10.77) becomes

$$\begin{Bmatrix} \dfrac{\partial N_i}{\partial x} \\[2ex] \dfrac{\partial N_i}{\partial y} \end{Bmatrix} = \dfrac{1}{\dfrac{\partial x}{\partial \xi}\dfrac{\partial y}{\partial \eta} - \dfrac{\partial x}{\partial \eta}\dfrac{\partial y}{\partial \xi}} \begin{bmatrix} \dfrac{\partial y}{\partial \eta} & -\dfrac{\partial y}{\partial \xi} \\[2ex] -\dfrac{\partial x}{\partial \eta} & \dfrac{\partial x}{\partial \xi} \end{bmatrix} \begin{Bmatrix} \dfrac{\partial N_i}{\partial \xi} \\[2ex] \dfrac{\partial N_i}{\partial \eta} \end{Bmatrix} \qquad (10.78)$$

The denominator in equation (10.78) is the determinant of the Jacobian. Equations (10.75) to (10.78) are general and apply to *all* quadrilateral elements. Their extension to hexahedral elements is straightforward.

We next consider the special case of a bilinear quadrilateral element, for which $N_{en} = 4$. The interpolation functions are

$$N_i = \frac{1}{4}(1 + \xi_i\xi)(1 + \eta_i\eta) \tag{10.79}$$

where the constants ξ_i and η_i are defined in Table 10.1. The derivatives of the interpolation functions with respect to ξ and η are thus

$$\frac{\partial N_i}{\partial \xi} = \frac{1}{4}\xi_i(1 + \eta_i\eta) \quad ; \quad \frac{\partial N_i}{\partial \eta} = \frac{1}{4}\eta_i(1 + \xi_i\xi) \tag{10.80}$$

The derivatives of x and y with respect to ξ and η follow from equations (10.75) and (10.76), viz.,

$$\frac{\partial x}{\partial \xi} = \frac{1}{4}\sum_{m=1}^{4} x_m^{(e)}\xi_m(1 + \eta_m\eta) = \frac{1}{4}(k_1 + k_3\eta) \tag{10.81}$$

$$\frac{\partial x}{\partial \eta} = \frac{1}{4}\sum_{m=1}^{4} x_m^{(e)}\eta_m(1 + \xi_m\xi) = \frac{1}{4}(k_2 + k_3\xi) \tag{10.82}$$

$$\frac{\partial y}{\partial \xi} = \frac{1}{4}\sum_{m=1}^{4} y_m^{(e)}\xi_m(1 + \eta_m\eta) = \frac{1}{4}(k_4 + k_6\eta) \tag{10.83}$$

$$\frac{\partial y}{\partial \eta} = \frac{1}{4}\sum_{m=1}^{4} y_m^{(e)}\eta_m(1 + \xi_m\xi) = \frac{1}{4}(k_5 + k_6\xi) \tag{10.84}$$

where

$$k_1 = \sum_{m=1}^{4} x_m^{(e)}\xi_m = x_1^{(e)} - x_2^{(e)} - x_3^{(e)} + x_4^{(e)} \tag{10.85}$$

$$k_2 = \sum_{m=1}^{4} x_m^{(e)} \eta_m = x_1^{(e)} + x_2^{(e)} - x_3^{(e)} - x_4^{(e)} \qquad (10.86)$$

$$k_3 = \sum_{m=1}^{4} x_m^{(e)} \xi_m \eta_m = x_1^{(e)} - x_2^{(e)} + x_3^{(e)} - x_4^{(e)} \qquad (10.87)$$

$$k_4 = \sum_{m=1}^{4} y_m^{(e)} \xi_m = y_1^{(e)} - y_2^{(e)} - y_3^{(e)} + y_4^{(e)} \qquad (10.88)$$

$$k_5 = \sum_{m=1}^{4} y_m^{(e)} \eta_m = y_1^{(e)} + y_2^{(e)} - y_3^{(e)} - y_4^{(e)} \qquad (10.89)$$

$$k_6 = \sum_{m=1}^{4} y_m^{(e)} \xi_m \eta_m = y_1^{(e)} - y_2^{(e)} + y_3^{(e)} - y_4^{(e)} \qquad (10.90)$$

The Jacobian determinant in equation (10.78) is thus

$$\frac{\partial x}{\partial \xi} \frac{\partial y}{\partial \eta} - \frac{\partial x}{\partial \eta} \frac{\partial y}{\partial \xi} = \frac{1}{16} [(k_1 + k_3 \eta)(k_5 + k_6 \xi) - (k_2 + k_3 \xi)(k_4 + k_6 \eta)]$$

$$= \frac{1}{16} [(k_1 k_5 - k_2 k_4) + (k_1 k_6 - k_3 k_4)\xi + (k_3 k_5 - k_2 k_6)\eta] \quad (10.91)$$

We note that the bilinear term $\xi \eta$ drops out, rendering the Jacobian determinant linear in ξ and η. This finding will prove useful in quantifying element distortions in Section 10.6.

Finally, the derivatives of interpolation functions with respect to global coordinates are evaluated. From equation (10.78) we get

$$\frac{\partial N_i}{\partial x} = \frac{(k_5 + k_6 \xi)\xi_i(1 + \eta_i \eta) - (k_4 + k_6 \eta)\eta_i(1 + \xi_i \xi)}{(k_1 k_5 - k_2 k_4) + (k_1 k_6 - k_3 k_4)\xi + (k_3 k_5 - k_2 k_6)\eta} \qquad (10.92)$$

$$\frac{\partial N_i}{\partial y} = \frac{(k_1 + k_3 \eta)\eta_i(1 + \xi_i \xi) - (k_2 + k_3 \xi)\xi_i(1 + \eta_i \eta)}{(k_1 k_5 - k_2 k_4) + (k_1 k_6 - k_3 k_4)\xi + (k_3 k_5 - k_2 k_6)\eta} \qquad (10.93)$$

10.6   Element Distortions

When parent element domains are parametrically mapped into distorted configurations, care must be taken to avoid non-unique mappings. In such mappings, the one-to-one relationship between natural and global coordinates ceases to exist. Mathematically, non-unique mappings are characterized by a *change in sign* of the Jacobian determinant [605]. Physically, such mappings occur when an element is excessively distorted or folds back upon itself.

Some specific measures of element distortion are presented in this section. In addition, some guidelines are given that help avoid overly distorted elements.

One guideline that applies to *all* elements is that the node numbers comprising the element connectivity must be specified in an order that is consistent with that assumed in the parent domain. Failure to do this will produce a change of sign in the Jacobian determinant.

A second general guideline is that elements such as triangles and quadrilaterals that are derived in the plane may behave poorly if displaced out of plane (warped) to fit a curved surface [212].

10.6.1   Linear Triangular Elements

The most common parameters used to quantify the distortion of linear simplex elements are the interior vertex angles. In particular, no vertex angle should approach π [38], for this "maximum angle condition" leads to triangles with near-zero area (Figure 10.9b) and thus a near zero Jacobian determinant (recall equation (10.74) in Example 10.5).

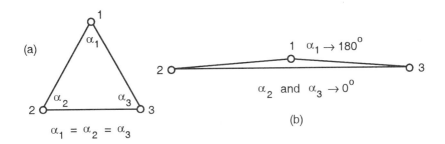

Figure 10.9: Linear Triangular Element: (a) Equilateral, and (b) Heavily Distorted

A second measure of distortion is the ratio of element area to the sum of the squares of the lengths of the element edges [326], normalized such that a value of 1.0 corresponds to an equilateral triangle (Figure 10.9a). The latter configuration represents the least distorted geometry for a linear triangle.

At all points in a linear triangle the Jacobian determinant is constant. The necessary condition for a proper one-to-one mapping is thus a positive Jacobian determinant (det $\mathbf{J}$) at a *single point* within the element.

To quantify the reduction of the Jacobian determinant with distortion, consider the element shown in Figure 10.10(a). The element area $A^{(e)}$ is equal to 2.0, and det $\mathbf{J} = 2A^{(e)} = 4.0$

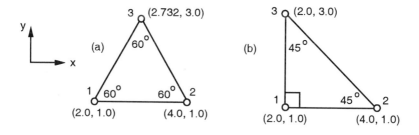

Figure 10.10: Hypothetical Triangular Element : (a) Equilateral, (b) Right

If the element is made into a right triangle with the same base and height (Figure 10.10b), $A^{(e)}$ will obviously be unchanged. If the element is next distorted by increasing the interior angle associated with node 1 (call this angle α_1) to 120 degrees, with the other two interior angles equal to 30 degrees and the coordinates of nodes 1 and 2 unchanged, $A^{(e)}$ is reduced by about 13 percent to 1.732. Finally, if the element is further distorted by increasing α_1 to 150 degrees with the other two interior angles equal to 15 degrees, $A^{(e)}$ further decreases to one-half of the original area. If α_1 is increased further, $A^{(e)}$ will even more rapidly approach zero.

10.6.2   Quadratic Triangular Elements

As in the case of linear triangles, the interior vertex angles in quadratic triangular elements should not approach π. In straight-sided configurations this would give an element whose area is approaching zero in the manner shown in Figure 10.9(b). For curved-sided configurations this would give locally zero or near-zero values of the Jacobian determinant (Figure 10.11b).

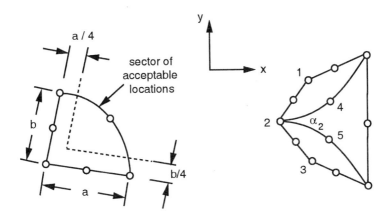

Figure 10.11: Quadratic Isoparametric Triangular Element: (a) Acceptable Locations of Midside Node Along Curved Side; (b) Unacceptable Vertex Angles: At Nodes 1, 3, 4 and 5 Approaching π; At Node 2 Approaching Zero

Along curved boundaries in the domain Ω, a commonly used element shape employs two straight sides, with centered midside nodes, and one curved side (Figure 10.11a). It can be shown [515] that the Jacobian determinant is non-zero in the element provided the midside node on the curved side lies anywhere in the indicated sector. If the element has straight sides, the midside nodes must lie in the middle half of the respective edges [515].

Recently Salem et al. [479, 480] developed a general element distortion metric for quadratic triangles that is based on the concept of *mid-node admissible spaces*. The degree of element distortion depends on the value of a "distortion parameter." This parameter is a function of the normalized Jacobian of the parent element domain, the degree to which an element edge is curved, and the offset of the midside node.

10.6.3 Higher-Order Triangular Elements

When checking the descriptions of higher-order triangular elements in an actual finite element computer program, the sign of the Jacobian determinant computed at the quadrature points is typically monitored to ensure that it does not change. If such a condition is encountered, computations are usually terminated, as this indicates the likelihood of an error in the input data or an overly distorted element.

10.6.4   Tetrahedral Elements

Relatively little work has been done to quantify the distortion of tetrahedral elements. Knupp [291] discussed the necessary and sufficient conditions for a positive Jacobian determinant in the linear tetrahedral element. Field [157] extended a complete Jacobian check of the quadratic triangle to the quadratic tetrahedral element.

10.6.5   Bilinear Quadrilateral Elements

The distortion of bilinear quadrilateral elements is typically quantified by the measures shown in Figure 10.12. A thorough discussion of these measures together with their relation to the Jacobian determinant through a set of shape parameters has been given by Robinson [467].

For isoparametric mappings of quadrilateral elements based on bilinear interpolation functions, the Jacobian determinant is known to be linear within the element (recall equation 10.91). Thus, if the Jacobian determinant has the *same sign* at all four corners of Ω^p, it cannot vanish inside. In the actual distorted element, the Jacobian determinant is non-zero if and only if the quadrilateral is *convex*; that is, all its interior vertex angles are than π [515].

The least distorted configuration for a bilinear quadrilateral element has vertex angles of $\pi/2$ and an aspect ratio (Figure 10.12a) of one; in short, the element is square. Even if right vertex angles are maintained, the element can become overly distorted if the aspect ratio a/b or b/a becomes excessive. A detailed investigation of the importance of aspect ratio in sensitivity testing of elements is given by Robinson [469].

If the taper in either of the two coordinate directions (Figure 10.12b and c) becomes excessive, the element will approach a triangular configuration. The corresponding distortion, quantified by the ration c/d or e/g, will likely lead to unpredictable results.

If an interior angle is less than, but close to π (Figure 10.13a), the element will likely be numerically ill-conditioned. It will facilitate the loss of significant digits during computation, and will likely lead to sporadic and unpredictable results.

If one of the interior vertex angles equals π, then the Jacobian determinant will equal to zero at the corresponding node (Figure 10.13b). If the angle exceeds π, the Jacobian determinant will change sign locally.

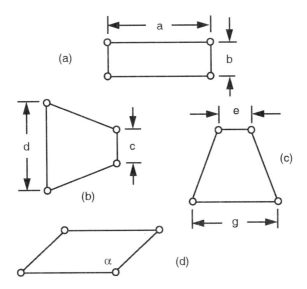

Figure 10.12: Distortion Measures for Bilinear Quadrilateral Elements: (a) Aspects Ratio; (b) Vertical Taper; (c) Horizontal Taper (d) Skew

Straight lines drawn within Ω^p will then go *outside* of the distorted element (Figure 10.13c).

If the element is a parallelogram, the Jacobian determinant is equal to one-quarter of the element area. As such, there is no mapping distortion. However, an extremely narrow parallelogram (Figure 10.13d) will likely be ill-conditioned because two of its angles approach π. In this skewed configuration (Figure 10.12d), the area of the element approaches zero. Consequently, in the analysis of stress-deformation problems the element would have a much greater stiffness than other, less distorted elements and would likely be a source of ill-conditioning in an assemblage of elements [100].

There is, however, no reason why the above extremes should ever be approached in a mesh. The elements used should not be too dissimilar from the parent domain. The interior vertex angles should stay within 20 or 30 degrees of a right angle; otherwise a triangular element would be preferred [100].

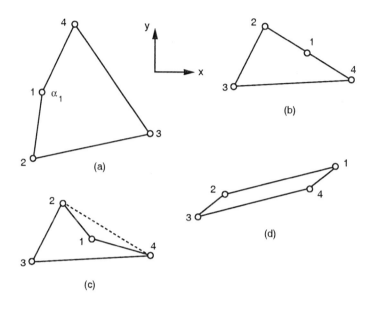

Figure 10.13: Bilinear Isoparametric Quadrilateral Element with Vertex Angle at Node 1: (a) Approaching π; (b) Equal to π; (c) Greater than π (d) Approaching Zero in a Parallelogram Configuration

10.6.6   Biquadratic Quadrilateral Elements

Extending his work on bilinear elements, Robinson [468] introduced a set of shape and distortion parameters for an eight-node serendipity quadrilateral element. These were expressed in terms of the shape polynomials of the element. In a similar manner, Barlow [54] investigated the errors due to shape distortions in the same element.

For eight- or nine-node isoparametric quadrilaterals, two basic conditions must be satisfied in order to ensure a non-zero Jacobian determinant everywhere in the element. First, as in the case of bilinear quadrilaterals, no interior vertex angle in the element should equal or exceed π.

Second, the midside nodes must be contained in the "middle half" of the distance between adjacent corner nodes (note the similarity of this guideline to the one-dimensional quadratic element considered in Example 10.3.). The middle third is, however, deemed safer [605]. This guideline is thought to be overly restrictive [479], since it would likely force an automatic meshing algorithm to generate an unnecessarily large number of elements in order to maintain low levels of element distortion. Further insight into computing Jacobian determinants for eight-node quadrilateral elements is given in [55].

Numerical experiments [222] also indicate that when the three nodes along a side of a nine-node Lagrangian element are used to model a circular arc, the difference in shape between the finite element and the actual domain is not significant unless the angle subtended at the center of curvature exceeds $\pi/2$. In practice there is of course no need to use such extremely distorted elements.

10.6.7 Hexahedral Elements

The quantification of distortion for hexahedral elements is not as straightforward as for quadrilaterals. For example, the invertibility of the isoparametric map for trilinear hexahedral elements is *not* guaranteed by checking the positivity of the Jacobian determinant at the eight element vertices or on the twelve element edges [291].

To study the errors due to shape distortion of twenty-node hexahedral elements, Barlow [54] used measures similar to ones developed Robinson [468]. However, no general guidelines to minimize distortions emerged.

In light of the absence of general distortion guidelines, when checking the descriptions of linear and higher-order hexahedral elements in an actual finite element computer program, the sign of the Jacobian determinant computed at the quadrature points is typically monitored to ensure that it does not change. If such a condition is encountered, computations are usually terminated, as this indicates the likelihood of an error in the input data or an overly distorted element.

Remark

1. To illustrate the potential importance of element distortions, it is timely to recount the case of the Sleipner off-shore platform [70]. In August of 1991, this concrete platform was completed in floating position in a Norwegian Fjord. When subjected to submergence testing, the platform suddenly sprang a leak and sank in less than twenty minutes. The rupture and implosion of the structure that ensued resulted in earthquake-like shock waves that were recorded throughout Norway. The only remains that could be found on the sea floor afterwards were a heap of concrete rubble and twisted reinforcement bars. The reason for the failure was traced to inaccurate shear stresses that were predicted using a rather coarse and somewhat distorted mesh of eight-node hexahedral "brick" elements. These stresses were extrapolated and the resulting distribution used in the selection of reinforcing steel, which proved incapable of carrying the submergence loads.

10.6.8   Further Insight Into Element Distortion

Even without attaining the extreme mappings presented above, geometric
distortions can significantly affect the general predictive capabilities of elements. To better illustrate this fact, consider a cantilever beam of unit
thickness loaded at its free end by an axial stress $\bar{\sigma} = 50.0$ and by a moment
of couple $\bar{M} = 100.0$ in the manner shown in Figure 10.14(a).

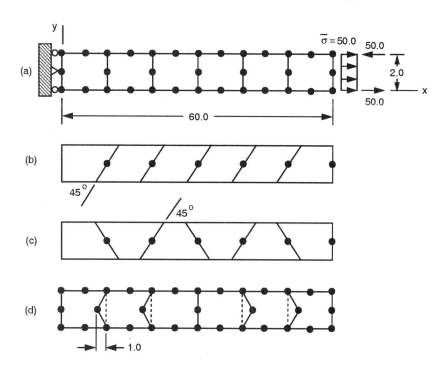

Figure 10.14: Cantilever Beam Loaded Axially and by a Moment: (a) Original Mesh, (b) "Parallelogram" Distortion, (c) "Trapezoidal" Distortion,
(d) Parabolic Distortion

From basic mechanics, the longitudinal displacement u of the beam is
known to be *linear* in x, viz.,

$$u = \frac{\bar{P}x}{EA} \tag{10.94}$$

where A is the cross-sectional area of the beam, E is the elastic modulus, and $\bar{P} = \bar{\sigma} A$. In analyzing the beam, an elastic modulus equal to 2.0×10^6 was used.

Using Bernoulli-Euler beam theory, the transverse displacement of the beam is *quadratic* in x, viz., [440]

$$v = \frac{\bar{M} x^2}{2EI} \tag{10.95}$$

where I is the moment of inertia about the z-axis. Also from beam theory, the axial strain ε_{11} is known to vary *linearly* with y. Since $\varepsilon_{11} = \partial u / \partial x$, u must thus be *quadratic* in y.

The approximation for u and v in an element has the usual form, viz.,

$$\hat{u}^{(e)}(x, y) = \sum_{m=1}^{N_{dof}} N_m(\xi, \eta) \hat{u}_m^{(e)} \tag{10.96}$$

$$\hat{v}^{(e)}(x, y) = \sum_{m=1}^{N_{dof}} N_m(\xi, \eta) \hat{v}_m^{(e)} \tag{10.97}$$

For a parametric mapping, the element geometry is interpolated in the manner described in Section 10.5, viz.,

$$x = \sum_{m=1}^{N_{pt}} G_m(\xi, \eta) x_m^{(e)} \tag{10.98}$$

$$y = \sum_{m=1}^{N_{pt}} G_m(\xi, \eta) y_m^{(e)} \tag{10.99}$$

where $(x_m^{(e)}, y_m^{(e)})$ are the global (x, y) coordinates of the point into which point m in (ξ, η) coordinates shall be mapped, and N_{pt} equals the number of points in the element that are used to define its geometry.

Consider the isoparametric eight-node quadrilateral element. For this biquadratic element, $N_{pt} = N_{dof} = 8$, and the interpolation functions G_m are identical to the N_m. As such, the basis functions for $\hat{u}^{(e)}$ and y are $(1, \xi, \eta, \xi^2, \xi\eta, \eta^2, \xi^2\eta, \xi\eta^2)$. The basis functions for y^2 will thus contain the terms ξ^4, η^4, $\xi^4\eta^2$, etc., that are clearly *not* present in the expression for $u^{(e)}$. It follows that states such as $u^{(e)} = y^2$ cannot be represented if the element is isoparametric.

We next consider a *subparametric* eight-node quadrilateral element. This is realized by keeping the edges of the element straight and the edge

Table 10.2: Comparison of Tip Displacements for Cantilever Beam

| | 8-Node Elements | | 9-Node Elements | |
Case	$\hat{u}$ $(\times 10^{-3})$	$\hat{v}$ $(\times 10^{-1})$	$\hat{u}$ $(\times 10^{-3})$	$\hat{v}$ $(\times 10^{-1})$
(a)	1.500	1.350	1.500	1.350
(b)	1.500	1.346	1.500	1.350
(c)	1.500	1.254	1.500	1.350
(d)	1.500	0.470	1.500	0.909

nodes centered (see Exercise 10.2). In this case the basis functions for y are $(1, \xi, \eta, \xi\eta)$. The resulting basis functions for y^2 thus include the following nine terms: $(1, \xi, \eta, \xi^2, \xi\eta, \eta^2, \xi^2\eta, \xi\eta^2, \xi^2\eta^2)$. Since the basis functions for $u^{(e)}$ are again $(1, \xi, \eta, \xi^2, \xi\eta, \eta^2, \xi^2\eta, \xi\eta^2)$, we see that the subparametric eight-node element cannot represent linear strain states exactly because of the absence of the $\xi^2\eta^2$ term.

The lowest order element that includes all of the above nine terms is one with nine nodes. Thus, an important advantage of this element over the eight-node element is the ability to represent linearly varying strain states exactly[5].

To quantify the above observations, the beam shown in Figure 10.14(a) was first analyzed using six eight-node elements. As expected, the linear variation in $\hat{u}$ and the quadratic variation in $\hat{v}$ were represented exactly. The exercise was next repeated using nine-node elements. The results were identical to those obtained for the eight-node elements. The values of $\hat{u}$ and $\hat{v}$ at the free end of the beam, for both element types, are listed in the first row of Table 10.2.

The elements were next distorted in the "parallelogram" manner shown in Figure 10.14(b). Although the eight-node element is subparametric, the value of $\hat{v}$ at the free end of the beam deviates from the exact value. On the other hand, the subparametric nine-node element is unaffected by the distortion and produces exact results. Since both the distorted eight- and nine-node elements can represent linear displacements exactly, the corresponding values of $\hat{u}$ at the free end are unaffected by the distortion.

The elements were next placed in the "trapezoidal" configuration shown in Figure 10.14(c). The eight-node element is again subparametric, and again the value of $\hat{v}$ at the free end differs from the exact value. The sub-

---

[5]Further details pertaining to both elements are given in Chapter 9.

parametric nine-node element is once again unaffected by the distortion and produces exact results. Since both the distorted eight- and nine-node elements can represent linear displacements exactly, the corresponding values of $\hat{u}$ at the free end are again unaffected by the distortion.

The final distortion affects only the interior mid-side nodes in the manner shown in Figure 10.14(d). Both the eight- and nine-node elements become isoparametric, and the values of $\hat{v}$ at the free end for both elements differ from the exact value. Not surprisingly, the values of $\hat{u}$ at the free end are again unaffected by the distortion.

We close this section by noting that the stiffening of distorted quadrilateral elements observed above has been investigated in the past by several authors. In particular, the eight-node quadrilateral element was studied by Stricklin et al. [517] and Bäcklund [48]. Gifford [199] studied the twelve-node serendipity element and found that it was affected by distortions in a manner similar to the eight-node element. He concluded that for acceptable results the acute angle defining the distortion in such elements should be less than $\pi/4$. Lee and Bathe [308] performed a comparison study of both serendipity and Lagrangian elements. They found that the latter are not affected by angular distortions (Figures 10.14b and c), while the former are. Both classes of elements are, however, affected by curved-edge distortions ((Figure 10.14d). Lee and Bathe [308] thus concluded that, when considering various forms of distortion, the nine- and sixteen-node Lagrangian elements are preferable to the eight- and twelve-node serendipity elements. However, for rectangular or parallelogram configurations, the computationally less expensive serendipity elements should be used. These conclusions are consistent with the findings of Cook [115], who limited himself to quadratic elements.

10.7 Computer Implementation Issues

Having discussed the topics of mapping and parametric elements, it is timely to address computer implementation issues related to the determination of element arrays. In the discussion of Chapter 7, the (i, j)th entry in the element properties ("stiffness") matrix was shown to be computed from

$$K_{ij}^{(e)} = \int_{\Omega^e} \mathbf{B}_i^T \mathbf{C} \mathbf{B}_j \, d\Omega \qquad (10.100)$$

where $\mathbf{B}_i$ and $\mathbf{B}_j$ denote submatrices of the "strain-displacement" transformation matrix, $\mathbf{C}$ is a matrix of material parameters, Ω^e denotes the element domain, and a superscript T denotes the operation of matrix transposition. The subscripts i and j range from 1 to N_{en}, where N_{en} represents the number of nodes associated with the element.

In addition, the ith entry in the element right hand side vector is computed from

$$f_i^{(e)} = \int_{\Omega^e} \mathbf{N}_i^T \mathbf{b} \, d\Omega \qquad (10.101)$$

where $\mathbf{N}_i$ is a submatrix of element interpolation functions, and $\mathbf{b}$ is a vector of "body force" components.[6]

To simplify the subsequent discussion, assume that the element possesses only *scalar* nodal unknowns (e.g., temperature, hydraulic head, etc.) for analysis of three-dimensional continua. Consequently, the total number of degrees of freedom associated with the element equals to N_{en}.

In light of the fact that, in general, numerical integration must be used, equations (10.100) and (10.101) are re-written as

$$K_{ij}^{(e)} = \sum_{n=1}^{N_{quad}} w_n \mathbf{B}_i^T(\xi_n, \eta_n, \zeta_n) \mathbf{C} \mathbf{B}_j(\xi_n, \eta_n, \zeta_n) \det \mathbf{J}(\xi_n, \eta_n, \zeta_n) \quad (10.102)$$

$$f_i^{(e)} = \sum_{n=1}^{N_{quad}} w_n N_i^T(\xi_n, \eta_n, \zeta_n) \, b \det J(\xi_n, \eta_n, \zeta_n) \qquad (10.103)$$

where N_{quad} represents the total number of quadrature points, w_n denotes a quadrature weight, and $\mathbf{J}$ is the Jacobian matrix defined in equation (10.7).

The form of equation (10.102) suggests the need for *three* "do loops," namely: one associated with the numerical integration variable n, and one associated with each of the subscripts i and j. Assuming a symmetric $\mathbf{K}^{(e)}$ and employing "pseudo-code," the numerical implementation of the element equations thus has the following general form:

---

[6] In heat conduction and flow through porous media applications, this would correspond to a vector of source/sink terms.

```
                         ! get quadrature points & weights

CALL integ_points (. . . nquad, . . .)

CALL integ_weights (. . . , nquad)

DO n=1,nquad

  CALL form_B ( . . . . )          ! compute B sub-matrices

  CALL elm_jacobian ( . . . )      ! get Jacobian determinant

  term = jac_det*weight(n)

                        ! compute body force vector

  CALL get_body_force ( . . . . , body_force, . . . . )

                        ! form [ C ]
  CALL form_C (. . . . . )

  DO i=1,nen
                        ! form ''stiffness'' matrix
    DO j=i,nen
      k_e(i,j) = k_e(i,j) + jac_det*weight(n)*b(i)*d(i,j)*b(j)
    END DO
                        ! form r.h.s. vector

    f_e(i) = f_e(i) + jac_det*weight(n)*shf(i)*body_force

  END DO

END DO
                              ! make matrix symmetric
CALL matrix_symmetric ( . ,k_e, . )
```

The discussion in this section has been somewhat general. This is by design, for the intent was to introduce the reader to the manner in which entries in the element arrays are determined numerically. A more detailed discussion, which addresses such issues as storage schemes and computational efficiency, is given in textbooks such as [252, 605].

10.8   Concluding Remarks

In this chapter, the subject of element mapping has been investigated. It was shown that distorted element configurations, such as those necessitated by the discretization of curved boundaries, can be realized by mapping simple parent domains into these configurations. As such, the natural coordinates discussed in Appendix C are mapped to global coordinates.

When parent element domains are parametrically mapped into distorted configurations, care must be taken to avoid non-unique mappings. In such mappings, the one-to-one relationship between natural and global coordinates ceases to exist. Mathematically, non-unique mappings are characterized by a change in sign of the determinant of the Jacobian matrix. In this chapter, some specific measures of element distortion were presented, along with guidelines that help avoid overly distorted elements.

To simplify the associated computations, element equations are evaluated in the parent domain. For all but the simplest elements, this requires numerical integration. In finite element applications the most efficient approach approach to numerical integration typically involves the use of Gaussian quadrature.

Since the element equations are evaluated in the parent domain, the element interpolation functions developed in Chapter 9 apply directly to distorted elements.

10.9 Exercises

10.1
Consider a four-node isoparametric element configured such that node 1 lies half-way along the straight line between nodes 2 and 4 (Figure 10.15).

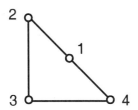

Figure 10.15: Triangular Element Formed by Degenerating a Four-Node Quadrilateral

Show that at node 1 the determinant of the Jacobian is zero (NOTE: the fact that the Jacobian determinant is zero at a node does not detrimentally affect the performance of this element).

10.2
A certain eight-node "serendipity" element has its midside nodes in the (x, y) space placed so as to lie at the midpoints of the straight lines joining the corner nodes (Figure 10.16).

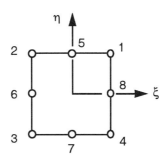

Figure 10.16: Subparametric Quadratic Serendipity Element

Show that the resulting isoparametric mapping between the element and the eight-node parent domain (i.e., the biunit square shown below) involves only the *bilinear* interpolation functions corresponding to the corner nodes (the element is thus effectively subparametric). Show this only for the mapping involving x-coordinates – the case for the y-coordinates follows by inspection.

The interpolation functions for the element are:

- For $i = 1$ to 4: $N_i = \dfrac{1}{4}(1 + \xi_i\xi)(1 + \eta_i\eta)(\xi_i\xi + \eta_i\eta - 1)$

- For $i = 5$ and 7: $N_i = \dfrac{1}{2}(1 - \xi^2)(1 + \eta_i\eta)$

- For $i = 6$ and 8: $N_i = \dfrac{1}{2}(1 + \xi_i\xi)(1 - \eta^2)$

Chapter 11

Finite Element Analysis of Scalar Field Problems

11.1 Introductory Remarks

In this chapter, the steady-state finite element analysis of problems involving *scalar* primary dependent variables is considered. Although the physical nature of such problems is rather diverse, the solution techniques used are quite similar. An interesting observation of scalar field problems is that they apply to many branches of engineering and physics.

11.2 General Governing Equations

The general equation governing scalar field problems is commonly obtained by combining two separate equations; viz.,

- A *conservation principle* or *balance equation*

$$\nabla^T \mathbf{q} + H\phi + Q = 0 \tag{11.1}$$

where ϕ is the primary dependent variable, $\mathbf{q} = \left\{ q_x \quad q_y \quad q_z \right\}^T$ is a vector of flux components, H and Q are constants or, in the most general case, functions of the spatial coordinates (x, y, z), and

$$\nabla = \left\{ \frac{\partial}{\partial x} \quad \frac{\partial}{\partial y} \quad \frac{\partial}{\partial z} \right\}^T \tag{11.2}$$

is the *gradient operator*. In the above equations, the superscript T denotes the operation of matrix transposition. The inner product $\nabla^T \mathbf{q}$ represents the *divergence* of the flux.

- A *constitutive relation*

$$\mathbf{q} = \mathbf{K} \nabla \phi \tag{11.3}$$

where $\mathbf{K}$ is a matrix of material parameters, which is typically *symmetric*. The symmetric nature of $\mathbf{K}$ renders the governing equations self-adjoint.

Equations (11.1) and (11.3) are typically combined by substituting the latter equation into the former to give

$$\nabla^T \left(\mathbf{K} \nabla \phi \right) + H\phi + Q = 0 \tag{11.4}$$

The boundary conditions associated with this equation consist of

- Essential boundary conditions

$$\phi = \bar{\phi} \quad \text{on} \quad \Gamma_1 \tag{11.5}$$

- Natural boundary conditions

$$\left(\mathbf{K} \nabla \phi \right)^T \mathbf{n} = \bar{q} \quad \text{on} \quad \Gamma_2 \tag{11.6}$$

where $\bar{q}$ is a known value of flux *normal* to the boundary. The vector $\mathbf{n}$ is the outward unit normal to that portion of the boundary Γ_2 where natural boundary conditions are specified (Figure 11.1).

- Mixed boundary conditions

$$\left(\mathbf{K} \nabla \phi \right)^T \mathbf{n} - h \left(\phi - \bar{\phi}_\infty \right) = 0 \quad \text{on} \quad \Gamma_3 \tag{11.7}$$

where h is a known coefficient and $\bar{\phi}_\infty$ denotes a known "far field" value of the primary dependent variable.

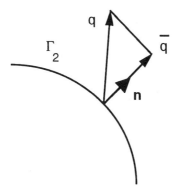

Figure 11.1: Schematic Illustration of Normal Component of Specified Boundary Flux

Remarks

1. When H is positive and $Q = 0$, equation (11.4) is known as the *Helmholtz* equation. If $H = 0$, it reduces to the *quasi-harmonic* or *Poisson's* equation. If, in addition, $Q = 0$, then the resulting expression becomes *Laplace's* equation.

2. In expanded form, equations (11.1), (11.3) and (11.4) become

$$\left\{\frac{\partial}{\partial x} \quad \frac{\partial}{\partial y} \quad \frac{\partial}{\partial z}\right\} \left\{\begin{matrix} q_x \\ q_y \\ q_z \end{matrix}\right\} = \frac{\partial q_x}{\partial x} + \frac{\partial q_y}{\partial y} + \frac{\partial q_z}{\partial z} \tag{11.8}$$

$$\left\{\begin{matrix} q_x \\ q_y \\ q_z \end{matrix}\right\} = \begin{bmatrix} k_{11} & k_{12} & k_{13} \\ k_{12} & k_{22} & k_{23} \\ k_{13} & k_{23} & k_{33} \end{bmatrix} \left\{\begin{matrix} \dfrac{\partial \phi}{\partial x} \\[2mm] \dfrac{\partial \phi}{\partial y} \\[2mm] \dfrac{\partial \phi}{\partial z} \end{matrix}\right\} \tag{11.9}$$

and

$$\frac{\partial}{\partial x}\left(k_{11}\frac{\partial\phi}{\partial x} + k_{12}\frac{\partial\phi}{\partial y} + k_{13}\frac{\partial\phi}{\partial z}\right) + \frac{\partial}{\partial y}\left(k_{12}\frac{\partial\phi}{\partial x} + k_{22}\frac{\partial\phi}{\partial y} + k_{23}\frac{\partial\phi}{\partial z}\right)$$

$$+\frac{\partial}{\partial z}\left(k_{13}\frac{\partial\phi}{\partial x} + k_{23}\frac{\partial\phi}{\partial y} + k_{33}\frac{\partial\phi}{\partial z}\right) + H\phi + Q = 0 \qquad (11.10)$$

3. From the discussion of boundary conditions in Section 1.3.5 of Chapter 1, it follows that for the present problem $\Gamma_1 \cup \Gamma_2 \cup \Gamma_3 = \Gamma$, and $\Gamma_1 \cap \Gamma_2 = \Gamma_1 \cap \Gamma_3 = \Gamma_2 \cap \Gamma_3 = \emptyset$. The symbols $\cup$, $\cap$ and $\emptyset$ denote *set union, set intersection* and *null set*, respectively.

4. Equation (11.4) is determined for an arbitrary set of global (x, y, z) coordinates and assumes an *anisotropic* material. It is always possible to determine another set of *local* or *material* coordinates (x', y', z') that represent *principal* material directions. With respect to these coordinates $\mathbf{K}$ is *diagonal*. In this case equation (11.4) simplifies to

$$\frac{\partial}{\partial x}\left(k_{11}\frac{\partial\phi}{\partial x}\right) + \frac{\partial}{\partial y}\left(k_{22}\frac{\partial\phi}{\partial y}\right) + \frac{\partial}{\partial z}\left(k_{33}\frac{\partial\phi}{\partial z}\right) + H\phi + Q = 0 \qquad (11.11)$$

For the special case of a material whose properties are independent of direction, $\mathbf{K} = k\mathbf{I}$, where $\mathbf{I}$ denotes the identity matrix and k represents the single material coefficient that characterizes the material. Such materials are said to be *isotropic*.

The fluxes measured in the two coordinate systems are related through simple coordinate transformations. To better understand this, we write the general constitutive relation of equation (11.3) in terms of material coordinates as

$$\mathbf{q}' = \begin{Bmatrix} q_{x'} \\ q_{y'} \\ q_{z'} \end{Bmatrix} = \mathbf{K}'\begin{Bmatrix} \dfrac{\partial\phi}{\partial x'} & \dfrac{\partial\phi}{\partial y'} & \dfrac{\partial\phi}{\partial z'} \end{Bmatrix}^T \qquad (11.12)$$

where $\mathbf{K}'$ is a *diagonal* matrix of material coefficients.

From equation (11.12) it is evident that the flux in each of the three primed coordinate directions is *uncoupled* from the other directions. Thus a gradient in the x'-direction produces a flux only in the x'-direction; i.e., parallel to the gradient, etc.

The flux vectors transform like displacements; thus

$$\mathbf{q} = \mathbf{R}^T\mathbf{q}' \qquad (11.13)$$

where

$$
\mathbf{R} = \begin{bmatrix} r_{11} & r_{12} & r_{13} \\ r_{21} & r_{22} & r_{23} \\ r_{31} & r_{32} & r_{33} \end{bmatrix} \tag{11.14}
$$

is an orthogonal transformation matrix, and r_{ij} denotes the cosine of the angle between the ith primed and the jth unprimed coordinate axes.

The fluxes in the two coordinate systems are related through the chain rule of differentiation, viz.,

$$
\left\{ \begin{array}{c} \dfrac{\partial \phi}{\partial x'} \\[2mm] \dfrac{\partial \phi}{\partial y'} \\[2mm] \dfrac{\partial \phi}{\partial z'} \end{array} \right\} = \begin{bmatrix} \dfrac{\partial x}{\partial x'} & \dfrac{\partial y}{\partial x'} & \dfrac{\partial z}{\partial x'} \\[2mm] \dfrac{\partial x}{\partial y'} & \dfrac{\partial y}{\partial y'} & \dfrac{\partial z}{\partial y'} \\[2mm] \dfrac{\partial x}{\partial z'} & \dfrac{\partial y}{\partial z'} & \dfrac{\partial z}{\partial z'} \end{bmatrix} \left\{ \begin{array}{c} \dfrac{\partial \phi}{\partial x} \\[2mm] \dfrac{\partial \phi}{\partial y} \\[2mm] \dfrac{\partial \phi}{\partial z} \end{array} \right\} = \mathbf{R} \left\{ \begin{array}{c} \dfrac{\partial \phi}{\partial x} \\[2mm] \dfrac{\partial \phi}{\partial y} \\[2mm] \dfrac{\partial \phi}{\partial z} \end{array} \right\} \tag{11.15}
$$

Substituting equation (11.12) into (11.13), and then equation (11.15) into the resulting expression, gives the final desired expression

$$
\mathbf{q} = \mathbf{R}^T \mathbf{K}' \mathbf{R} \left\{ \begin{array}{c} \dfrac{\partial \phi}{\partial x} \\[2mm] \dfrac{\partial \phi}{\partial y} \\[2mm] \dfrac{\partial \phi}{\partial z} \end{array} \right\} = \mathbf{K} \left\{ \begin{array}{c} \dfrac{\partial \phi}{\partial x} \\[2mm] \dfrac{\partial \phi}{\partial y} \\[2mm] \dfrac{\partial \phi}{\partial z} \end{array} \right\} \tag{11.16}
$$

For the special case of an *isotropic* material, equation (11.16) and the definition of an orthogonal matrix gives

$$
\mathbf{K}' = \mathbf{R}\mathbf{K}\mathbf{R}^T = \mathbf{R}\,(k\mathbf{I})\,\mathbf{R}^T = k\mathbf{R}\mathbf{R}^T = k\mathbf{I} = \mathbf{K} \tag{11.17}
$$

If the material coefficients are specified with respect to principal material directions, they can thus be easily transformed to the global coordinates. Once the fluxes are known with respect to the global coordinates, the corresponding values related to the local or material coordinates are obtained from the inverse of equation (11.13), that is, $\mathbf{q}' = \mathbf{R}\mathbf{q}$.

For the special case of a *two-dimensional* anisotropic domain (Figure 11.2), it is implicitly assumed that one of the three principal axes is perpendicular to the plane of the body. The associated transformation matrix is thus

$$\mathbf{R} = \begin{bmatrix} \cos\alpha & \sin\alpha \\ -\sin\alpha & \cos\alpha \end{bmatrix} \qquad (11.18)$$

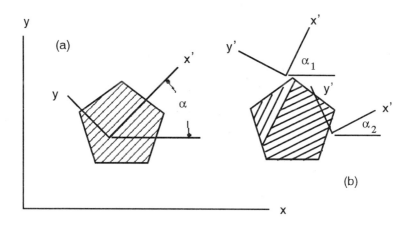

Figure 11.2: Schematic Illustration of Anisotropic Two-Dimensional Solution Domain: Principal Material Axes not Aligned with Global Axes, (a) Single, (b) Dual Material Orientation

Thus, if the global coordinate axes are different from the (primed) principal material axes, then

$$\begin{Bmatrix} q_{x'} \\ q_{y'} \end{Bmatrix} = \mathbf{R} \begin{Bmatrix} q_x \\ q_y \end{Bmatrix} \qquad (11.19)$$

and

$$\mathbf{K} = \mathbf{R}^T \mathbf{K}' \mathbf{R}$$
$$= \begin{bmatrix} \left(k_{11'}\cos^2\alpha + k_{22'}\sin^2\alpha\right) & \cos\alpha\sin\alpha\left(k_{11'} - k_{22'}\right) \\ \cos\alpha\sin\alpha\left(k_{11'} - k_{22'}\right) & \left(k_{11'}\sin^2\alpha + k_{22'}\cos^2\alpha\right) \end{bmatrix} \qquad (11.20)$$

For the special case of *isotropic* materials $k_{11'} = k_{22'} = k$. Equation (11.20) thus reduces to the expected result $\mathbf{K} = k\mathbf{I} = \mathbf{K}'$.

5. If, as shown in Figure 11.2(a), the principal material axes have the *same* orientation throughout the domain, then the global $x - y$ axes can, of course, be defined to be parallel to the principal axes. Depending on the size of the problem, this can significantly reduce the associated computational effort.

To supplement the previous general development, we provide the following specific examples.

Example 11.1: Heat Transfer by Convection and Conduction

One of the most commonly studied scalar field problems is that of heat transfer by convection and conduction. Beginning in the mid-sixties, this problem was analyzed using the finite element method [549, 571].

For a three-dimensional material element, the balance of energy is obtained from the first law of thermodynamics, viz.,

$$\nabla^T \mathbf{q} - S = \frac{\partial q_x}{\partial x} + \frac{\partial q_y}{\partial y} + \frac{\partial q_z}{\partial z} - S = 0 \qquad (11.21)$$

where $\mathbf{q} = \begin{Bmatrix} q_x & q_y & q_z \end{Bmatrix}^T$ is the vector of heat flux components (units: $Et^{-1}L^{-2}$), $S = S(x, y, z)$ represents the heat generated per unit volume (units: $Et^{-1}L^{-3}$).

The constitutive relation, written with respect to principal material axes, is generalized Fourier's law, viz.,

$$\mathbf{q} = -\mathbf{K}\nabla T = - \begin{bmatrix} k_{11} & 0 & 0 \\ 0 & k_{22} & 0 \\ 0 & 0 & k_{33} \end{bmatrix} \begin{Bmatrix} \dfrac{\partial T}{\partial x} \\[2mm] \dfrac{\partial T}{\partial y} \\[2mm] \dfrac{\partial T}{\partial z} \end{Bmatrix} \qquad (11.22)$$

where $\mathbf{K}$ is a matrix of thermal conductivity coefficients (units: $EL^{-1}T^{-1}$), and T is the temperature.

Substituting equation (11.22) into equation (11.21) gives the equation governing steady state heat transfer, viz.,

$$\nabla^T \left(\mathbf{K}\nabla T \right) + S = 0 \qquad (11.23)$$

The associated boundary conditions are: Prescribed temperature (essential boundary condition)

$$T = \bar{T} \quad \text{on} \quad \Gamma_1 \qquad (11.24)$$

prescribed surface heat flux per unit area (natural boundary condition)

$$-\left(\mathbf{K}\nabla T\right)^T \mathbf{n} = \bar{q} \quad \text{on} \quad \Gamma_2 \qquad (11.25)$$

and convection heat transfer (mixed boundary condition)

$$-\left(\mathbf{K}\nabla T\right) \mathbf{n} = h\left(T - \bar{T}_\infty\right) \quad \text{on} \quad \Gamma_3 \qquad (11.26)$$

where h is a heat transfer coefficient (units: $EL^{-1}T^{-1}$), T represents the temperature along the boundary (an unknown), and T_∞ represents the ambient temperature (a known). The sign convention for the heat flux per unit area $\mathbf{q}$ is such that positive values are associated with heat flowing *into* the body. An thorough discussion of other aspects related to sign convention is presented by Segerlind [488].

Example 11.2: Transverse Displacement of Elastic Membrane Resting on Elastic Foundation

This problem is the two-dimensional analogue of the one-dimensional "string" resting on an elastic (Winkler) foundation, which is considered as an exercise in Chapter 7. As in the one-dimensional problem, *infinitesimal* displacements are assumed. Force equilibrium in the transverse (z) direction gives the following relation

$$\nabla^T \mathbf{f} + kw = \frac{\partial f_x}{\partial x} + \frac{\partial f_y}{\partial y} + kw = F \qquad (11.27)$$

where f_x is the transverse component of the x-component of membrane tension, per unit length in the y-direction (units: FL^{-1}), f_y is the transverse component of the y-component of membrane tension, per unit length in the x-direction (units: FL^{-1}), k is the elastic stiffness per unit area of the foundation (units: FL^{-3}), w is the transverse displacement (units: L), and F is the applied transverse force per unit area of the membrane (units: FL^{-2}).

Consistent with the assumption of infinitesimal displacements, the primary "stiffness" resisting the transverse displacement is due to the transverse component of the membrane tension. The constitutive relation, with respect to the principal material axes, is thus

$$\mathbf{f} = -\mathbf{T}\nabla w = -\begin{bmatrix} T_1 & 0 \\ 0 & T_2 \end{bmatrix} \left\{ \begin{array}{c} \dfrac{\partial w}{\partial x} \\[2mm] \dfrac{\partial w}{\partial y} \end{array} \right\} \tag{11.28}$$

where T_1 is the tension in the x-direction, per unit length in the y-direction, and T_2 is the tension in the y-direction, per unit length in the x-direction (units: FL^{-1}). Substituting equation (11.28) into (11.27) gives the governing equation

$$-\nabla^T (\mathbf{T}\nabla w) + kw = F \tag{11.29}$$

The associated boundary conditions are prescribed transverse displacements (essential boundary condition)

$$w = \bar{w} \quad \text{on} \quad \Gamma_1 \tag{11.30}$$

or prescribed transverse components of tension (natural boundary condition)

$$-(\mathbf{T}\nabla w)^T \mathbf{n} = \bar{F} \text{on} \quad \Gamma_2 \tag{11.31}$$

Example 11.3: Two-Dimensional Irrotational Fluid Flow

A fluid flow is said to be *irrotational* if the vorticity or spin tensor vanishes everywhere, viz.,

$$\nabla \times \mathbf{v} = \left(\frac{\partial v_z}{\partial y} - \frac{\partial v_y}{\partial z} \right) \mathbf{e}_1 + \left(\frac{\partial v_x}{\partial z} - \frac{\partial v_z}{\partial x} \right) \mathbf{e}_2$$
$$+ \left(\frac{\partial v_y}{\partial x} - \frac{\partial v_x}{\partial y} \right) \mathbf{e}_3 = \mathbf{0} \tag{11.32}$$

where e_1, e_2 and e_3 are unit vectors parallel to the x, y, and z coordinate axes, respectively. It follows that the criterion for irrotationality can thus be written as

$$\frac{\partial v_z}{\partial y} - \frac{\partial v_y}{\partial z} = 0 \quad ; \quad \frac{\partial v_x}{\partial z} - \frac{\partial v_z}{\partial x} = 0 \quad ; \quad \frac{\partial v_y}{\partial x} - \frac{\partial v_x}{\partial y} = 0 \qquad (11.33)$$

A significant result related to irrotational flow is the theorem that states that if velocity components at all points in a region of flow can be expressed as continuous partial derivatives of a scalar *velocity potential* ϕ in the manner

$$\mathbf{v} = -\nabla \phi(x, y, z, t) \qquad (11.34)$$

then the flow must be irrotational. The converse of this theorem is likewise true. That is, any irrotational flow may be expressed as the gradient of a scalar velocity potential [490].

Next consider the special case of two-dimensional *incompressible* flow. It can be shown that for such a flow

$$v_x = -\frac{\partial \psi}{\partial y} \quad \text{and} \quad v_y = \frac{\partial \psi}{\partial x} \qquad (11.35)$$

where $\psi(x, y)$ represents an arbitrary *stream function* [490]. If this fluid is further assumed to be irrotational, it follows from equations (11.34) and (11.35) that

$$\frac{\partial \psi}{\partial y} = \frac{\partial \phi}{\partial x} \quad \text{and} \quad \frac{\partial \psi}{\partial x} = -\frac{\partial \phi}{\partial y} \qquad (11.36)$$

thus giving a relationship between the stream function and the velocity potential.[1] Note that for a two-dimensional irrotational fluid, the irrotationality condition is described by the third of equation (11.33). Substituting equations (11.35) into this equation gives Laplace's equation, viz.,

$$\frac{\partial^2 \psi}{\partial x^2} + \frac{\partial^2 \psi}{\partial y^2} = \nabla^2 \psi = 0 \qquad (11.37)$$

---

[1] Lines of constant ψ are thus perpendicular to lines of constant ϕ; there is no flow perpendicular to a streamline.

where ∇^2 represents the Laplacian operator. The associated boundary conditions are

$$\psi = \bar{\psi} \quad \text{on} \quad \Gamma_1 \quad \text{or} \quad \frac{\partial \psi}{\partial n} = \bar{v}_n \quad \text{on} \quad \Gamma_2 \qquad (11.38)$$

where $\Gamma_1 \cup \Gamma_2 = \Gamma$ and $\Gamma_1 \cap \Gamma_2 = \emptyset$.

For two-dimensional irrotational incompressible flow, the *velocity potential* ϕ can likewise be shown to satisfy Laplace's equation [490]. Thus $\nabla^2 \phi = 0$, indicating that like the stream function, the velocity potential is also *harmonic*. The associated velocity components are thus given by

$$v_x = -\frac{\partial \phi}{\partial x} \quad \text{and} \quad v_y = \frac{\partial \phi}{\partial y} \qquad (11.39)$$

The associated boundary conditions are

$$\phi = \bar{\phi} \quad \text{on} \quad \Gamma_1 \quad \text{or} \quad \frac{\partial \phi}{\partial n} = \bar{v}_n \quad \text{on} \quad \Gamma_2 \qquad (11.40)$$

where again $\Gamma_1 \cup \Gamma_2 = \Gamma$ and $\Gamma_1 \cap \Gamma_2 = \emptyset$.

In the previous development the incompressible and irrotational fluid has tacitly been assumed to be *ideal* or *inviscid*. As such, no friction occurs between the fluid and a surface, thus simplifying the solution.

11.3 Historical Note

Although the finite element method initially developed largely through structural mechanics applications, it quickly expanded into problems involving non-structural applications. The application to problems involving scalar primary dependent variables was greatly facilitated by Zienkiewicz and Cheung [585], who showed that the method applied to all field problems governed by quasi-harmonic differential forms. This was supported by subsequent application of the method to problems in heat conduction [549, 571], torsion of irregularly shaped cross-sections [228], seepage problems [586], certain fluid flow problems governed by a quasi-linear Poisson equation [573], and three-dimensional field problems [588].

11.4   Development of General Finite Element Equations

The general finite element equations shall be derived using the five-step approach introduced in Chapter 7.

* **Step 1:** Identification of the Primary Dependent Variable

The primary dependent variable is the scalar quantity ϕ. Assuming an irreducible formulation, it follows that associated with each node in the element will be one degree of freedom.

* **Step 2:** Definition of the Gradient and Constitutive Relations

The general constitutive relation applicable to this class of problems is given by equation 11.3.

* **Step 3:** Identification of the Element Equations

The element equations shall be derived using a weighted residual approach. The development thus begins with the *strong form* of the scalar field problem given by equation (11.4).

Weak Form

To initiate the development of the weak form of this equation, we multiply it by an arbitrary weighting function w and integrate over the domain Ω, viz.,

$$\int_\Omega w \left[\nabla^T \left(\mathbf{K} \nabla \phi \right) + H\phi + Q \right] d\Omega = 0 \qquad (11.41)$$

Integrating the first term in equation (11.41) by parts gives

$$\int_\Omega w \left[\nabla^T \left(\mathbf{K} \nabla \phi \right) \right] d\Omega = \oint_\Gamma w \left(\mathbf{K} \nabla \phi \right)^T \mathbf{n} \, d\Gamma - \int_\Omega \nabla^T w \left(\mathbf{K} \nabla \phi \right) d\Omega \quad (11.42)$$

Since $\oint_\Gamma = \oint_{\Gamma_1} + \oint_{\Gamma_2} + \oint_{\Gamma_3}$, and in light of equations (11.6) and (11.7), the boundary integral in equation (11.42) becomes

$$\oint_{\Gamma} w \, (\mathbf{K}\nabla\phi)^T \, \mathbf{n} \, d\Gamma = \oint_{\Gamma_1} w \, (\mathbf{K}\nabla\phi)^T \, \mathbf{n} \, d\Gamma$$

$$+ \oint_{\Gamma_2} w\bar{q} \, d\Gamma + \oint_{\Gamma_3} wh \left(\phi - \bar{\phi}_\infty \right) \, d\Gamma \tag{11.43}$$

In addition, as shown in Section 4.7 of Chapter 3, the weighting function w must vanish on Γ_1, thus further simplifying equation (11.43).

Substituting the resulting expression into equation (11.42), and the resulting expression into equation (11.41), the weak form of the problem becomes

$$\int_\Omega \nabla^T w \, (\mathbf{K}\nabla\phi) \, d\Omega - \int_\Omega w \left(H\phi + Q \right) \, d\Omega$$

$$- \oint_{\Gamma_2} w\bar{q} \, d\Gamma - \oint_{\Gamma_3} wh \left(\phi - \bar{\phi}_\infty \right) \, d\Gamma = 0 \tag{11.44}$$

This represents the weak form of the general equation governing scalar field problems. The solution of equation (11.44) is known to be identical to the solution of equation (11.4) [252]. Before proceeding to the weighted residual or discrete form of the governing equations, it is insightful to investigate the above integration by parts for *two-dimensional* problems.

Denote by α the angle that the outward unit normal $\mathbf{n}$ to the boundary makes with the positive x axis (Figure 11.3).

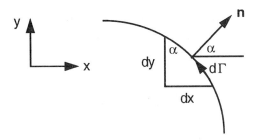

Figure 11.3: Portion of Two-Dimensional Boundary with Outward Unit Normal Vector

Thus

$$\cos \alpha = \frac{dy}{d\Gamma} \quad , \quad \sin \alpha = -\frac{dx}{d\Gamma} \tag{11.45}$$

and that the scalar components of **n** are

$$n_1 = |n| \cos \alpha = \frac{dy}{d\Gamma} \quad , \quad n_2 = |n| \sin \alpha = -\frac{dx}{d\Gamma} \tag{11.46}$$

Assuming a diagonal **K**, it follows that

$$(\mathbf{K}\nabla\phi)^T \mathbf{n} = \left(\left\{ \frac{\partial \phi}{\partial x} \quad \frac{\partial \phi}{\partial y} \right\} \begin{bmatrix} k_{11} & 0 \\ 0 & k_{22} \end{bmatrix} \right) \bullet \langle n_1 e_1 + n_2 e_2 \rangle$$

$$= k_{11} \frac{\partial \phi}{\partial x} n_1 + k_{22} \frac{\partial \phi}{\partial y} n_2 = k_{11} \frac{\partial \phi}{\partial x} \cos \alpha - k_{22} \frac{\partial \phi}{\partial y} \sin \alpha \tag{11.47}$$

In the case of an *isotropic* material, $k_{11} = k_{22} = k$, and equation (11.47) reduces to

$$(\mathbf{K}\nabla\phi)^T \mathbf{n} = k \left(\frac{\partial \phi}{\partial x} \cos \alpha - \frac{\partial \phi}{\partial y} \sin \alpha \right) = k \frac{\partial \phi}{\partial n} \tag{11.48}$$

where $\partial\phi/\partial n$ is the derivative of ϕ normal to the boundary, viz.,

$$\frac{\partial \phi}{\partial n} = \mathbf{n} \bullet \nabla\phi \tag{11.49}$$

Next consider the integration by parts of the first term in equation (11.41), now specialized for two-dimensions. Since a diagonal **K** has been assumed, this term is now written as

$$\int_\Omega w \left[\nabla^T (\mathbf{K}\nabla\phi) \right] d\Omega$$

$$= \int_\Omega w \left[\frac{\partial}{\partial x} \left(k_{11} \frac{\partial \phi}{\partial x} \right) + \frac{\partial}{\partial y} \left(k_{22} \frac{\partial \phi}{\partial y} \right) \right] d\Omega \tag{11.50}$$

Applying Green's theorem to the respective terms gives

$$\int_{\Omega} w \left[\frac{\partial}{\partial x} \left(k_{11} \frac{\partial \phi}{\partial x} \right) \right] d\Omega = \oint_{\Gamma} w k_{11} \frac{\partial \phi}{\partial x} n_1 \, d\Gamma - \int_{\Omega} \frac{\partial w}{\partial x} k_{11} \frac{\partial \phi}{\partial x} d\Omega \quad (11.51)$$

$$\int_{\Omega} w \left[\frac{\partial}{\partial y} \left(k_{22} \frac{\partial \phi}{\partial y} \right) \right] d\Omega = \oint_{\Gamma} w k_{22} \frac{\partial \phi}{\partial y} n_2 \, d\Gamma - \int_{\Omega} \frac{\partial w}{\partial y} k_{22} \frac{\partial \phi}{\partial y} d\Omega \quad (11.52)$$

where the line integrals are evaluated around the boundary Γ in a *counterclockwise* direction (Figure 11.3). Combining equations (11.51) and (11.52) gives

$$\int_{\Omega} w \left[\nabla^T (\mathbf{K} \nabla \phi) \right] d\Omega = \oint_{\Gamma} w \left(k_{11} \frac{\partial \phi}{\partial x} n_1 + k_{22} \frac{\partial \phi}{\partial y} n_2 \right) d\Gamma$$
$$- \int_{\Omega} \left(\frac{\partial w}{\partial x} k_{11} \frac{\partial \phi}{\partial x} + \frac{\partial w}{\partial y} k_{22} \frac{\partial \phi}{\partial y} \right) d\Omega \quad (11.53)$$

In light of equation (11.47), the first term in equation (11.53) is simply

$$\oint_{\Gamma} w \left(k_{11} \frac{\partial \phi}{\partial x} n_1 + k_{22} \frac{\partial \phi}{\partial y} n_2 \right) d\Gamma = \oint_{\Gamma} w \, (\mathbf{K} \nabla \phi)^T \, \mathbf{n} \, d\Gamma \quad (11.54)$$

Finally, the second term in equation (11.53) is manipulated in the following manner

$$\int_{\Omega} \left(\frac{\partial w}{\partial x} k_{11} \frac{\partial \phi}{\partial x} + \frac{\partial w}{\partial y} k_{22} \frac{\partial \phi}{\partial y} \right) d\Omega = \int_{\Omega} \left\{ \frac{\partial w}{\partial x} \quad \frac{\partial w}{\partial y} \right\} \left\{ \begin{array}{c} k_{11} \dfrac{\partial \phi}{\partial x} \\[2mm] k_{22} \dfrac{\partial \phi}{\partial y} \end{array} \right\} d\Omega$$
$$= \int_{\Omega} \nabla^T w \, (\mathbf{K} \nabla \phi) \, d\Omega \quad (11.55)$$

Equations (11.55) and (11.54) are precisely the first and third terms, respectively, in equation (11.42), though with opposite signs.

Weighted Residual (Discrete) Form

In the previous equations, no approximations have as yet been made with respect to the primary dependent variable — the solution to the weak form of the governing equation is known to be identical to the solution of the strong form [252]. In anticipation of the subsequent finite element formulation, we assume the usual approximation for ϕ over a typical element domain Ω^e, viz.,

$$\hat{\phi}^{(e)}(x,y,z) = \sum_{m=1}^{N_{dof}} N_m(x,y,z)\hat{\phi}_m^{(e)} = \mathbf{N}\hat{\phi}_n^{(e)} \tag{11.56}$$

where $\mathbf{N}$ is a $(1 * N_{dof})$ vector of piecewise defined element interpolation functions, and $\hat{\phi}_n^{(e)}$ is a $(N_{dof} * 1)$ vector of nodal values of $\hat{\phi}^{(e)}$. For irreducible elements such as those assumed in this chapter, $N_{dof} = N_{en}$, where N_{en} equals the number of nodes in the element.

The arbitrary weighting functions w must be replaced by the finite set of prescribed functions w_i that vanish along Γ_1. Adopting a Galerkin weighted residual statement of the problem, it follows that $w_i = N_i$. Thus

$$\hat{w}(x,y,z) = \sum_{i=1}^{N_{en}} N_i(x)\beta_i^{(e)} = \mathbf{N}\beta \tag{11.57}$$

where the $\beta_i^{(e)}$ are arbitrary unknown coefficients.

Substituting equations (11.56) and (11.57), along with derivatives of $\phi_m^{(e)}$ and $\hat{w}$ into equation (11.44) gives

$$\int_{\Omega^e} \beta^T \mathbf{B}^T \mathbf{K} \mathbf{B} \hat{\phi}_n^{(e)} \, d\Omega - \int_{\Omega^e} \beta^T \mathbf{N}^T H \mathbf{N} \hat{\phi}_n^{(e)} \, d\Omega - \int_{\Omega^e} \beta^T \mathbf{N}^T Q \, d\Omega$$

$$- \oint_{\Gamma_2} \beta^T \mathbf{N}^T \bar{q} \, d\Gamma - \oint_{\Gamma_3} \beta^T \mathbf{N}^T h \left(\mathbf{N}\hat{\phi}_n^{(e)} - \bar{\phi}_\infty \right) d\Gamma = 0 \tag{11.58}$$

where $\mathbf{B} \equiv \nabla \mathbf{N}$. The matrix $\mathbf{B}$ is commonly partitioned as

$$\mathbf{B} = \begin{bmatrix} \mathbf{B_1} & \mathbf{B_2} & \cdots & \mathbf{B_{N_{en}}} \end{bmatrix} \tag{11.59}$$

with each submatrix $\mathbf{B_i}$ of size $(N_{rowb} * 1)$. Since the primary dependent variable is a scalar, $N_{rowb} = N_{sdim}$, where N_{sdim} is the spatial dimension of the problem. Consequently, the size of $\mathbf{B}$ is $(N_{sdim} * N_{en})$.

For *arbitrary* $\beta_i^{(e)}$, the element equations become

$$\mathbf{K}^{(e)} \hat{\phi}_n^{(e)} = \mathbf{f}^{(e)} \tag{11.60}$$

where

$$\mathbf{K}^{(e)} = \int_{\Omega^e} \mathbf{B}^T \mathbf{K} \mathbf{B} \, d\Omega - \int_{\Omega^e} \mathbf{N}^T H \mathbf{N} \, d\Omega - \int_{\Gamma_3} \mathbf{N}^T h \mathbf{N} \, d\Omega \tag{11.61}$$

and

$$\mathbf{f}^{(e)} = \int_{\Omega^e} \mathbf{N}^T Q \, d\Omega + \oint_{\Gamma_2} \mathbf{N}^T \bar{q} \, d\Gamma - \oint_{\Gamma_3} \mathbf{N}^T h \bar{\phi}_\infty \, d\Gamma \tag{11.62}$$

Investigating more closely the first integrand in equation 11.61, we find that

$$\mathbf{B_i}^T \mathbf{K} \mathbf{B_j} = \left\{ \frac{\partial N_i}{\partial x} \quad \frac{\partial N_i}{\partial y} \quad \frac{\partial N_i}{\partial z} \right\} \begin{bmatrix} k_{11} & k_{12} & k_{13} \\ k_{12} & k_{22} & k_{23} \\ k_{13} & k_{23} & k_{33} \end{bmatrix} \begin{Bmatrix} \dfrac{\partial N_i}{\partial x} \\[2mm] \dfrac{\partial N_i}{\partial y} \\[2mm] \dfrac{\partial N_i}{\partial z} \end{Bmatrix}$$

$$= \left(\frac{\partial N_i}{\partial x} k_{11} + \frac{\partial N_i}{\partial y} k_{12} + \frac{\partial N_i}{\partial z} k_{13} \right) \frac{\partial N_j}{\partial x}$$

$$+ \left(\frac{\partial N_i}{\partial x} k_{12} + \frac{\partial N_i}{\partial y} k_{22} + \frac{\partial N_i}{\partial z} k_{23} \right) \frac{\partial N_j}{\partial y}$$

$$+ \left(\frac{\partial N_i}{\partial x} k_{13} + \frac{\partial N_i}{\partial y} k_{23} + \frac{\partial N_i}{\partial z} k_{33} \right) \frac{\partial N_j}{\partial z} \tag{11.63}$$

where $i, j = 1, 2, \cdots, N_{en}$. If $\mathbf{K}$ is diagonal, $k_{12} = k_{13} = k_{23} = 0$, and equation (11.63) simplifies accordingly. The two-dimensional counterpart of equation (11.63) is

$$\mathbf{B_i}^T \mathbf{K} \mathbf{B_j} = \left(\frac{\partial N_i}{\partial x} k_{11} + \frac{\partial N_i}{\partial y} k_{12} \right) \frac{\partial N_j}{\partial x}$$

$$+ \left(\frac{\partial N_i}{\partial x} k_{12} + \frac{\partial N_i}{\partial y} k_{22} \right) \frac{\partial N_j}{\partial y} \tag{11.64}$$

• **Step 4:** Selection of Interpolation Functions

We begin this step by observing that the highest order derivative of the interpolation functions appearing in the element equations is one; using the notation of Section 1.3.4, $s = 1$. It follows that any of the elements discussed in Chapter 9 can thus be used in finite element analyses of scalar field problems.

• **Step 5:** Specialization of Approximate Element Equations

Depending on the specific element chosen, the associated interpolation functions and their derivatives must appropriately be substituted into the equations (11.61) and (11.62).

11.5   Torsion of Straight, Prismatic Bars

In this section we focus on the specific problem of torsion of straight, prismatic bars. This classical problem of the theory of elasticity has been widely used by engineers and mathematicians alike to evaluate the accuracy of approximate solution techniques. We begin with a discussion of the strong form of the problem.

11.5.1   The Solution of Saint-Venant

The exact solution of the problem of torsion of circular shafts, developed by Coulomb [119], was subsequently applied by Navier [403] to bars having non-circular cross-sections. However, the assumption that the cross-sections of the bar remain plane and rotate without any distortion during twist did not agree with experimental measurements [143].

The correct solution to the problem of non-circular bars twisted by couples applied only at the ends was obtained by Saint-Venant [478]. He proposed a semi-inverse method of solution in which assumptions are made concerning the deformation of the twisted bar. The assumed deformations must next be shown to satisfy the equations of equilibrium and the boundary conditions. Finally, from Kirchhoff's uniqueness theorem for elasticity solutions [538], it follows that the assumptions made at the outset are valid and the solution obtained is the exact one. The only restrictions are that the twisting moments on the ends are applied as shear stresses in exactly the manner required by the solution.

Consider a straight prismatic bar of arbitrary cross-section subjected to equal and opposite twisting moments M_T applied at its ends in the manner shown in Figure 11.4. The origin of the rectangular Cartesian coordinate system is located at the center of twist of the cross-section; i.e., the point about which the cross-section rotates during twisting (at the center of twist the displacements $u = v = 0$). The location of the center of twist is a function of the shape of the cross-section.

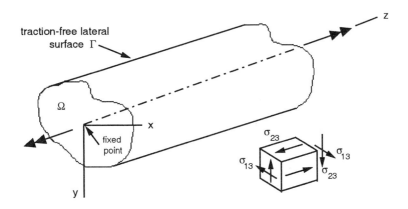

Figure 11.4: Prismatic Bar of Arbitrary Cross-Section Subjected to Twisting Moment

The rotation of the bar is assumed to be identical to the case for circular shafts, that is $u = -\theta z y$ and $v = \theta z x$, where the product θz represents the angle of rotation of a cross-section at a distance z from the origin. The angle of twist per unit length, θ, is assumed to remain constant.

If the prismatic bar is non-circular in cross-section, initially plane cross-sections experience out-of-plane deformation or *warping*. To account for this, Saint-Venant assumed that the warping of cross-sections is independent of axial location, and is defined by a warping function $\psi(x, y)$. The axial displacement is thus assumed to be $w = \theta \psi$, where ψ remains to be determined. In keeping with the solution of Coulomb, for circular shafts $w = 0$.

The material is assumed to be homogeneous, isotropic and linear elastic. In light of the assumed displacement field, the infinitesimal strain-displacement relations give the following strain components:

$$\varepsilon_{11} = \frac{\partial u}{\partial x} = 0 \quad ; \quad \varepsilon_{22} = \frac{\partial v}{\partial y} = 0 \quad ; \quad \varepsilon_{33} = \frac{\partial w}{\partial z} = 0 \qquad (11.65)$$

$$\gamma_{12} = \frac{\partial u}{\partial y} + \frac{\partial v}{\partial x} = 0 \tag{11.66}$$

$$\gamma_{13} = \frac{\partial u}{\partial z} + \frac{\partial w}{\partial x} = -\theta y + \theta \frac{\partial \psi}{\partial x} = \theta \left(\frac{\partial \psi}{\partial x} - y \right) \tag{11.67}$$

$$\gamma_{23} = \frac{\partial v}{\partial z} + \frac{\partial w}{\partial y} = \theta x + \theta \frac{\partial \psi}{\partial y} = \theta \left(\frac{\partial \psi}{\partial y} + x \right) \tag{11.68}$$

where γ_{12}, γ_{13} and γ_{23} denote engineering measures of shearing strain.

From the constitutive relations for an isotropic, linear elastic material, it follows that, in the absence of initial strains and stresses, the corresponding components of stress are

$$\sigma_{11} = \sigma_{22} = \sigma_{33} = \sigma_{12} = 0 \tag{11.69}$$

$$\sigma_{13} = G\gamma_{13} = G\theta \left(\frac{\partial \psi}{\partial x} - y \right) \tag{11.70}$$

$$\sigma_{23} = G\gamma_{23} = G\theta \left(\frac{\partial \psi}{\partial y} + x \right) \tag{11.71}$$

where G represents the elastic shear modulus.

The warping function $\psi(x,y)$ must next be determined such that the equations of equilibrium are satisfied. In the absence of body forces, the only equilibrium equation that is not trivially satisfied is

$$\frac{\partial \sigma_{13}}{\partial x} + \frac{\partial \sigma_{23}}{\partial y} = 0 \tag{11.72}$$

Substituting equations (11.70) and (11.71) gives

$$\frac{\partial}{\partial x} \left[G\theta \left(\frac{\partial \psi}{\partial x} - y \right) \right] + \frac{\partial}{\partial y} \left[G\theta \left(\frac{\partial \psi}{\partial y} + x \right) \right] = \frac{\partial^2 \psi}{\partial x^2} + \frac{\partial^2 \psi}{\partial y^2} = 0 \tag{11.73}$$

The warping function ψ is thus seen to be a harmonic function that satisfies Laplace's equation.

The final requirement of the warping function is that it satisfy the boundary conditions. At the boundary of the cross-section, the component of the stress in the direction of the unit outward normal **n** to the boundary must vanish, and the resultant shear stress must be directed along the tangent to the boundary. Failure to do so would contradict the initial assumption that the lateral surface is free from traction. Consequently this leads to the following relation [538]:

$$\left(\frac{\partial\psi}{\partial x} - y\right)\frac{dy}{ds} - \left(\frac{\partial\psi}{\partial y} + x\right)\frac{dx}{ds} = 0 \qquad (11.74)$$

where equations (11.70) and (11.71) have been used.

The torsion of straight, prismatic bars thus consists of the determination of a warping function $\psi(x, y)$ that satisfies the equilibrium equation (11.73) and the boundary conditions (11.74). Note that the twisting moment M_T does not explicitly appear in the solution.

The Saint-Venant solution gives the *same* stress distribution in all cross-sections. It therefore requires that the external forces applied at the ends of the bar be distributed in a consistent manner. Only under such conditions does the solution become the exact one for the problem. However, if the Saint-Venant principle is applied to the problem, the application of statically equivalent systems of forces at the ends will only affect the solution in the vicinity of the ends; at a sufficient distance from the ends the solution would be expected to be identical to that of Saint-Venant.

11.5.2   The Solution of Prandtl

An alternate solution, which has the advantage of leading to a simpler boundary condition, was proposed by Prandtl [442]. Similar to the solution of Saint-Venant, this solution is semi-inverse, only now it begins with an assumed stress distribution. In particular, the only non-zero components of stress are assumed to be σ_{13} and σ_{23}. Assuming negligible body forces, the equilibrium equations reduce to

$$\frac{\partial\sigma_{13}}{\partial z} = 0 \quad ; \quad \frac{\partial\sigma_{23}}{\partial z} = 0 \qquad (11.75)$$

along with equation (11.72).

Equations (11.75) indicate that σ_{13} and σ_{23} do not vary with the longitudinal coordinate z. Since they are independent of z, the expressions for

σ_{13} and σ_{23} in the Saint-Venant solution (equations 11.70 and 11.71) satisfy equations (11.75).

Since an isotropic, linear elastic material has been assumed, it follows from the constitutive relations that

$$\varepsilon_{11} = \varepsilon_{22} = \varepsilon_{33} = \gamma_{12} = 0 \qquad (11.76)$$

$$\gamma_{13} = \frac{1}{G}\sigma_{13} \quad ; \quad \gamma_{23} = \frac{1}{G}\sigma_{23} \qquad (11.77)$$

Of the six compatibility equations, five give non-trivial results [538]. These indicate that γ_{13} and γ_{23} (and, through the constitutive relations, σ_{13} and σ_{23}) must be functions only of x and y. This is completely in agreement with the assumptions of Saint-Venant.

Letting C be a constant, the equation

$$\frac{\partial \gamma_{13}}{\partial y} - \frac{\partial \gamma_{23}}{\partial x} = C \qquad (11.78)$$

must likewise hold. This equation is typically solved by introducing the so-called "Prandtl stress function" $\Phi(x, y)$ [442], such that

$$\sigma_{13} = \frac{\partial \Phi}{\partial y} \quad ; \quad \sigma_{23} = -\frac{\partial \Phi}{\partial x} \qquad (11.79)$$

Equations (11.79) satisfy the equilibrium equations (11.75) and (11.72). Using equations (11.77) in equation (11.78) gives

$$\frac{1}{G}\left(\frac{\partial^2 \Phi}{\partial y^2} + \frac{\partial^2 \Phi}{\partial x^2}\right) = C \qquad (11.80)$$

Thus, if the compatibility equations are to be satisfied, the Prandtl stress function Ψ must satisfy the Poisson's equation (11.80). To evaluate the constant C, we return to the expressions for shear stresses proposed by Saint-Venant (equations 11.70 and 11.71), and relate these to Ψ, viz.,

$$\sigma_{13} = G\theta\left(\frac{\partial \psi}{\partial x} - y\right) = \frac{\partial \Phi}{\partial y} \qquad (11.81)$$

$$\sigma_{23} = G\theta \left(\frac{\partial \psi}{\partial y} + x \right) = -\frac{\partial \Phi}{\partial x} \qquad (11.82)$$

To eliminate Ψ, differentiate equation (11.81) with respect to y and equation (11.82) with respect to x. Subtracting the resulting equations gives

$$\frac{\partial^2 \Phi}{\partial x^2} + \frac{\partial^2 \Phi}{\partial y^2} = -2G\theta \qquad (11.83)$$

The constant C in equation (11.80) is thus equal to -2θ where, as before, θ is the angle of twist per unit length. This result indirectly proves that the Saint-Venant solution satisfies the compatibility equations.

Focusing on the boundary conditions, substitution of equations (11.81) and (11.82) into equation (11.74) gives

$$\frac{\partial \Phi}{\partial y} \frac{dy}{ds} + \frac{\partial \Phi}{\partial x} \frac{dx}{ds} = \frac{\partial \Phi}{\partial s} = 0 \qquad (11.84)$$

Along a portion ds of the boundary, the stress function Φ must thus be constant. Typically, it is simply chosen to be *zero* [538].

Although it does not explicitly appear in the previous solution, the twisting moment M_T can easily be related to Ψ. From a consideration of the equilibrium conditions at the ends of the bar, it can be shown that [89]

$$M_T = 2 \iint \Phi \, dx \, dy \qquad (11.85)$$

Insight into equation (11.85) shows that one-half of the torque is due to σ_{13} and the other one-half to σ_{23}.

11.5.3 Finite Element Equations

Comparing equation (11.83) to equation (11.4), it is evident that the former is obtained by setting $\mathbf{K} = \mathbf{I}$, $H = 0$, and $Q = 2G\theta$ in the latter. Finally, we note that in the Prandtl solution to the torsion problem, only homogeneous essential boundary conditions are specified.

In light of the above observations, it follows that the element coefficient matrix of equation (11.61) reduces to

$$\mathbf{K}^{(e)} = \int\limits_{A^{(e)}} \mathbf{B}^T \mathbf{B}\, t\, dA \tag{11.86}$$

where $A^{(e)}$ denotes the cross-sectional area of the element. Since the member is assumed to be prismatic, we analyze a typical cross-section, implying that the element thickness t is equal to unity.

The right-hand side vector of equation (11.62) likewise reduces to

$$\mathbf{f}^{(e)} = 2G\theta \int\limits_{A^{(e)}} \mathbf{N}^T\, t\, dA \tag{11.87}$$

where t is again equal to unity.

Example 11.4: Torsion of a Rectangular Cross-Section

Consider a typical rectangular cross-section such as that shown in Figure 11.5.

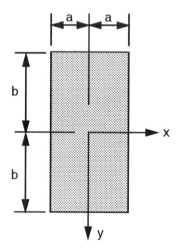

Figure 11.5: Prismatic Rectangular Cross-Section

The exact solution for the Prandtl stress function is [538]

$$\Phi = \frac{32G\theta\,a^2}{\pi^3} \sum_{n=1,3,5,\cdots}^{\infty} \frac{1}{n^3}(-1)^{\frac{n-1}{2}} \left[1 - \frac{\cosh\left(\frac{n\pi y}{2a}\right)}{\cosh\left(\frac{n\pi b}{2a}\right)}\right] \cos\frac{n\pi x}{2a} \qquad (11.88)$$

with the maximum value occurring at $x = y = 0$. The maximum shear stress, which occurs at $x = \pm a$ and $y = 0$, is

$$|\sigma|_{\max} = 2Ga\theta - \frac{16G\theta a}{\pi^2} \sum_{n=1,3,5,\cdots}^{\infty} \frac{1}{n^2 \cosh\left(\frac{n\pi b}{2a}\right)} \qquad (11.89)$$

Consider the specific case of $b = 3$ and $a = 2$. The elastic shear modulus is taken to be $G = 8.0\text{x}10^6 \; N/cm^2$, and the angle of twist per unit length is assumed to be $\theta = 1.745\text{x}10^{-4} \; rad/cm$. The maximum values for the Prandtl stress function and the shear stress are thus

$$\Phi_{\max} = 4500\,N/cm \quad \text{and} \quad |\sigma|_{\max} = 4733\,N/cm^2 \qquad (11.90)$$

In discretizing the domain, advantage is taken of the double symmetry of the cross-section. It can be shown [100] that along the symmetry axes, the boundary condition is $\partial\Phi/\partial n = 0$.

Using a standard finite element program for heat transfer analyses, the following thermal conductivities are specified: $k_{11} = k_{22} = 1.0$, $k_{12} = 0.0$. The source term is $S = 2G\theta = 2(8.0\text{x}10^6)(1.745\text{x}10^{-4}) = 2792.5 \; N/cm^3$ and applies to all elements. In interpreting the secondary dependent variables, note that $\hat{\sigma}_{23} = \hat{q}_x$ and $\hat{\sigma}_{13} = -\hat{q}_y$, where q_x and q_y represent the heat flux per unit area (computed using Fourier's law) in the global x and y coordinate directions, respectively.

We first analyze the problem using meshes of four-node quadrilateral elements. The first two meshes are shown in Figure 11.6.

The results of the finite element analyses using four-node quadrilaterals are summarized in Table 11.1. In all cases, the values of Φ_{max} occurred at the origin. The number of "free" degrees of freedom is equal to the total number of degrees of freedom minus the essential nodal specifications ($\Phi = 0$) made along the boundary. The percent relative errors are shown in parentheses below the respective results.

Remarks

1. As evident from the above results, convergence to the exact solution of both the primary dependent variable (as evidenced by Φ_{max}) and the secondary dependent variable (as evidenced by $|\sigma|_{\max}$) is observed.

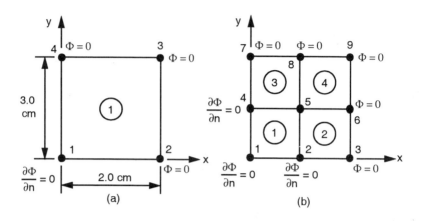

Figure 11.6: Meshes of Four-Node Quadrilaterals: (a) One and (b) Four Element

2. The primary dependent variable converges more rapidly than does the secondary dependent variable.

3. The *location* of $|\sigma|_{\max}$ changes because fluxes are computed at the element centers, while the exact maximum shear stress occurs on the boundary of Ω.

We next investigate the performance of quadratic elements. In particular, uniform meshes consisting of one and four eight-node quadrilaterals are used. The results of the corresponding analyses are summarized in Table 11.2. The above remarks pertaining to the linear elements likewise hold for the quadratic elements.

Comparing the results shown in Tables 11.1 and 11.2, it is evident that the quadratic elements are somewhat more efficient than the linear ones.

Example 11.5: Torsion of a Quarter Circle

We next consider the torsion of a quarter-circle cross-section with radius equal to 3 cm. Since the solution domain will be discretized using bilinear (four-node) quadrilaterals, the problem illustrates the notion domain discretization error. In addition, since no exact solution exists for this problem, the accuracy of the results can only be assessed by progressive refinement of the mesh.

Although symmetry considerations allow us analyze one-half of the domain (that is, one-eighth of a circle), we choose to instead analyze the entire

Table 11.1: Torsion of Rectangular Section: Summary of Results Obtained using Four-Node Quadrilateral Elements

Number of Elements	Free d.o.f.	Φ_{max} (N/cm)	$\|\sigma\|_{\max}$ (N/cm^2)	(x, y)
1	1	5800 (28.9)	1450 (69.4)	(1.000,1.500)
4	4	4736 (5.2)	3298 (30.3)	(1.500,0.750)
16	16	4556 (1.2)	4025 (15.0)	(1.750,3.750)
64	64	4516 (0.4)	4382 (7.4)	(1.875,0.188)
256	256	4506 (0.1)	4559 (3.7)	(1.938,0.094)

Table 11.2: Torsion of Rectangular Section: Summary of Results Obtained using Eight-Node Quadrilateral Elements

Number of Elements	Free d.o.f.	Φ_{max} (N/cm)	$\|\sigma\|_{\max}$ (N/cm^2)	(x, y)
1	3	4089 (9.1)	3616 (23.6)	(0.634,2.366)
4	12	4495 (0.1)	4130 (12.7)	(1.789,0.317)
16	96	4502 (0.04)	4435 (6.3)	(1.894,0.159)

domain. The reasons for this are twofold: first to avoid the possibility of moderately distorted quadrilaterals, and second to verify the symmetry of the solution.

Since the entire quarter circle is considered, the boundary conditions along all external edges are $\Phi = 0$. The product $2G\theta$ is again taken equal to 2792.5 $N/cm^3 0$.

The first mesh used contains 4 elements along each edge, for a total of 12 elements (Figure 11.7). The domain discretization error along the curved boundary is evident. The corresponding contour plot of the Prandtl stress function Φ is shown in Figure 11.8. Due to the coarseness of the mesh, the distribution of the stress function would be expected to be rather approximate. Note that in keeping with the homogeneous essential boundary condition of $\Phi = 0$, the contours are approximately parallel to the boundaries of the quarter circle. Also note that, as expected, the solution is symmetric with respect to the line bisecting the quarter circle. This serves as a check on the solution.

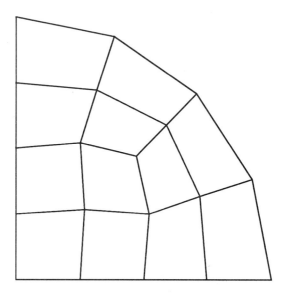

Figure 11.7: Quarter Circle Discretized by Mesh of 12 Four-Node Quadrilaterals

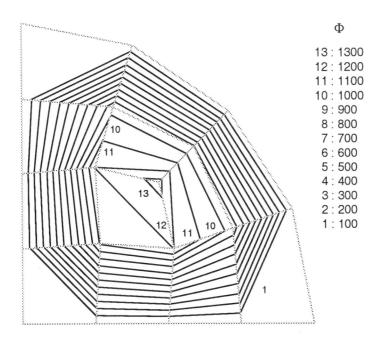

Φ

13 : 1300
12 : 1200
11 : 1100
10 : 1000
9 : 900
8 : 800
7 : 700
6 : 600
5 : 500
4 : 400
3 : 300
2 : 200
1 : 100

Figure 11.8: Contours of Prandtl Stress Function Corresponding to 12 Element Mesh

The maximum value of Φ is $1330\,N/cm$, and occurs at the point (1.414, 1.414). The maximum shear stress magnitude is $1483\,N/cm^2$, and occurs at the points (0.388, 1.154) and (1.154, 0.388). As expected, the two locations of maximum shear stress magnitude are symmetric with respect to the line bisecting the quarter circle.

The discretization is refined to 8 elements per side, which results in a total of 48 total elements. The discretization error along the curved boundary will thus be reduced. The maximum value of Φ is now $1375\,N/cm$, and occurs at the point (1.163, 1.163). The maximum shear stress magnitude is $1979\,N/cm^2$, and occurs at the points (0.195, 1.341) and (1.341, 0.195).

The discretization is again refined to 16 elements per side, resulting in a total of 192 total elements. The maximum value of Φ is now $1380\,N/cm$, and occurs at the point (1.306, 1.306). The maximum shear stress magnitude is $2234\,N/cm^2$, and occurs at the points (0.097, 1.425) and (1.425, 0.097). Assessing the results of the three analyses, it appears that the solution for the primary dependent variable Φ is converging. The same cannot be said, with great certainty, concerning the maximum shear stress.

As such, we refine the mesh one more time to 32 elements per side and 768 total elements. The corresponding mesh is shown in Figure 11.9. The discretization error along the curved boundary has obviously been significantly reduced. The contour plot of Φ is shown in Figure 11.10. The maximum value of Φ is now $1383\,N/cm$, and occurs at the point (1.238, 1.238). The maximum shear stress magnitude is $2367\,N/cm^2$, and occurs at the points (0.049, 1.369) and (1.369, 0.049).

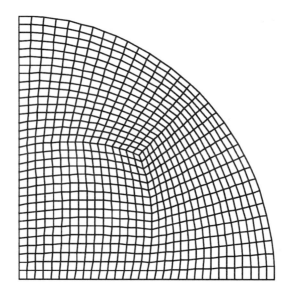

Figure 11.9: Quarter Circle Discretized by Mesh of 768 Four-Node Quadrilaterals

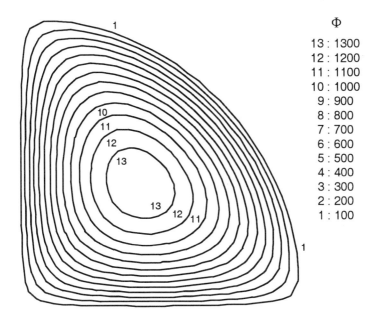

Φ

13 : 1300
12 : 1200
11 : 1100
10 : 1000
9 : 900
8 : 800
7 : 700
6 : 600
5 : 500
4 : 400
3 : 300
2 : 200
1 : 100

Figure 11.10: Contours of Prandtl Stress Function Corresponding to 768 Element Mesh

11.6   Flow Through Porous Geologic Media

Closely related to the problem of irrotational fluid flow (see Example 11.3) is the steady-state flow through porous geologic media. Assuming the solid phase to be rigid and the pores of the medium to be saturated with a homogeneous and incompressible fluid, the flow is governed by the equation

$$\nabla^T (\mathbf{K}\nabla\phi) + Q = 0 \qquad (11.91)$$

where ϕ is the piezometric head (units: L), Q represents a source (recharge) or sink (pumping) per unit volume (units: t^{-1}), and $\mathbf{K}$ is the permeability or hydraulic conductivity tensor (units: Lt^{-1}).

Underlying the derivation of equation (11.91) is the assumption that the velocity of flow through the porous geologic media is related to the gradient of ϕ by Darcy's law [131]. This in turn assumes the flow through the porous medium to be laminar. Since the permeability coefficients are inversely proportional to the viscosity of the pore fluid, it follows that the assumption of an ideal fluid is not valid for this problem. Although all six independent k_{ij} values appear in $\mathbf{K}$, seldom can all these be determined for a specific porous medium. Instead, only the *diagonal* entries are typically known.

The boundary conditions associated with equation (11.91) consist of either specified values of piezometric head (i.e., $\phi = \bar{\phi}$) on Γ_1 or specified values (typically zero) of seepage velocity $v = -k \partial \phi / \partial n$ on Γ_2 where $\Gamma_1 \cup \Gamma_2 = \Gamma$ and $\Gamma_1 \cap \Gamma_2 = \emptyset$.

In the field of *groundwater flow*, two important problems are governed by equation (11.91). The first is the calculation of either the drawdown of groundwater due to a well that is removing fluid from an aquifer, or the influx of fluid due to a pump recharging an aquifer. For a two-dimensional flow, with k_{11} and k_{22} the only non-zero permeability coefficients, the governing equation is

$$\frac{\partial}{\partial x}\left(k_{11}\frac{\partial \phi}{\partial x}\right) + \frac{\partial}{\partial y}\left(k_{22}\frac{\partial \phi}{\partial y}\right) + Q = 0 \qquad (11.92)$$

The associated boundary conditions consist of known values of ϕ or the seepage of fluid into or out of the aquifer along a boundary. The seepage is described by the following natural boundary condition

$$\bar{q} = -\left(k_{11}\frac{\partial \phi}{\partial x}\cos\alpha + k_{22}\frac{\partial \phi}{\partial y}\sin\alpha\right) \qquad (11.93)$$

where α is measured counterclockwise from the global x axis (see Figure 11.3). The fluid velocity components (secondary dependent variables) are computed using generalized Darcy's Law, viz.,

$$v_x = -k_{11}\frac{\partial \phi}{\partial x} \quad ; \quad v_y = -k_{22}\frac{\partial \phi}{\partial y} \qquad (11.94)$$

The second general problem governed by equation (11.91) is the *steady-state seepage* of groundwater under dams, sheet pile structures, etc. An early application of the finite element method to the solution of the seepage problem is given in [586].

In such problems, the "source term" Q is typically zero. The associated boundary conditions generally consist of known values of ϕ on Γ_1 or of zero seepage conditions $\partial \phi / \partial n = 0$ on Γ_2 along impermeable boundaries.

For example, consider the hypothetical sheet pile wall shown in Figure 11.11. The associated mathematical model is shown in Figure 11.12. We note the following important considerations related to this model:

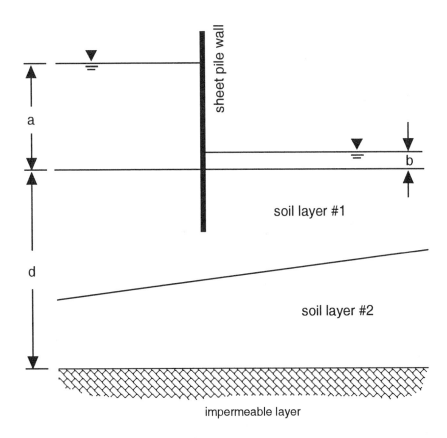

Figure 11.11: Hypothetical Sheet Pile Wall

- An element boundary must be aligned with the (inclined) material interface between the two soils.

- The largest change in direction of flow through the soil mass will occur in the vicinity of the sheet pile wall. Consequently, the mesh in this region should be rather fine, as compared to that farther away from the wall.

- The lateral extent of the mesh must be selected judiciously. We assume that the mesh in the vicinity of the sheet pile wall is sufficiently fine and is not varied as the size of the mesh is changed laterally. The lateral extent of the mesh is then deemed sufficient if:

1. The primary dependent variable ϕ and the flow velocity in the vicinity of the sheet pile wall do not change appreciably as the size of the mesh is progressively increased, and

2. The primary dependent variable ϕ at the left and right boundaries approaches values specified along the top boundary. Recall that the boundary condition along the left and right boundaries is a natural (Neumann) boundary condition. The value of ϕ is thus an unknown along these boundaries.

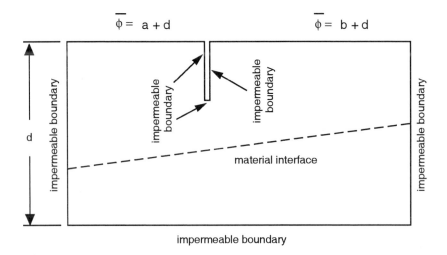

Figure 11.12: Mathematical Model of Hypothetical Sheet Pile Wall

In concluding this section, it is important to note that in both types of problems discussed above, the porous media is assumed to be free from deformation due to mechanical loads. If coupled flow-deformation problems are to be solved, the governing equations increase in complexity, and the resulting finite element formulation becomes *mixed*. Further details pertaining to mixed formulations are given in [252, 605].

11.7 Exercises

11.1

Using a finite element computer program suitable for analyzing scalar field problems (e.g., a heat transfer program), analyze the following cross-sections. Using at least three ever finer meshes, estimate the maximum value of the Prandtl stress function Φ and the maximum magnitude of the shear stress. If an exact solution exists, compare the approximate results with the exact. Discuss your results.

For all analyses, use the following values: $G = 1.154 \times 10^7$ psi and $\theta = 4.433 \times 10^{-4}$ radians per inch. Unless otherwise noted, all dimensions shown below are in units of inches.

• **Triangular Cross-Section**

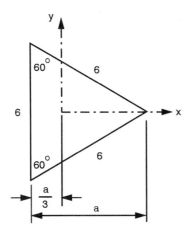

Figure 11.13: Triangular Cross-Section for Torsional Analysis

Exact Solution [538]:

$$\Phi = -G\theta \left[\frac{1}{2} \left(x^2 + y^2 \right) - \frac{x}{2a} \left(x^2 - 3y^2 \right) - \frac{2}{27}a^2 \right]$$

$$\sigma_{13} = -G\theta y \left(1 + \frac{3x}{a} \right) \quad ; \quad \sigma_{23} = G\theta \left[x - \frac{3}{2a} \left(x^2 - y^2 \right) \right]$$

$$|\sigma|_{\max} = \frac{G\theta a}{2} \quad \text{at } x = -\frac{a}{3}, y = 0$$

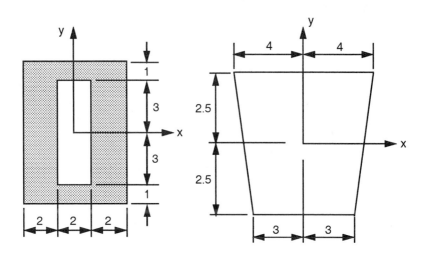

Figure 11.14: Cross-Sections for Torsional Analysis

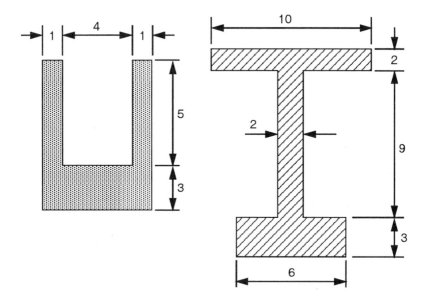

Figure 11.15: Additional Cross-Sections for Torsional Analysis

• **Solid Elliptical Cross-Section**

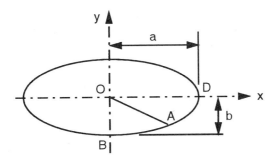

Figure 11.16: Solid Elliptical Cross-Section for Torsional Analysis

Exact Solution [538]:
The Prandtl stress function for this problem is

$$\Phi = \frac{-G\theta a^2 b^2}{(a^2 + b^2)} \left(\frac{x^2}{a^2} + \frac{y^2}{b^2} - 1 \right)$$

The shear stresses are

$$\sigma_{13} = -\frac{2a^2 G\theta}{a^2 + b^2} y \quad ; \quad \sigma_{23} = \frac{2b^2 G\theta}{a^2 + b^2} x$$

Notes
1. The resulting shear stress along any radius OA has a constant direction that coincides with the direction of the tangent to the boundary at point A.
2. Along OB, the stress component σ_{13} is zero; along OD, σ_{13} is zero.
3. The maximum shear stress occurs on the boundary, at the ends of the minor axis. The magnitude is equal to

$$|\sigma|_{\max} = |\sigma_{13} \left(y = \pm b \right)| = \frac{2a^2 bG\theta}{a^2 + b^2}$$

4. Setting $a = b$, the solution reduces to Coulomb's solution for a circular cross-section.
5. For this problem, let $a = 3$ and $b = 2$ inches.

11.2

Use a computer program with two-dimensional heat conduction elements to determine the highest temperature in the plane body shown in Figure 11.17. What is this value, and where does it occur? The numerical values listed in Figure 11.17 are assumed to have consistent units.

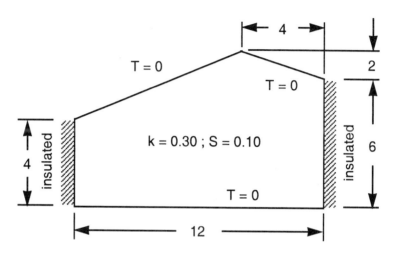

Figure 11.17: Plane Body for Heat Conduction Analysis

11.3

The cross-section of an insulated undersea oil pipeline is shown in Figure 11.17. The thermal properties of the structural support material are $k_{11} = k_{22} = 0.20 \, cal/^{\circ}C - cm - s$. For the insulation material $k_{11} = k_{22} = 0.020 \, cal/^{\circ}C - cm - s$. The pipe is assumed to be full of oil; treat this as an essential boundary condition of $\bar{T} = 85^{\circ}C$. The thickness of the pipe wall is assumed negligible. In addition, it is assumed that there is no heat transfer into the soil; that is, this boundary is fully insulated. Finally, the temperature of the sea water surrounding the pipeline is $5^{\circ}C$.

Using a computer program with two-dimensional heat conduction elements, determine the temperature distribution in the cross-section. Present your results in the form of temperature contours. In addition, on a sketch of the cross-section, plot (to a reasonable scale) the heat flux vectors at the center of each element. To reduce computational effort, use whatever symmetry conditions are appropriate.

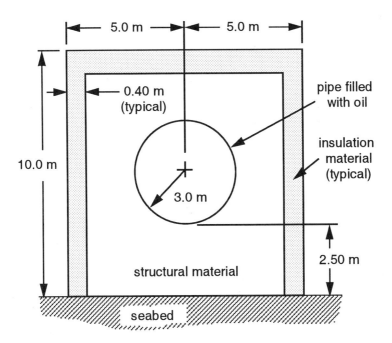

Figure 11.18: Cross-Section of Undersea Oil Pipeline

11.4

Using a computer program with two-dimensional scalar elements, analyze the sheet pile wall shown in Figure 11.19. Present your results in the form of contours of hydraulic head. On a separate figure, plot (to a reasonable scale) the velocity vectors at the center of each element.

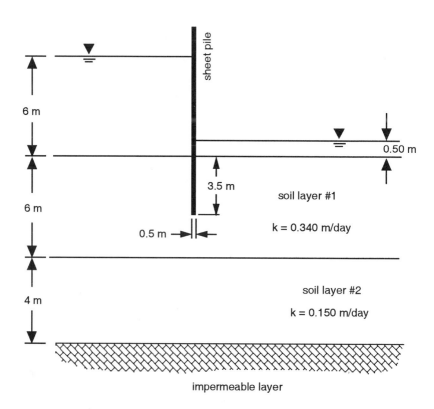

Figure 11.19: Sheet Pile Wall

Chapter 12

Finite Element Analysis in Linear Elastostatics

12.1 Introductory Remarks

In this chapter finite element equations are developed for analysis of linear elastic bodies under static conditions. As noted in Chapter 6, some of the earliest applications of the finite element method were in this area of solid mechanics.

A basic understanding of linear elasticity by the reader is assumed. Further information pertaining to the subject is available in classical textbooks on the subject [310, 504, 529, 538], as well as in ones devoted to the finite element analysis of elastic bodies [104]. Historical accounts of the theory of elasticity are available in books by Todhunter and Pearson [540] and Timoshenko [539], and in an article by Truesdell [544].

12.2 Development of General Finite Element Equations

Using the five-step approach introduced in Chapter 7, a general irreducible finite element formulation is developed herein. In subsequent sections, this formulation is specialized for three-dimensional, and then for various two-dimensional idealizations.

Although equations associated with the theory of elasticity are commonly cast in tensor form, one involving vectors and matrices is more conducive to numerical analysis. Consequently, such a form is used herein.

- **Step 1:** Identification of the Primary Dependent Variable

Since an irreducible, displacement based formulation is adopted herein, displacement components constitute the primary dependent variables. For a given element these components are approximated as

$$\hat{\mathbf{u}}^{(e)} = \mathbf{N}\hat{\mathbf{u}}_n^{(e)} \tag{12.1}$$

where $\hat{\mathbf{u}}^{(e)}$ is a $(N_{sdim} * 1)$ vector of approximate displacements, $\mathbf{N}$ is a $(N_{sdim} * N_{eddof})$ matrix of element interpolation functions, and $\hat{\mathbf{u}}_n^{(e)}$ is a $(N_{eddof} * 1)$ vector containing the unknown nodal values of displacement components.

- **Step 2:** Definition of the Gradient and Constitutive Relations

The displacements and displacement gradients are assumed to be sufficiently small such that no distinction need be made between Lagrangian and Eulerian descriptions. Consequently, the total strains are defined by

$$\varepsilon = \mathbf{S}\mathbf{u} \tag{12.2}$$

where ε is a $(N_{rowb} * 1)$ vector of strain components, and $\mathbf{S}$ is a $(N_{rowb} * N_{sdim})$ matrix of linear operators.

Making use of equation (12.1), the strains for a given element are approximated in the following manner:

$$\hat{\varepsilon} = \mathbf{S}\hat{\mathbf{u}}^{(e)} = \mathbf{S}\mathbf{N}\hat{\mathbf{u}}_n^{(e)} = \mathbf{B}\hat{\mathbf{u}}_n^{(e)} \tag{12.3}$$

where $\hat{\varepsilon}$ is a $(N_{rowb} * 1)$ vector of approximate strain components, and $\mathbf{B}$ is the $(N_{rowb} * N_{eddof})$ strain-displacement matrix.

The material is assumed to be *homogeneous, linearly elastic,* and either *anisotropic* (due to some technological process such as rolling of metals, etc., the properties of the material are different in different directions), or *isotropic* (the elastic properties are the same in all directions).

The constitutive relations for an anisotropic linear elastic material are commonly referred to as *generalized Hooke's law.* For such a material, the constitutive relations, in "direct" form, are given by

$$\varepsilon = \mathbf{A}\left(\sigma - \sigma^0\right) + \varepsilon^0 \tag{12.4}$$

where σ is a $(N_{rowb} * 1)$ vector of stress components, $\mathbf{A}$ is a $(N_{rowb} * N_{rowb})$ matrix of compliance coefficients characterizing the material, σ^0 is the $(N_{rowb} * 1)$ vector of initial (strain independent) stresses (that is, a pre-stressing present before deformations are allowed), and ε^0 is the $(N_{rowb} * 1)$ vector of initial strains (that is, those due to shrinkage, crystal growth, moisture induced swelling, thermal effects, etc.; in short, any stress independent strain).

Explicit expressions for the elastic compliance coefficients populating $\mathbf{A}$, in terms of "engineering" constants are given in [310].

The constitutive relations, in "inverse" form, are given by

$$\sigma = \mathbf{C}\left(\varepsilon - \varepsilon^0\right) + \sigma^0 \tag{12.5}$$

where $\mathbf{C}$, which is the inverse of $\mathbf{A}$, is a $(N_{rowb} * N_{rowb})$ matrix of elastic coefficients that characterize the material.

Thermal Strains

In order to illustrate the relation between initial strains and initial stresses, consider the specific case of thermal strains/stresses. In standard uncoupled, quasi-static thermoelasticity, the initial strain vector is given by [184]

$$\varepsilon^0 = \alpha_t \left(T - T_0\right) \tag{12.6}$$

where α_t is a $(N_{rowb} * 1)$ vector of coefficients of thermal expansion, T denotes the temperature, and T_0 is the uniform temperature at which the material is strain (stress)-free.

The constitutive relations, written in inverse form without initial stresses, are specialized for initial thermal strains, viz.,

$$\sigma = \mathbf{C}\left(\varepsilon - \alpha_t \left(T - T_0\right)\right) = \mathbf{C}\varepsilon - \beta\left(T - T_0\right) \tag{12.7}$$

where $\beta = \mathbf{C}\alpha_t$. The initial strain can thus alternately be viewed as an initial stress equal to $-\beta\left(T - T_0\right)$.

Assuming the initial strains and stresses to be known *exactly*, the approximate stresses are computed from the following expression:

$$\hat{\sigma} = \mathbf{C}\left(\hat{\varepsilon} - \varepsilon^0\right) + \sigma^0 = \mathbf{C}\mathbf{B}\hat{u}_n^{(e)} - \mathbf{C}\varepsilon^0 + \sigma^0 \tag{12.8}$$

Specific forms of **C**, which depend upon the number of planes of elastic symmetry exhibited by the material, are presented in subsequent sections.

- **Step 3:** Identification of the Element Equations

For problems in linear elasticity, the element equations can be obtained either through a variational principle or by employing a weighted residual statement. Both approaches shall be presented herein.

The element equations are first derived using a *variational* approach. This involves choosing suitable nodal displacements that make stationary the approximate functional $\hat{I}$. If the interpolation functions used in a piecewise definition of the nodal displacements satisfy the convergence criteria discussed in Chapter 6, the functional may be written as a sum of individual functionals defined over all the elements, that is,

$$\hat{I} = \sum_{e=1}^{N_e} \hat{I}^{(e)} \tag{12.9}$$

where N_e equals the total number of elements. Thus, instead of working with the functional defined over the entire solution domain Ω, only the functionals associated with the individual elements need be considered. Recalling that an irreducible formulation is assumed, the element functional will be stationary provided that

$$\frac{\partial \hat{I}}{\partial \hat{u}_i^{(e)}} = 0 \quad , \quad \frac{\partial \hat{I}}{\partial \hat{v}_i^{(e)}} = 0 \quad , \quad \frac{\partial \hat{I}}{\partial \hat{w}_i^{(e)}} = 0 \quad ; \quad i = 1, 2, \cdots, N_{en} \tag{12.10}$$

where $\hat{u}_i^{(e)}$, $\hat{v}_i^{(e)}$ and $\hat{w}_i^{(e)}$ represent displacement components in the x, y and z directions, respectively, corresponding to node i in element e. This is a set of N_{eddof} equations that characterize the behavior of a typical element.

For displacement-based formulations the functional is equal to the total potential energy, which consists of the sum of the strain energy and the potential due to the applied loading. Making use of the constitutive relation given by equation (12.5), the strain energy for a typical element is written in terms of the strain vector as

$$U^{(e)} = \frac{1}{2} \int_{\Omega^{(e)}} \varepsilon^T \mathbf{C} \varepsilon \, d\Omega - \int_{\Omega^{(e)}} \varepsilon^T \mathbf{C} \varepsilon^0 \, d\Omega + \int_{\Omega^{(e)}} \varepsilon^T \sigma^0 \, d\Omega$$

$$+ \int_{\Omega^{(e)}} \varepsilon^{0^T} \left(\mathbf{C} \varepsilon^0 - \sigma^0 \right) d\Omega \tag{12.11}$$

The potential energy associated with the external loading is

$$V_a^{(e)} = - \int_{\Omega^{(e)}} \mathbf{u}^T \mathbf{b} \, d\Omega - \int_{\Gamma_2} \mathbf{u}^T \bar{\mathbf{t}} \, d\Gamma \tag{12.12}$$

where Γ_2 represents that portion of the surface along which a traction (a natural boundary condition) is specified, and b and $\bar{t}$ are $(N_{sdim} * 1)$ vectors of body force components and surface tractions, respectively.

The total potential energy (the functional), is thus $I^{(e)} = U^{(e)} + V_a^{(e)}$. The approximation to $I^{(e)}$ is thus

$$\hat{I}^{(e)} = \frac{1}{2} \int_{\Omega^{(e)}} \hat{\mathbf{u}}_n^{(e)T} \mathbf{B}^T \mathbf{C} \mathbf{B} \hat{\mathbf{u}}_n^{(e)} \, d\Omega - \int_{\Omega^{(e)}} \hat{\mathbf{u}}_n^{(e)T} \mathbf{B}^T \mathbf{C} \, \varepsilon^0 \, d\Omega$$

$$+ \int_{\Omega^{(e)}} \hat{\mathbf{u}}_n^{(e)T} \mathbf{B}^T \sigma^0 \, d\Omega + \int_{\Omega^{(e)}} \varepsilon^{0T} \left(\mathbf{C} \, \varepsilon^0 - \sigma^0 \right) \, d\Omega$$

$$- \int_{\Omega^{(e)}} \hat{\mathbf{u}}_n^{(e)T} \mathbf{N}^T \mathbf{b} \, d\Omega - \int_{\Gamma_2} \hat{\mathbf{u}}_n^{(e)T} \mathbf{N}^T \bar{\mathbf{t}} \, d\Gamma \tag{12.13}$$

where equation (12.3) has been used.

To ensure equilibrium, for variations of admissible displacements the total potential energy must be stationary. Using equation (12.10) leads to the following set of element equations:

$$\mathbf{K}^{(e)} \hat{\mathbf{u}}_n^{(e)} = \mathbf{f}^{(e)} \tag{12.14}$$

where

$$\mathbf{K}^{(e)} = \int_{\Omega^{(e)}} \mathbf{B}^T \mathbf{C} \mathbf{B} \, d\Omega \tag{12.15}$$

$$\mathbf{f}^{(e)} = \int_{\Omega^{(e)}} \mathbf{B}^T \mathbf{C} \varepsilon^0 \, d\Omega - \int_{\Omega^{(e)}} \mathbf{B}^T \sigma^0 \, d\Omega$$

$$+ \int_{\Omega^{(e)}} \mathbf{N}^T \mathbf{b} \, d\Omega + \int_{\Gamma_2} \mathbf{N}^T \bar{\mathbf{t}} \, d\Gamma \tag{12.16}$$

The Weighted Residual Method Applied to Linear Elasticity

In the previous development, the integral statement necessary for deriving element equations associated with linear elastostatics was obtained using a variational approach. In this note the same equations will be derived using the method of weighted residuals. The result will be shown to be a statement of the principle of *virtual work*.

Multiplying each of the equations of static equilibrium [538] by arbitrary weighting functions and integrating over the volume of the body gives

$$\delta u \left[\int_V \left(\frac{\partial \sigma_{11}}{\partial x} + \frac{\partial \sigma_{21}}{\partial y} + \frac{\partial \sigma_{31}}{\partial z} + b_1 \right) dV \right] = 0 \qquad (12.17)$$

$$\delta v \left[\int_V \left(\frac{\partial \sigma_{12}}{\partial x} + \frac{\partial \sigma_{22}}{\partial y} + \frac{\partial \sigma_{32}}{\partial z} + b_2 \right) dV \right] = 0 \qquad (12.18)$$

$$\delta w \left[\int_V \left(\frac{\partial \sigma_{13}}{\partial x} + \frac{\partial \sigma_{23}}{\partial y} + \frac{\partial \sigma_{33}}{\partial z} + b_3 \right) dV \right] = 0 \qquad (12.19)$$

where δu, δv and δw represent the arbitrary weighting functions.

The divergence theorem is next used to integrate by parts those terms containing partial derivatives. For example,

$$\delta u \int_V \left(\frac{\partial \sigma_{11}}{\partial x} \right) dV = - \int_V \sigma_{11} \frac{\partial (\delta u)}{\partial x} dV + \int_\Gamma \delta u \, \sigma_{11} \, n_1 \, d\Gamma \qquad (12.20)$$

where Γ denotes the surface area of the body and n_1 represents the unit outward normal parallel to the x coordinate axis. Equations (12.17) to (12.19) become

$$- \int_V \left[\sigma_{11} \frac{\partial (\delta u)}{\partial x} + \sigma_{21} \frac{\partial (\delta u)}{\partial y} + \sigma_{31} \frac{\partial (\delta u)}{\partial z} \right] dV + \int_V \delta u \, b_1 dV$$

$$+ \int_\Gamma \delta u \left[\sigma_{11} \, n_1 + \sigma_{21} \, n_2 + \sigma_{31} \, n_3 \right] d\Gamma = 0 \qquad (12.21)$$

$$-\int_V \left[\sigma_{12} \frac{\partial(\delta v)}{\partial x} + \sigma_{22} \frac{\partial(\delta v)}{\partial y} + \sigma_{32} \frac{\partial(\delta v)}{\partial z} \right] dV + \int_V \delta v \, b_2 dV$$

$$+ \int_\Gamma \delta v \left[\sigma_{12} \, n_1 + \sigma_{22} \, n_2 + \sigma_{32} \, n_3 \right] d\Gamma = 0 \qquad (12.22)$$

$$-\int_V \left[\sigma_{13} \frac{\partial(\delta w)}{\partial x} + \sigma_{23} \frac{\partial(\delta w)}{\partial y} + \sigma_{33} \frac{\partial(\delta w)}{\partial z} \right] dV + \int_V \delta w \, b_3 dV$$

$$+ \int_\Gamma \delta w \left[\sigma_{13} \, n_1 + \sigma_{23} \, n_2 + \sigma_{33} \, n_3 \right] d\Gamma = 0 \qquad (12.23)$$

Equations (12.21) to (12.23) represent the weak form of a weighted residual statement of the governing (static equilibrium) differential equations. Since they are homogeneous, these equations can be combined into a single equation, viz.,

$$-\int_V \left\{ \sigma_{11} \frac{\partial(\delta u)}{\partial x} + \sigma_{12} \left[\frac{\partial(\delta u)}{\partial y} + \frac{\partial(\delta v)}{\partial x} \right] + \sigma_{13} \left[\frac{\partial(\delta u)}{\partial z} + \frac{\partial(\delta w)}{\partial x} \right] \right.$$

$$\left. + \sigma_{22} \frac{\partial(\delta v)}{\partial y} + \sigma_{23} \left[\frac{\partial(\delta v)}{\partial z} + \frac{\partial(\delta w)}{\partial y} \right] + \sigma_{33} \frac{\partial(\delta w)}{\partial z} \right\} dV$$

$$+ \int_V \left(\delta u \, b_1 + \delta v \, b_2 + \delta w \, b_3 \right) dV$$

$$+ \int_\Gamma \left\{ \delta u \left[\sigma_{11} \, n_1 + \sigma_{21} \, n_2 + \sigma_{31} \, n_3 \right] + \delta v \left[\sigma_{12} \, n_1 + \sigma_{22} \, n_2 + \sigma_{32} \, n_3 \right] \right.$$

$$\left. + \delta w \left[\sigma_{13} \, n_1 + \sigma_{23} \, n_2 + \sigma_{33} \, n_3 \right] \right\} d\Gamma = 0 \qquad (12.24)$$

where the symmetry of the stress tensor has been taken into account. The six stress components appearing in the equilibrium equations are functions of the displacement components u, v and w. If the arbitrary weighting functions δu, δv and δw are taken to be *virtual displacement increments* consistent with the essential boundary conditions, it follows that the virtual strain increments will be given by

$$\{\delta\varepsilon\} = \left\{\begin{array}{c} \delta\varepsilon_{11} \\ \delta\varepsilon_{22} \\ \delta\varepsilon_{33} \\ \delta\gamma_{12} \\ \delta\gamma_{13} \\ \delta\gamma_{23} \end{array}\right\} = \left\{\begin{array}{c} \frac{\partial(\delta u)}{\partial x} \\[4pt] \frac{\partial(\delta v)}{\partial y} \\[4pt] \frac{\partial(\delta w)}{\partial z} \\[4pt] \frac{\partial(\delta u)}{\partial y} + \frac{\partial(\delta v)}{\partial x} \\[4pt] \frac{\partial(\delta u)}{\partial z} + \frac{\partial(\delta w)}{\partial x} \\[4pt] \frac{\partial(\delta v)}{\partial z} + \frac{\partial(\delta w)}{\partial y} \end{array}\right\} = \mathbf{S} \left\{\begin{array}{c} \delta u \\ \delta v \\ \delta w \end{array}\right\} \tag{12.25}$$

where infinitesimal displacements and displacement gradients have again been assumed and γ_{12}, γ_{13} and γ_{23} denote virtual "engineering" shear strains. Equation (12.24) thus becomes

$$\int_V \{\sigma_{11}\delta\varepsilon_{11} + \sigma_{12}\delta\gamma_{12} + \sigma_{13}\delta\gamma_{13} + \sigma_{22}\delta\varepsilon_{22} + \sigma_{23}\delta\gamma_{23} + \sigma_{33}\delta\varepsilon_{33}\}\, dV$$

$$- \int_V \{b_1\delta u_1 + b_2\delta u_2 + b_3\delta u_3\}\, dV$$

$$+ \int_{\Gamma_2} \{\bar{t}_1\, \delta u_1 + \bar{t}_2\, \delta u_2 + \bar{t}_3\, \delta u_3\}\, d\Gamma = 0 \tag{12.26}$$

In matrix form, this equation is written as

$$\int_V \delta\boldsymbol{\varepsilon}^T\boldsymbol{\sigma}\, dV - \int_V \delta\mathbf{u}^T\mathbf{b}\, dV - \int_{\Gamma_2} \delta\mathbf{u}^T\bar{\mathbf{t}}\, d\Gamma = 0 \tag{12.27}$$

where

$$\boldsymbol{\sigma} = \left\{\begin{array}{cccccc} \sigma_{11} & \sigma_{22} & \sigma_{33} & \sigma_{12} & \sigma_{13} & \sigma_{23} \end{array}\right\}^T \tag{12.28}$$

$$\mathbf{b} = \left\{\begin{array}{ccc} b_1 & b_2 & b_3 \end{array}\right\}^T \tag{12.29}$$

and

$$
\bar{\mathbf{t}} = \begin{Bmatrix} \bar{t}_1 \\ \bar{t}_2 \\ \bar{t}_3 \end{Bmatrix} = \begin{Bmatrix} \sigma_{11}n_1 + \sigma_{21}n_2 + \sigma_{31}n_3 \\ \sigma_{12}n_1 + \sigma_{22}n_2 + \sigma_{32}n_3 \\ \sigma_{13}n_1 + \sigma_{23}n_2 + \sigma_{33}n_3 \end{Bmatrix}
$$

$$
= \begin{bmatrix} \sigma_{11} & \sigma_{12} & \sigma_{13} \\ \sigma_{21} & \sigma_{22} & \sigma_{23} \\ \sigma_{31} & \sigma_{32} & \sigma_{33} \end{bmatrix} \begin{Bmatrix} n_1 \\ n_2 \\ n_3 \end{Bmatrix} \tag{12.30}
$$

is a vector containing the x, y and z components of the surface traction specified along a portion of the boundary Γ_2. The vectors $\delta\varepsilon$ and $\delta\mathbf{u}$ are as defined in equation (12.25). Equation (12.26) is a statement of the principle of *virtual work*, which is valid for linear as well as non-linear constitutive relations. This statement is seen to be nothing more than a weak form of the weighted residual statement of the governing (equilibrium) differential equations.

To realize the desired element equations, we note that

$$
\delta\hat{\mathbf{u}}^{(e)} = \mathbf{N}\delta\hat{\mathbf{u}}_n^{(e)} \tag{12.31}
$$

$$
\delta\varepsilon^{(e)} = \mathbf{B}\delta\hat{\mathbf{u}}_n^{(e)} \tag{12.32}
$$

The approximation to equation (12.27) thus becomes

$$
\int_{V^{(e)}} \delta\hat{\mathbf{u}}_n^{(e)T} \mathbf{B}^T \hat{\sigma} \, dV - \int_{V^{(e)}} \delta\hat{\mathbf{u}}_n^{(e)T} \mathbf{N}^T \mathbf{b} \, dV
$$

$$
- \int_{\Gamma_2} \delta\hat{\mathbf{u}}_n^{(e)T} \mathbf{N}^T \bar{\mathbf{t}} \, d\Gamma = 0 \tag{12.33}
$$

where the integration is understood to apply only to a single element domain and its surface. Substituting equation (12.8) for $\hat{\sigma}$, and noting that the $\delta\hat{\mathbf{u}}_n^{(e)}$ are arbitrary, gives results that are *identical* to equations (12.15) and (12.16).

• **Step 4:** Selection of Interpolation Functions

As evident from the element equations, irreducible formulations for elastostatics require only C^0 continuous interpolation functions. Consequently, any of the elements discussed in Chapter 9 are applicable to such formulations.

• **Step 5:** Specialization of Approximate Element Equations

The specialization of the element equations depends upon the specific element type being considered. As such, the last step in the development of the element equations shall be carried out in the subsequent sections, where the general formulation is specialized.

12.3   Three-Dimensional Idealizations

For three-dimensional analyses $N_{sdim} = 3$ and $N_{rowb} = 6$. The approximate displacements for a given element are

$$\hat{\mathbf{u}}^{(e)} = \begin{Bmatrix} \hat{u} \\ \hat{v} \\ \hat{w} \end{Bmatrix} = \mathbf{N}\hat{\mathbf{u}}_{\mathbf{n}}^{(e)} \tag{12.34}$$

where $\hat{\mathbf{u}}_{\mathbf{n}}^{(e)}$ is the vector of unknown displacement degrees of freedom u, v and w present at each node.

The vectors of total and initial strains are

$$\varepsilon = \begin{Bmatrix} \varepsilon_{11} & \varepsilon_{22} & \varepsilon_{33} & \gamma_{12} & \gamma_{13} & \gamma_{23} \end{Bmatrix}^T \tag{12.35}$$

$$\varepsilon^0 = \begin{Bmatrix} \varepsilon_{11}^0 & \varepsilon_{22}^0 & \varepsilon_{33}^0 & \gamma_{12}^0 & \gamma_{13}^0 & \gamma_{23}^0 \end{Bmatrix}^T \tag{12.36}$$

where *engineering* measures of shear strains have been used. The strain-displacement equations, written in explicit form with respect to a rectangular Cartesian coordinate system, are as follows

$$\varepsilon_{11} = \frac{\partial u}{\partial x} \quad , \quad \varepsilon_{22} = \frac{\partial v}{\partial y} \quad , \quad \varepsilon_{33} = \frac{\partial w}{\partial z} \tag{12.37}$$

$$\gamma_{12} = \frac{\partial u}{\partial y} + \frac{\partial v}{\partial x} \quad , \quad \gamma_{13} = \frac{\partial u}{\partial z} + \frac{\partial w}{\partial x} \quad , \quad \gamma_{23} = \frac{\partial v}{\partial z} + \frac{\partial w}{\partial y} \tag{12.38}$$

The matrix of linear operators is thus

$$
S = \begin{bmatrix}
\partial/\partial x & 0 & 0 \\
0 & \partial/\partial y & 0 \\
0 & 0 & \partial/\partial z \\
\partial/\partial y & \partial/\partial x & 0 \\
\partial/\partial z & 0 & \partial/\partial x \\
0 & \partial/\partial z & \partial/\partial y
\end{bmatrix}
\tag{12.39}
$$

The corresponding stress vectors are

$$
\boldsymbol{\sigma} = \left\{ \sigma_{11} \quad \sigma_{22} \quad \sigma_{33} \quad \sigma_{12} \quad \sigma_{13} \quad \sigma_{23} \right\}^T
\tag{12.40}
$$

$$
\boldsymbol{\sigma}^0 = \left\{ \sigma_{11}^0 \quad \sigma_{22}^0 \quad \sigma_{33}^0 \quad \sigma_{12}^0 \quad \sigma_{13}^0 \quad \sigma_{23}^0 \right\}^T
\tag{12.41}
$$

The matrix of element interpolation functions is written in terms of N_{en} submatrices as

$$
\mathbf{N} = \begin{bmatrix} \mathbf{N}_1 & \mathbf{N}_2 & \cdots & \mathbf{N}_{N_{en}} \end{bmatrix}
\tag{12.42}
$$

Each of the $(3 * 3)$ submatrices has the following general form:

$$
\mathbf{N}_i = \begin{bmatrix}
N_i & 0 & 0 \\
0 & N_i & 0 \\
0 & 0 & N_i
\end{bmatrix} = N_i \mathbf{I} \quad ; \quad i = 1, 2, \cdots, N_{en}
\tag{12.43}
$$

where $\mathbf{I}$ is the identity matrix.

The strain-displacement matrix is likewise written in terms of submatrices as

$$
\mathbf{B} = \mathbf{S} \begin{bmatrix} \mathbf{N}_1 & \mathbf{N}_2 & \cdots & \mathbf{N}_{N_{en}} \end{bmatrix} = \begin{bmatrix} \mathbf{B}_1 & \mathbf{B}_2 & \cdots & \mathbf{B}_{N_{en}} \end{bmatrix}
\tag{12.44}
$$

where each of the $(6 * 3)$ submatrices has the general form

$$
\mathbf{B}_i = \begin{bmatrix}
\partial N_i/\partial x & 0 & 0 \\
0 & \partial N_i/\partial y & 0 \\
0 & 0 & \partial N_i/\partial z \\
\partial N_i/\partial y & \partial N_i/\partial x & 0 \\
\partial N_i/\partial z & 0 & \partial N_i/\partial x \\
0 & \partial N_i/\partial z & \partial N_i/\partial y
\end{bmatrix} \quad ; \quad i = 1, 2, \cdots, N_{en}
\tag{12.45}
$$

From the point of view of material characterization, the analysis of bodies made of anisotropic materials is quite complicated, especially in three dimensions. This is because of the need to experimentally determine the entries in the matrices $\mathbf{A}$ or $\mathbf{C}$ appearing in equations (12.4) and (12.5), respectively. In the most general case, these matrices contain *twenty-one* independent entries.

Example 12.1: Constitutive Relation for Cellulose Fibers

Cellulose, a basic building block of most plant life forms, is used in many industrial applications. Cellulose fibers are quite anisotropic. This anisotropy has been quantified by Tashiro and Kobayashi [528], who presented the following material coefficients related to the global x, y and z axes (and not necessarily to the orientation of the fibers):

$$\mathbf{C} = \begin{bmatrix} 18.07 & 3.84 & 0.60 & 0 & 0 & 0.05 \\ 3.84 & 18.38 & 0.35 & 0 & 0 & 3.72 \\ 0.60 & 0.35 & 167.97 & 0 & 0 & -0.02 \\ 0 & 0 & 0 & 6.08 & -5.03 & 0 \\ 0 & 0 & 0 & -5.03 & 6.29 & 0 \\ 0.05 & 3.72 & -0.02 & 0 & 0 & 21.77 \end{bmatrix} GPa \qquad (12.46)$$

From the relative magnitude of the above stiffness parameters, it is evident that cellulose fibers exhibit fairly pronounced anisotropy.

To make better sense of material idealizations based on entries in $\mathbf{C}$, we partition this matrix into $(3 * 3)$ submatrices as

$$\mathbf{C} = \begin{bmatrix} \mathbf{C_{11}} & \mathbf{C_{12}} \\ \mathbf{C_{21}} & \mathbf{C_{22}} \end{bmatrix} \qquad (12.47)$$

where, due to symmetry, $\mathbf{C_{12}} = \mathbf{C_{21}}^T$.

Non-zero entries in $\mathbf{C_{11}}$ mean that the normal stresses σ_{11}, σ_{22} and σ_{33} are functions of the strains ε_{11}, ε_{22} and ε_{33}. If, in addition, $\mathbf{C_{12}} \neq \mathbf{0}$, the normal stresses are also functions of the shear strains γ_{12}, γ_{13} and γ_{23}. If $\mathbf{C_{12}} \neq \mathbf{0}$, then $\mathbf{C_{21}} \neq \mathbf{0}$, implying that the shear stresses σ_{12}, σ_{13} and σ_{23} are functions of ε_{11}, ε_{22} and ε_{33}. Finally, if $\mathbf{C_{22}} \neq \mathbf{0}$, the shear stresses are also functions of the shear strains.

Analyzing $\mathbf{C_{22}}$ more closely, it is evident that a diagonal form of this submatrix means that shear stresses are only functions of the corresponding shear strains. Once non-zero diagonal entries are present in $\mathbf{C_{22}}$, the shear stresses become functions of two or more shear strains.

Fortunately, many of the important engineering materials possess some internal structure that exhibits certain symmetries, and thus simplifies the composition of $\mathbf{A}$ or $\mathbf{C}$. Some special cases of elastic symmetry are summarized below.

Special Case 1: Materials Possessing a Single Plane of Elastic Symmetry

A plane of elastic symmetry is defined in the following manner: every two directions that are symmetrical with respect to this plane are equivalent in terms of elastic properties. A direction normal to the plane of elastic symmetry (here assumed to be the z-axis) is termed the *principal direction of elasticity*. In this case only one principal direction passes through a point in the body. If such a plane of elastic symmetry passes through each point in a body, then the constitutive relations of equations (12.4) and (12.5) simplify. More precisely, these relations are obtained by requiring that the coefficients a_{ij} remain unchanged when the z-axis is inverted. Consequently, the compliance matrix has the following entries:

$$\mathbf{A} = \begin{bmatrix} a_{11} & a_{12} & a_{13} & a_{14} & 0 & 0 \\ a_{12} & a_{22} & a_{23} & a_{24} & 0 & 0 \\ a_{13} & a_{23} & a_{33} & a_{34} & 0 & 0 \\ a_{14} & a_{24} & a_{34} & a_{44} & 0 & 0 \\ 0 & 0 & 0 & 0 & a_{55} & a_{56} \\ 0 & 0 & 0 & 0 & a_{56} & a_{66} \end{bmatrix} \tag{12.48}$$

Due to their symmetry, the constitutive relations now involve only *thirteen* elastic constants.

Special Case 2: Materials Possessing Three Planes of Elastic Symmetry (Orthotropic Materials)

Consider a material through each point of which pass *three* mutually perpendicular planes of elastic symmetry. If similar planes are parallel at all points in the material, then taking (x, y, z) coordinate axes normal to these planes (i.e., along the principal directions) it follows that there should be no interaction between the various shear components or between the shear and normal components. Consequently, the compliance matrix has the following entries:

$$\mathbf{A} = \begin{bmatrix} a_{11} & a_{12} & a_{13} & 0 & 0 & 0 \\ a_{12} & a_{22} & a_{23} & 0 & 0 & 0 \\ a_{13} & a_{23} & a_{33} & 0 & 0 & 0 \\ 0 & 0 & 0 & a_{44} & 0 & 0 \\ 0 & 0 & 0 & 0 & a_{55} & 0 \\ 0 & 0 & 0 & 0 & 0 & a_{66} \end{bmatrix} \qquad (12.49)$$

Employing the notation of Lekhnitskii [310], we introduce the engineering constants E_i, G_{ij} and ν_{ij} (no summation on repeated indices is implied). These constants are related to the compliance coefficients in the following manner: $a_{11} = 1/E_1$, $a_{12} = -\nu_{21}/E_2$, $a_{13} = -\nu_{31}/E_3$, $a_{21} = -\nu_{12}/E_1$, $a_{22} = 1/E_2$, $a_{23} = -\nu_{32}/E_3$, $a_{31} = -\nu_{13}/E_1$, $a_{32} = -\nu_{23}E_2$, $a_{33} = 1/E_3$, $a_{44} = 1/G_{12}$, $a_{55} = 1/G_{13}$, and $a_{66} = 1/G_{23}$.

Of the twelve elastic constants entering the above equations, only *nine* are independent. More precisely, the symmetry of $\mathbf{A}$ implies that

$$a_{12} = a_{21} \quad \Rightarrow \quad \frac{\nu_{21}}{E_2} = \frac{\nu_{12}}{E_1} \quad \text{or} \quad E_1\nu_{21} = E_2\nu_{12} \qquad (12.50)$$

$$a_{13} = a_{31} \quad \Rightarrow \quad \frac{\nu_{31}}{E_3} = \frac{\nu_{13}}{E_1} \quad \text{or} \quad E_1\nu_{31} = E_3\nu_{13} \qquad (12.51)$$

$$a_{23} = a_{32} \quad \Rightarrow \quad \frac{\nu_{32}}{E_3} = \frac{\nu_{23}}{E_2} \quad \text{or} \quad E_2\nu_{32} = E_3\nu_{23} \qquad (12.52)$$

In inverse form, the non-zero entries in $\mathbf{C}$ are

$$c_{11} = \frac{E_1(1 - \nu_{23}\nu_{32})}{D} \quad ; \quad c_{12} = c_{21} = \frac{E_1(\nu_{21} + \nu_{31}\nu_{23})}{D} \qquad (12.53)$$

$$c_{13} = c_{31} = \frac{E_1(\nu_{31} + \nu_{21}\nu_{32})}{D} \quad ; \quad c_{22} = \frac{E_2(1 - \nu_{13}\nu_{31})}{D} \qquad (12.54)$$

$$c_{23} = c_{32} = \frac{E_2(\nu_{32} + \nu_{12}\nu_{31})}{D} \quad ; \quad c_{33} = \frac{E_3(1 - \nu_{12}\nu_{21})}{D} \qquad (12.55)$$

$$c_{44} = G_{12} \quad ; \quad c_{55} = G_{13} \quad ; \quad c_{66} = G_{23} \qquad (12.56)$$

where $D = 1 - \nu_{12}\nu_{21} - \nu_{13}\nu_{31} - \nu_{23}\nu_{32} - 2\nu_{12}\nu_{31}\nu_{23}$. Examples of orthotropic materials include fiber reinforced composite beams, wood (as a first approximation), sheet metal and honeycomb.

Special Case 3: Materials Possessing a Plane of Isotropy (Transversely Isotropic Materials)

Such materials possess the following properties: through all points there pass parallel planes of elastic symmetry in which all directions are elastically equivalent (i.e., planes of isotropy). Thus at each point there exists one principal direction and an infinite number of principal directions in a plane normal to the first direction. Taking the z-axis normal to the plane of isotropy, with the x and y axes directed arbitrarily in this plane, it follows that the compliance matrix has the following entries:

$$\mathbf{A} = \begin{bmatrix} a_{11} & a_{12} & a_{13} & 0 & 0 & 0 \\ a_{12} & a_{22} & a_{23} & 0 & 0 & 0 \\ a_{13} & a_{23} & a_{33} & 0 & 0 & 0 \\ 0 & 0 & 0 & 2(a_{11} - a_{12}) & 0 & 0 \\ 0 & 0 & 0 & 0 & a_{55} & 0 \\ 0 & 0 & 0 & 0 & 0 & a_{66} \end{bmatrix} \tag{12.57}$$

The following *five* material constants are introduced: E_1 (the elastic modulus for tension or compression in the plane of isotropy), E_2 (the elastic modulus for tension or compression normal to the plane of isotropy), ν_1 (Poisson's ratio[1] characterizing transverse contraction in the plane of isotropy), ν_2 (Poisson's ratio associated with tension applied in a direction normal to the plane of isotropy), and G_2 (the shear modulus for the plane of isotropy and any plane perpendicular to it). The shear modulus G_1 equals to $0.5E_1/(1 + \nu_1)$ and is thus *not* independent.

The compliance coefficients are related to the above five material constants in the following manner: $a_{11} = a_{22} = /E_1$, $a_{12} = -\nu_1/E_1$, $a_{13} = a_{23} = -\nu_2 E_2$, $a_{33} = 1/E_2$, $a_{44} = 2(a_{11} - a_{12}) = 2(1 + \nu_1)/E_1$, $a_{55} = a_{66} = 1/G_2$. In inverse form, the non-zero entries in $\mathbf{C}$ are

$$c_{11} = c_{22} = \frac{E_1 \left[E_2 - E_1 (\nu_2)^2 \right]}{(1 + \nu_1) \left[E_2 (1 - \nu_1) - 2E_1 (\nu_2)^2 \right]} \tag{12.58}$$

$$c_{12} = \frac{E_1 \left[E_1 (\nu_2)^2 + E_2 \nu_1 \right]}{(1 + \nu_1) \left[E_2 (1 - \nu_1) - 2E_1 (\nu_2)^2 \right]} \tag{12.59}$$

---

[1] Named in honor of S. D. Poisson (1781-1840).

$$c_{13} = c_{23} = \frac{E_1 E_2 \nu_2}{E_2(1 - \nu_1) - 2E_1(\nu_2)^2} \tag{12.60}$$

$$c_{33} = \frac{(E_2)^2 (1 - \nu_1)}{E_2(1 - \nu_1) - 2E_1(\nu_2)^2} \tag{12.61}$$

$$c_{44} = \frac{E_1}{2(1 + \nu_1)} \quad ; \quad c_{55} = c_{66} = G_2 \tag{12.62}$$

An example of a transversely isotropic material is a unidirectional fiber reinforced thermoplastic composite. In this case the z-axis would be taken to coincide with the fiber direction.

Special Case 4: Isotropic Linear Elastic Material

The simplest linearly elastic material, for which the elastic behavior is *independent* of the orientation of the coordinate axes, is called *isotropic*. A more precise definition would be: a material whose constitutive relations are unaltered under orthogonal transformations of coordinates. For an isotropic linear elastic material the constitutive relations simplify in that only *two* constants are required to completely describe the material behavior. For example, the elastic (Young's) modulus E and Poisson's ratio ν. Setting $E_1 = E_2 = E$ along with $\nu_1 = \nu_2 = \nu$, and noting that now $G_2 = E/2(1 + \nu)$, the compliance matrix for a transversely isotropic reduces to

$$\mathbf{A} = \frac{1}{E} \begin{bmatrix} 1 & -\nu & -\nu & 0 & 0 & 0 \\ -\nu & 1 & -\nu & 0 & 0 & 0 \\ -\nu & -\nu & 1 & 0 & 0 & 0 \\ 0 & 0 & 0 & 2(1+\nu) & 0 & 0 \\ 0 & 0 & 0 & 0 & 2(1+\nu) & 0 \\ 0 & 0 & 0 & 0 & 0 & 2(1+\nu) \end{bmatrix} \tag{12.63}$$

where it is noted that now $a_{44} = a_{55} = a_{66} = 2(a_{11} - a_{12})$. In inverse form, the non-zero entries in $\mathbf{C}$ are

$$c_{11} = c_{22} = c_{33} = \frac{E(1 - \nu)}{(1 + \nu)(1 - 2\nu)} = \lambda + 2\mu = K + \frac{4}{3}G \tag{12.64}$$

$$c_{12} = c_{13} = c_{23} = \frac{E\nu}{(1+\nu)(1-2\nu)} = \lambda = K - \frac{2}{3}G \qquad (12.65)$$

$$c_{44} = c_{55} = c_{66} = \frac{E}{2(1+\nu)} = \mu = G \qquad (12.66)$$

In the above equations, E is the elastic (Young's) modulus, ν is Poisson's ratio, λ and μ are the Lamé constants, $G = \mu$ is the elastic shear modulus, and K is the elastic bulk modulus. The relationships between the various elastic moduli associated with isotropic materials are summarized in Section 12.8.

Further details pertaining to special material characterizations are available in the books by Love [331] and Lekhnitskii [310], to name a few. We next outline the element equations for a specific three-dimensional element.

Example 12.2: Eight-Node, Isoparametric Hexahedral Element

To specialize the previous discussion, consider the eight-node, isoparametric ("brick") element (see Figure 9.8). As mentioned in Chapter 9, this trilinear element is the first member of both the Lagrangian and serendipity element families. For this element, $N_{sdim} = 3$ and $N_{en} = 8$. Since three displacement degrees of freedom are present at each node, $N_{eddof} = 3*8 = 24$. Finally, since six components of stress and strain are involved in the analysis, $N_{rowb} = 6$.

The approximate displacements in a given element are given by equation (12.34), with

$$\hat{\mathbf{u}}_n^{(e)} = \left\{ \hat{u}_1^{(e)} \quad \hat{v}_1^{(e)} \quad \hat{w}_1^{(e)} \quad \cdots \quad \hat{u}_8^{(e)} \quad \hat{v}_8^{(e)} \quad \hat{w}_8^{(e)} \right\}^T \qquad (12.67)$$

and

$$\mathbf{N} = \begin{bmatrix} \mathbf{N}_1 & \mathbf{N}_2 & \cdots & \mathbf{N}_8 \end{bmatrix} \qquad (12.68)$$

Each submatrix of element interpolation functions has the general form of equation (12.43), viz.,

$$\mathbf{N}_i = N_i \mathbf{I} \quad ; \quad i = 1, 2, \cdots, 8 \qquad (12.69)$$

where the interpolation functions are as derived in Section 9.2.3, viz.,

$$N_i = \frac{1}{8}(1 + \xi_i \xi)(1 + \eta_i \eta)(1 + \zeta_i \zeta) \qquad (12.70)$$

with $\xi_i, \eta_i, \zeta_i = \pm 1$.

The matrix of linear operators is given by equation (12.39). The strain-displacement matrix is written in terms of submatrices as

$$\mathbf{B} = \mathbf{S} \begin{bmatrix} \mathbf{N}_1 & \mathbf{N}_2 & \cdots & \mathbf{N}_8 \end{bmatrix} = \begin{bmatrix} \mathbf{B}_1 & \mathbf{B}_2 & \cdots & \mathbf{B}_8 \end{bmatrix} \qquad (12.71)$$

where each of the submatrices has the general form given in equation (12.45).

The first derivatives of interpolation functions are computed in the manner described in Chapter 10, viz.,

$$\begin{Bmatrix} \dfrac{\partial N_i}{\partial x} \\[2mm] \dfrac{\partial N_i}{\partial y} \\[2mm] \dfrac{\partial N_i}{\partial z} \end{Bmatrix} = \mathbf{J}^{-1} \begin{Bmatrix} \dfrac{\partial N_i}{\partial \xi} \\[2mm] \dfrac{\partial N_i}{\partial \eta} \\[2mm] \dfrac{\partial N_i}{\partial \zeta} \end{Bmatrix} \quad ; \quad i = 1, 2, \cdots, 8 \qquad (12.72)$$

where $\mathbf{J}$ is the Jacobian matrix, and

$$\frac{\partial N_i}{\partial \xi} = \frac{1}{8}\xi_i(1 + \eta_i \eta)(1 + \zeta_i \zeta) \qquad (12.73)$$

$$\frac{\partial N_i}{\partial \eta} = \frac{1}{8}\eta_i(1 + \xi_i \xi)(1 + \zeta_i \zeta) \qquad (12.74)$$

$$\frac{\partial N_i}{\partial \zeta} = \frac{1}{8}\zeta_i(1 + \xi_i \xi)(1 + \eta_i \eta) \qquad (12.75)$$

Since the element is *isoparametric*, the parametric mapping between between the natural coordinates (ξ, η, ζ) and the global coordinates (x, y, z) is given by

$$x = \sum_{m=1}^{N_{pt}} x_m^{(e)} N_m(\xi, \eta, \zeta) \tag{12.76}$$

$$y = \sum_{m=1}^{N_{pt}} y_m^{(e)} N_m(\xi, \eta, \zeta) \tag{12.77}$$

$$z = \sum_{m=1}^{N_{pt}} z_m^{(e)} N_m(\xi, \eta, \zeta) \tag{12.78}$$

where $N_{pt} = N_{en}$ and $(x_m^{(e)}, y_m^{(e)}, z_m^{(e)})$ are the global coordinates of node m in element e.

The $(24 * 24)$ element stiffness matrix is partitioned into $(3 * 3)$ submatrices in the following manner:

$$\mathbf{K}^{(e)} = \begin{bmatrix} \mathbf{K}_{11}^{(e)} & \cdots & \mathbf{K}_{18}^{(e)} \\ \vdots & \ddots & \vdots \\ \mathbf{K}_{81}^{(e)} & \cdots & \mathbf{K}_{88}^{(e)} \end{bmatrix} \tag{12.79}$$

A typical submatrix is computed as

$$\mathbf{K}_{ij}^{(e)} = \int_{\Omega^e} \mathbf{B_i}^T \mathbf{C} \mathbf{B_j} \, d\Omega$$

$$= \sum_{n=1}^{N} \mathbf{B_i}^T(\xi_n, \eta_n, \zeta_n) \mathbf{C} \mathbf{B_j}(\xi_n, \eta_n, \zeta_n) \, w_n \, |\mathbf{J}| \tag{12.80}$$

where $i, j = 1, 2, \cdots, 8$ and N equals the total number of quadrature points.

The determination of the element load vector, given by equation (12.16), is straightforward. The only point worth noting is that the vector of body force components (units of FL^{-3}) is

$$\mathbf{b} = \{ b_x \quad b_y \quad b_z \}^T \tag{12.81}$$

We next consider some two-dimensional idealizations used in elastic analyses. Since the domain for such idealizations is two-dimensional, $N_{sdim} = 2$. In addition, the number of stress and strain components involved (and thus N_{rowb}) also reduces.

12.4   Plane Stress Idealizations

If a thin plate is loaded by forces applied at the boundary, parallel to the plane of the plate and distributed uniformly over the thickness (Figure 12.1), the stress component σ_{33} will be negligible (often zero) through the thickness of the plate. The shear stresses σ_{13} and σ_{23} will likewise be negligible. Under these assumptions, it follows from the balance of linear momentum that under static conditions the body force in the z-direction must be zero. The state of stress is thus completely specified by σ_{11}, σ_{22} and σ_{12}, and is referred to as *plane stress*. These components of stress are assumed to be functions only of x and y.

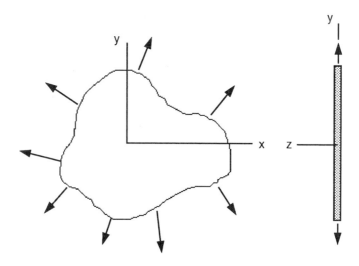

Figure 12.1: Schematic Illustration of Plane Stress

For plane stress idealizations $N_{sdim} = 2$ and $N_{rowb} = 3$. The approximate displacements for a given element are

$$\hat{\mathbf{u}}^{(e)} = \left\{ \begin{array}{c} \hat{u} \\ \hat{v} \end{array} \right\} = \mathbf{N}\hat{\mathbf{u}}_{\mathbf{n}}^{(e)} \tag{12.82}$$

where $\hat{\mathbf{u}}_{\mathbf{n}}^{(e)}$ is the vector of unknown displacement degrees of freedom u and v present at each node.

The vectors of total and initial strains are

$$\varepsilon = \left\{ \varepsilon_{11} \quad \varepsilon_{22} \quad \gamma_{12} \right\}^T \tag{12.83}$$

$$\varepsilon^0 = \left\{ \varepsilon_{11}^0 \quad \varepsilon_{22}^0 \quad \gamma_{12}^0 \right\}^T \tag{12.84}$$

where *engineering* measures of shear strains have again been used. The strain component ε_{33} is not independent and is given by

$$\varepsilon_{33} = \frac{-c_{13}\left(\varepsilon_{11} - \varepsilon_{11}^0\right) - c_{23}\left(\varepsilon_{22} - \varepsilon_{22}^0\right) - c_{34}\left(\gamma_{12} - \gamma_{12}^0\right)}{c_{33}}$$
$$-\frac{\sigma_{33}^0}{c_{33}} + \varepsilon_{33}^0 \tag{12.85}$$

The matrix of linear operators is thus

$$\mathbf{S} = \begin{bmatrix} \partial/\partial x & 0 \\ 0 & \partial/\partial y \\ \partial/\partial y & \partial/\partial x \end{bmatrix} \tag{12.86}$$

The corresponding stress vectors are

$$\boldsymbol{\sigma} = \left\{ \sigma_{11} \quad \sigma_{22} \quad \sigma_{12} \right\}^T \tag{12.87}$$

$$\boldsymbol{\sigma}^0 = \left\{ \sigma_{11}^0 \quad \sigma_{22}^0 \quad \sigma_{12}^0 \right\}^T \tag{12.88}$$

The matrix of element interpolation functions is again written in terms of N_{en} submatrices in the manner described by equation (12.42). However, each of the submatrices now has a size of $(2 * 2)$ and the general form

$$\mathbf{N}_i = \begin{bmatrix} N_i & 0 \\ 0 & N_i \end{bmatrix} = N_i \mathbf{I} \quad ; \quad i = 1, 2, \cdots, N_{en} \tag{12.89}$$

The strain-displacement matrix is again described by equation (12.44). However, each submatrix now has a size of $(3 * 2)$ and the general form

$$\mathbf{B}_i = \begin{bmatrix} \partial N_i/\partial x & 0 \\ 0 & \partial N_i/\partial y \\ \partial N_i/\partial y & \partial N_i/\partial x \end{bmatrix} \quad ; \quad i = 1, 2, \cdots, N_{en} \tag{12.90}$$

The constitutive matrices $\mathbf{A}$ and $\mathbf{C}$ are $(2 * 2)$ in size. In the most general case, these matrices contain *six* independent entries.

Special Case 5: Anisotropic Linear Elastic Material

Under plane stress conditions, $c_{i5} = c_{i6} = 0$ $(i = 1, 2, \cdots, 6)$. From equation (12.5) ε_{33} is seen not to be independent, viz.,

$$\varepsilon_{33} = \frac{-c_{13}(\varepsilon_{11} - \varepsilon_{11}^0) - c_{23}(\varepsilon_{22} - \varepsilon_{22}^0) - c_{34}(\gamma_{12} - \gamma_{12}^0)}{c_{33}} + \varepsilon_{33}^0 \quad (12.91)$$

where σ_{33}^0 has been assumed to be zero. Substituting for ε_{33} into the general expressions for σ_{11}, σ_{22} and σ_{12} and simplifying the resulting expressions, gives the following entries in $\mathbf{C}$:

$$c_{11}^* = c_{11} - \frac{c_{13}c_{13}}{c_{33}} \quad ; \quad c_{12}^* = c_{12} - \frac{c_{13}c_{23}}{c_{33}} \quad (12.92)$$

$$c_{13}^* = c_{14} - \frac{c_{13}c_{34}}{c_{33}} \quad ; \quad c_{22}^* = c_{22} - \frac{c_{23}c_{23}}{c_{33}} \quad (12.93)$$

$$c_{23}^* = c_{24} - \frac{c_{23}c_{34}}{c_{33}} \quad ; \quad c_{33}^* = c_{44} - \frac{c_{34}c_{34}}{c_{33}} \quad (12.94)$$

The corresponding vector of initial stresses is

$$\sigma^0 = \left\{ \begin{array}{c} \sigma_{11}^0 \\ \sigma_{22}^0 \\ \sigma_{12}^0 \end{array} \right\} - \frac{\sigma_{33}^0}{c_{33}} \left\{ \begin{array}{c} c_{13} \\ c_{23} \\ c_{34} \end{array} \right\} \quad (12.95)$$

If σ_{33}^0 is zero, the initial stresses simplify accordingly. The initial stress vector associated with initial strains is simply $\sigma^0 = -\mathbf{C}\varepsilon^0$.

Special Case 6: Orthotropic Linear Elastic Material

The elastic constitutive property matrix for an orthotropic linear elastic material under conditions of plane stress is derived by first writing the constitutive relations in direct form (see Special Case 2). The associated compliance coefficients are: $a_{11} = 1/E_1$, $a_{12} = -\nu_{21}/E_2$, $a_{13} = a_{23} = 0$, $a_{21} = -\nu_{12}/E_1$, $a_{22} = 1/E_2$, and $a_{33} = 1/G_{12}$. Inverting the constitutive relations gives the following non-zero entries in $\mathbf{C}$:

$$c_{11}^* = \frac{E_1}{1 - \nu_{12}\nu_{21}} \quad ; \quad c_{12}^* = \frac{E_1\nu_{21}}{1 - \nu_{12}\nu_{21}} \tag{12.96}$$

$$c_{22}^* = \frac{E_2}{1 - \nu_{12}\nu_{21}} \quad ; \quad c_{33}^* = G_{12} \tag{12.97}$$

The corresponding vector of initial stresses is

$$\boldsymbol{\sigma}^0 = \left\{ \begin{array}{c} \sigma_{11}^0 \\ \sigma_{22}^0 \\ \sigma_{12}^0 \end{array} \right\} - \frac{\sigma_{33}^0}{c_{33}} \left\{ \begin{array}{c} c_{13} \\ c_{23} \\ 0 \end{array} \right\} = \left\{ \begin{array}{c} \sigma_{11}^0 \\ \sigma_{22}^0 \\ \sigma_{12}^0 \end{array} \right\} - \sigma_{33}^0 \left\{ \begin{array}{c} \frac{E_1(\nu_{31} + \nu_{21}\nu_{32})}{E_3(1 - \nu_{12}\nu_{21})} \\ \frac{E_1(\nu_{32} + \nu_{12}\nu_{31})}{E_3(1 - \nu_{12}\nu_{21})} \\ 0 \end{array} \right\} \tag{12.98}$$

If σ_{33}^0 is zero, the initial stresses simplify accordingly. The initial stress vector associated with initial strains is simply

$$\boldsymbol{\sigma}^0 = -\mathbf{C}\boldsymbol{\varepsilon}^0 = - \left\{ \begin{array}{c} c_{11}^*\varepsilon_{11}^0 + c_{12}^*\varepsilon_{22}^0 \\ c_{12}^*\varepsilon_{11}^0 + c_{22}^*\varepsilon_{22}^0 \\ 0 \end{array} \right\} \tag{12.99}$$

Next consider a material whose preferred directions (described by the x' and y' axes) are not aligned with the global x and y coordinate axes (Figure 12.2).

The constitutive property matrix associated with the x' and y' axes is denoted by $\mathbf{C}'$. This matrix is transformed to the global $x - y$ coordinates through the following matrix triple product

$$\mathbf{C} = \mathbf{R}^T\mathbf{C}'\mathbf{R} \tag{12.100}$$

where

$$\mathbf{R} = \begin{bmatrix} (\cos\alpha)^2 & (\sin\alpha)^2 & \cos\alpha\sin\alpha \\ (\sin\alpha)^2 & (\cos\alpha)^2 & -\cos\alpha\sin\alpha \\ -2\cos\alpha\sin\alpha & 2\cos\alpha\sin\alpha & (\cos\alpha)^2 - (\sin\alpha)^2 \end{bmatrix} \tag{12.101}$$

is an orthogonal transformation matrix.

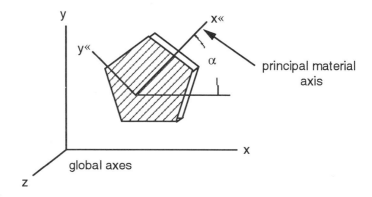

Figure 12.2: Orthotropic Plane Stress Body: Material Axes not Aligned with Global Axes

---

Special Case 7: Isotropic Linear Elastic Material

For an isotropic, linear elastic material $E_1 = E_2 = E$, $\nu_1 = \nu_2 = \nu$ and the shear modulus $G_2 = E/2(1 + \nu)$. The matrix $\mathbf{C}$ contains the following non-zero entries:

$$c_{11}^* = c_{22}^* = \frac{E}{1 - \nu^2} = \frac{4\mu(\lambda + \mu)}{\lambda + 2\mu} \qquad (12.102)$$

$$c_{12}^* = \frac{E\nu}{1 - \nu^2} = \frac{2\mu\lambda}{\lambda + 2\mu} \qquad (12.103)$$

$$c_{33}^* = \frac{E}{2(1 + \nu)} = \mu \qquad (12.104)$$

The corresponding initial stress vector is

$$\sigma^0 = \left\{ \begin{matrix} \sigma_{11}^0 \\ \sigma_{22}^0 \\ \sigma_{12}^0 \end{matrix} \right\} + \frac{\nu \sigma_{33}^0}{1 - \nu} \left\{ \begin{matrix} 1 \\ 1 \\ 0 \end{matrix} \right\} = \left\{ \begin{matrix} \sigma_{11}^0 \\ \sigma_{22}^0 \\ \sigma_{12}^0 \end{matrix} \right\} + \frac{\lambda \sigma_{33}^0}{\lambda + 2\mu} \left\{ \begin{matrix} 1 \\ 1 \\ 0 \end{matrix} \right\} \qquad (12.105)$$

12.5 Generalized Plane Strain Idealizations

Consider a homogeneous elastic body having general rectilinear anisotropy and bounded by a cylindrical surface. The length of the body is assumed infinite. Its cross-section is arbitrary, either finite or infinite, and either simply- or multiply connected. The origin of the coordinate system is placed at an arbitrary point within the cross-section. The z-axis is taken parallel to the generators, and the x- and y-axes are placed in accordance with the shape of the particular cross-section (Figure 12.3). The body is assumed to be loaded by surface and body forces acting in the planes of the cross-section, that is, normal to the generators, and not varying along its length.

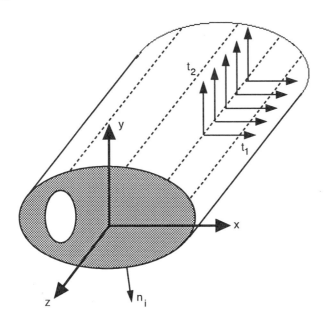

Figure 12.3: Schematic Illustration of Body Idealized Under Generalized Plane Strain Conditions

Clearly, for the body described above, the stresses and deformations of all cross-sections are only functions of the x and y coordinates. In an isotropic or anisotropic body at each point of which there exists a plane of elastic symmetry normal to the generators, the cross-sections remain plane. The deformation is then said to be "plane strain" (to be discussed later in this section). If, however, in the more general case none of the planes of elastic symmetry are parallel to the $x - y$ plane, and particularly in the general case of anisotropy, the deformation is no longer plane; the cross-sections will now warp, but all identically. This kind of deformation, commonly referred to as *generalized plane strain* [310], is considered herein.

Based upon the above discussion, the strain and stress fields will be functions of only x, y and time. It is evident that although these quantities are independent of z, unlike the case of plane strain, quantities such as ε_{33} and the shearing strains γ_{13} and γ_{23} do not, in general, vanish.

From the compatibility equations for linear systems [538], it can be shown that

$$\frac{\partial^2 \varepsilon_{33}}{\partial x^2} = \frac{\partial^2 \varepsilon_{33}}{\partial y^2} = 0 \qquad (12.106)$$

As such, under generalized plane strain conditions ε_{33} must be a linear function of x and y, viz.,

$$\varepsilon_{33} = A(t)x + B(t)y + C(t) \qquad (12.107)$$

where A, B and C are arbitrary functions of time. If no torsional loading is applied to the body, it follows that

$$u = -\frac{A(t)}{2}(z)^2 + u^*(x, y, t) \qquad (12.108)$$

$$v = -\frac{B(t)}{2}(z)^2 + v^*(x, y, t) \qquad (12.109)$$

$$w = [A(t)x + B(t)y + C(t)]\, z + w^*(x, y, t) \qquad (12.110)$$

where u^*, v^* and w^* represent the parts of the displacements u, v and w that are independent of z.

Next suppose the body shown in Figure 12.3 has, at each point, a plane of elastic symmetry normal to the generators (i.e., normal to the z-axis). In this case, all the equations of elasticity can be satisfied by assuming that the displacement w is constant, and by considering the displacements u and v to be functions only of x and y. It follows that

$$\varepsilon_{33} = \frac{\partial w}{\partial z} = 0 \qquad (12.111)$$

and

$$\gamma_{13} = \frac{\partial u}{\partial z} + \frac{\partial w}{\partial x} = 0 \quad ; \quad \gamma_{23} = \frac{\partial v}{\partial z} + \frac{\partial w}{\partial y} = 0 \qquad (12.112)$$

Assuming some degree of isotropy, these equations imply that the shear stresses $\sigma_{13} = \sigma_{31}$ and $\sigma_{23} = \sigma_{32}$ are likewise zero. From the general constitutive relations in "inverse" form (equation 12.5), it follows that

$$\sigma_{33} = c_{13}\varepsilon_{11} + c_{23}\varepsilon_{22} + c_{34}\gamma_{12} + \sigma_{33}^{0} \qquad (12.113)$$

The state of stress is thus completely specified by σ_{11}, σ_{22} and by σ_{12}, with these components functions only of x and y.

When one and only one of the principal strains at a point in a continuum is zero, the deformation field in all planes perpendicular to the direction of the zero principal strain is identical. As such, the state of *plane strain* is also referred to as a state of *plane deformation*. Conditions of plane deformation are typically assumed when one coordinate direction (e.g., the z direction) of a body is quite large compared to the other two, and the cross-sectional area of the body remains unchanged. Examples of such bodies include dams, culverts, tunnels, etc.

Some specific material symmetries for plane strain idealizations are next presented. Further details pertaining to finite element analyses for generalized plane strain idealizations can be found in [127].

Special Case 8: Anisotropic Linear Elastic Material

Under a state of plane strain $c_{15} = c_{16} = c_{25} = c_{26} = c_{35} = c_{36} = c_{45} = c_{46} = c_{56} = 0$ and $\varepsilon_{33} = 0$. From equation (12.5) it follows that

$$\sigma_{11} = c_{11}(\varepsilon_{11} - \varepsilon_{11}^{0}) + c_{12}(\varepsilon_{22} - \varepsilon_{22}^{0}) - c_{13}\varepsilon_{33}^{0} + c_{14}(\gamma_{12} - \gamma_{12}^{0})$$
$$+ \sigma_{11}^{0} \qquad (12.114)$$

$$\sigma_{22} = c_{12}(\varepsilon_{11} - \varepsilon_{11}^{0}) + c_{22}(\varepsilon_{22} - \varepsilon_{22}^{0}) - c_{23}\varepsilon_{33}^{0} + c_{24}(\gamma_{12} - \gamma_{12}^{0})$$
$$+ \sigma_{22}^{0} \qquad (12.115)$$

and

$$\sigma_{12} = c_{14}(\varepsilon_{11} - \varepsilon_{11}^0) + c_{24}(\varepsilon_{22} - \varepsilon_{22}^0) - c_{34}\varepsilon_{33}^0 + c_{44}(\gamma_{12} - \gamma_{12}^0)$$
$$+ \sigma_{12}^0 \qquad (12.116)$$

Rearranging the above expressions and writing them in matrix form gives,

$$\begin{Bmatrix} \sigma_{11} \\ \sigma_{22} \\ \sigma_{12} \end{Bmatrix} = \begin{bmatrix} c_{11} & c_{12} & c_{14} \\ c_{12} & c_{22} & c_{24} \\ c_{14} & c_{24} & c_{44} \end{bmatrix} \begin{Bmatrix} \varepsilon_{11} - \varepsilon_{11}^{(0)} \\ \varepsilon_{22} - \varepsilon_{22}^{(0)} \\ \gamma_{12} - \gamma_{12}^{(0)} \end{Bmatrix} + \begin{Bmatrix} \sigma_{11}^{(0)} - c_{13}\varepsilon_{33}^{(0)} \\ \sigma_{22}^{(0)} - c_{23}\varepsilon_{33}^{(0)} \\ \sigma_{12}^{(0)} - c_{34}\varepsilon_{33}^{(0)} \end{Bmatrix} \qquad (12.117)$$

With ε_{11}, ε_{22} and γ_{12} known, σ_{33} is computed from

$$\sigma_{33} = c_{13}(\varepsilon_{11} - \varepsilon_{11}^0) + c_{23}(\varepsilon_{22} - \varepsilon_{22}^0) - c_{33}\varepsilon_{33}^0$$
$$+ c_{34}(\gamma_{12} - \gamma_{12}^0) + \sigma_{33}^0 \qquad (12.118)$$

Special Case 9: Transversely Isotropic Linear Elastic Material

The z-axis is taken normal to the plane of isotropy, with the x and y axes directed arbitrarily in this plane. Using the material constants E_1, E_2, G_2, ν_1 and ν_2 previously introduced in conjunction with a general transversely isotropic material idealization (Special Case 3), and noting that in this case $\varepsilon_{33} = 0$ by definition of plane strain, the elements of the elastic constitutive property matrix are

$$c_{11} = \frac{E_1 \left[E_2 - E_1(\nu_2)^2 \right]}{(1 + \nu_1) \left[E_2(1 - \nu_1) - 2E_1(\nu_2)^2 \right]} \qquad (12.119)$$

$$c_{12} = \frac{E_1 E_2 \nu_2}{E_2(1 - \nu_1) - 2E_1(\nu_2)^2} \qquad (12.120)$$

$$c_{22} = \frac{(E_2)^2 (1 - \nu_1)}{E_2(1 - \nu_1) - 2E_1(\nu_2)^2} \qquad (12.121)$$

$$c_{13} = c_{23} = 0.0 \quad ; \quad c_{33} = G_2 \qquad (12.122)$$

Consider next a material whose preferred material directions (x' and y') are not aligned with the global x and y coordinate axes (recall Figure 12.2). The elements of the constitutive property matrix $\mathbf{C}'$ associated with the x'-y' axes are defined by the above equations for c_{11} to c_{33}. The $\mathbf{C}'$ matrix is transformed to the global x and y axes through the following matrix triple product:

$$\mathbf{C} = \mathbf{R}^T \mathbf{C} \mathbf{R} \tag{12.123}$$

where the transformation matrix is defined by

$$\mathbf{R} = \begin{bmatrix} (\cos\alpha)^2 & (\sin\alpha)^2 & \cos\alpha\sin\alpha \\ (\sin\alpha)^2 & (\cos\alpha)^2 & -\cos\alpha\sin\alpha \\ -2\cos\alpha\sin\alpha & 2\cos\alpha\sin\alpha & (\cos\alpha)^2 - (\sin\alpha)^2 \end{bmatrix} \tag{12.124}$$

The angle α is the measured counterclockwise from the global x-axis to the x'-axis.

Special Case 10: Isotropic Linear Elastic Material

For an isotropic, linear elastic material $E_1 = E_2 = E$, $\nu_1 = \nu_2 = \nu$, and the shear modulus $G_2 = 0.5E/(1 + \nu)$. From the general equations for a transversely isotropic material (Special Case 3), it thus follows that for a plane strain idealization,

$$\mathbf{C} = \frac{E}{(1+\nu)(1-2\nu)} \begin{bmatrix} 1-\nu & \nu & 0 \\ \nu & 1-\nu & 0 \\ 0 & 0 & 0.5(1-2\nu) \end{bmatrix}$$

$$= \begin{bmatrix} (\lambda+2\mu) & \lambda & 0 \\ \lambda & (\lambda+2\mu) & 0 \\ 0 & 0 & \mu \end{bmatrix} \tag{12.125}$$

$$\sigma^{(0)} = \left\{ \begin{matrix} \sigma_{11}^{(0)} - \nu\varepsilon_{33}^{(0)} \\ \sigma_{22}^{(0)} - \nu\varepsilon_{33}^{(0)} \\ \sigma_{12}^{(0)} \end{matrix} \right\} = \left\{ \begin{matrix} \sigma_{11}^{(0)} - \lambda\varepsilon_{33}^{(0)} \\ \sigma_{22}^{(0)} - \lambda\varepsilon_{33}^{(0)} \\ \sigma_{12}^{(0)} \end{matrix} \right\} \tag{12.126}$$

with

$$\sigma_{33} = \nu\left[\left(\sigma_{11} - \sigma_{11}^{(0)}\right) + \left(\sigma_{22} - \sigma_{22}^{(0)}\right)\right] + \sigma_{33}^{(0)} - E\varepsilon_{33}^{(0)} \tag{12.127}$$

The vector of initial thermal strains, and associated vector of initial stresses, are now given by

$$\varepsilon^{(0)} = \alpha_t(T - T_0) \left\{1 \quad 1 \quad 0\right\}^T \qquad (12.128)$$

$$\sigma^{(0)} = -\mathbf{C}\varepsilon^{(0)} = -\frac{E\alpha_t}{1 - 2\nu}(T - T_0) \left\{1 \quad 1 \quad 0\right\}^T \qquad (12.129)$$

Example 12.3: Element Arrays for a Three-Node Triangular Element

To specialize the discussion of plane stress and plane strain idealizations, consider the two-dimensional simplex element. For this element $N_{sdim} = 2$ and $N_{en} = 3$. Since two displacement degrees of freedom are present at each node (Figure 12.4), $N_{eddof} = 2 * 3 = 6$. Finally, since only three components of stress and strain are involved in the analysis, $N_{rowb} = 3$.

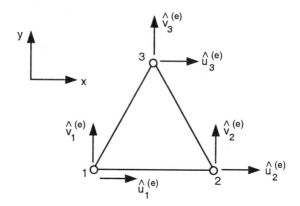

Figure 12.4: Typical Linear Triangular Element

The approximate displacements in a given element are given by equation (12.34), with

$$\hat{\mathbf{u}}_n^{(e)} = \left\{\hat{u}_1^{(e)} \quad \hat{v}_1^{(e)} \quad \hat{u}_2^{(e)} \quad \hat{v}_2^{(e)} \quad \hat{u}_3^{(e)} \quad \hat{v}_3^{(e)}\right\}^T \qquad (12.130)$$

and

$$\mathbf{N} = \begin{bmatrix} \mathbf{N}_1 & \mathbf{N}_2 & \mathbf{N}_3 \end{bmatrix} \qquad (12.131)$$

Each submatrix of element interpolation functions has the general form of equation (12.43), viz.,

$$\mathbf{N}_i = N_i \begin{bmatrix} 1 & 0 \\ 0 & 1 \end{bmatrix} \quad ; \quad i = 1, 2, 3 \qquad (12.132)$$

where, $A^{(e)}$ is the area of the element, and, as shown in Section 9.4.2,

$$N_i = \frac{1}{2A^{(e)}} \left(a_i + b_i x + c_i y \right) \qquad (12.133)$$

where

$$a_1 = x_2 y_3 - x_3 y_2 \quad , \quad b_1 = y_2 - y_3 \quad , \quad c_1 = x_3 - x_2 \qquad (12.134)$$
$$a_2 = x_3 y_1 - x_1 y_3 \quad , \quad b_2 = y_3 - y_1 \quad , \quad c_2 = x_1 - x_3 \qquad (12.135)$$
$$a_3 = x_1 y_2 - x_2 y_1 \quad , \quad b_3 = y_1 - y_2 \quad , \quad c_3 = x_2 - x_1 \qquad (12.136)$$

For *either* plane stress or plane strain, the matrix of linear operators is

$$\mathbf{S} = \begin{bmatrix} \partial/\partial x & 0 \\ 0 & \partial/\partial y \\ \partial/\partial y & \partial/\partial x \end{bmatrix} \qquad (12.137)$$

The strain-displacement matrix is written in terms of submatrices as

$$\mathbf{B} = \mathbf{S} \begin{bmatrix} \mathbf{N}_1 & \mathbf{N}_2 & \mathbf{N}_3 \end{bmatrix} = \begin{bmatrix} \mathbf{B}_1 & \mathbf{B}_2 & \mathbf{B}_3 \end{bmatrix} \qquad (12.138)$$

Recalling from Example 10.5 that

$$\frac{\partial N_i}{\partial x} = \frac{b_i}{2A^{(e)}} \quad ; \quad \frac{\partial N_i}{\partial y} = \frac{c_i}{2A^{(e)}} \qquad (12.139)$$

the submatrices $\mathbf{B_i}$ gave the following general form:

$$\mathbf{B_i} = \frac{1}{2A^{(e)}} \begin{bmatrix} b_i & 0 \\ 0 & c_i \\ c_i & b_i \end{bmatrix} \qquad (12.140)$$

The element stiffness matrix is again computed using Equation (12.15), viz.,

$$\mathbf{K}^{(e)} = \int_{\Omega^{(e)}} \mathbf{B}^T \mathbf{C} \mathbf{B} \, d\Omega \qquad (12.141)$$

where $\Omega^{(e)} = t \, A^{(e)}$. For plane strain idealizations, the element thickness t is equal to unity. Since $\mathbf{B}$ and $\mathbf{C}$ are constant and, provided that t is constant, the integrand in equation (12.141) simply yields the element area $A^{(e)}$.

Thus, for a *plane stress* idealization,

$$\mathbf{K}^{(e)} = \frac{t}{4A^{(e)}} \frac{E}{1-\nu^2} \mathbf{B}^T \begin{bmatrix} 1 & \nu & 0 \\ \nu & 1 & 0 \\ 0 & 0 & \dfrac{(1-\nu)}{2} \end{bmatrix} \mathbf{B} \qquad (12.142)$$

Similarly, for a *plane strain* idealization

$$\mathbf{K}^{(e)} = \frac{E}{4A^{(e)}(1+\nu)(1-2\nu)} \mathbf{B}^T \begin{bmatrix} (1-\nu) & \nu & 0 \\ \nu & (1-\nu) & 0 \\ 0 & 0 & \dfrac{(1-2\nu)}{2} \end{bmatrix} \mathbf{B}$$
$$(12.143)$$

where

$$\mathbf{B} = \begin{bmatrix} b_1 & 0 & b_2 & 0 & b_3 & 0 \\ 0 & c_1 & 0 & c_2 & 0 & c_3 \\ c_1 & b_1 & c_2 & b_2 & c_3 & b_3 \end{bmatrix} \qquad (12.144)$$

The determination of the element load vector, given by equation (12.16), is straightforward. The only point worth noting is that the vector of body force components (units of FL^{-3}) is

$$\mathbf{b} = \{ b_x \quad b_y \}^T \qquad (12.145)$$

12.6 Axisymmetric Idealizations

We begin the discussion of axisymmetry with the three-dimensional infinitesimal strain state, written in cylindrical coordinates, viz.,

$$\varepsilon_{rr} = \frac{\partial u}{\partial r} \quad ; \quad \varepsilon_{zz} = \frac{\partial v}{\partial z} \quad ; \quad \varepsilon_{\theta\theta} = \frac{u}{r} + \frac{1}{r}\frac{\partial w}{\partial \theta} \tag{12.146}$$

$$\gamma_{rz} = \frac{\partial u}{\partial z} + \frac{\partial v}{\partial r} \tag{12.147}$$

$$\gamma_{r\theta} = \frac{1}{r}\frac{\partial u}{\partial \theta} + \frac{\partial w}{\partial r} - \frac{w}{r} \tag{12.148}$$

$$\gamma_{z\theta} = \frac{\partial w}{\partial z} + \frac{1}{r}\frac{\partial v}{\partial \theta} \tag{12.149}$$

where u, v and w now represent displacements in the radial r, axial z, and circumferential θ directions, respectively.

For convenience, the strains are partitioned into submatrices in the following manner:

$$\varepsilon = \left\{ \begin{matrix} \varepsilon_1 \\ \varepsilon_2 \end{matrix} \right\} = \mathbf{S} \left\{ \begin{matrix} u \\ v \\ w \end{matrix} \right\} \tag{12.150}$$

where

$$\varepsilon_1 = \left\{ \begin{matrix} \varepsilon_{rr} & \varepsilon_{zz} & \varepsilon_{\theta\theta} & \gamma_{rz} \end{matrix} \right\}^T \quad ; \quad \varepsilon_2 = \left\{ \begin{matrix} \gamma_{r\theta} & \gamma_{z\theta} \end{matrix} \right\}^T \tag{12.151}$$

Under conditions of axisymmetry, the geometry of the body, the material, the loading and the boundary conditions are independent of the circumferential coordinate θ. Setting derivatives with respect to θ equal to zero, the strain equations simplify accordingly. Noting that $N_{sdim} = 3$ and $N_{rowb} = 6$, the $(6 * 3)$ matrix of linear operators is partitioned in the following manner:

$$\mathbf{S} = \begin{bmatrix} \mathbf{S_{11}} & \mathbf{0} \\ \mathbf{0} & \mathbf{S_{11}} \end{bmatrix} \tag{12.152}$$

where

$$\mathbf{S_{11}} = \begin{bmatrix} \partial/\partial r & 0 \\ 0 & \partial/\partial z \\ 1/r & 0 \\ \partial/\partial r & \partial/\partial z \end{bmatrix} \quad ; \quad \mathbf{S_{22}} = \begin{bmatrix} \partial/\partial r - 1/r \\ \partial/\partial z \end{bmatrix} \tag{12.153}$$

Since the material constants are independent of θ, it follows that $c_{15} = c_{16} = c_{25} = c_{26} = c_{35} = c_{36} = c_{45} = c_{46} = c_{56} = 0$. The resulting constitutive relations are thus "uncoupled," viz.,

$$\begin{Bmatrix} \sigma_1 \\ \sigma_2 \end{Bmatrix} = \begin{bmatrix} \mathbf{C_{11}} & 0 \\ 0 & \mathbf{C_{22}} \end{bmatrix} \begin{Bmatrix} \varepsilon_1 \\ \varepsilon_2 \end{Bmatrix} \tag{12.154}$$

where

$$\sigma_1 = \begin{Bmatrix} \sigma_{rr} & \sigma_{zz} & \sigma_{\theta\theta} & \sigma_{rz} \end{Bmatrix}^T \tag{12.155}$$

$$\sigma_2 = \begin{Bmatrix} \sigma_{r\theta} & \sigma_{z\theta} \end{Bmatrix}^T \tag{12.156}$$

and

$$\mathbf{C_{11}} = \begin{bmatrix} c_{11} & c_{12} & c_{13} & c_{14} \\ c_{12} & c_{22} & c_{23} & c_{24} \\ c_{13} & c_{23} & c_{33} & c_{34} \\ c_{14} & c_{24} & c_{34} & c_{44} \end{bmatrix} \quad ; \quad \mathbf{C_{22}} = \begin{bmatrix} c_{55} & 0 \\ 0 & c_{66} \end{bmatrix} \tag{12.157}$$

For finite element analyses, the matrix of element interpolation functions is again given by equations (12.42) and (12.43). The $(6 * N_{eddof})$ strain-displacement matrix is partitioned in the following manner:

$$\mathbf{B} = \begin{bmatrix} \mathbf{B_{11}} & \mathbf{B_{12}} & \cdots & \mathbf{B_{1,N_{en}}} \\ \mathbf{B_{21}} & \mathbf{B_{22}} & \cdots & \mathbf{B_{2,N_{en}}} \end{bmatrix} \tag{12.158}$$

where the submatrices have the following general form:

$$\mathbf{B_{1i}} = \begin{bmatrix} \partial N_i/\partial r & 0 & 0 \\ 0 & \partial N_i/\partial z & 0 \\ N_i/r & 0 & 0 \\ \partial N_i/\partial z & \partial N_i/\partial r & 0 \end{bmatrix} \tag{12.159}$$

and

$$\mathbf{B_{2i}} = \begin{bmatrix} 0 & 0 & \partial N_i/\partial r - N_i/r \\ 0 & 0 & \partial N_i/\partial z \end{bmatrix} \qquad (12.160)$$

where $i = 1, 2, \cdots, N_{en}$. The approximate element strains are thus computed from

$$\hat{\varepsilon}^{(e)} = \left\{ \begin{matrix} \hat{\varepsilon}_1^{(e)} \\ \hat{\varepsilon}_2^{(e)} \end{matrix} \right\} = \begin{bmatrix} \mathbf{B_{11}} & \mathbf{B_{12}} & \cdots & \mathbf{B_{1,N_{en}}} \\ \mathbf{B_{21}} & \mathbf{B_{22}} & \cdots & \mathbf{B_{2,N_{en}}} \end{bmatrix} \hat{\mathbf{u}}_\mathbf{n}^{(e)} \qquad (12.161)$$

where

$$\hat{\varepsilon}_1^{(e)} = \left\{ \begin{matrix} \hat{\varepsilon}_{rr}^{(e)} \\ \hat{\varepsilon}_{zz}^{(e)} \\ \hat{\varepsilon}_{\theta\theta}^{(e)} \\ \hat{\gamma}_{rz}^{(e)} \end{matrix} \right\} = \left\{ \begin{matrix} \partial \hat{u}^{(e)}/\partial r \\ \partial \hat{v}^{(e)}/\partial z \\ \hat{u}^{(e)}/r \\ \partial \hat{u}^{(e)}/\partial z + \partial \hat{v}^{(e)}/\partial r \end{matrix} \right\} \qquad (12.162)$$

and

$$\hat{\varepsilon}_2^{(e)} = \left\{ \begin{matrix} \hat{\gamma}_{r\theta}^{(e)} \\ \hat{\gamma}_{z\theta}^{(e)} \end{matrix} \right\} = \left\{ \begin{matrix} \partial \hat{w}^{(e)}/\partial r - \hat{w}^{(e)}/r \\ \partial \hat{w}^{(e)}/\partial z \end{matrix} \right\} \qquad (12.163)$$

The vector of nodal unknowns is now

$$\hat{\mathbf{u}}_\mathbf{n}^{(e)} = \left\{ \hat{u}_1^{(e)} \quad \hat{v}_1^{(e)} \quad \hat{w}_1^{(e)} \quad \cdots \quad \hat{u}_{(N_{en})}^{(e)} \quad \hat{v}_{(N_{en})}^{(e)} \quad \hat{w}_{(N_{en})}^{(e)} \right\}^T \qquad (12.164)$$

The rather sparse structure of $\mathbf{B}$ and $\mathbf{C}$ gives rise to a particularly simple form for the element stiffness matrix. Recalling that this matrix is computed from equation (12.15), and writing $\mathbf{B}$ and $\mathbf{C}$ in partitioned form leads to the following general $(3 * 3)$ submatrix of $\mathbf{K}^{(e)}$:

$$\mathbf{K_{mn}^{(e)}} = \int_{\Omega^{(e)}} \left(\mathbf{B_{1m}}^T \mathbf{C_{11}} \mathbf{B_{1n}} + \mathbf{B_{2m}}^T \mathbf{C_{22}} \mathbf{B_{2n}} \right) d\Omega \qquad (12.165)$$

where $m, n = 1, 2, \cdots, N_{en}$, and the submatrices $\mathbf{C_{11}}$ and $\mathbf{C_{22}}$ are defined in equation (12.157).

Multiplication of the above submatrices leads to the following results

$$\mathbf{B_{1m}}^T \mathbf{C_{11}} \mathbf{B_{1n}} = \begin{bmatrix} \kappa_{11} & \kappa_{12} & 0 \\ \kappa_{21} & \kappa_{22} & 0 \\ 0 & 0 & 0 \end{bmatrix} \tag{12.166}$$

where

$$\kappa_{11} = c_{11} N_{m,r} N_{n,r} + c_{13} \frac{1}{r} (N_{m,r} N_n + N_{n,r} N_m)$$
$$+ c_{14}(N_{m,r} N_{n,z} + N_{m,z} N_{n,r}) + c_{33} \frac{1}{r^2} N_m N_n$$
$$+ c_{34} \frac{1}{r} (N_m N_{n,z} + N_n N_{m,z}) + c_{44} N_{m,z} N_{n,z} \tag{12.167}$$

$$\kappa_{12} = c_{12} N_{m,r} N_{n,z} + c_{14} N_{m,r} N_{n,r} + c_{23} \frac{1}{r} N_{n,z} N_m$$
$$+ c_{24} N_{m,z} N_{n,z} + c_{34} \frac{1}{r} N_{n,r} N_m + c_{44} N_{m,z} N_{n,r} \tag{12.168}$$

$$\kappa_{21} = c_{12} N_{m,z} N_{n,r} + c_{14} N_{m,r} N_{n,r} + c_{23} \frac{1}{r} N_{m,z} N_n$$
$$+ c_{24} N_{m,z} N_{n,z} + c_{34} \frac{1}{r} N_{m,r} N_n + c_{44} N_{m,r} N_{n,z} \tag{12.169}$$

$$\kappa_{22} = c_{22} N_{m,z} N_{n,z} + c_{24}(N_{m,z} N_{n,r} + N_{m,r} N_{n,z})$$
$$+ c_{44} N_{m,r} N_{n,r} \tag{12.170}$$

In the above equations, $N_{m,r} \equiv \partial N_m / \partial r$. In a similar manner,

$$\mathbf{B_{2m}}^T \mathbf{C_{22}} \mathbf{B_{2n}} = \begin{bmatrix} 0 & 0 & 0 \\ 0 & 0 & 0 \\ 0 & 0 & \lambda_{33} \end{bmatrix} \tag{12.171}$$

where

$$\lambda_{33} = c_{55} \left(N_{m,r} - \frac{N_m}{r} \right) \left(N_{n,r} - \frac{N_n}{r} \right) + c_{66} N_{m,z} N_{n,z} \tag{12.172}$$

Two specific classes of axisymmetric problems are next considered.

Problem Class 1: $w = w(r, z)$, $u = v = 0$

This problem class corresponds to the *torsion* of axisymmetric bodies such as shafts of varying circular cross-section.[2] The state of general axisymmetry degenerates to one for which $N_{sdim} = 1$ and one displacement degree of freedom, the *circumferential* displacement w, is present at each node. The vector of nodal unknowns becomes

$$\hat{\mathbf{u}}_n^{(e)} = \left\{ \hat{w}_1^{(e)} \quad \hat{w}_2^{(e)} \quad \cdots \quad \hat{w}_{(N_{en})}^{(e)} \right\}^T \qquad (12.173)$$

In the case of axisymmetric torsion the only non-zero strains are: $\gamma_{13} \equiv \gamma_{r\theta}$ and $\gamma_{23} \equiv \gamma_{z\theta}$, implying that $N_{rowb} = 2$. The corresponding stress components are: $\sigma_{13} \equiv \sigma_{r\theta}$ and $\sigma_{23} \equiv \sigma_{z\theta}$. The approximate element strains are related to the nodal displacements according to equation (12.163).

The general strain-displacement matrix reduces to

$$\mathbf{B} = \begin{bmatrix} \mathbf{B_1} & \mathbf{B_2} & \cdots & \mathbf{B_{N_{en}}} \end{bmatrix} \qquad (12.174)$$

where a typical submatrix has the form

$$\mathbf{B_i} = \begin{bmatrix} \dfrac{\partial N_i}{\partial r} - \dfrac{N_i}{r} \\ \dfrac{\partial N_i}{\partial z} \end{bmatrix} \quad ; \quad i = 1, 2, \cdots, N_{en} \qquad (12.175)$$

A typical submatrix in $\mathbf{K}_{mn}^{(e)}$ now has size $(1 * 1)$ and is computed from

$$\mathbf{K}_{mn}^{(e)} = \int_{\Omega^{(e)}} \left(\mathbf{B_m}^T \mathbf{C_{22}} \mathbf{B_n} \right) d\Omega = \int_{\Omega^{(e)}} \lambda_{33} \, d\Omega \qquad (12.176)$$

where $m, n = 1, 2, \cdots, N_{en}$, and λ_{33} is as defined above.

For the finite element analysis of such problems, numerical quadrature is adequate and introduces no additional errors. The same order of quadrature should be used as for plane stress and plane strain idealizations.

---

[2]Further details pertaining to the solution of this problem are given in Article 49 of [504].

Problem Class 2: $w = 0$, $u = u(r, z)$, $v = v(r, z)$

This problem class corresponds to *torsionless* axisymmetric deformation. The state of general axisymmetry degenerates to one for which $N_{sdim} = 2$ and two displacement degrees of freedom, u (radial) and v (axial), are present at each node. The vector of nodal unknowns becomes

$$\hat{\mathbf{u}}_n^{(e)} = \left\{ \hat{u}_1^{(e)} \quad \hat{v}_1^{(e)} \quad \cdots \quad \hat{u}_{(N_{en})}^{(e)} \quad \hat{v}_{(N_{en})}^{(e)} \right\}^T \tag{12.177}$$

The only non-zero strains are: $\varepsilon_{11} \equiv \varepsilon_{rr}$, $\varepsilon_{22} \equiv \varepsilon_{zz}$, $\varepsilon_{33} \equiv \varepsilon_{\theta\theta}$ and $\gamma_{12} \equiv \gamma_{rz}$, implying that $N_{rowb} = 4$. The corresponding stress components are: $\sigma_{11} \equiv \sigma_{rr}$, $\sigma_{22} \equiv \sigma_{zz}$, $\sigma_{33} \equiv \sigma_{\theta\theta}$ and $\sigma_{12} \equiv \sigma_{rz}$. The approximate element strains are related to the nodal displacements according to equation (12.162).

It is evident that any radial displacement u automatically induces a strain in the circumferential (θ) direction. Since the stresses in this direction are non-zero, it follows that this (fourth) component of strain and stress must be considered in the analysis. In this manner the torsionless axisymmetric case differs from that of plane stress/plane strain.

The general strain-displacement matrix reduces to

$$\mathbf{B} = \begin{bmatrix} \mathbf{B_1} & \mathbf{B_2} & \cdots & \mathbf{B_{N_{en}}} \end{bmatrix} \tag{12.178}$$

where a typical submatrix has the form

$$\mathbf{B_i} = \begin{bmatrix} \partial N_i/\partial r & 0 \\ 0 & \partial N_i/\partial z \\ N_i/r & 0 \\ \partial N_i/\partial z & \partial N_i/\partial r \end{bmatrix} \quad ; \quad i = 1, 2, \cdots, N_{en} \tag{12.179}$$

A typical submatrix in $\mathbf{K}_{mn}^{(e)}$ now has size $(2*2)$ and is computed from

$$\mathbf{K}_{mn}^{(e)} = \int_{\Omega^{(e)}} \left(\mathbf{B_m}^T \mathbf{C_{11}} \mathbf{B_n} \right) d\Omega = \int_{\Omega^{(e)}} \begin{bmatrix} \kappa_{11} & \kappa_{12} \\ \kappa_{21} & \kappa_{22} \end{bmatrix} d\Omega \tag{12.180}$$

where $m, n = 1, 2, \cdots, N_{en}$, and κ_{11}, κ_{12}, κ_{21} and κ_{22} are as previously defined.

12.7 Computation of Equivalent Nodal Loads

In elastostatic finite element analyses employing irreducible elements, the only permissible form of loading is through suitable concentrated nodal forces. Thus body forces, forces due to initial stresses and/or strains, and forces due prescribed surface tractions (pressures) must thus be converted to equivalent nodal loads. Recall that these forces enter the element equations through the right-hand side vector given by equation (12.16), viz.,

$$\mathbf{f}^{(e)} = \int\limits_{\Omega^{(e)}} \mathbf{B}^T \mathbf{C} \boldsymbol{\varepsilon}^0 \, d\Omega - \int\limits_{\Omega^{(e)}} \mathbf{B}^T \boldsymbol{\sigma}^0 \, d\Omega$$

$$+ \int\limits_{\Omega^{(e)}} \mathbf{N}^T \mathbf{b} \, d\Omega + \int\limits_{\Gamma_2} \mathbf{N}^T \bar{\mathbf{t}} \, d\Gamma \qquad (12.181)$$

In this section, we focus on the last term in the above equation, where $\bar{\mathbf{t}}$ is a vector of applied surface tractions, with units of stress (FL^{-2}). In three-dimensional idealizations, the surface tractions are applied along surfaces; in two-dimensional idealizations they are applied along element edges.

For simplicity, consider a two-dimensional idealization. A distributed edge loading (traction), in either the normal or tangential direction, may be prescribed along any element interface (Figure 12.5). Such distributed loads have units of stress.

To facilitate the discussion, let N_p equal the number of nodes present along a loaded element edge. To be consistent with the order of node numbering employed in defining the element connectivity, during specification of the edge loading these N_p nodes should likewise be specified in a *counterclockwise* sequence.

A traction normal to an edge is assumed to be positive if, in conjunction with a counterclockwise node number specification, it is directed *into* the element. A tangential load is assumed to be positive if it acts in a counter-clockwise direction with respect to the loaded element (Figure 12.5). Such definitions are necessary in order to avoid confusion when distributed loads are specified along the interface between two elements, such as that shown in Figure 12.6. Assuming a counterclockwise node number specification, if the applied tractions are assumed to act on element 1, they are considered to be positive. If they instead are taken as acting on element 2, these same tractions are negative.

The distributed loads need not be constant but can vary along the element edge. In the case of a bilinear element N_p equals 2, implying that

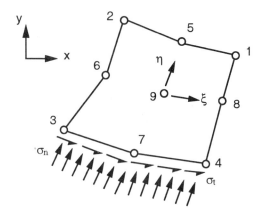

Figure 12.5: Normal and Tangential Distributed Edge Loads Applied to a Biquadratic Lagrange Element

at least a linear loading distribution can be accommodated. In the case of a biquadratic element N_p equals 3. The loading can thus, at best, vary parabolically.

Denoting the values of the normal (σ_n) and tangential (σ_t) tractions associated with node i ($i = 1, 2, \cdots N_p$) by $\sigma_n^{(i)}$ and $\sigma_t^{(i)}$, respectively, the following general expression is written:

$$\hat{\sigma}_n = \sum_{j=1}^{N_p} \sigma_n^{(j)} N_j(\psi) \tag{12.182}$$

$$\hat{\sigma}_t = \sum_{j=1}^{N_p} \sigma_t^{(j)} N_j(\psi) \tag{12.183}$$

where the N_j are standard interpolation functions used to parametrically define the edge tractions, and ψ denotes the curvilinear natural coordinate parallel to the loaded element edge.

12.7.1   Plane Stress and Plane Strain Idealizations

Consider a typical incremental length ds of the loaded edge such as that shown in Figure 12.7.

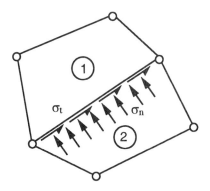

Figure 12.6: Edge Loads Specified Along a Typical Element Interface

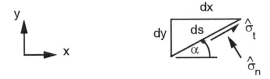

Figure 12.7: Trigonometric Relations for Element Tractions Applied Along Γ_2

Under conditions of plane stress or plane strain, the incremental force components in the global x and y directions are

$$dQ_x = (\hat{\sigma}_t \, ds \, \cos\alpha - \hat{\sigma}_n \, ds \, \sin\alpha) \, t = (\hat{\sigma}_t \, dx - \hat{\sigma}_n \, dy) \, t \quad (12.184)$$

$$dQ_y = (\hat{\sigma}_t \, ds \, \sin\alpha + \hat{\sigma}_n \, ds \, \cos\alpha) \, t = (\hat{\sigma}_n \, dx + \hat{\sigma}_t \, dy) \, t \quad (12.185)$$

where t is the thickness of the two-dimensional body (under plane strain conditions, t is equal to unity).

Since computation of the equivalent nodal forces requires integration along the loaded edge, we make use of the following relations:

$$dx = \frac{\partial x}{\partial s} ds \quad ; \quad dy = \frac{\partial y}{\partial s} \tag{12.186}$$

The incremental force components are thus written as

$$dQ_x = \left(\hat{\sigma}_t \frac{\partial x}{\partial s} ds - \hat{\sigma}_n \frac{\partial y}{\partial s} \right) t\, ds \tag{12.187}$$

$$dQ_y = \left(\hat{\sigma}_n \frac{\partial x}{\partial s} ds + \hat{\sigma}_t \frac{\partial y}{\partial s} \right) t\, ds \tag{12.188}$$

The increment in work done by the edge tractions is

$$d\hat{W} = \hat{u}\, dQ_x + \hat{v}\, dQ_y \tag{12.189}$$

where $\hat{u}$ and $\hat{v}$ are the approximate displacements in the global x and y directions, respectively. The contribution of the distributed loads to the total potential energy will, of course, be the *negative* of $d\hat{W}$.

Substituting for dQ_x and dQ_y into equation (12.189), using the standard finite element approximations for $\hat{u}$ and $\hat{v}$, and integrating along the element edge gives the following expression for total work done by the edge tractions, viz.,

$$\hat{W} = \int_s \left[\left(\sum_{i=1}^{N_p} N_i \hat{u}_i^{(e)} \right) \left(\hat{\sigma}_t \frac{\partial x}{\partial s} ds - \hat{\sigma}_n \frac{\partial y}{\partial s} \right) \right.$$
$$\left. + \left(\sum_{i=1}^{N_p} N_i \hat{v}_i^{(e)} \right) \left(\hat{\sigma}_n \frac{\partial x}{\partial s} ds + \hat{\sigma}_t \frac{\partial y}{\partial s} \right) \right] t\, ds \tag{12.190}$$

The equivalent nodal force components at node i ($i = 1, 2, \cdots N_p$) are thus

$$\hat{Q}_x^{(i)} = \frac{\partial \hat{W}}{\partial \hat{u}_i^{(e)}} = \int_s \left[N_i \left(\hat{\sigma}_t \frac{\partial x}{\partial s} - \hat{\sigma}_n \frac{\partial y}{\partial s} \right) \right] t\, ds \tag{12.191}$$

$$\hat{Q}_y^{(i)} = \frac{\partial \hat{W}}{\partial \hat{v}_i^{(e)}} = \int_s \left[N_i \left(\hat{\sigma}_n \frac{\partial x}{\partial s} + \hat{\sigma}_t \frac{\partial y}{\partial s} \right) \right] t\, ds \tag{12.192}$$

Expressing σ_n and σ_t by equations (12.182) and (12.183), respectively, gives the final form of the equivalent nodal forces, viz.,

$$\hat{Q}_x^{(i)} = \int_{-1}^{1} N_i \left[\left(\sum_{j=1}^{N_p} \sigma_t^{(j)} N_j \right) \frac{\partial x}{\partial s} - \left(\sum_{j=1}^{N_p} \sigma_n^{(j)} N_j \right) \frac{\partial y}{\partial s} \right] t \, |\mathbf{J}| \, d\psi \quad (12.193)$$

$$\hat{Q}_y^{(i)} = \int_{-1}^{1} N_i \left[\left(\sum_{j=1}^{N_p} \sigma_n^{(j)} N_j \right) \frac{\partial x}{\partial s} + \left(\sum_{j=1}^{N_p} \sigma_t^{(j)} N_j \right) \frac{\partial y}{\partial s} \right] t \, |\mathbf{J}| \, d\psi \quad (12.194)$$

where $|\mathbf{J}|$ represents the determinant of the Jacobian matrix corresponding to the mapping from global (x, y) coordinates through s to natural ψ coordinates. It is understood that derivatives of x and y with respect to s shall be expressed through the chain rule of differentiation as $\partial x/\partial s = (\partial x/\partial \psi)(\partial \psi/\partial s)$ and $\partial y/\partial s = (\partial y/\partial \psi)(\partial \psi/\partial s)$, respectively.

In computer programs employing parametric mapping of arbitrarily shaped element domains, the calculation of equivalent nodal loads must be performed numerically [244]. Using standard Gauss-Legendre quadrature, equations (12.193) and (12.194) become

$$\hat{Q}_x^{(i)} = \sum_{k=1}^{N} w_k N_i(\psi_k) \left[\left(\sum_{j=1}^{N_p} N_j(\psi_k) \sigma_t^{(j)} \right) \frac{\partial x}{\partial s}(\psi_k) \right.$$
$$\left. - \left(\sum_{j=1}^{N_p} N_j(\psi_k) \sigma_n^{(j)} \right) \frac{\partial y}{\partial s}(\psi_k) \right] t \, |\mathbf{J}| \quad (12.195)$$

$$\hat{Q}_y^{(i)} = \sum_{k=1}^{N} w_k N_i(\psi_k) \left[\left(\sum_{j=1}^{N_p} N_j(\psi_k) \sigma_n^{(j)} \right) \frac{\partial x}{\partial s}(\psi_k) \right.$$
$$\left. + \left(\sum_{j=1}^{N_p} N_j(\psi_k) \sigma_t^{(j)} \right) \frac{\partial y}{\partial s}(\psi_k) \right] t \, |\mathbf{J}| \quad (12.196)$$

where N equals the number of quadrature points used, the w_k are weights associated with the numerical quadrature, ψ_k denotes the natural coordinates of the quadrature points, and the derivatives of x and y with respect to s are again expressed through the chain rule of differentiation. Explicit values of N for linear and quadratic elements are determined in the following examples.

Example 12.4: Linear elements ($N_p = 2$)

In the case of linear elements such as 3-node triangles or 4-node quadri-laterals, the loaded edge has the configuration shown in Figure 12.8.

Figure 12.8: Linear Element Edge and Associated "Parent" Domain

The relation between natural ψ and global s coordinates is

$$\psi = \frac{2(s - s_c)}{h} \tag{12.197}$$

where s_c is the midpoint of the element edge, and

$$h = \sqrt{(x_2 - x_1)^2 + (y_2 - y_1)^2} \tag{12.198}$$

is the length of the edge. The interpolation functions are the usual linear expressions, viz.,

$$N_1 = \frac{1}{2}(1 - \psi) \quad , \quad N_2 = \frac{1}{2}(1 + \psi) \tag{12.199}$$

For an isoparametric mapping

$$x = \sum_{i=1}^{2} x_i N_i(\psi) \quad ; \quad y = \sum_{i=1}^{2} y_i N_i(\psi) \tag{12.200}$$

and

$$\frac{\partial x}{\partial s} = \frac{\partial x}{\partial \psi}\frac{\partial \psi}{\partial s} = \frac{2}{h}\sum_{i=1}^{2} x_i \frac{\partial N_i}{\partial \psi} = \frac{1}{h}(x_2 - x_1) \tag{12.201}$$

$$\frac{\partial y}{\partial s} = \frac{\partial y}{\partial \psi}\frac{\partial \psi}{\partial s} = \frac{2}{h}\sum_{i=1}^{2} y_i \frac{\partial N_i}{\partial \psi} = \frac{1}{h}(y_2 - y_1) \tag{12.202}$$

Equations (12.195) and (12.196) contain products of N_i, N_j and either $\partial x/\partial s$ or $\partial y/\partial s$. Since the interpolation functions are linear in ψ, and since these partial derivatives are constants, it follows that the expressions for $\hat{Q}_x^{(i)}$ and $\hat{Q}_y^{(i)}$ are *quadratic* in ψ. To exactly integrate such functions, Gauss-Legendre quadrature requires that N equal 2. Further details pertaining to numerical quadrature are given in Appendix F.

Consider the specific case of a shear stress applied along the edge of the linear rectangular element shown in Figure 12.9. Here $h = 2b$, $\partial x/\partial s = 0$, $\partial y/\partial s = 1$ and $|\mathbf{J}| = b$. The shear stress distribution can, at most, be *trapezoidal.*

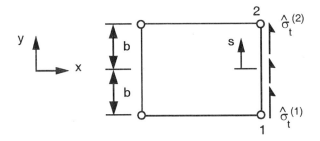

Figure 12.9: Shear Stress Applied to a Rectangular Linear Element

From equations (12.195) and (12.196) it is clear that $\hat{Q}_x^{(i)} = 0$, and that

$$\hat{Q}_y^{(i)} = \int_{-1}^{1} N_i \left(\sum_{j=1}^{N_p} \sigma_t^{(j)} N_j \right) tb \, d\psi \qquad (12.203)$$

or

$$\left\{ \begin{matrix} \hat{Q}_y^{(1)} \\ \hat{Q}_y^{(2)} \end{matrix} \right\} = \int_{-1}^{1} \begin{bmatrix} N_1 N_1 & N_1 N_2 \\ N_1 N_2 & N_2 N_2 \end{bmatrix} \left\{ \begin{matrix} \sigma_t^{(1)} \\ \sigma_t^{(2)} \end{matrix} \right\} tb \, d\psi \qquad (12.204)$$

Evaluating the integrals exactly (identical results would be obtained by using two-point Gauss-Legendre quadrature), gives the following results for the general case of a trapezoidal shear stress distribution:

$$\left\{ \begin{matrix} \hat{Q}_y^{(1)} \\ \hat{Q}_y^{(2)} \end{matrix} \right\} = \frac{tb}{3} \begin{bmatrix} 2 & 1 \\ 1 & 2 \end{bmatrix} \left\{ \begin{matrix} \sigma_t^{(1)} \\ \sigma_t^{(2)} \end{matrix} \right\} \qquad (12.205)$$

For the special case of a *triangular* distribution, $\sigma_t^{(1)} = 0$ and $\sigma_t^{(2)} = \bar{\sigma}_t$. It follows that

$$\left\{ \begin{array}{c} \hat{Q}_y^{(1)} \\ \hat{Q}_y^{(2)} \end{array} \right\} = \frac{tb\bar{\sigma}_t}{3} \left\{ \begin{array}{c} 1 \\ 2 \end{array} \right\} \qquad (12.206)$$

We note that, even though the nodal values $\sigma_t^{(1)} = 0$, the corresponding equivalent nodal force is *non-zero*. The check for this case is that the total force in the y direction $(\hat{Q}_y^{(1)} + \hat{Q}_y^{(2)})$ is equal to $\bar{\sigma}_t$ multiplied by the area over which it is applied $0.5(2bt)$. This is indeed seen to be the case.

Finally, for the special case of a *constant* shear stress distribution, $\sigma_t^{(1)} = \sigma_t^{(2)} = \bar{\sigma}_t$, we have the more "expected" result

$$\left\{ \begin{array}{c} \hat{Q}_y^{(1)} \\ \hat{Q}_y^{(2)} \end{array} \right\} = tb\bar{\sigma}_t \left\{ \begin{array}{c} 1 \\ 1 \end{array} \right\} \qquad (12.207)$$

The check for this case is that the total force in the y direction is equal to $\bar{\sigma}_t$ multiplied by the area over which it is applied $(2bt)$. This is indeed seen to be the case.

Example 12.5: Quadratic elements $(N_p = 3)$

In the case of quadratic elements such as 6-node triangles or 8- or 9-node quadrilaterals, the loaded edge has the configuration shown in Figure 12.10.

Figure 12.10: Quadratic Element Edge and Associated "Parent" Domain

The relation between natural ψ and global s coordinates is again defined by equations (12.197) and (12.198).

The interpolation functions are the usual quadratic expressions, viz.,

$$N_1 = \frac{1}{2}\psi(\psi - 1) \quad , \quad N_2 = \frac{1}{2}\psi(\psi + 1) \quad , \quad N_3 = 1 - \psi^2 \qquad (12.208)$$

For an isoparametric mapping

$$\frac{\partial x}{\partial s} = \frac{2}{h} \sum_{i=1}^{3} x_i \frac{\partial N_i}{\partial \psi} = \frac{1}{h} [(2\psi - 1)x_1 + (2\psi + 1)x_2 - 4\psi x_3] \quad (12.209)$$

$$\frac{\partial y}{\partial s} = \frac{2}{h} \sum_{i=1}^{3} y_i \frac{\partial N_i}{\partial \psi} = \frac{1}{h} [(2\psi - 1)y_1 + (2\psi + 1)y_2 - 4\psi y_3] \quad (12.210)$$

Equations (12.195) and (12.196) contain products of N_i, N_j and either $\partial x/\partial s$ or $\partial y/\partial s$. Since the interpolation functions are quadratic in ψ, and since these partial derivatives are linear, it follows that the expressions for $\hat{Q}_x^{(i)}$ and $\hat{Q}_y^{(i)}$ are *fifth-order* in ψ. To exactly integrate such functions, Gauss-Legendre quadrature requires that N equal 3.

Consider the specific case of a normal stress applied along the edge of the quadratic (subparametric) rectangular element shown in Figure 12.11. Once again $h = 2b$, $\partial x/\partial s = 0$, $\partial y/\partial s = 1$, and $|\mathbf{J}| = b$. The normal stress distribution can, at most, be *parabolic*.

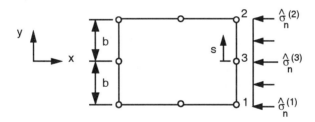

Figure 12.11: Shear Stress Applied to a Quadratic Rectangular Element

From equations (12.195) and (12.196) it is clear that $\hat{Q}_y^{(i)} = 0$, and that

$$\hat{Q}_x^{(i)} = -\int_{-1}^{1} N_i \left(\sum_{j=1}^{N_p} \sigma_n^{(j)} N_j \right) tb\, d\psi \quad (12.211)$$

or

$$\begin{Bmatrix} \hat{Q}_x^{(1)} \\ \hat{Q}_x^{(2)} \\ \hat{Q}_x^{(3)} \end{Bmatrix} = -\int_{-1}^{1} \begin{bmatrix} N_1 N_1 & N_1 N_2 & N_1 N_3 \\ N_1 N_2 & N_2 N_2 & N_2 N_3 \\ N_1 N_3 & N_2 N_3 & N_3 N_3 \end{bmatrix} \begin{Bmatrix} \sigma_n^{(1)} \\ \sigma_n^{(2)} \\ \sigma_n^{(3)} \end{Bmatrix} tb\, d\psi \quad (12.212)$$

Evaluating the integrals exactly (identical results would be obtained by using three-point Gauss-Legendre quadrature), gives the following results for the general case of a trapezoidal shear stress distribution:

$$\begin{Bmatrix} \hat{Q}_x^{(1)} \\ \hat{Q}_x^{(2)} \\ \hat{Q}_x^{(3)} \end{Bmatrix} = -\frac{tb}{15} \begin{bmatrix} 4 & -1 & 2 \\ -1 & 4 & 2 \\ 2 & 2 & 16 \end{bmatrix} \begin{Bmatrix} \sigma_n^{(1)} \\ \sigma_n^{(2)} \\ \sigma_n^{(3)} \end{Bmatrix} \qquad (12.213)$$

For the special case of a *parabolic* normal stress distribution, $\sigma_n^{(1)} = \sigma_n^{(2)} = 0$ and $\sigma_n^{(3)} = \bar{\sigma}_n$. It follows that

$$\begin{Bmatrix} \hat{Q}_x^{(1)} \\ \hat{Q}_x^{(2)} \\ \hat{Q}_x^{(3)} \end{Bmatrix} = -\frac{tb\bar{\sigma}_n}{15} \begin{Bmatrix} 2 \\ 2 \\ 16 \end{Bmatrix} \qquad (12.214)$$

Thus, even though the nodal values $\sigma_n^{(1)} = \sigma_n^{(2)} = 0$, the corresponding equivalent nodal forces are *non-zero*. The check for this case is that the total force in the x direction $(\hat{Q}_x^{(1)} + \hat{Q}_x^{(2)} + \hat{Q}_x^{(3)})$ is equal to $\bar{\sigma}_n$ multiplied by the area over which it is applied $2(2bt)/3$. This is indeed seen to be the case.

For a *trapezoidal* normal stress distribution, $\sigma_n^{(3)} = 0.5(\sigma_n^{(1)} + \sigma_n^{(2)})$. The corresponding equivalent nodal forces are

$$\begin{Bmatrix} \hat{Q}_x^{(1)} \\ \hat{Q}_x^{(2)} \\ \hat{Q}_x^{(3)} \end{Bmatrix} = -\frac{tb}{3} \begin{Bmatrix} \sigma_n^{(1)} \\ \sigma_n^{(2)} \\ 2(\sigma_n^{(1)} + \sigma_n^{(2)}) \end{Bmatrix} \qquad (12.215)$$

The check for this case is left to the reader.

For a *triangular* normal stress distribution, $\sigma_n^{(1)} = \bar{\sigma}_n$, $\sigma_n^{(2)} = 0$, and $\sigma_n^{(3)} = 0.5\bar{\sigma}_n$. The corresponding equivalent nodal forces are

$$\begin{Bmatrix} \hat{Q}_x^{(1)} \\ \hat{Q}_x^{(2)} \\ \hat{Q}_x^{(3)} \end{Bmatrix} = -\frac{tb\bar{\sigma}_n}{3} \begin{Bmatrix} 1 \\ 0 \\ 2 \end{Bmatrix} \qquad (12.216)$$

The check for this case is left to the reader.

Finally, for the special case of a *constant* normal stress distribution, $\sigma_n^{(1)} = \sigma_n^{(2)} = \sigma_n^{(3)} = \bar{\sigma}_n$, we have the following, not necessarily intuitive, result:

$$\left\{\begin{array}{c} \hat{Q}_x^{(1)} \\ \hat{Q}_x^{(2)} \\ \hat{Q}_x^{(3)} \end{array}\right\} = -\frac{tb\bar{\sigma}_n}{3} \left\{\begin{array}{c} 1 \\ 1 \\ 4 \end{array}\right\} \tag{12.217}$$

The check for this case is that the total force in the x direction is equal to $\bar{\sigma}_n$ multiplied by the area over which it is applied ($2bt$). This is indeed seen to be the case.

12.7.2 Axisymmetric Idealizations

For problems idealized assuming conditions of torsionless axisymmetry, the thickness t, appearing in the previous equations, is replaced by the radial coordinate r. Equations (12.195) and (12.196) thus become

$$\hat{Q}_x^{(i)} = \sum_{k=1}^{N} w_k N_i(\psi_k) \left[\left(\sum_{j=1}^{N_p} N_j(\psi_k)\sigma_t^{(j)} \right) \frac{\partial x}{\partial s}(\psi_k) \right.$$
$$\left. - \left(\sum_{j=1}^{N_p} N_j(\psi_k)\sigma_n^{(j)} \right) \frac{\partial y}{\partial s}(\psi_k) \right] r\, |\mathbf{J}| \tag{12.218}$$

$$\hat{Q}_y^{(i)} = \sum_{k=1}^{N} w_k N_i(\psi_k) \left[\left(\sum_{j=1}^{N_p} N_j(\psi_k)\sigma_n^{(j)} \right) \frac{\partial x}{\partial s}(\psi_k) \right.$$
$$\left. + \left(\sum_{j=1}^{N_p} N_j(\psi_k)\sigma_t^{(j)} \right) \frac{\partial y}{\partial s}(\psi_k) \right] r\, |\mathbf{J}| \tag{12.219}$$

where r is function of the natural coordinate ψ. For example, for linear elements ($N_p = 2$), the radial coordinate varies linearly with ψ, viz.,

$$r(\psi) = \frac{1}{2}(1-\psi)\,r_1 + \frac{1}{2}(1+\psi)\,r_2 \tag{12.220}$$

where r_1 and r_2 represent the radial coordinates of the element nodes along the loaded edge. For quadratic elements ($N_p = 3$), the radial coordinate varies quadratically with ψ; viz.,

$$r(\psi) = \frac{1}{2}\psi\,(\psi-1)\,r_1 + \frac{1}{2}\psi\,(\psi+1)\,r_2 + \left(1-\psi^2\right) r_3 \tag{12.221}$$

where r_1, r_2 and r_3 represent the radial coordinates of the element nodes along the loaded edge.

12.7.3   Potential Errors Along Curved Boundaries

As noted in Section 7.7, a geometric error may be introduced when a distributed flux is applied to a curved boundary. If the boundary is discretized using straight-sided elements, the flux, which is applied normal to the boundary, will be oriented incorrectly to some extent. A solution to this problem is to use curved-sided elements (i.e., quadratic and higher order) [100].

12.8   Relations Between Moduli

For *isotropic* elastic idealizations, the material constants are related in the following manner:

$$\lambda = \frac{2\mu\nu}{1 - 2\nu} = \frac{\mu(E - 2\mu)}{3 - E} = K - \frac{2}{3}\mu = \frac{E\nu}{(1 + \nu)(1 - 2\nu)}$$
$$= \frac{3K\nu}{1 + \nu} = \frac{3K(3K - E)}{9K - E} \tag{12.222}$$

$$\mu = G + \frac{\lambda(1 - 2\nu)}{2\nu} = \frac{3}{2}(K - \lambda) = \frac{E}{2(1 + \nu)}$$
$$= \frac{3K(1 - 2\nu)}{2(1 + \nu)} = \frac{3KE}{9K - E} \tag{12.223}$$

$$\nu = \frac{\lambda}{2(\lambda + \mu)} = \frac{\lambda}{(3K - \lambda)} = \frac{E}{2\mu} - 1 = \frac{3K - 2\mu}{2(3K + \mu)} = \frac{3K - E}{6K}$$
$$= -\frac{1}{4}\left(\frac{E}{\lambda} + 1\right) \pm \frac{1}{4}\sqrt{\left(\frac{E}{\lambda} + 1\right)^2 + 8} \tag{12.224}$$

$$E = \frac{\mu(3\lambda + 2\mu)}{\lambda + \mu} = \frac{\lambda(1 + \nu)(1 - 2\nu)}{\nu} = \frac{9K(K - \lambda)}{3K - \lambda} = 2\mu(1 + \nu)$$
$$= \frac{9K\mu}{3K + \mu} = 3K(1 - 2\nu) \tag{12.225}$$

$$K = \lambda + \frac{2}{3}\mu = \frac{\lambda(1 + \nu)}{3\nu} = \frac{2\mu(1 + \nu)}{3(1 - 2\nu)} = \frac{\mu E}{3(3\mu - E)}$$
$$= \frac{E}{3(1 - 2\nu)} \tag{12.226}$$

12.9   Exercises

All of the following exercises should be analyzed using a finite element program that can perform two-dimensional analyses of linear elastic continua.

12.1

This problem looks into the simulation of beam flexure. Analyze the beam of unit thickness shown in Figure 12.12. Compare the numerical solution with the standard (Bernoulli-Euler) beam theory solution. In particular, compare the transverse displacement v at the mid-span of the beam (i.e., at $x = L/2$ and $y = 0$).

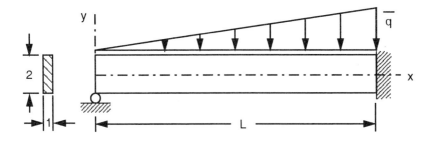

Figure 12.12: Propped Cantilever Beam

For all analyses performed, assume $\bar{q} = 10.0$ (units of force per distance), and *plane stress* conditions with elastic modulus equal to 20×10^5, and Poisson's ratio equal to 0.20. First assume $L = 8.0$ (span to depth ratio = 4). Then repeat the analyses for $L = 20$ (span to depth ratio = 10). For each value of L, please perform the following analyses:

- Using 2 (vertical) by 8 (horizontal), 4 by 16, and 8 by 32 meshes of four-node quadrilateral elements, analyze the beam.

- Reanalyze the beam using 1 by 2, and 2 by 4 meshes of eight-node serendipity elements, and then nine-node Lagrangian elements. How do results generated using these biquadratic elements compare to the four-node quadrilaterals? How does the performance of the two biquadratic elements compare to each other?

For each value of L, present your results in a single plot having normalized displacement as the ordinate versus the total number of unconstrained degrees of freedom (in log scale). The ordinate is simply the absolute value of the approximate transverse displacement at mid-span divided by the value predicted by Bernoulli-Euler beam theory. The latter value can easily be shown to be

$$v(x) = \frac{1}{120}\frac{q}{EI}\left(\frac{x^5}{L} + L^3x - 2Lx^3\right)$$

where positive v and q act in the positive y-coordinate direction. In the above equation, I denotes the moment of inertia (second area moment) of the cross-section. On the aforementioned plots, for each respective element type, draw a separate curve. Discuss your findings in detail. Further details pertaining to the performance of quadrilateral elements in bending simulations are available in [116].

12.2

This exercise looks into the simulation of beam flexure. Analyze the simply supported beam of unit thickness shown in Figure 12.13. Note that due to symmetry about the centerline, only one-half of the beam needs to be modeled (with appropriate nodal specifications made along the axis of symmetry). For all analyses performed, assume $c = 1.0$, $L = 8.0$, and plane stress conditions with $E = 10,000$, and $\nu = 0.20$. Let $\bar{q} = 1.0$.

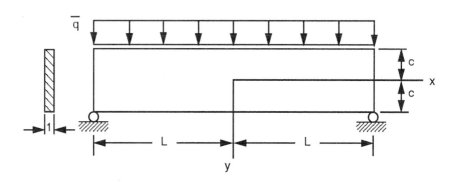

Figure 12.13: Simply Supported Beam

- Using 2 (vertical) by 4 (horizontal), 4 by 8, and 8 by 16 meshes of four-node quadrilateral elements, analyze the beam. Compare the numerical solution with both the standard (Bernoulli-Euler) beam theory solution, and that given by the theory of elasticity (see Article 22 in [538]). In particular, compare the stresses σ_{11}, σ_{22} and

σ_{12} acting through the thickness at $x = L/2$, and the transverse displacement v at the mid-span of the beam (i.e., at $x = y = 0$). The determination of stresses may require interpolation.

- Reanalyze the beam using 1 by 2 and 2 by 4 meshes of eight-node serendipity elements, and then nine-node Lagrangian elements. How do results generated using these biquadratic elements compare to the four-node quadrilaterals? How does the performance of the two biquadratic elements compare to each other?

- As a final, albeit approximate, assessment of the computational effort associated with the respective solutions, plot (as abscissa) the total number of unconstrained nodal (displacement) degrees of freedom versus $(\hat{v})_{y=0}/\delta$. Here δ represents the midspan displacement associated given by the theory of elasticity solution.

Discuss your findings in detail. Further details pertaining to the performance of quadrilateral elements in bending simulations are available in [116].

12.3

The purpose of this exercise is to evaluate the robustness of elements in the incompressible limit. An element is considered to be "robust" if its performance is not sensitive to physical parameters entering into the governing differential equations. The performance of many standard elements in axisymmetric, plane strain, and three-dimensional analyses of linear elastic continua is known to be sensitive to Poisson's ratio ν approaching the incompressible limit of 0.5. Indeed, for values of ν near 0.5 the energy stored by a unit volumetric strain is many orders greater than the energy stored by a unit deviatoric strain [252]. Thus elements exhibiting a strong coupling between volumetric and deviatoric strains often produce poor results in the nearly incompressible range. The problem of analyzing incompressible, or nearly incompressible media is beyond the scope of this introductory text. However, the behavior of "standard" elements in the incompressible limit is important to know even at the introductory level.

Consider the propped cantilever beam shown in Figure 12.12, with $L = 20$. Assuming *plane strain* conditions, along with the elastic modulus and load listed in Exercise 12.1, please do the following:

- Using 2 (vertical) by 8 (horizontal), 4 by 16, and 8 by 32 meshes of four-node quadrilateral elements, analyze the beam for ν equal to 0.0, 0.2, 0.4, 0.49 and 0.4999. In a figure plot the ratio of the computed transverse displacement at the middle of the beam to the exact value (with respect to Bernoulli-Euler beam theory) versus ν. Discuss your results. At which value of ν do result seem to significantly degrade?

- Repeat the above analysis using a two-element mesh of eight-node and nine-node quadrilateral elements. How does the performance of these elements compare to that of the four-node quadrilaterals?

Remark

1. The performance of triangular elements in the incompressible limit is quite poor; consequently, we do not consider them in this exercise.

12.4

Consider a thin, flat plate with a small circular hole drilled through it (Figure 12.14). We desire to ascertain the effect of the hole on the stress distribution in the plate. From Saint-Venant's principle we can conclude that the stress distribution shall be affected only in that portion of the plate in the immediate vicinity of the hole. Due to symmetry we need only to analyze the problem shown in Figure 12.15.

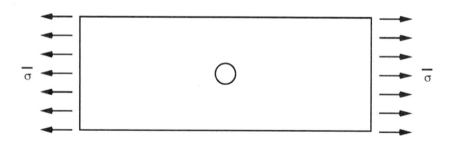

Figure 12.14: Plate with Circular Hole

Assume plane stress conditions, and an isotropic elastic material characterized by an elastic modulus of 30.0×10^6, and Poisson's ratio equal to 0.30. Please do the following:

- Using a coarse mesh of four-node quadrilateral elements, obtain a preliminary solution. Represent your results in the form of plots of σ_{11} along ab and σ_{22} along cd.

- Repeat the above analysis, only using a refined mesh. Plot the same quantities (on the same figures) as for the coarse mesh. Compare the results of both analyses with the elasticity solution (see Article 35 in [538]). Discuss your results in detail.

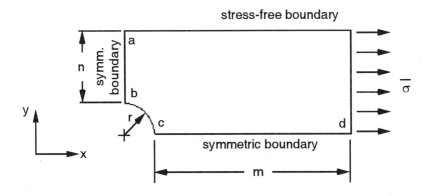

Figure 12.15: Mathematical Model of Plate with Circular Hole

12.5

Solve the thick-walled cylinder problem shown in Figure 12.16, subjected to a non-zero internal pressure. Assume an elastic modulus equal to 1000, and Poisson's ratio of 0.25. Use meshes of 1 (vertical) by 2 (horizontal), 1 by 4 and, 1 by 8 four-node quadrilateral elements, and 1 by 1, 1 by 2 and, 1 by 3 eight-node quadrilateral elements. On four separate diagrams, plot the radial displacement (u_r) and the three stress components σ_{rr}, σ_{zz} and $\sigma_{\theta\theta}$ versus radius (r) for each of the meshes. Represent these values by *discrete* symbols. On these same figures, plot the elasticity solution for the problem (see Article 28 in [538]). Depict the latter using *continuous* lines. Discuss your findings.

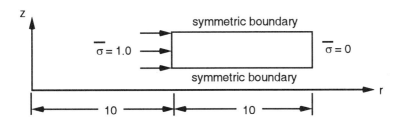

Figure 12.16: Mathematical Model of Thick-Walled Cylinder Problem

12.6

Consider the problem of a thin "strip" (really a beam with different length to depth ratios), loaded by concentrated forces.[3] The physical problem is shown in Figure 12.17.

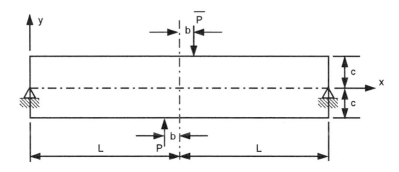

Figure 12.17: Thin Strip Loaded by Concentrated Forces

We desire to investigate the effect of concentrated forces on a solution. In particular, we are interested in the region affected by the force, and the resulting stress distributions. In the analysis, use the following geometric values: $L = 10.0$, $b = 0.50$, $c = 5.0$, and a unit thickness. Assume an isotropic, linear elastic material with $E = 30.0 \times 10^6$ and $\nu = 0.30$. Finally, use an applied load $\bar{P} = 50,000$, and assume plane stress conditions.

Using four-node quadrilaterals throughout, determine the following approximate stress distributions:

- σ_{12} versus y at $x = L$ (use interpolation as appropriate). How does this shear stress distribution compare to the parabolic distribution derived for beams? In the original work of Filon, this distribution was represented on a plot of $2c\sigma_{12}/P$ (abscissa) versus y/c (ordinate).

- σ_{22} versus x at $y = 0$ over the length of the beam. Discuss the localization of σ_{22} in light of the concentrated force. This can be represented as a plot of normalized quantities such as x/c (abscissa) versus $c\sigma_{22}/2P$ (ordinate).

---

[3]This problem was originally analyzed by L. N. G. Filon, *Transactions of the Royal Society*, (London), ser. A, **201**: 67 (1903). Further information is given in Article 24 of [538].

12.7
We wish to analyze the classical problem of an elastic half-space subjected to a pressure loading (see Article 36 in [538]). In 1892, Flamant solved this problem by using the three-dimensional solution of J. V. Boussinesq.[4] The problem is shown in Figure 12.18.

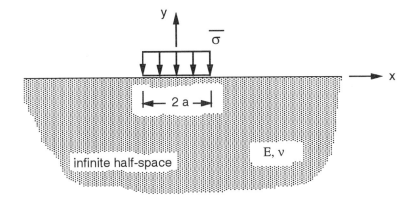

Figure 12.18: Elastic Half-Space Subjected to Pressure Loading

Without using "infinite" elements, we cannot of course extend the finite element mesh out to infinity. Thus we analyze instead the problem shown in Figure 12.19. We desire the stress distribution (don't worry about the displacements) of the second problem to be a good (say within one or two percent) approximation to the first. Using the same elements throughout, determine the minimum values for the dimensions b and d so that the above approximation is realized.

Assume $\bar{\sigma} = 1.0$, $E = 10,000$, $\nu = 0.20$ and $a = 1.0$. Explain the criterion used to decide when the two solutions are essentially the same. For two subsequent meshes, plot the stress contours for σ_{22} at two or three depths in the neighborhood of the applied stress.

---

[4] Joseph Valentin Boussinesq (1842-1929) is considered by many to be one of the most distinguished pupils of Saint-Venant. He was responsible for advances in hydrodynamics, optics, thermodynamics, and the theory of elasticity. Alfred-Aimé Flamant was another pupil of Saint-Venant.

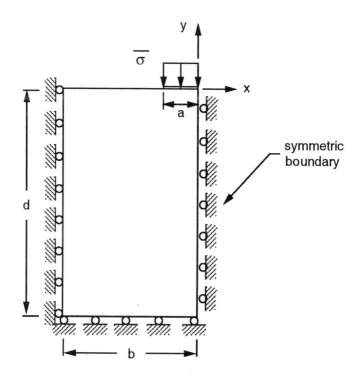

Figure 12.19: Mathematical Model of Elastic Half-Space Subjected to Pressure Loading

Chapter 13

Implementation, Modeling, and Related Issues

13.1 Introductory Remarks

After introducing the finite element method in Chapter 6, the subsequent chapters have presented details concerning the formulation of element equations (Chapter 7), the basic steps in performing finite element analyses and examples of some simple steady-state finite element analyses (Chapter 8), the development of element interpolation functions (Chapter 9), the notion of element mapping (Chapter 10), and finally applications to scalar field and elastostatic problems (Chapters 11 and 12, respectively). Beginning with Chapter 8, various aspects related to the numerical implementation of the finite element method have been discussed in rather general terms.

This chapter focuses on complete finite element programs and on related issues such as the role of finite element analysis in engineering design, mesh generation, modeling techniques and the associated errors. The treatment is not meant to be exhaustive, as several texts have been written that are devoted to programming and implementation of the finite element method [244, 8, 502]. For the interested reader, historical overviews of early program development in the area of structural analysis are given in [186, 238, 110, 370].

13.2   Role of Modeling and Analysis in Engineering Design

As noted in Section 1.5, in recent years computing power has increased significantly, making computational resources more easily available to analysts. Significant improvements in computational performance, coupled with the low price of random access memory and disk storage, have allowed engineers to analyze larger and more sophisticated problems. The role of mathematical modeling and finite element analysis in the design process is graphically summarized in Figure 13.1.

The makeup of finite element analysis programs and the computational aspects related to the development of mathematical models for finite element analyses are discussed in the subsequent sections of this chapter.

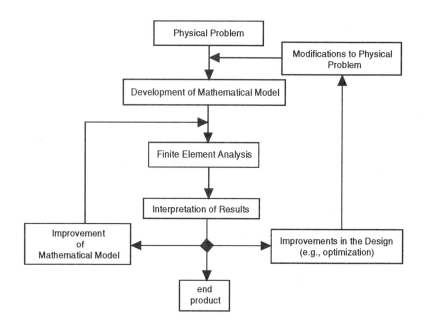

Figure 13.1: Schematic Overview of Mathematical Modeling and Finite Element Analysis in the Design Process

13.3 Phases of a Finite Element Analysis

Every finite element analysis is comprised of the following three general phases: (1) the preprocessing phase; (2) the solution phase; and, (3) the postprocessing phase. The general tasks performed in each of the three phases are listed below.[1]

1. Preprocessing Phase

   - Specification of analysis options
   - Generation of mesh
   - Specification of materials
   - Specification of nodal constraints
   - Check for "reasonableness" of model created

2. Solution Phase

   - Generation of all element arrays
   - Assembly of all element arrays
   - Application of nodal specifications
   - Solution of global equations
   - Computation of secondary dependent variables
   - Storage of results in a manner suitable for subsequent postprocessing

3. Postprocessing Phase

   - Hard copy print out (for small problems)
   - Graphics (deformed geometry, contour plots, vector fields, etc.)

Relating the present discussion to Figure 13.1, we note that the preprocessing phase of the analysis is associated with the development of the mathematical model. The finite element solution represents the solution phase of the analysis. Finally, the postprocessing phase of the analysis greatly facilitates the interpretation of the results.

---

[1]Such a breakdown of tasks is certainly not unique. It is also assumed that, in accordance with good programming practice, basic array management tasks such as allocation and initialization are performed.

13.4   Outline of a Finite Element Program

In the past, the three phases of a finite element analysis were often contained in a *single* program. A number of finite element analysis programs, particularly introductory ones or special purpose codes, still adhere to such design. It is thus worthwhile to outline their structure, which consists of the following phases[2]:

1. Input of Data

   - Initialization of arrays
   - Reading of input data (with minimal input from the analyst; as much information as possible is generated). This includes:
     - General solution control information
     - Material constants
     - Nodal coordinates
     - Element connectivity
     - Nodal specifications
     - Output control (i.e., control over which nodal and element results shall be displayed – such output control reduces the often voluminous amounts of data generated during finite element analyses)
   - Checking if input is syntactically correct.
   - Checking if input data is "reasonable," for example:
     - Check if length of one-dimensional line elements is zero
     - Check for zero Jacobian derivatives at each quadrature point in each element
     - Check for excessive element distortions
   - Checking if adequate memory has been allocated.
   - Generation of any unknown nodal coordinates.
   - "Echo" printing of the input data.

2. Development of Global Arrays

   - Forming all element arrays.
   - Accounting for nodal specifications (at element or global level).
   - Assembly of element arrays to form global arrays.

---

[2] This outline is by no means unique.

3. Solution of Global Equations

4. Computation of Secondary Dependent Variables

5. Display of Results

The greater availability of fast and relatively inexpensive computational resources has allowed engineers and physicists to analyze larger and more sophisticated problems. In the context of finite element analyses, this usually translates into the development of larger models with increasingly complex shape and high levels of mesh refinement. As the models studied increase in complexity and size, more and more effort is spent on the development of the mathematical model prior to analysis; that is, on the preprocessing phase. Likewise, a substantial amount of time is also spent in the postprocessing phase viewing and interpreting the typically large body of results generated by the finite element analysis.

In appreciation of the need for powerful yet easy to use pre- and postprocessing software, a *modular* approach to finite element analyses has developed over time. In particular, general "stand-alone" pre- and postprocessors have been written to complement specialized analysis programs. Such a modular design is founded on the rationale that each phase of the analysis process should draw upon the expertise of a specific software development team with a specific design objective.

The preprocessor should be an efficient, powerful, yet flexible program that completely defines the geometry and element connectivity of the mathematical model, assigns material types to groups of elements, and possibly assigns nodal specifications. Ideally the preprocessor possesses robust graphical capabilities that can be displayed on all popular computing platforms (possessing reasonable computational resources, of course). The end product of the preprocessor is a file(s), with known format, that contain all of the aforementioned data.

Before leaving the subject of preprocessors, it is timely to note that the rise in popularity of computer-aided design (CAD) and computer-aided design and drafting (CADD) application software has led to their use in developing finite element meshes. Although such use of CAD and CADD software is still somewhat in its infancy, further expansion in this area is likely.

The purpose of the analysis program, or "computing engine," is to analyze a mathematical model. Ideally, the analysis program should not have to do any development of the mathematical model, for this should have all been done by the more efficient preprocessor. Likewise, the analysis program should not be assigned the task of graphically displaying results. This task is relegated to the more efficient stand alone postprocessor. Thus, the analysis program simply prompts for the name of a file(s) containing the description of a complete mathematical model, and for the name of a file(s) into which results should be written.

Finally, the purpose of the postprocessor is to conveniently display the results of the analysis for scrutiny by the analyst. Typically this entails the use of interactive graphics. Similar to the preprocessor, the postprocessor must be an efficient, powerful, yet flexible program. It must afford the user ample opportunity to ascertain the results of the finite element analysis. Examples of graphical capabilities include deformed mesh plots, contour plots, time-history plots, plots of vector fields, etc.

The only drawback of the above modular approach to finite element analysis is the disparity of formats associated with the input/output files created in each phase of the analysis. The remedy to this problem is the data file translator. A translator is a piece of software whose sole task is to read a data file, with known format, and to write (translate) this data into another format. Typically, a translator is written for a specific *pair* of programs; namely, a program whose output will serve as input to the translator, and a program whose input will be the output of the translator. Ideally translation should be done automatically, with minimal input from the user. The role of translators in the modular analysis procedure is summarized in Figure 13.2.

The "preprocessor to analysis" translator reads the output from a specific preprocessor. It then converts this data into the format recognized by the analysis program. In a similar fashion, the "analysis to postprocessor" translator reads the output of the analysis program and converts this data into a format recognized by a specific postprocessor.

Having described the workings of finite element analysis programs in a general sense, attention is next turned to some specific details. The first topic reviewed is the construction of finite element meshes.

13.5   Meshing Guidelines Revisited

Beginning in Chapter 8, general guidelines have been given to help in deciding how to discretize of the domain. We now reiterate many of these guidelines, and add some others.

13.5.1   Element Types

With respect to the selection of elements, the accuracy of finite element analyses is known to depend on the type of element used and on the distribution of elements. In a given analysis, a variety of element shapes could,

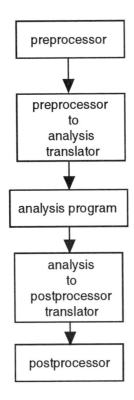

Figure 13.2: Representation of "Modular" Finite Element Analyses

in theory, be used. In practice, only a few different types of elements are typically employed. The specific element types used depend upon the spatial extent of the domain and upon the idealization chosen in the analysis. Some common element types have been introduced in Chapter 8; the associated interpolation functions have been derived in Chapter 9.

13.5.2 Placement of Elements

The following general guidelines for placement of elements have been presented in Chapter 8:

1. Element boundaries should be placed along all exterior and interior boundaries of the body being discretized, in a manner that minimizes the domain discretization error.

2. Since material and geometric parameters are typically assumed to be constant within an element, element interfaces must coincide with material interfaces.

3. Nodes should be placed at any location in Ω at which a value of the primary dependent variable(s) is desired.

4. Elements should not be overly distorted.[3]

Even in simple meshes, it is common practice to place the bulk of the elements in locations where the gradient of the primary dependent variables would be expected (or is known) to be high.

13.5.3   Individual Element Shapes

The subject of element distortions was discussed at length in Section 10.6. We recall that different elements have different sensitivities to shape distortion. Consequently, few general guidelines apply to all elements. None the less, the following points are worth reiterating:

- In quadrilateral elements, the aspect ratio should be kept near unity (Figure 13.3a).

- The vertex angles in quadrilateral elements should be near 90 degrees, thus minimizing taper (Figure 13.3b and c) and skew (Figure 13.3d). In triangular elements the vertex angles should be near 60 degrees.

- The edges in higher-order elements should be straight.

- Edge nodes should be uniformly distributed.

These guidelines are not without exceptions. For example:

- Elements with large aspect ratios can be used in regions of Ω where gradients are very small.

- Moderately distorted quadrilateral elements with vertex angles greater than 90 (but less than 180) degrees can still give accurate results.

- Element sides can be curved to fit a curved boundary; sides of such elements interior to the mesh should be straight, however.

- Placing edge nodes at the quarter points along an edge allows an element to simulate a stress singularity at the corner [221].

---

[3] A quantification of element distortion and the associated consequences, have been discussed in Section 10.6 of Chapter 10.

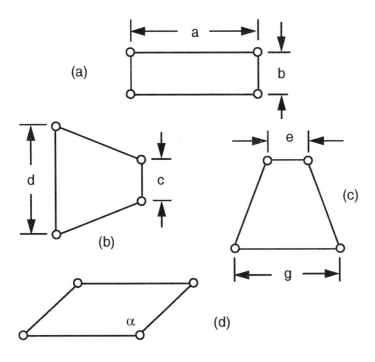

Figure 13.3: Distortion Measures for Bilinear Quadrilateral Elements: (a) Aspects Ratio; (b) Vertical Taper; (c) Horizontal Taper (d) Skew

This brings to light the important observation that, in many physical applications, elements with special properties are required to achieve maximum accuracy. For example, certain *singular* or *localization* elements for modeling point and line singularities have been developed [221, 7, 57, 541, 508, 251, 422, 67]. A second example are so-called *infinite* elements [72, 73, 61, 595, 128, 374, 148] that allow for accurate representation of semi-infinite domains without resorting to boundary solution procedures. The final example are *interface* elements that facilitate the modeling of relative displacements, contact stresses, etc. along material interfaces [202, 232, 196, 282, 140, 63, 193, 485, 276].

Detailed discussions of these specialized elements are beyond the scope of this introductory text. For further information, the interested reader is thus directed to the references cited above.

13.5.4   Element Combinations

When combining elements of different types, care must be exercised to ensure that *compatible* configurations be used. Recall that, in accordance with the compatibility criterion, the interpolation functions must be C^{s-1} continuous across all element interfaces. This will be true provided the elements have the same edge or face nodes and coordinates, and if the values of the primary dependent variable along the common edge or face are defined by the same interpolation functions [58].

Such a condition will be realized provided elements have corresponding nodes connected; that is, corner nodes to corner nodes and edge nodes to edge nodes (Figure 13.4a). If this is not done (Figure 13.4b to d), then interelement continuity is generally destroyed, since the degrees of freedom defining the approximate solution on the side of one element are *not* the same degrees of freedom defining the approximation on the side of the adjacent element [100].

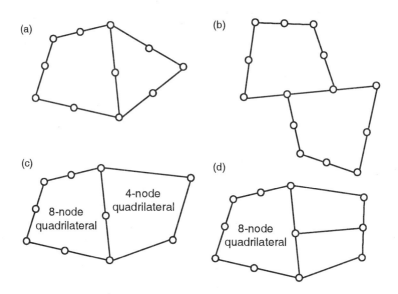

Figure 13.4: Element Combinations: (a) Compatible, and (b), (c), (d) Incompatible

There are two ways in which to properly combine elements of different orders. The first is through the use of *transition elements*. A detailed discussion of such elements was given in Section 9.6. A second approach is to impose a linear *constraint equation* that constrains the edge degrees of freedom of the higher-order element to equal the approximate solution on the side of the lower-order element.

Improper element connections do not necessarily condemn the complete finite element analysis. Provided the surrounding mesh is satisfactory, spurious gradients caused by local meshing errors usually die out rather than propagate [116].

13.5.5 Changes in Mesh Density

Since finite element meshes are rarely uniform, some comments concerning changes in mesh density are in order. There are four different approaches for changing the density of a mesh. These can be used separately or in combination.

The first approach uses only one type of element and gradually increases or decreases its size. For present-day mesh generators, the automatic creation of such meshes is a trivial matter.

The second approach for changing mesh density is very similar to the first, but is not restricted to elements of the same type. For example, the mesh shown in Figure 9.24(b) uses triangular elements as transitions from smaller to larger quadrilaterals. This type of mesh tends to be more irregular than that shown in Figure 9.24(a), particularly if the mesh density is changed in more than one direction.

The third approach to changing mesh density is through the use of transition elements. Such elements have been discussed in Section 9.6.

The final approach to changing mesh density is through the use of special constraint equations that constrain specific degrees of freedom in higher-order elements so as to make them compatible with lower-order elements to which they are connected. For example, in the case of a quadratic element connected to a linear one, the constraint equation would constrain the midside degree of freedom associated with the quadratic element to equal the element interpolation function on the side of the linear element.

If a mesh is graded, this should be done in a manner that produces no great discrepancy in size between adjacent elements. Abrupt changes in mesh refinement from very fine to very coarse should be avoided, as the error in the immediate vicinity of the fine mesh is constrained to being comparable to the error in the coarse mesh [100]. The graded mesh shown in Figure 13.5(a) is thus preferable over that shown in Figure 13.5(b).

Even in properly graded meshes, elements should not be refined so much that the size of the smallest one is very much smaller than the largest one. Such discrepancies tend to produce ill-conditioned global properties matrices $\mathbf{K}$. The degree of ill-conditioning is very problem-dependent and, at

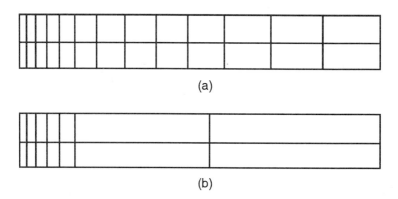

(a)

(b)

Figure 13.5: Hypothetical Graded Meshes: (a) Smoothly Graded, and (b) Abruptly Graded

best, only rough guidelines have been proposed. Cook et al. [116] suggest that, in going from one element to the next, the ratio of the elastic modulus to the element volume should not change by more than a factor of three. In the opinion of Burnett [100], ratios of 100 to 1 in linear dimensions (thus 10,000 to 1 in area) between the largest and smallest element can sometimes be troublesome in "medium-size problems (a few hundred degrees of freedom) on low-precision (32-bit) computers." In an effort to lessen the arbitrary aspects of mesh grading, optimization techniques have been to the process [427].

13.6   Mesh Generation

As evident from the previous discussion, the accuracy of a finite element solution is strongly influenced by the quality of the mesh used to discretize the domain Ω. In addition, it is evident that the description of all but the simplest meshes involves the specification of appreciable amounts of data.

In light of these observations, it comes as no surprise that the generation of high quality meshes is a very active area of research. Mesh generation is complicated by the realization that a mesh must satisfy almost contradictory requirements:

- It must conform to the shape of the simulated domain.

- It must be composed of elements that are the "right size and shape, avoiding excessive distortion (e.g., overly small or overly large interior angles.

- Over relatively short distances it will typically have to grade from large elements (coarse discretization) to small elements (fine discretization).

For finite element meshing applications, two-dimensional domains may be subdivided into triangular or quadrilateral elements. This applies equally to surfaces in three dimensions, sometimes referred to as "2.5-dimensional meshes [194]. Three-dimensional domains (volumes) are primarily subdivided into tetrahedral or hexahedral elements.

There has long been a debate among finite element analysts as to what element shapes produce the most accurate results. There is the often-held position that quadrilateral and hexahedral shaped elements have superior performance to triangle and tetrahedral shaped elements when comparing an equivalent number of degrees of freedom. Use of hexahedral elements can also vastly reduce the number of elements and consequently computational effort. In addition, hexahedral and quadrilateral elements are more suited for non-linear analysis as well as situations where alignment of elements is important to the physics of the problem, such as in computational fluid dynamics or simulation of composite materials [423].

13.6.1 Types of Meshes

There are basically two types of meshes: *structured* and *unstructured*. A mesh can be entirely structured or entirely unstructured, or a combination of the two (Figure 13.6).

Structured Meshes

A structured mesh has connectivity of the "finite difference type [194]; that is, the intersection of nodal lines is orthogonal. Consequently, all interior nodes of the mesh have an equal number of adjacent elements. In Figure 13.6, that portion of the mesh consisting of rectangular elements is structured.

Structured meshing is commonly referred to as "grid generation [536], and is particularly well suited for domains defined by rectangular or square boundaries. The meshes generated by a structured grid generator are typically composed of exclusively quadrilateral or hexahedral elements. Algorithms employed generally involve complex iterative smoothing techniques that attempt to align elements with boundaries or physical domains. Structured grid generators are most commonly used within the field of com-

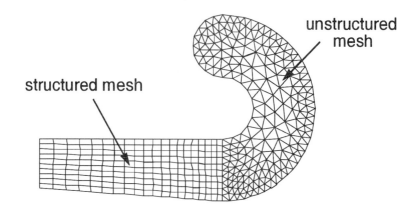

Figure 13.6: Mesh Containing Both Structured and Unstructured Portions (after Bickerton [74])

putational fluid dynamics (CFD), where strict alignment of elements can be required by the analysis code or necessary to capture physical phenomenon [424].

Unstructured Meshes

In unstructured meshes, any number of elements may meet at a single node. For example, in Figure 13.6, that portion of the mesh consisting of triangular elements is unstructured. In two and three-dimensional unstructured meshes, triangular and tetrahedral elements, respectively, are by far the most common element types used. Less common in unstructured meshes is the use of quadrilateral and hexahedral elements.

While there exists some overlap between structured and unstructured mesh generation technologies, the main feature that distinguishes the two fields is the use of iterative smoothing algorithms employed by structured grid generators [424].

A wide variety of schemes for generating two- and three-dimensional meshes has been developed. Some of these schemes are designed to handle specific geometries, while others can be used in a more general context. Manual, semi-automatic and fully automatic schemes exist that lead to structured, unstructured, or mixed meshes. Some major mesh generation schemes is now briefly reviewed. Further details pertaining to this subject can be found in any of a number of references on the subject such as [194, 149, 150, 318, 253].

13.6.2 Overview of Structured Meshing Schemes

In this category are grid generation schemes that apply to structured meshes.

Semi-Automatic Mesh Generation

In this approach, the domain is broken into simple subregions such as triangles, rectangles, pentagons, etc. The nodes along the boundary of each subregion are generated. Once the analyst decides upon the mesh density, an algorithm is used to locate the coordinates of all nodes lying in the interior of the subregions. Algorithms are then used to "compress nodes common to two adjacent subregions. One such a semi-automatic approach is the "isoparametric" scheme originally developed by Herrmann [231] and subsequently refined [273] and incorporated into a mesh generation preprocessor for mixed meshes [409].

Transport Mapping

Under this heading are methods that construct the mesh of the real domain as a mapping, by a suitable transformation, of a mesh of an elementary geometry such as a triangle, quadrilateral, etc. [218].

Explicit Solution of Partial Differential Equations

This scheme is similar to transport mapping, only the mapping function is not given initially but is computed by solving partial differential equations in order to satisfy certain useful properties (e.g., element density, orthogonality, etc.). This technique is commonly used in commercial finite volume programs [536].

13.6.3 Overview of Unstructured Meshing Schemes

In this category are grid generation schemes that apply to unstructured meshes.

Quadtree/Octree Schemes

The schemes in this category are based on the deformation and local modification of an easily obtainable grid. They produce meshes of triangular or tetrahedral elements.

The primary scheme in this category is of type "quadtree" or "octree," for two- and three-dimensional domains, respectively [577, 578, 492]. In such a scheme, the domain is enclosed in a quadrilateral or quadrilateral parallelepipedon that is split into boxes. These boxes are recursively subdivided by decomposition based on a quaternary "tree" (for 2-space) or octal "tree" (for 3-space) until the desired resolution is achieved.

Delaunay Triangulation

Possibly the most popular techniques for triangular and tetrahedral meshes are those utilizing the Delaunay criterion [135]. This criterion is related to the work of Dirichlet [141], who proposed a mechanism to connect an arbitrary set of points in space. The associated geometric construction is commonly referred to as a Voronoï diagram [550]. If all pairs of points that have some segment of a Voronoï boundary in common are joined, a triangulation of the set of points is generated. This triangulation is known as the Delaunay triangulation.

By itself, the Delaunay criterion is not an algorithm for generating a mesh – it merely provides the means by which to connect a set of existing points in space. The use of this criterion for developing schemes to triangulate a set of vertices [552] led to the subsequent development of meshing schemes [49, 195, 458, 553]. Starting from the boundary of Ω, Voronoï-Delaunay based schemes create internal nodes and elements. The boundary is described either globally (e.g., by an analytical definition), or in a discrete manner (e.g., as a list of edges or triangular faces).

Advancing Front

Another very popular family of triangle and tetrahedral meshing schemes are based on the advancing front method [327, 328, 330]. Similar to the Voronoï-Delaunay based schemes, the advancing front method derives the mesh, element by element, from the boundary data. For example, tetrahedra are built progressively inward from the triangulated surface. An active front is maintained where new tetrahedra are formed.

Mapped Meshing

This scheme is used to generate unstructured quadrilateral and hexahedral meshes. When the geometry of Ω is appropriate (e.g., opposite edges of areas to be meshed must have equal numbers of divisions), mapped meshing will generally produce high quality results [424].

"Sweeping" is another class of mapped hexahedral meshing. In this scheme, a quadrilateral mesh is swept through space along a curve. Using the same topology as the quadrilateral mesh, layers of hexahedra are formed at specified intervals. Further details pertaining to mapped meshing, along with applicable references, are given in [424].

Unstructured Quadrilateral Meshing

Schemes for generating unstructured quadrilateral meshes can be divided into two general categories: *indirect* and *direct*. In an indirect scheme, Ω is first discretized into a set of triangles. Appropriate algorithms are then used to convert the triangles into quadrilaterals. In a direct scheme, a surface is directly meshed into quadrilaterals without first generating a triangular mesh.

Although many direct generation schemes have been developed, they tend to fall into one of two categories [424]. In the first category are schemes that rely on some form of decomposition of Ω into simpler regions that can subsequently be resolved into elements. The second category includes those schemes that utilyze a moving front method for locating nodes and elements. Further details pertaining to unstructured quadrilateral meshes, along with applicable references, are given in [424].

Unstructured Hexahedral Meshing

Similar to quadrilateral meshing, unstructured hexahedral meshing schemes can be indirect and direct. In an indirect scheme, Ω is first discretized into a set of tetrahedral elements. These are subsequently subdivided into a number of hexahedra. However, the quality of the resulting meshes tends to be rather poor [424], thus precluding wide use of such indirect schemes.

For this reason, several direct unstructured hexahedral meshing schemes have been developed. These include the grid-based approach [486], medial surface methods [312, 444, 445], "plastering" [82], and "whisker weaving" [530]. Further details pertaining to these schemes, along with applicable references, are given in [424].

13.6.4 Overview of General Meshing Schemes

In this category are schemes that apply to *both* structured and unstructured meshes.

Manual Mesh Generation

For problems with very simple geometries, it is feasible to have the user explicitly define all nodes and elements comprising the mesh. Once the mesh becomes at all complex, this scheme quickly loses its practicality.

Combinations of Schemes

Under this heading are schemes that use a composition of meshes obtained by any of the previous schemes. In such schemes, the problem is split into a set of simpler subproblems. Using one or more of the previous classes of schemes, these are then meshed. The final result is then obtained by transformation and combination of the partial results [194].

13.7   Mesh Renumbering Schemes

As mentioned in Section 8.5.1, one of the most effective ways in which to reduce the storage and computational effort associated with the solution of the global equations is to number the nodes or elements in as optimal a manner as possible.

Over the years, many algorithms have been developed that automatically renumber a mesh in order to maximize efficiency. These algorithms are *solver-specific*. Thus no single strategy is best for all objectives or all programs. For example, in the case of banded solvers, the algorithms renumber the mesh to minimize the bandwidth (or, for symmetric systems, the half-bandwidth) of the global $\mathbf{K}$ [10, 9, 129, 207, 114, 6, 197, 449]. For profile solvers, the nodes are renumbered in a manner that reduces the profile (and may leave the bandwidth unchanged) [499]. Finally, for frontal solvers the wave front is minimized by renumbering the elements [457, 62, 501].

The procedures for nodal and element renumbering both require a mesh with initial node and element numbers assigned. This starting mesh is typically created using a generation scheme such as those alluded to above. Renumbering algorithms use graph theory and its terminology [129]. They do not guarantee an optimum numbering, or even an improvement over the original numbering.

Renumbering algorithms require the creation of a table of nodal connectivities. There is thus some computational effort expended in generating this table that goes along with the actual renumbering. The automatic renumbering is, however, cost-effective, especially if the same numbering is used in repeated solutions, as is the case in transient or non-linear analyses [116].

Further details pertaining to many of the commonly used renumbering algorithms are available in comparison papers [198, 319, 154] and certain finite element texts [430, 218]. Detailed discussions of the advantages and disadvantages of most popular renumbering algorithms are given by George [194] and Carey [101].

13.8 Sources of Error

The following sources of error are associated with the use of computer programs to perform finite element analyses.

13.8.1 Programming Errors

Programming errors refer to potential unintentional errors that have been made in writing the analysis program. A method for assessing errors associated with the numerical implementation of elements, the so-called "patch test," has been discussed in Appendix D.

13.8.2 Errors in Input Data

Probably the most obvious source of *unintentional* error, and often the easiest to avoid, is that associated with errors in the description of the mathematical model. Such errors come under the heading of "human errors" or possibly "blunders." Obviously, if errors are made in describing: (1) the geometry of the solution domain, (2) the element data, (3) the material idealizations, (4) the nodal specifications, (5) solution control information, etc., then the incorrect mathematical model shall be analyzed. Some errors in input data are, however, unavoidable. For example, a certain degree of roundoff error is to be expected in describing the mathematical model, though this error is usually not significant.

13.9 Concluding Remarks

This chapter has focused on issues related to the numerical implementation of the finite element method and on techniques used to create valid mathematical models. The role of the finite element method in engineering design, a topic that was first introduced in Section 1.5, was reviewed.

The three major phases of any finite element analysis were described, followed by a general outline of a finite element analysis program. Although traditionally such programs have contained all three of the aforementioned phases, this monolithic software model has been giving way to a more modular approach. The impetus for this trend is the widespread availability of robust pre- and postprocessing software.

General guidelines for properly discretizing the solution domain, first presented in Chapter 8, were reviewed and expanded. This was followed by a brief overview of the major mesh generation schemes currently in use.

The chapter concluded with a discussion of errors associated with the use of computer programs to perform finite element analyses.

Appendix A

Mathematical Potpourri

A.1 Indicial Notation

A tensor of any order, and/or its components, may be represented clearly and concisely by the use of indicial notation . This convention was believed to have been introduced by Einstein. In this notation, letter indices, either subscripts or superscripts, are appended to the *generic* or *kernel* letter representing the tensor quantity of interest; e.g. A_{ij}, B_{ijk}, δ_{ij}, a_{kl}, etc. Some benefits of using indicial notation include: (1) economy in writing; and, (2) compatibility with computer languages (e.g. easy correlation with "do loops"). Some rules for using indicial notation follow.

Index rule

In a given term, a letter index may occur no more than twice.

Range Convention

When an index occurs unrepeated in a term, that index is understood to take on the values $1, 2, \cdots, N$ where N is a specified integer that, depending on the space considered, determines the range of the index.

Summation Convention

When an index appears twice in a term, that index is understood to take on all the values of its range, and the resulting terms are *summed*. For example, $A_{kk} = A_{11} + A_{22} + \cdots + A_{NN}$.

Free Indices

By virtue of the range convention, unrepeated indices are free to take the values over the range, that is, $1, 2, \cdots, N$. These indices are thus termed "free." The following items apply to free indices:

- Any equation must have the same free indices in each term.

- The tensorial rank of a given term is equal to the number of free indices.

- $N^{(\text{no. of free indices})}$ = number of components represented by the symbol.

Dummy Indices

In the summation convention, repeated indices are often referred to as "dummy" indices, since their replacement by any other letter not appearing as a free index does not change the meaning of the term in which they occur.

In the following equations, the repeated indices are thus "dummy" indices: $A_{kk} = A_{mm}$ and $a_{ik} b_{kl} = a_{in} b_{nl}$.

In the equation $E_{ij} = e_{im} e_{mj}$ i and j represent free indices and m is a dummy index. Assuming $N = 3$ and using the range convention, it follows that $E_{ij} = e_{i1} e_{1j} + e_{i2} e_{2j} + e_{i3} e_{3j}$.

Care must be taken to avoid breaking grammatical rules in the indicial "language." For example, the expression $\mathbf{a} \bullet \mathbf{b} = (a_k e_k) \bullet (b_k e_k)$ is erroneous since the summation on the dummy indices is ambiguous. To avoid such ambiguity, a dummy index can *only* be paired with *one other* dummy index in an expression. A good rule to follow is use separate dummy indices for each implied summation in an expression.

Contraction of Indices

Contraction refers to the process of summing over a pair of repeated indices. This *reduces* the order of a tensor by *two*.

For example, contracting the indices of A_{ij} (a second-order tensor) leads to A_{kk} (a zeroth-order tensor or scalar). Contracting the indices of B_{ijk} (a third-order tensor) leads to B_{ikk} (a first-order tensor). Contracting the indices of C_{ijkl} (a fourth-order tensor) leads to C_{ijmm} (a second-order tensor).

Comma Subscript Convention

A subscript comma followed by a subscript index "i" indicates partial differentiation with respect to each coordinate x_i. Thus,

$$\phi_{,m} \equiv \frac{\partial \phi}{\partial x_m} \quad ; \quad a_{i,j} \equiv \frac{\partial a_i}{\partial x_j} \quad ; \quad C_{ij,kl} \equiv \frac{\partial C_{ij}}{\partial x_k \partial x_l} \quad ; \quad \text{etc.} \tag{A.1}$$

- If i remains a free index, differentiation of a tensor with respect to i produces a tensor of order one *higher*. For example

$$A_{j,i} = \frac{\partial A_j}{\partial x_i} \tag{A.2}$$

- If i is a dummy index, differentiation of a tensor with respect to i produces a tensor of order one *lower*. For example

$$V_{m,m} = \frac{\partial V_m}{\partial x_m} = \frac{\partial V_1}{\partial x_1} + \frac{\partial V_2}{\partial x_2} + \cdots + \frac{\partial V_N}{\partial x_N} \tag{A.3}$$

A.2 Classification of PDEs

The classification of a partial differential equation (PDE) is an important concept in that the general theory and methods of solution typically apply only to a given class of equations. In addition, each class of equations demands a particular set of initial and boundary conditions that must be specified to obtain a well-posed problem that is suitable for an analytical or a numerical solution. Six basic classifications exist [155]:

Order of the PDE

The order of a partial differential equation is equal to the order of the *highest partial derivative* in the equation.

Number of Variables

The number of variables is equal to the number of *independent* variables.

Linearity

PDEs are either *linear* or *nonlinear*. In linear partial differential equations the dependent variable and all its derivatives appear in a linear fashion. More precisely, a second-order linear partial differential equation in two variables has the following general form:

$$A\frac{\partial^2 \phi}{\partial x^2} + B\frac{\partial^2 \phi}{\partial x \partial y} + C\frac{\partial^2 \phi}{\partial y^2} + D\frac{\partial \phi}{\partial x} + E\frac{\partial \phi}{\partial y} + F\phi - G = 0 \qquad (A.4)$$

where A, B, C, D, E, F and G can be constants or given functions of the independent variables x and y.

Kinds of Coefficients

If the coefficients A, B, C, D, E and F in equation (A.4) are constants, then this equation is said to have *constant coefficients*. Otherwise, the coefficients are said to be *variable*. In the case of partial differential equations with variable coefficients, the type of the equation can change from point to point in the solution domain.

Homogeneity

If the right-hand side $G(x,y)$ is identically zero for all x and y, equation (A.4) is said to be *homogeneous*. Otherwise the equation is said to be *nonhomogeneous*.

Basic Types of PDEs

All linear partial differential equations are either *parabolic, hyperbolic,* or *elliptic*. Referring to equation (A.4), this classification is based on the sign of the quantity $B^2 - 4AC$. This classification is related to the nature of characteristics, which are lines along which a disturbance or information can propagate. Along a characteristic, a partial differential equation in two dimensions reduces to an ordinary differential equation [265].

If $B^2 - 4AC > 0$, the partial differential equation is said to be *hyperbolic*. A hyperbolic partial differential equation has two real and distinct characteristics, which are often employed to obtain the solution. In this case, a disturbance propagates at finite speed over a finite region, bounded by the two families of characteristics. Hyperbolic partial differential equations typically describe vibrating systems and wave motion. Some examples are given below.

- Wave equation: $\ddot{u} = c^2 \nabla^2 u$

- Wave equation with friction: $\ddot{u} = c^2 \nabla^2 u - h\dot{u}$

- Transmission line equation: $\ddot{u} = c^2 \frac{\partial^2 u}{\partial x^2} - h\dot{u} - ku$

If $B^2 - 4AC = 0$, the partial differential equation is said to be *parabolic*. For parabolic partial differential equations, the two families of characteristics merge, giving rise to an infinite propagation speed and information flow in one direction. The solution at a point thus only depends upon the results obtained along this direction up to the point, and not on results in the region beyond [265]. Parabolic partial differential equations typically describe heat flow and diffusion processes. Some examples are given below.

- Diffusion equation: $\dot{u} = \alpha^2 \nabla^2 u$

- Diffusion equation with source: $\dot{u} = \alpha^2 \nabla^2 u + f(x_1, x_2, x_3, t)$

- One-dimensional diffusion-convection equation:

$$\dot{u} = \alpha^2 \frac{\partial^2 u}{\partial x^2} - k \frac{\partial u}{\partial x}$$

If $B^2 - 4AC < 0$, the partial differential equation is said to be *elliptic*. For ellipticpartial differential equations, complex characteristics are obtained, and, as a result, no directional restrictions arise; the disturbance propagates in all directions. At a given point, the solution is affected by disturbances at every other point in the region. There are no preferred directions, and the solution must be obtained over the entire region simultaneously [265]. Elliptic partial differential equations typically describe steady state phenomena. Some examples are given below.

- Laplace's equation: $\nabla^2 u = 0$

- Poisson's equation: $\nabla^2 u = k$

- Helmholtz's equation: $\nabla^2 u + \lambda^2 u = 0$

- Schrödinger's equation: $\nabla^2 u + k(E - V)u = 0$

A.3 Some Finite Difference Formulas

Consider a function f(x) that is analytic[1] in the neighborhood of a point $x = c$. For convenience and brevity, we adopt the following subscript notation:

$$f(x)|_{x=c} \equiv f_j \quad ; \quad f(x)|_{x=c+h} \equiv f_{j+1} \tag{A.5}$$

---

[1] If a function $f(x)$ possesses a derivative at $x = c$ and at every point in some neighborhood of c, then $f(x)$ is said to be *analytic* at c and c is called a *regular point* of the function [575].

A.3.1   Forward Differences in One Dimension

Using suitable Taylor series expansions, the following forward difference approximations $O(h)$ are derived:

$$f'(x_j) = \frac{f_{j+1} - f_j}{h} \tag{A.6}$$

$$f''(x_j) = \frac{f_{j+2} - 2f_{j+1} + f_j}{h^2} \tag{A.7}$$

$$f'''(x_j) = \frac{f_{j+3} - 3f_{j+2} + 3f_{j+1} - f_j}{h^3} \tag{A.8}$$

$$f^{iv}(x_j) = \frac{f_{j+4} - 4f_{j+3} + 6f_{j+2} - 4f_{j+1} + f_j}{h^4} \tag{A.9}$$

By including more terms in the Taylor series expansion, expressions of greater accuracy may be obtained. These are summarized below for the case of approximations $O(h)^2$

$$f'(x_j) = \frac{-f_{j+2} + 4f_{j+1} - 3f_j}{2h} \tag{A.10}$$

$$f''(x_j) = \frac{-f_{j+3} + 4f_{j+2} - 5f_{j+1} + 2f_j}{h^2} \tag{A.11}$$

$$f'''(x_j) = \frac{-3f_{j+4} + 14f_{j+3} - 24f_{j+2} + 18f_{j+1} - 5f_j}{2h^3} \tag{A.12}$$

$$h^4 f^{iv}(x_j) = -2f_{j+5} + 11f_{j+4} - 24f_{j+3}$$
$$+26f_{j+2} - 14f_{j+1} + 3f_j \tag{A.13}$$

A.3.2   Backward Differences in One Dimension

Some backward difference approximations $O(h)$ are summarized below:

$$f'(x_j) = \frac{f_j - f_{j-1}}{h} \tag{A.14}$$

$$f''(x_j) = \frac{f_j - 2f_{j-1} + f_{j-2}}{h^2} \tag{A.15}$$

$$f'''(x_j) = \frac{f_j - 3f_{j-1} + 3f_{j-2} - f_{j-3}}{h^3} \tag{A.16}$$

$$f^{iv}(x_j) = \frac{f_j - 4f_{j-1} + 6f_{j-2} - 4f_{j-3} + f_{j-4}}{h^4} \tag{A.17}$$

Some higher-order backward difference approximations $O(h)^2$ are summarized below:

$$f'(x_j) = \frac{3f_j - 4f_{j-1} + f_{j-2}}{2h} \tag{A.18}$$

$$f''(x_j) = \frac{2f_j - 5f_{j-1} + 4f_{j-2} - f_{j-3}}{h^2} \tag{A.19}$$

$$f'''(x_j) = \frac{5f_j - 18f_{j-1} + 24f_{j-2} - 14f_{j-3} + 3f_{j-4}}{2h^3} \tag{A.20}$$

$$h^4 f^{iv}(x_j) = 3f_j - 14f_{j-1} + 26f_{j-2}$$
$$-24f_{j-3} + 11f_{j-4} - 2f_{j-5} \tag{A.21}$$

A.3.3 Central Differences in One Dimension

By suitably manipulating forward and backward Taylor series expansions, the following central difference approximations $O(h)^2$ are obtained:

$$f'(x_j) = \frac{f_{j+1} - f_{j-1}}{2h} \tag{A.22}$$

$$f''(x_j) = \frac{f_{j+1} - 2f_j + f_{j-1}}{h^2} \tag{A.23}$$

$$f'''(x_j) = \frac{f_{j+2} - 2f_{j+1} + 2f_{j-1} - f_{j-2}}{2h^3} \tag{A.24}$$

$$f^{iv}(x_j) = \frac{f_{j+2} - 4f_{j+1} + 6f_j - 4f_{j-1} + f_{j-2}}{h^4} \tag{A.25}$$

Some higher-order central difference approximations $O(h)^4$ are summarized below:

$$f'(x_j) = \frac{-f_{j+2} + 8f_{j+1} - 8f_{j-1} + f_{j-2}}{12h} \tag{A.26}$$

$$f''(x_j) = \frac{-f_{j+2} + 16f_{j+1} - 30f_j + 16f_{j-1} - f_{j-2}}{12h^2} \tag{A.27}$$

$$8h^3 f'''(x_j) = -f_{j+3} + 8f_{j+2} - 13f_{j+1}$$
$$+13f_{j-1} - 8f_{j-2} + f_{j-3} \tag{A.28}$$

$$6h^4 f^{iv}(x_j) = -f_{j+3} + 12f_{j+2} - 39f_{j+1}$$
$$+56f_j - 39f_{j-1} + 12f_{j-2} - f_{j-3} \tag{A.29}$$

Further information pertaining to finite differences can be found in standard numerical analysis textbooks such as [166, 249, 265, 396, 452, 481].

Remarks

1. The coefficients associated with any finite difference approximation must add up to zero so that if a function is constant over the interval under consideration, the first (and subsequent) derivative must be exactly zero.

2. Error terms such as $O(h)^2$ mean that halving h reduces the truncation error by one-quarter.

A.4   Self-Adjoint Operators

Consider the general boundary-value problem described by

$$L(\phi) - f = 0 \quad \text{in} \quad \Omega \tag{A.30}$$

and

$$B(\phi) - r = 0 \quad \text{on} \quad \Gamma \tag{A.31}$$

The linear system defined by the operators L and B is said to be *self-adjoint* if, for any two functions ϕ_1 and ϕ_2 (sufficiently differentiable in Ω) that satisfy homogeneous boundary conditions, we have

$$\int_\Omega \phi_1 L(\phi_2) d\Omega = \int_\Omega \phi_2 L(\phi_1) d\Omega \tag{A.32}$$

where the integration is carried out over the domain associated with the problem. The validation of equation (A.32) usually involves integration by parts, or the application of a formalized form of integration by parts such as Green's theorem.

Example 1: Bernoulli-Euler Beam Theory

The governing differential equation for a beam of length h is written in the form

$$\frac{d^2}{dx^2}\left(EI\frac{d^2v}{dx^2}\right) - q = 0 \tag{A.33}$$

where E is the elastic modulus, I is the second moment of area, and $q(x)$ represents an applied distributed load per unit length along the beam. Equation (A.33) implies that, with respect to equation(A.30), $\phi = v$, $L = \frac{d^2}{dx^2}\left(EI\frac{d^2}{dx^2}\right)$, and $f = q(x)$.

Consider two different transverse deflections, v_1 and v_2, both of which satisfy the following *homogeneous* boundary conditions at $x = 0$ and at $x = h$.

$$v\big|_{x=0} = 0 \tag{A.34}$$

$$\frac{dv}{dx}\bigg|_{x=0} = 0 \tag{A.35}$$

$$EI\frac{d^2v}{dx^2}\bigg|_{x=h} = 0 \tag{A.36}$$

$$\frac{d}{dx}\left(EI\frac{d^2v}{dx^2}\right)\bigg|_{x=h} = 0 \tag{A.37}$$

To determine whether the differential operator L is self-adjoint, we must show that

$$\int_0^h v_1 \frac{d^2}{dx^2}\left(EI\frac{d^2v_2}{dx^2}\right)\,dx = \int_0^h v_2 \frac{d^2}{dx^2}\left(EI\frac{d^2v_1}{dx^2}\right)\,dx \tag{A.38}$$

We begin with the first quantity in equation (A.38). Integration by parts gives

$$\int_0^h v_1 \frac{d^2}{dx^2}\left(EI\frac{d^2v_2}{dx^2}\right)\,dx$$

$$= \left[v_1 \frac{d}{dx}\left(EI\frac{d^2v_2}{dx^2}\right)\right]_0^h - \int_0^h \frac{dv_1}{dx}\frac{d}{dx}\left(EI\frac{d^2v_2}{dx^2}\right)\,dx \tag{A.39}$$

From equations (A.34) and (A.37) it follows that the first term on the right hand side will drop out. Integrating by parts a second time gives

$$\int_0^h v_1 \frac{d^2}{dx^2}\left(EI\frac{d^2 v_2}{dx^2}\right)\ dx$$

$$= -\left[\frac{dv_1}{dx}\left(EI\frac{d^2 v_2}{dx^2}\right)\right]_0^h + \int_0^h \frac{d^2 v_1}{dx^2}\left(EI\frac{d^2 v_2}{dx^2}\right)\ dx \qquad (A.40)$$

From equations (A.35) and (A.36) it follows that the first term on the right hand side will once again drop out. Integrating by parts a third time gives

$$\int_0^h v_1 \frac{d^2}{dx^2}\left(EI\frac{d^2 v_2}{dx^2}\right)\ dx$$

$$= \left[EI\frac{d^2 v_1}{dx^2}\left(\frac{dv_2}{dx}\right)\right]_0^h - \int_0^h \frac{d}{dx}\left(EI\frac{d^2 v_1}{dx^2}\right)\frac{dv_2}{dx}\ dx \qquad (A.41)$$

From equations (A.35) and (A.36) it follows that the first term on the right hand side will once again drop out. Integrating by parts one last time gives

$$\int_0^h v_1 \frac{d^2}{dx^2}\left(EI\frac{d^2 v_2}{dx^2}\right)\ dx$$

$$= -\left[v_2 \frac{d}{dx}\left(EI\frac{d^2 v_1}{dx^2}\right)\right]_0^h + \int_0^h v_2 \frac{d^2}{dx^2}\left(EI\frac{d^2 v_1}{dx^2}\right)\ dx \qquad (A.42)$$

From equations (A.34) and (A.37) it follows that the first term on the right hand side will again drop out, leaving equation(A.38). The operator L is thus self-adjoint.

Appendix B

Some Notes on Heat Flow

B.1 Introductory Remarks

This appendix contains a brief overview of heat flow in materials. For further details, the interested reader is directed to books on the subject such as those by Carslaw and Jaeger [105], Özişik [425] and White [558], to name a few.

B.2 Fourier's Law of Heat Conduction

The flow of heat in a material was quantified by Fourier,[1] who noted that the amount of heat to be conducted was proportional to the area normal to the flow, the temperature difference per length parallel to the direction of flow, and the duration for which the temperature difference was maintained [169]. The constant of proportionality is a property of the material.

Assuming the temperature varies only in the x_1-direction, Fourier's heat conduction law is written as

$$\tilde{q}_1 = \frac{dQ}{dt} = -k_{11} A_1 \frac{dT}{dx_1} \tag{B.1}$$

where Q is the heat (units of E), $\tilde{q}_1$ is the heat transfer rate or "heat flux" (units of Et^{-1}), k_{11} is a material transport property called the *thermal conductivity* (units of $Et^{-1}L^{-1}T^{-1}$), A_1 is the cross-sectional area normal to the direction in which heat flows (units of L^2) and T is the temperature (units of T).

---

[1] Jean Baptiste Joseph Fourier (1768-1830) was a French mathematical physicist.

Fourier's Law is valid for all common solids, liquids and gases. Equation (B.1) is also written in the form

$$q_1 = \frac{\tilde{q}_1}{A_1} = -k_{11} \frac{dT}{dx_1} \tag{B.2}$$

where q_1 is the heat flux per unit area.

Generalizing equation (B.1) for a three-dimensional anisotropic continuum, the relation between the heat flux per unit area vector and the temperature gradient is

$$\begin{Bmatrix} q_1 \\ q_2 \\ q_3 \end{Bmatrix} = - \begin{bmatrix} k_{11} & k_{12} & k_{13} \\ k_{21} & k_{22} & k_{23} \\ k_{31} & k_{32} & k_{33} \end{bmatrix} \begin{Bmatrix} \partial T/dx_1 \\ \partial T/dx_2 \\ \partial T/dx_3 \end{Bmatrix} \tag{B.3}$$

or

$$\mathbf{q} = -\mathbf{K}\nabla T \tag{B.4}$$

where

$$\nabla = \left\{ \frac{\partial}{\partial x_1} \quad \frac{\partial}{\partial x_2} \quad \frac{\partial}{\partial x_3} \right\}^T \tag{B.5}$$

is the gradient operator.

Equation (B.3) indicates that in anisotropic materials a given heat flux component is linearly dependent upon a combination of the temperature gradients in that direction and in the other two orthogonal directions.

The conductivity coefficients can be shown to obey the reciprocity relation $k_{ij} = k_{ji}$, implying that the matrix of thermal conductivities is *symmetric* [425]. Furthermore, according to irreversible thermodynamics [446], the diagonal thermal conductivities k_{11}, k_{22} and k_{33} are *positive*. Finally, the magnitude of the off-diagonal conductivities is limited by the relation $k_{ii}k_{jj} - k_{ij}k_{ij} > 0$; $i \neq j$, where no summation over repeated indices is implied.

B.3   Heat Conduction Equations

In the conduction of heat, work effects are negligible [558] and, as a result, the first law of thermodynamics reduces to

$$\frac{dQ}{dt} = \frac{dE}{dt} \tag{B.6}$$

where Q and E denote the heat and energy associated with the body. In words, the net heat conducted into the body is equal to the energy increase of the body.

B.3.1   Three–Dimensional Heat Conduction

Consider the infinitesimal element of an anisotropic material shown in Figure B.1. The net heat conducted into the material element is

$$d\tilde{q}_{in} = \tilde{q}_1 + (-\tilde{q}_1^+) + \tilde{q}_2 + (-\tilde{q}_2^+) + \tilde{q}_3 + (-\tilde{q}_3^+) \tag{B.7}$$

Assuming a continuous material,[2] it follows that all properties will vary smoothly enough to use the calculus. The outgoing flux terms are expanded in a first-order Taylor series, viz.,

$$\tilde{q}_1^+ = \tilde{q}_1 + \frac{\partial \tilde{q}_1}{\partial x_1} \Delta x_1 \tag{B.8}$$

$$\tilde{q}_2^+ = \tilde{q}_2 + \frac{\partial \tilde{q}_2}{\partial x_2} \Delta x_2 \tag{B.9}$$

$$\tilde{q}_3^+ = \tilde{q}_3 + \frac{\partial \tilde{q}_3}{\partial x_3} \Delta x_3 \tag{B.10}$$

Equation (B.7) thus reduces to

$$d\tilde{q}_{in} = -\frac{\partial \tilde{q}_1}{\partial x_1} \Delta x_1 - \frac{\partial \tilde{q}_2}{\partial x_2} \Delta x_2 - \frac{\partial \tilde{q}_3}{\partial x_3} \Delta x_3 \tag{B.11}$$

---

[2]The fundamental assumption of a continuum implies that the material contains no voids or defects; consequently, all physical quantities must be described by a unique continuous function. For a solid with or without heat sources (sinks) and subject to suitable thermal boundary conditions, at any given location and time in the solid there is thus a unique temperature.

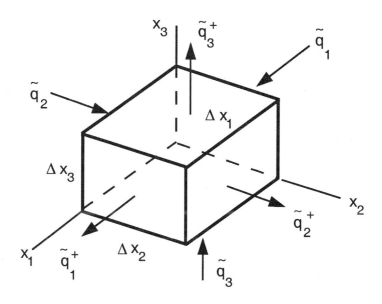

Figure B.1: Illustration of Heat Flow in a Three-Dimensional Body

From generalized Fourier's Law, equation (B.3), it follows that

$$\tilde{q}_1 = -\left(k_{11}\frac{\partial T}{\partial x_1} + k_{12}\frac{\partial T}{\partial x_2} + k_{13}\frac{\partial T}{\partial x_3} \right) \Delta x_2 \Delta x_3 \qquad \text{(B.12)}$$

$$\tilde{q}_2 = -\left(k_{12}\frac{\partial T}{\partial x_1} + k_{22}\frac{\partial T}{\partial x_2} + k_{23}\frac{\partial T}{\partial x_3} \right) \Delta x_1 \Delta x_3 \qquad \text{(B.13)}$$

$$\tilde{q}_3 = -\left(k_{13}\frac{\partial T}{\partial x_1} + k_{23}\frac{\partial T}{\partial x_2} + k_{33}\frac{\partial T}{\partial x_3} \right) \Delta x_1 \Delta x_2 \qquad \text{(B.14)}$$

where $k_{ij} = k_{ji}$. Substituting equations (B.12) to (B.14) into equation (B.11) gives

$$d\tilde{q}_{in} = \frac{\partial}{\partial x_1}\left(k_{11}\frac{\partial T}{\partial x_1} + k_{12}\frac{\partial T}{\partial x_2} + k_{13}\frac{\partial T}{\partial x_3}\right)\Delta x_1 \Delta x_2 \Delta x_3$$

$$+\frac{\partial}{\partial x_2}\left(k_{12}\frac{\partial T}{\partial x_1} + k_{22}\frac{\partial T}{\partial x_2} + k_{23}\frac{\partial T}{\partial x_3}\right)\Delta x_1 \Delta x_2 \Delta x_3$$

$$+\frac{\partial}{\partial x_3}\left(k_{13}\frac{\partial T}{\partial x_1} + k_{23}\frac{\partial T}{\partial x_2} + k_{33}\frac{\partial T}{\partial x_3}\right)\Delta x_1 \Delta x_2 \Delta x_3 \qquad \text{(B.15)}$$

The net heat conduction is thus seen to be proportional to the elemental volume.

The heat generated within the element is simply $S\Delta x_1 \Delta x_2 \Delta x_3$, where $S = S(x_1, x_2, x_3)$ represents the heat generated per unit volume (units of $Et^{-1}L^{-3}$). For example, S accounts for internal heat production due to such causes as dielectric or induction heating, radioactive decay, absorption from radiation, chemical reactions, etc.

The increase in element energy is given by $\rho(\Delta x_1 \Delta x_2 \Delta x_3)c_v\,(\partial T/\partial t)$, where ρ is the mass density of the material (units of ML^{-3}), c_v is the specific heat at constant volume (units of $EM^{-1}T^{-1}$), and t represents time.

Substituting equation (B.15) and the above expressions for heat generated and increase in energy into equation (B.6), and canceling the term $(\Delta x_1 \Delta x_2 \Delta x_3)$ gives

$$\frac{\partial}{\partial x_1}\left(k_{11}\frac{\partial T}{\partial x_1} + k_{12}\frac{\partial T}{\partial x_2} + k_{13}\frac{\partial T}{\partial x_3}\right)$$

$$+\frac{\partial}{\partial x_2}\left(k_{12}\frac{\partial T}{\partial x_1} + k_{22}\frac{\partial T}{\partial x_2} + k_{23}\frac{\partial T}{\partial x_3}\right)$$

$$+\frac{\partial}{\partial x_3}\left(k_{13}\frac{\partial T}{\partial x_1} + k_{23}\frac{\partial T}{\partial x_2} + k_{33}\frac{\partial T}{\partial x_3}\right) + S = \rho c_v \frac{\partial T}{\partial t} \qquad \text{(B.16)}$$

In condensed matrix form equation (B.16) becomes

$$\nabla^T\left(\mathbf{K}\nabla T\right) + S = \rho c_v \frac{\partial T}{\partial t} \qquad \text{(B.17)}$$

where the gradient operator has been defined in equation (B.5), and $\mathbf{K}$ is the matrix of thermal conductivity coefficients defined in equation (B.4).

Equation (B.16) is the governing differential equation for heat conduction in an anisotropic solid continuum. If any or all of the physical properties ρ, c_v and k_{ij} are functions of temperature, the problem becomes *non-linear*.

For an *isotropic* material, $k_{11} = k_{22} = k_{33} = k$, and all the remaining (off-diagonal) conductivity coefficients are zero. Equation (B.16) thus simplifies to

$$\frac{\partial}{\partial x_1}\left(k\frac{\partial T}{\partial x_1}\right) + \frac{\partial}{\partial x_2}\left(k\frac{\partial T}{\partial x_2}\right) + \frac{\partial}{\partial x_3}\left(k\frac{\partial T}{\partial x_3}\right) + S = \rho c_v \frac{\partial T}{\partial t} \qquad \text{(B.18)}$$

Furthermore, if k is constant, equation (B.18) further simplifies to

$$\alpha \nabla^2 T + \frac{S}{\rho c_v} = \frac{\partial T}{\partial t} \qquad \text{(B.19)}$$

where $\alpha = k/(\rho c_v)$ represents the *thermal diffusivity* of the material. Finally, for steady state heat conduction, the right hand side of equation (B.18) and equation (B.19) is set equal to zero.

B.3.2   Two-Dimensional Heat Conduction

Consider a body of unit thickness that is insulated along its upper and lower surfaces (i.e., parallel to the plane of the paper), or a body that is "infinitely" long in the thickness direction. In either case, a uniform temperature through the thickness is assumed, implying that no heat flows in this direction. In general, the material is assumed to be inhomogeneous and anisotropic in the plane of heat flow (taken here to be the $x_1 - x_2$ plane). It is evident that the three-dimensional heat conduction problem now reduces to two-dimensional one. The heat flux vector is thus written as

$$\begin{Bmatrix} q_1 \\ q_2 \end{Bmatrix} = - \begin{bmatrix} k_{11} & k_{12} \\ k_{12} & k_{22} \end{bmatrix} \begin{Bmatrix} \partial T/\partial x_1 \\ \partial T/\partial x_2 \end{Bmatrix} \qquad \text{(B.20)}$$

where $k_{ij} = k_{ij}(x_1, x_2)$. Equation (B.16) therefore simplifies to

$$\frac{\partial}{\partial x_1}\left(k_{11}\frac{\partial T}{\partial x_1} + k_{12}\frac{\partial T}{\partial x_2}\right) + \frac{\partial}{\partial x_2}\left(k_{12}\frac{\partial T}{\partial x_1} + k_{22}\frac{\partial T}{\partial x_2}\right)$$
$$+ S(x_1, x_2) = \rho c_v \frac{\partial T}{\partial t} \qquad \text{(B.21)}$$

If the material is *isotropic*, $k_{12} = k_{21} = 0$, $k_{11} = k_{22} = k$, and equation (B.21) reduces to

$$k\nabla^2 T + S(x_1, x_2) = \rho c_v \frac{\partial T}{\partial t} \tag{B.22}$$

where ∇ is now the gradient operator in two dimensions.

Finally, for *steady state* heat conduction, the right hand side of equation (B.21) or (B.22) is set equal to zero.

B.3.3 One-Dimensional Heat Conduction

Consider the insulated rod shown below in Figure B.2. As evident from this figure, heat can only flow in a direction parallel to the x_1-axis.

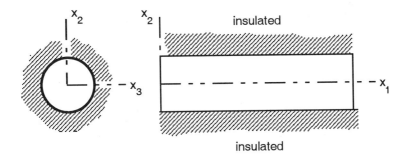

Figure B.2: Schematic Illustration of an Insulated Rod

In this case the general heat conduction problem reduces to a one-dimensional one. Equation (B.16) thus simplifies to

$$\frac{d}{dx_1}\left(k_{11}\frac{dT}{dx_1}\right) + S(x_1) = \rho c_v \frac{dT}{dt} \tag{B.23}$$

If the material has a *constant* conductivity coefficient, equation (B.23) reduces to

$$k_{11}\frac{d^2 T}{dx_1^2} + S(x_1) = \rho c_v \frac{dT}{dt} \tag{B.24}$$

Finally, for the case of steady state heat conduction, the right-hand side of equation (B.23) or (B.24) is set equal to zero.

B.4   Boundary and Initial Conditions

Consider a typical domain Ω bounded by a closed curve Γ in the manner shown in Figure B.3.

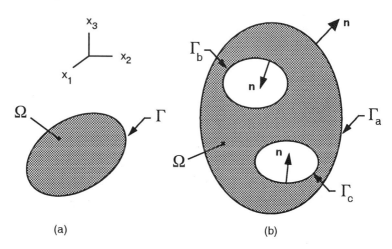

(a)   (b)

Figure B.3: Schematic Illustration of a Typical Solution Domain and its Boundary: (a) Simply Connected, (b) Multiply Connected ($\Gamma = \Gamma_a + \Gamma_b + \Gamma_c$)

The governing differential equation (B.16) describing heat conduction in such a continuum is first-order in time and second-order in space. Thus, in general it requires the specification of an *initial condition* and appropriate *boundary conditions*.

If the problem is time dependent, the entire temperature field (the primary dependent variable) must be specified as an initial condition at some particular time, viz.,

$$T = \bar{T}_o(x_1, x_2, x_3) \quad \text{for } t = t_0, \quad \forall x \in \Omega \qquad \text{(B.25)}$$

The most common case is that of a uniform temperature $T = \bar{T}_o$ everywhere at $t = 0$.

Regardless of whether or not the problem is time dependent, appropriate boundary conditions must also be specified at every point along Γ for all times. Three types of boundary conditions are possible, namely:

1. Prescribed temperature along the boundary (*essential* or *Dirichlet* boundary condition):

$$T = \bar{T} \quad \text{on} \quad \Gamma_1 \qquad (B.26)$$

where the bar indicates a known value of temperature.

2. Prescribed boundary heat flux (*natural* or *Neumann* boundary condition):

$$q_n = -k\frac{\partial T}{\partial n} = \bar{q} \quad \text{on} \quad \Gamma_2 \qquad (B.27)$$

For insulated boundaries, $\bar{q} = 0$.

3. Prescribed convection to surroundings (*mixed* or *Robin* boundary condition):

$$q_n = -k\frac{\partial T}{\partial n} = \beta(T_b - T_\infty) \quad \text{on} \quad \Gamma_3 \qquad (B.28)$$

In equations (B.27) and (B.28) n denotes the outward normal coordinate from the boundary, k denotes the value of an appropriate conductivity coefficient at boundary point, and β represents the heat transfer coefficient. Recall that $\Gamma = \Gamma_1 \cup \Gamma_2 \cup \Gamma_3$ and $\Gamma_1 \cap \Gamma_2 = \Gamma_1 \cap \Gamma_3 = \Gamma_2 \cap \Gamma_3 = \emptyset$.

Appendix C

Local and Natural Coordinate Systems

C.1 Introductory Remarks

Although finite element equations could be developed in terms of *global* coordinates, it is typically more economical to carry out the associated computations using coordinates that rely on the element geometry for their definition. The notion of such coordinate systems may appear to be a formality, introduced to emphasize the fact that individual finite elements are to be considered disjoint for the purpose of developing local approximations $\hat{\phi}^{(e)}$. However, all finite element equations involve integration, and many of these integrals cannot be evaluated analytically. The difficulties associated with evaluating integrals numerically can often be lessened by writing the integral in a new coordinate system. The objective of this appendix is to discuss several *local* and *natural* coordinate systems that facilitate the numerical integrations associated with finite element equations.

C.2 Local Coordinate Systems

Consider a typical one-dimensional element such as that shown in Figure C.1(a). The number of nodes in the element is immaterial. The global y-coordinate of node i is denoted by y_i, and $h^{(e)}$ is the element length. The global coordinate of any point in the element is given by

$$y = y_i + \bar{y} \qquad (\text{C.1})$$

where $\bar{y}$ represents the *local* coordinate. It is evident that the definition of the local coordinate is dependent on the geometry of the specific element under consideration.

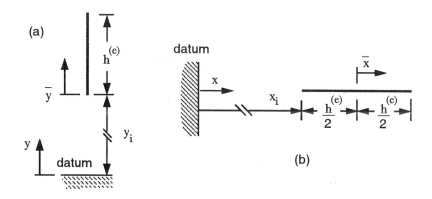

Figure C.1: Local Coordinate System : (a) Vertical Orientation, (b) Horizontal Orientation

Another local coordinate system is shown in Figure C.1(b). Here the global coordinate of any point in the element is given by

$$x = x_i + \frac{h^{(e)}}{2} + \bar{x} \qquad (C.2)$$

The local coordinate system is thus defined by

$$\bar{x} = x - x_i - \frac{h^{(e)}}{2} \qquad (C.3)$$

Often it is preferable to non-dimensionalize local coordinates. This leads to the topic of *natural* coordinate systems.

C.3 Natural Coordinate Systems

A natural coordinate system is a local system that specifies a point within the element by a dimensionless number whose magnitude typically does not exceed unity [488]. Very often the specific limits of a natural coordinate system are chosen so as to facilitate the application of numerical integration formulas.

C.3.1 One-Dimensional Natural Coordinate Systems

As a first example of a natural coordinate system in one dimension, consider the vertically oriented local coordinate system shown in Figure C.1(a). This system is non-dimensionalized by dividing by the length of the element, viz.,

$$\lambda = \frac{\bar{y}}{h^{(e)}} = \frac{y - y_i}{y_j - y_i} \tag{C.4}$$

where λ represents the natural coordinate, which varies from $\lambda(y = y_i) = 0$ to $\lambda(y = y_j) = 1$.

As a second example of a natural coordinate system, consider the horizontally oriented one-dimensional local coordinate system shown in Figure C.1(b). Denote the coordinate of the element mid-point by x_c. The natural coordinate system is then defined by

$$\xi = \frac{2(x - x_c)}{x_j - x_i} = \frac{2\left[x - 0.5(x_i + x_j)\right]}{h^{(e)}} \quad ; \quad -1 \le \xi \le 1 \tag{C.5}$$

It follows that at $x = x_i$:

$$\xi(x_i) = \frac{2(x_i - x_c)}{x_j - x_i} = \frac{2\left[x_i - 0.5(x_i + x_j)\right]}{h^{(e)}} = \frac{2\left[-0.5h^{(e)}\right]}{h^{(e)}} = -1 \tag{C.6}$$

at $x = x_c$:

$$\xi(x_c) = \frac{2(x_c - x_c)}{x_j - x_i} = 0 \tag{C.7}$$

and at $x = x_j$:

$$\xi(x_j) = \frac{2(x_j - x_c)}{x_j - x_i} = \frac{2\left[x_j - 0.5(x_i + x_j)\right]}{h^{(e)}} = \frac{2\left[0.5h^{(e)}\right]}{h^{(e)}} = +1 \tag{C.8}$$

The above results can be extended to the case of x_c located at any point within the element.

C.3.2   Natural Coordinates for Triangular Elements

The location in a triangular finite element of a point p, having global coordinates (x_p, y_p), is realized by defining the three natural coordinates[1] L_1, L_2, and L_3 shown in Figure C.2(a). Each of the L_i is equal to the ratio of a perpendicular distance s_i from one side, to the altitude h_i associated with the same side. In Figure C.2(b) this definition is schematically shown for the case of L_1.

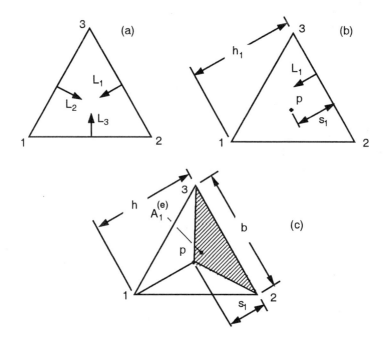

Figure C.2: (a) Natural Coordinates for a Triangular Element, (b) Definition of Natural Coordinate L_1, (c) Subarea of a Triangular Element

It follows that each of the L_i varies between unity (at vertex i) and zero (along the line connecting the other two vertices). The lines of constant

---

[1] The terminology "natural coordinates" appears to have been first used in the context of triangular elements by Argyris [23].

L_i are shown in Figure C.3. Each of these lines is parallel to the side from which the particular L_i is measured.

To find explicit expressions for the L_i note that the original Cartesian coordinates of any point within or on the triangular element should be linearly related to the natural coordinates, viz.,

$$x = L_1 x_1 + L_2 x_2 + L_3 x_3 \tag{C.9}$$

$$y = L_1 y_1 + L_2 y_2 + L_3 y_3 \tag{C.10}$$

where (x_i, y_i) are the coordinates of vertex i ($i = 1, 2, 3$) in the triangle. In addition, at any point in the triangle the L_i must also sum to one, viz.,

$$L_1 + L_2 + L_3 = 1 \tag{C.11}$$

implying that only two of the natural coordinates are independent.

Equations (C.9) to (C.11) are next written in matrix form:

$$\begin{Bmatrix} 1 \\ x \\ y \end{Bmatrix} = \begin{bmatrix} 1 & 1 & 1 \\ x_1 & x_2 & x_3 \\ y_1 & y_2 & y_3 \end{bmatrix} \begin{Bmatrix} L_1 \\ L_2 \\ L_3 \end{Bmatrix} \tag{C.12}$$

Inverting equation (C.12) gives the following expressions for the natural coordinates:

$$\begin{Bmatrix} L_1 \\ L_2 \\ L_3 \end{Bmatrix} = \frac{1}{2A^{(e)}} \begin{bmatrix} a_1 & a_2 & a_3 \\ b_1 & b_2 & b_3 \\ c_1 & c_2 & c_3 \end{bmatrix} \begin{Bmatrix} 1 \\ x \\ y \end{Bmatrix} \tag{C.13}$$

where

$$a_1 = x_2 y_3 - x_3 y_2 \quad , \quad b_1 = y_2 - y_3 \quad , \quad c_1 = x_3 - x_2 \tag{C.14}$$
$$a_2 = x_3 y_1 - x_1 y_3 \quad , \quad b_2 = y_3 - y_1 \quad , \quad c_2 = x_1 - x_3 \tag{C.15}$$
$$a_3 = x_1 y_2 - x_2 y_1 \quad , \quad b_3 = y_1 - y_2 \quad , \quad c_3 = x_2 - x_1 \tag{C.16}$$

The area of the element is equal to the determinant of the coefficient matrix in equation (C.12), viz.,

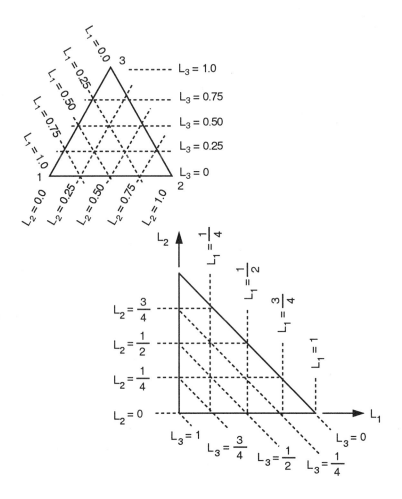

Figure C.3: Lines of Constant L_1, L_2, L_3 : Two Commonly Used Representations

$$A^{(e)} = \frac{1}{2} \begin{vmatrix} 1 & 1 & 1 \\ x_1 & x_2 & x_3 \\ y_1 & y_2 & y_3 \end{vmatrix} = \frac{1}{2}(a_1 + a_2 + a_3) \qquad \text{(C.17)}$$

The natural coordinates can also be derived in an *alternate* manner. From Figure C.2(c) the ratio of the shaded subregion to the area of the entire triangle is

$$\frac{A_1^{(e)}}{A^{(e)}} = \frac{\frac{1}{2}bs_1}{\frac{1}{2}bh} = \frac{s_1}{h} \qquad \text{(C.18)}$$

which is, by definition, equal to L_1 (see Figure C.2b). Similarly,

$$L_2 = \frac{A_2^{(e)}}{A^{(e)}} \quad \text{and} \quad L_3 = \frac{A_3^{(e)}}{A^{(e)}} \qquad \text{(C.19)}$$

The fact that $A_1^{(e)} + A_2^{(e)} + A_3^{(e)} = A^{(e)}$ confirms equation (C.11). In light of equations (C.18) and (C.19), the L_i are sometimes referred to as the *area coordinates*[2] associated with triangular elements.

C.3.3 Natural Coordinates for Tetrahedral Elements

The derivation of natural coordinates for tetrahedral elements follows very closely to that associated with triangular elements. Consider a typical tetrahedral element such as that shown in Figure C.4.

Define the natural coordinates L_1, L_2, L_3, and L_4. To find explicit expressions for the L_i note that the original Cartesian coordinates of any point within or on the triangular element should be linearly related to the natural coordinates, viz.,

$$x = L_1 x_1 + L_2 x_2 + L_3 x_3 + L_4 x_4 \qquad \text{(C.20)}$$
$$y = L_1 y_1 + L_2 y_2 + L_3 y_3 + L_4 y_4 \qquad \text{(C.21)}$$
$$z = L_1 z_1 + L_2 z_2 + L_3 z_3 + L_4 z_4 \qquad \text{(C.22)}$$

where (x_i, y_i, z_i) are the coordinates of vertex i ($i = 1, 2, 3, 4$) in the tetrahedron. In addition, we note that

---

[2] This terminology appears to have first been used by Zienkiewicz and Cheung [587].

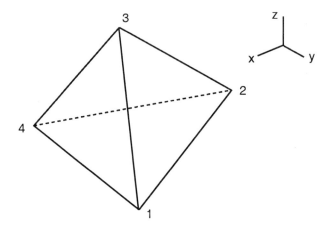

Figure C.4: Typical Tetrahedral Element

$$L_1 + L_2 + L_3 + L_4 = 1 \tag{C.23}$$

implying that only three of the natural coordinates are independent.
Writing these equations in matrix form gives

$$\begin{Bmatrix} 1 \\ x \\ y \\ z \end{Bmatrix} = \begin{bmatrix} 1 & 1 & 1 & 1 \\ x_1 & x_2 & x_3 & x_4 \\ y_1 & y_2 & y_3 & y_4 \\ z_1 & z_2 & z_3 & z_4 \end{bmatrix} \begin{Bmatrix} L_1 \\ L_2 \\ L_3 \\ L_4 \end{Bmatrix} \tag{C.24}$$

Inverting the coefficient matrix leads to the desired natural coordinates

$$\begin{Bmatrix} L_1 \\ L_2 \\ L_3 \\ L_4 \end{Bmatrix} = \frac{1}{6V^{(e)}} \begin{bmatrix} M_{11} & -M_{21} & M_{31} & -M_{41} \\ -M_{12} & M_{22} & -M_{32} & M_{42} \\ M_{13} & -M_{23} & M_{33} & -M_{43} \\ -M_{14} & M_{24} & -M_{34} & M_{44} \end{bmatrix} \begin{Bmatrix} 1 \\ x \\ y \\ z \end{Bmatrix} \tag{C.25}$$

where $V^{(e)}$ represents the volume of the tetrahedral element, which is simply the determinant of the coefficient matrix in equation (C.24), viz.,

$$V^{(e)} = \frac{1}{6} \begin{vmatrix} 1 & 1 & 1 & 1 \\ x_1 & x_2 & x_3 & x_4 \\ y_1 & y_2 & y_3 & y_4 \\ z_1 & z_2 & z_3 & z_4 \end{vmatrix} \tag{C.26}$$

and the M_{ij} denote minors of the coefficient matrix, viz.,

$$M_{11} = \begin{vmatrix} x_2 & x_3 & x_4 \\ y_2 & y_3 & y_4 \\ z_2 & z_3 & z_4 \end{vmatrix} , \quad M_{12} = \begin{vmatrix} x_1 & x_3 & x_4 \\ y_1 & y_3 & y_4 \\ z_1 & z_3 & z_4 \end{vmatrix} , \quad \cdots \tag{C.27}$$

In light of equations (C.27), a row expansion of equation (C.26) gives

$$\begin{aligned} 6V^{(e)} &= M_{11} - M_{12} + M_{13} - M_{14} \\ &= [x_2 (y_3 z_4 - y_4 z_3) + x_3 (y_4 z_2 - y_2 z_4) + x_4 (y_2 z_3 - y_3 z_2)] \\ &\quad - [x_1 (y_3 z_4 - y_4 z_3) + x_3 (y_4 z_1 - y_1 z_4) + x_4 (y_1 z_3 - y_3 z_1)] \\ &\quad + [x_1 (y_2 z_4 - y_4 z_2) + x_2 (y_4 z_1 - y_1 z_4) + x_4 (y_1 z_2 - y_2 z_1)] \\ &\quad - [x_1 (y_2 z_3 - y_3 z_2) + x_2 (y_3 z_1 - y_1 z_3) + x_3 (y_1 z_2 - y_2 z_1)] \end{aligned} \tag{C.28}$$

To cast the natural coordinates into a general form similar to that used in conjunction with triangular elements, define

$$a_1 = M_{11} \; ; \; a_2 = -M_{12} \; ; \; a_3 = M_{13} \; ; \; a_4 = -M_{14} \tag{C.29}$$
$$b_1 = -M_{21} \; ; \; b_2 = M_{22} \; ; \; b_3 = -M_{23} \; ; \; b_4 = M_{24} \tag{C.30}$$
$$c_1 = M_{31} \; ; \; c_2 = -M_{32} \; ; \; c_3 = M_{33} \; ; \; c_4 = -M_{34} \tag{C.31}$$
$$d_1 = -M_{41} \; ; \; d_2 = M_{42} \; ; \; d_3 = -M_{43} \; ; \; d_4 = M_{44} \tag{C.32}$$

The natural coordinates, which are sometimes also referred to as *volume coordinates*, are then written as

$$L_i = \frac{1}{6V^{(e)}} (a_i + b_i x + c_i y + d_i z) \quad ; \quad i = 1, 2, 3, 4 \tag{C.33}$$

where

$$6V^{(e)} = a_1 + a_2 + a_3 + a_4 \tag{C.34}$$

Remark

   1. *General* equations for natural coordinates associated with linear, triangular, tetrahedral or any N-dimensional simplex element have been derived by Fried [177].

C.3.4   Natural Coordinates for Rectangular Elements

Consider the rectangular element shown in Figure C.5. The global coordinates of the centroid of the element area are denoted by (x_c, y_c).

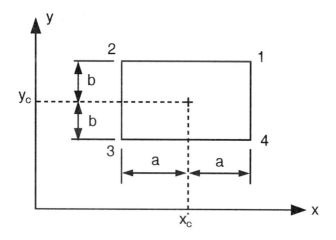

Figure C.5: Typical Rectangular Element

   The natural coordinates parallel to the global x and y-axes are defined in the following manner:

$$\xi = \frac{x - x_c}{a} \tag{C.35}$$

$$\eta = \frac{y - y_c}{b} \tag{C.36}$$

where $-1 \leq \xi, \eta \leq 1$. The first use of such coordinates for the more general case of quadrilateral elements is attributed to Taig and Kerr [527].

C.3.5 Natural Coordinates for Rectangular Prisms

The natural coordinates for rectangular prisms are a direct extension of those developed for rectangular elements. Consider the rectangular prism shown in Figure C.6.

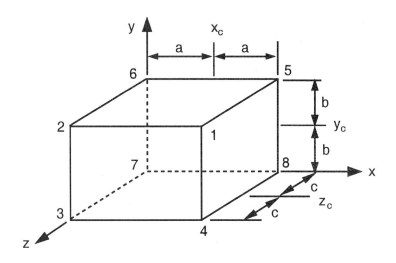

Figure C.6: Typical Rectangular Prism Element

Denoting the coordinates of the centroid of the element volume by (x_c, y_c, z_c), and letting the length of the element parallel to the global z-axis be $2c$, the natural coordinates parallel to the global x, y and z-axes will be

$$\xi = \frac{x - x_c}{a} \tag{C.37}$$

$$\eta = \frac{y - y_c}{b} \tag{C.38}$$

$$\zeta = \frac{z - z_c}{c} \tag{C.39}$$

respectively.

C.4 Exercises

C.1

Consider the three-node quadratic one-dimensional element shown in Figure C.7.

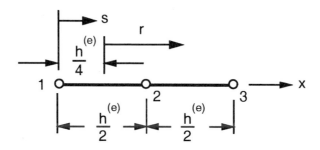

Figure C.7: Quadratic Element with Two Natural Coordinate Systems

The element interpolation functions relative to the s local coordinate system are

$$N_1 = \left(1 - \frac{2s}{h^{(e)}}\right)\left(1 - \frac{s}{h^{(e)}}\right)$$

$$N_2 = \frac{4s}{h^{(e)}}\left(1 - \frac{s}{h^{(e)}}\right)$$

$$N_3 = \frac{2s}{(h^{(e)})^2}\left(s - \frac{h^{(e)}}{2}\right)$$

Write these interpolation functions in terms of the r local coordinate system, where

$$-\frac{h^{(e)}}{4} \leq r \leq \frac{3h^{(e)}}{4}$$

Verify that the interpolation functions satisfy the basic property that $N_i(r_j) = \delta_{ij}$; $i, j = 1, 2, 3$, where δ_{ij} is the Kronecker delta.

C.2

Revisit the one-dimensional linear (simplex) element of Example 7.6, which is redrawn in Figure C.8.

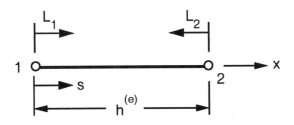

Figure C.8: One-Dimensional Simplex Element

Denoting by x the global longitudinal element axis, the interpolation functions were shown to be

$$N_1 = \frac{x_2 - x}{h^{(e)}} \quad , \quad N_2 = \frac{x - x_1}{h^{(e)}}$$

- First rewrite N_1 and N_2 in terms of the s coordinate system, where $0 \le s \le h^{(e)}$.

- Next, noting that the natural coordinates L_1 and L_2 are defined as the length ratios

$$L_1 = \frac{h^{(e)} - s}{h^{(e)}} \quad \text{and} \quad L_2 = \frac{s}{h^{(e)}}$$

where $L_1 + L_2 = 1$, rewrite the element interpolation functions in terms of L_1 and L_2.

Appendix D

The Patch Test

D.1 Introductory Remarks

In light of the discussion of conforming elements in Chapter 7, it is important to note that: (1) In some instances it is difficult to find interpolation functions that will automatically be continuous across Γ^e and, (2) The performance of conforming elements in the analysis of certain problems (e.g., the simulation of beam bending) is less than satisfactory; this in turn leads to the use of "non-conforming" elements that violate certain continuity requirements across Γ^e. However, if, in the limit as the size of the elements decreases, continuity across Γ^e is restored, the results will still tend towards the correct answer and convergence will be realized [605].

A rather simple test of the successful implementation and convergence of elements is the so-called "patch test." The development of the patch test is attributed to Irons and his coworkers [60, 255, 261], with subsequent contributions by others [19, 533].

The patch test involves the appropriate specification, on a rather arbitrary "patch" of elements, of nodal values of the primary and secondary dependent variables, and allows the analyst to assess whether an element satisfies the completeness criterion.[1]

Since the original presentation of the patch test lacked mathematical rigor, it is sometimes referred to as the "engineering version" of the test [252]. Subsequently, a mathematical basis for the patch test has been

---

[1] Concerning axisymmetric idealizations, the presence of terms containing $1/r$ in the element equations (via **B**) will prevent exact results from being obtained in a simple patch test. Thus, a uniform radial traction on the outer surface of a solid cylinder cannot be expected to yield the correct constant strain state. However, as the mesh is refined, the exact results will be approached.

established [513, 515], the test was shown to be contained in the variational formulations of the finite element method at the assembly level [172], and a fair amount of discussion has been generated regarding the necessity and sufficiency of this test [520, 262, 533] and its extensions [598, 604]. As a result, passing the the patch test is known to give a *sufficient* and *necessary* condition for convergence [533, 605].

The two guidelines for constructing a multi-element patch are:

1. The patch should contain *at least one node that does not lie along a boundary.* Some examples of patches of linear elements are shown below in Figure D.1. The extension to three-dimensional elements such as tetrahedra and hexahedra is straightforward.

2. Since some elements pass the patch test in certain special configurations but not in others, it is important to use an *irregular geometry* in constructing a patch.

A complete application of the patch test involves *three* versions of the test [605]. These are summarized below.

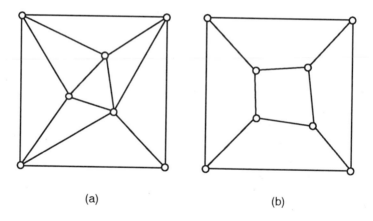

(a)   (b)

Figure D.1: Sample Patch of Linear Elements: (a) Triangular and (b) Quadrilateral

D.2   Test A

The nodal values of the primary dependent variables $\hat{\phi}_m^{(e)}$ are set in accord with monomials or independent polynomials of a form appropriate for the element considered (for example, in the case of a linear element in two dimensions, the monomials would consist of 1, x and y). The gradients of $\hat{\phi}_m^{(e)}$ are then verified as being exact in accordance with the assumed monomial or polynomial. Since $\hat{\phi}_m^{(e)}$ is specified at all the nodes (that is, essential specifications are made exclusively), *Test A* does not require explicit use of the coefficient matrix **K**; consequently, this test only verifies that the programming steps used to compute the secondary dependent variables contain no errors. Furthermore, *Test A* ascertains only the satisfaction of the basic differential equation and not of the natural (Neumann) nodal specifications.

D.3   Test B

The boundary nodal values of $\hat{\phi}_m^{(e)}$ are set in accord with monomials or independent polynomials of an appropriate form. The solution at the interior node(s) is then verified, as are the values of the gradients of $\hat{\phi}_m^{(e)}$ within the elements. Unlike *Test A*, *Test B* ascertains the accuracy of **K**. Similar to *Test A*, *Test B* does not verify the correctness of the natural nodal specifications. Tests A and *B* are but *necessary* conditions for convergence of a formulation; they do not establish sufficient conditions for convergence. A check must likewise be made of the formulation involving natural nodal specifications.

D.4   Test C

In this test only a *minimum* number of essential (Dirichlet) nodal specifications are made; i.e., only enough values of $\hat{\phi}_m^{(e)}$ necessary to obtain a physically valid solution (e.g., prevent rigid body motion in a stress-deformation problem; specify a single value of temperature in a heat transfer problem, etc.). At the remaining boundary nodes, natural specifications (e.g., tractions or heat fluxes) are made. A solution is then sought for the remaining nodal values of $\hat{\phi}_m^{(e)}$ and is compared with the exact basic solution assumed. *Test C* has the advantage that it can detect *spurious mechanisms* such as deformations that have zero strain energy. In such cases, **K** will be singular (within roundoff error). The nodal values $\hat{\phi}_m^{(e)}$ may or may not be very large; this depends on whether the "tractions" excite the spurious mechanism [370]. In "Test C form," the patch test is not only necessary but is also sufficient for convergence [605].

It is possible to reduce *Test C* to a *single* element patch [533]. However, since it lacks interior nodes, the single element test is less powerful than a

multi-element patch test. The single element test can determine whether
fluxes are correctly evaluated from the nodal values $\hat{\phi}_m^{(e)}$, but it cannot
determine whether adjacent elements are compatible [370]. In summary,
both single-element and multi-element patch tests are required to ensure
that an element will converge and is stable.

Example 1: Patch Test Applied to Linear Quadrilateral Elements

Consider the "patch" of four-node quadrilateral elements shown in Figure D.2. The nodal coordinates are listed in the second and third columns
of Table D.1.

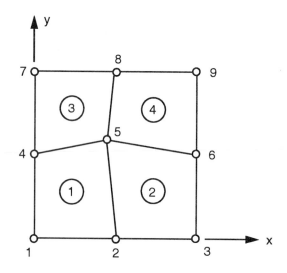

Figure D.2: Simple Patch of Four-Node Quadrilateral Elements

The material is assumed to be isotropic, linear elastic and is quantified
by an elastic modulus $E = 1000$, and by a Poisson's ratio ν of 0.30. A state
of *plane stress* is assumed. The displacement field in the global x and y
coordinate directions is taken to be $u = 0.002x$, $v = -0.0004y$, respectively.
Under the assumption of infinitesimal displacements and displacement gradients, this leads to the following strain state:[2]

---

[2]Further details pertaining to these and other equations associated with the theory
of elasticity are given in Chapter 12.

Table D.1: Nodal Coordinates and Specifications Associated with Patch
Test A

node	x	y	$\bar{u}$	$\bar{v}$
1	0.0	0.0	0.0	0.0
2	1.0	0.0	2.000×10^{-3}	0.0
3	2.0	0.0	4.000×10^{-3}	0.0
4	0.0	1.0	0.0	-4.000×10^{-4}
5	0.9	1.2	1.800×10^{-3}	-4.800×10^{-4}
6	2.0	1.0	4.000×10^{-3}	-4.000×10^{-4}
7	0.0	2.0	0.0	-8.000×10^{-4}
8	1.0	2.0	2.000×10^{-3}	-8.000×10^{-4}
9	2.0	2.0	4.000×10^{-3}	-8.000×10^{-4}

$$\varepsilon_{11} = \frac{\partial u}{\partial x} = 2.000 \times 10^{-3} \quad ; \quad \varepsilon_{22} = \frac{\partial v}{\partial y} = -4.000 \times 10^{-4} \tag{D.1}$$

$$\gamma_{12} = \frac{\partial u}{\partial y} + \frac{\partial v}{\partial x} = 0.0 \tag{D.2}$$

$$\varepsilon_{33} = \frac{-\nu}{1-\nu} (\varepsilon_{11} + \varepsilon_{22}) = -6.857 \times 10^{-4} \tag{D.3}$$

The associated stress state is

$$\sigma_{11} = \frac{E}{1-\nu^2} (\varepsilon_{11} + \nu\varepsilon_{22}) = 2.066 \times 10^{0} \tag{D.4}$$

$$\sigma_{22} = \frac{E}{1-\nu^2} (\nu\varepsilon_{11} + \varepsilon_{22}) = 2.198 \times 10^{-1} \tag{D.5}$$

$$\sigma_{12} = \frac{E}{2(1+\nu)} \gamma_{12} = 0.0 \tag{D.6}$$

with $\sigma_{33} = 0$ due to the assumption of plane stress.

For *Test A*, the nodal displacements ₍essential boundary conditions) listed in the last two columns of Table D.1 are thus specified. Performing an analysis of *Test A* using the *APES* computer program [274] yields the following results:

```
============================================================
| E L E M E N T   S T R A I N S  &  S T R E S S E S |
============================================================

        x1      eps-11     eps-22     eps-33     gam-12
        x2      sig-11     sig-22     sig-33     sig-12
     --------   ---------  ---------  ---------  ---------

--> element no.   1 (type = Q4P0  ) :
     4.75E-01   2.000E-03 -4.000E-04 -6.857E-04  1.793E-11
     5.50E-01   2.066E+00  2.198E-01  0.000E+00  6.896E-09

--> element no.   2 (type = Q4P0  ) :
     1.47E+00   2.000E-03 -4.000E-04 -6.857E-04  2.509E-11
     5.50E-01   2.066E+00  2.198E-01  0.000E+00  9.651E-09

--> element no.   3 (type = Q4P0  ) :
     4.75E-01   2.000E-03 -4.000E-04 -6.857E-04 -3.395E-11
     1.55E+00   2.066E+00  2.198E-01  0.000E+00 -1.306E-08

--> element no.   4 (type = Q4P0  ) :
     1.47E+00   2.000E-03 -4.000E-04 -6.857E-04 -1.982E-11
     1.55E+00   2.066E+00  2.198E-01  0.000E+00 -7.622E-09


===============================================
| N O D A L   D I S P L A C E M E N T S |
===============================================

                    coordinates          displacements
                    -----------          -------------
      node
    number      x1          x2          u1          u2
    ------   ----------  ----------  ----------  ----------
         1   1.000E-21  -2.000E-21   2.062E-23   2.284E-24
         2   1.000E+00   4.000E-21   2.000E-03   5.117E-24
         3   2.000E+00  -2.000E-21   4.000E-03   2.295E-24
         4  -2.000E-21   1.000E+00   3.296E-23  -4.000E-04
```

5	9.000E-01	1.200E+00	1.800E-03	-4.800E-04
6	2.000E+00	1.000E+00	4.000E-03	-4.000E-04
7	1.000E-21	2.000E+00	1.987E-23	-8.000E-04
8	1.000E+00	2.000E+00	2.000E-03	-8.000E-04
9	2.000E+00	2.000E+00	4.000E-03	-8.000E-04

The results indicate that standard four-node quadrilateral passes *Test A*. For *Test B*, the input is identical to that associated with Test A, except that at the interior node (node 5), no displacements are specified. Using the *APES* computer program, the results obtained are identical to those shown above. Since the displacements listed for node 5, which have now been computed and not assumed a priori, are in accordance with the monomials assumed, it follows that the element also passes *Test B*.

Finally, for *Test C*, a minimal number of nodal displacements are specified. In addition, distributed normal tractions are specified as shown in Figure D.3. The associated results follow.

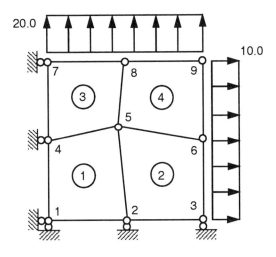

Figure D.3: Nodal Specifications Associated with Patch Test C

```
=========================================================
| E L E M E N T   S T R A I N S  &  S T R E S S E S |
=========================================================


        x1       eps-11     eps-22     eps-33     gam-12
        x2       sig-11     sig-22     sig-33     sig-12
      --------   ---------  ---------  ---------  ---------


--> element no.   1 (type = Q4P0  ) :
     4.75E-01  4.000E-03  1.700E-02 -9.000E-03 -1.735E-18
     5.50E-01  1.000E+01  2.000E+01  0.000E+00 -6.672E-16

--> element no.   2 (type = Q4P0  ) :
     1.47E+00  4.000E-03  1.700E-02 -9.000E-03  2.168E-19
     5.50E-01  1.000E+01  2.000E+01  0.000E+00  8.340E-17

--> element no.   3 (type = Q4P0  ) :
     4.75E-01  4.000E-03  1.700E-02 -9.000E-03  3.470E-18
     1.55E+00  1.000E+01  2.000E+01  0.000E+00  1.334E-15

--> element no.   4 (type = Q4P0  ) :
     1.47E+00  4.000E-03  1.700E-02 -9.000E-03 -8.674E-18
     1.55E+00  1.000E+01  2.000E+01  0.000E+00 -3.336E-15


==============================================
| N O D A L   D I S P L A C E M E N T S |
==============================================


             coordinates          displacements
             -----------          -------------

   node
  number    x1         x2          u1          u2
  ------  ---------- ----------  ----------  ----------
       1   1.000E-21 -2.000E-21   9.981E-23   2.078E-22
       2   1.000E+00  4.000E-21   4.000E-03   4.657E-22
       3   2.000E+00 -2.000E-21   8.000E-03   2.088E-22
       4  -2.000E-21  1.000E+00   1.596E-22   1.700E-02
       5   9.000E-01  1.200E+00   3.600E-03   2.040E-02
       6   2.000E+00  1.000E+00   8.000E-03   1.700E-02
       7   1.000E-21  2.000E+00   9.616E-23   3.400E-02
       8   1.000E+00  2.000E+00   4.000E-03   3.400E-02
       9   2.000E+00  2.000E+00   8.000E-03   3.400E-02
```

From the above results, it is evident that the four-node quadrilateral element also passes *Test C*. The patch test is thus completed.

D.5 Concluding Remarks

The basic aspects of the "engineering" patch test have been discussed. Once an element has been shown to pass all three forms of the test, convergence is assured as the size of the elements tends to zero.

In certain problems (e.g., nearly incompressible or inextensible elasticity), convergence is very slow unless a very large number of elements are used. For such problems, Zienkiewicz and Taylor [605] recommend the use of higher order patch tests involving the use of polynomial solutions of the differential equation and associated boundary conditions. The order of terms employed is larger than the basic polynomials used in a "lower order" patch test.

For further details pertaining to the patch test, the interested reader is directed to Chapter 5 in the book by MacNeal [370] and to Chapter 10 in Zienkiewicz and Taylor's book [605]. Specifics of actual two- and three-dimensional element patches as part of a set of "benchmark" problems are given by MacNeal and Harder [371].

Appendix E

Solution of Linear Systems of Equations

Depending on the kind and number of elements used in a mathematical model and on the manner in which the nodes are numbered, the time for solution of the equilibrium equations associated with a finite element analysis can represent a large percentage of the total solution time. Thus, if inappropriate solution techniques are used, the total computational cost can be quite large. In addition, if inappropriate (unstable) numerical procedures are used, a meaningful solution may be unattainable.

This appendix briefly reviews some techniques commonly used to solve the set of simultaneous linear equations that arise in steady state finite element analyses. The matrix *global* equations, generated from the element equations through the assembly process, have the following general form:

$$\mathbf{K}\hat{\boldsymbol{\phi}}_n = \mathbf{f} \tag{E.1}$$

where the global property matrix $\mathbf{K}$ is a square matrix with known coefficients, $\hat{\boldsymbol{\phi}}_n$ is the vector of nodal unknowns, and $\mathbf{f}$ is the vector of nodal "forcing" parameters.

The solution of simultaneous linear algebraic equations is typically realized using two different classes of methods: *direct* and *iterative* (indirect). Although both classes of methods have their advantages, for the size and types of problems considered in this introductory text, direct solvers are typically preferred. As such, in this appendix emphasis is placed on direct solution methods. The treatment is not meant to be exhaustive; further details pertaining to direct and iterative solvers are available in references cited herein.

E.1   Direct Methods

For non-singular $\mathbf{K}$, direct methods will, within machine accuracy, always produce a solution $\hat{\phi}_n$ to equation (E.1). Direct methods are known to:

- Be robust and to give predictable results, though not necessarily in as efficient a manner as possible. The requisite number of operations is known a priori.

- Possess optimal generality (e.g., the ability to easily handle multiple right hand side vectors) and automation (unlike some iterative solution techniques, no auxiliary grids are required).

- Exhibit rather poor scalability.

Direct methods are variants of basic Gauss elimination [191]. We thus first consider this approach, as applied to general non-singular coefficient matrices. This is followed by a discussion of ways in which the efficiency of the method can be improved when applied to the finite element method. Further details pertaining to direct methods are available in most numerical analysis texts, as well as in more classical references such as [377, 566].

E.1.1   Basic Gauss Elimination

The basic (sometimes referred to as "naive") Gauss elimination solution procedure for systems of simultaneous linear equations shall be described by solving the following linear system of equations:

$$
\begin{bmatrix}
6 & -1 & 2 & 4 \\
-2 & 4 & 1 & 0 \\
1 & 2 & 4 & -5 \\
-1 & -2 & 1 & 7
\end{bmatrix}
\begin{Bmatrix}
\hat{\phi}_1 \\
\hat{\phi}_2 \\
\hat{\phi}_3 \\
\hat{\phi}_4
\end{Bmatrix}
=
\begin{Bmatrix}
26 \\
9 \\
-3 \\
26
\end{Bmatrix}
\tag{E.2}
$$

- Step 1:
We seek to obtain zero elements in the first column of $\mathbf{K}$. To do this, we subtract $(-2/6)$ times row (equation) 1 from row (equation) 2, $(1/6)$ times row 1 from row 3, and $(-1/6)$ times row 1 from row 4. The resulting equations are shown below:

$$
\begin{bmatrix}
6 & -1 & 2 & 4 \\
0 & 11/3 & 5/3 & 4/3 \\
0 & 13/6 & 11/3 & -17/3 \\
0 & -13/6 & 4/3 & 23/3
\end{bmatrix}
\begin{Bmatrix}
\hat{\phi}_1 \\
\hat{\phi}_2 \\
\hat{\phi}_3 \\
\hat{\phi}_4
\end{Bmatrix}
=
\begin{Bmatrix}
26 \\
53/3 \\
-22/3 \\
91/3
\end{Bmatrix}
\tag{E.3}
$$

- Step 2:

The elements below the diagonal in the second column of **K** must next be set equal to zero. This is achieved by subtracting $(13/6)/(11/3) = 13/22$ times row 2 from row 3, and $(-13/6)/(11/3) = -13/22$ times row 2 from row 4. The resulting equations are shown below:

$$\begin{bmatrix} 6 & -1 & 2 & 4 \\ 0 & 11/3 & 5/3 & 4/3 \\ 0 & 0 & 59/22 & -71/11 \\ 0 & 0 & 51/22 & 93/11 \end{bmatrix} \begin{Bmatrix} \hat{\phi}_1 \\ \hat{\phi}_2 \\ \hat{\phi}_3 \\ \hat{\phi}_4 \end{Bmatrix} = \begin{Bmatrix} 26 \\ 53/3 \\ -391/22 \\ 897/22 \end{Bmatrix} \qquad (E.4)$$

- Step 3:

The elements below the diagonal in the third column of **K** must next be set equal to zero. This is achieved by subtracting $(51/22)/(59/22) = 51/59$ times row 3 from row 4. The resulting equations are shown below:

$$\begin{bmatrix} 6 & -1 & 2 & 4 \\ 0 & 11/3 & 5/3 & 4/3 \\ 0 & 0 & 59/22 & -71/11 \\ 0 & 0 & 0 & 828/59 \end{bmatrix} \begin{Bmatrix} \hat{\phi}_1 \\ \hat{\phi}_2 \\ \hat{\phi}_3 \\ \hat{\phi}_4 \end{Bmatrix} = \begin{Bmatrix} 26 \\ 53/3 \\ -391/22 \\ 3312/59 \end{Bmatrix} \qquad (E.5)$$

The coefficient matrix has now been reduced to *upper triangular form*. This portion of the solution is commonly referred to as *forward elimination*.

The unknowns are now computed using the process of *backward substitution*. Beginning with the fourth (last) equation (row):

$$x_4 = (3312/59)/(828/59) = 4.000 \qquad (E.6)$$

Proceeding next to the third equation we solve for x_3 in the following manner:

$$x_3 = \frac{(-391/22) - (-71/11)x_4}{59/22} = 3.000 \qquad (E.7)$$

Similarly,

$$x_2 = \frac{(53/3) - (4/3)x_4 - (5/3)x_3}{11/3} = 2.000 \qquad (E.8)$$

and finally

$$x_1 = \frac{26 - 4x_4 - 2x_3 - (-1)x_2}{6} = 1.000 \tag{E.9}$$

Remarks

1. Gauss elimination assumes that in step i, the ith diagonal element or "pivot" in the current coefficient matrix is non-zero. In the backward substitution portion of the solution, division by the diagonal elements is once again performed.

2. For many finite element applications, including those considered in this text, $\mathbf{K}$ is diagonally dominant. This alleviates the need for using pivoting strategies (such as row, column or complete pivoting). If such strategies are of interest, the reader is referred to texts such as [99, 30] that discuss this subject in detail.

E.1.2   Formal Statement of Basic Gauss Elimination

Using appropriate matrix operations, the basic Gauss elimination procedure is now formalized. Further details pertaining to this formal statement of the procedure are given in [509].

We first define an *elementary lower triangular matrix* of order n and index p, denoted by $\mathbf{M_k}$, as

$$\mathbf{M_p} = \begin{bmatrix} 1 & 0 & \cdots & & \cdots & \cdots & 0 \\ 0 & \ddots & 0 & & \cdots & \cdots & 0 \\ \vdots & 0 & 1 & 0 & \cdots & 0 \\ 0 & 0 & -m_{p+1,p} & 1 & \cdots & 0 \\ \vdots & \vdots & \vdots & 0 & \ddots & 0 \\ 0 & 0 & -m_{n,p} & 0 & \cdots & 1 \end{bmatrix} \tag{E.10}$$

where $m_{p+1,p} = K_{p+1,p}/K_{p,p}$ and K_{ij} is the (i,j)th entry in $\mathbf{K}$.

The reduction of the coefficient matrix $\mathbf{K}$ of order n to upper triangular form $\mathbf{U}$ can be written in the following form [509]:

$$\mathbf{M_{n-1}M_{n-2}}\cdots\mathbf{M_1K} = \mathbf{U} \tag{E.11}$$

This can easily be verified for the set of equations (E.2). In this case,

$$\mathbf{M_1} = \begin{bmatrix} 1 & 0 & 0 & 0 \\ 2/6 & 1 & 0 & 0 \\ -1/6 & 0 & 1 & 0 \\ 1/6 & 0 & 0 & 1 \end{bmatrix} \quad ; \quad \mathbf{M_2} = \begin{bmatrix} 1 & 0 & 0 & 0 \\ 0 & 1 & 0 & 0 \\ 0 & -13/22 & 1 & 0 \\ 0 & 13/22 & 0 & 1 \end{bmatrix} \tag{E.12}$$

$$\mathbf{M_3} = \begin{bmatrix} 1 & 0 & 0 & 0 \\ 0 & 1 & 0 & 0 \\ 0 & 0 & 1 & 0 \\ 0 & 0 & -51/59 & 1 \end{bmatrix} \tag{E.13}$$

and the reduced form given in equation (E.5) would be obtained from

$$\mathbf{M_3 M_2 M_1 K}\hat{\phi}_n = \mathbf{M_3 M_2 M_1 f} \tag{E.14}$$

In the actual computer implementation of standard Gauss elimination, the multiplications involving the $\mathbf{M_p}$ are not formally carried out. Instead, $\mathbf{K}$ is reduced to upper triangular form through direct modifications to the elements of $\mathbf{K}$ through the *elementary row operations* employed in steps 1 to 3 in the above example.

E.1.3 Triangular Decomposition

The inverse of an elementary lower triangular matrix $\mathbf{M_p}$ is obtained by simply reversing the signs of the off-diagonal elements. Denoting this inverse by $\mathbf{M_p}^{-1} \equiv \mathbf{L_p}$, we multiply, in succession, equation (E.11) by $\mathbf{L_{n-1}L_{n-2}\cdots L_1}$. This leads to the following result:

$$\mathbf{K} = \mathbf{L_1 L_2 \cdots L_{n-2} L_{n-1} U} \tag{E.15}$$

Recalling that the product of two lower triangular matrices yields a lower triangular matrix, we write

$$\mathbf{K} = \mathbf{LU} \tag{E.16}$$

The coefficient matrix is thus decomposed into the product of a *unit* lower triangular matrix and an upper triangular one. This process is called *Doolittle decomposition*, and is just basic Gauss elimination with non-zero coefficients stored in the lower triangular portion of the modified $\mathbf{K}$.

Remark

  1. The decomposition of $\mathbf{K}$ to LU form does *not* include the right-hand side vector $\mathbf{f}$.

Once $\mathbf{K}$ is decomposed to LU form, the solution of equation (E.1) is carried out using the following two-step procedure:

  1. The equations $\mathbf{Lz} = \mathbf{f}$ are first solved for the intermediate vector $\mathbf{z}$. Since $\mathbf{L}$ is a unit lower triangular matrix, this solution involves only *forward substitution*, beginning with the first equation.

  2. The solution vector $\hat{\boldsymbol{\phi}}_n$ is then obtained by solving the equations $\mathbf{U}\hat{\boldsymbol{\phi}}_n = \mathbf{z}$. Since $\mathbf{U}$ is an upper triangular matrix, this solution involves only *backward substitution*.

The primary reason for using a LU decomposition is the ease with which the solution for *multiple* right hand side vectors is realized. In this solution process, $\mathbf{K}$ is reduced to LU form *only once*. Then, given a right hand side vector $\mathbf{f}$, the above two-step procedure is used to solve for $\hat{\boldsymbol{\phi}}_n$. Then, given another vector $\mathbf{f}$, the solution process is repeated without the need for refactoring $\mathbf{K}$.

A common variant of LU decomposition is realized by expressing $\mathbf{U}$ as the product $\mathbf{U} = \mathbf{D}\tilde{\mathbf{U}}$, where $\mathbf{D}$ is a diagonal matrix containing the diagonal (pivot) elements of $\mathbf{U}$ and $\tilde{\mathbf{U}}$ is a *unit* upper triangular matrix. The coefficient matrix is thus decomposed into LDU form.

Different LU decompositions can be obtained depending on how the the diagonal matrix $\mathbf{D}$ is treated in the LDU decomposition. Three variants are commonly used, namely:

  1. The first variant associates $\mathbf{D}$ with the upper triangular part to give the factorization $\mathbf{K} = \mathbf{L}(\mathbf{D}\tilde{\mathbf{U}})$. This is seen to be the aforementioned *Doolittle* decomposition. Recall that $\mathbf{L}$ is unit lower triangular.

  2. The second variant associates $\mathbf{D}$ with the lower triangular part to give the factorization $\mathbf{K} = (\mathbf{L}\mathbf{D})\tilde{\mathbf{U}}$. This is known as *Crout* decomposition, and involves a unit upper triangular.

  3. Finally, when $\mathbf{K}$ is *symmetric*, the decomposition has the form $\mathbf{K} = \mathbf{L}\mathbf{D}\mathbf{L}^T$. If the diagonal elements of $\mathbf{K}$ are positive, then the diagonal matrix

$$\tilde{\mathbf{D}} = \operatorname{diag}\left[(\delta_{11})^{\frac{1}{2}} \quad (\delta_{22})^{\frac{1}{2}} \quad \cdots \quad (\delta_{nn})^{\frac{1}{2}}\right] \qquad \text{(E.17)}$$

exists. The coefficient matrix can then be decomposed in the following manner:

$$\mathbf{K} = \left(\mathbf{L}\tilde{\mathbf{D}}\right)\left(\mathbf{L}\tilde{\mathbf{D}}\right)^{T} = \tilde{\mathbf{L}}\tilde{\mathbf{L}}^{T} \tag{E.18}$$

This third variant is known as the *Cholesky* decomposition of $\mathbf{K}$, and requires a symmetric, positive definite coefficient matrix.

Implementation Notes

1. In the actual computer implementation of Gauss elimination employing LU decomposition, the vector $\mathbf{z}$ is effectively calculated at the same time as the elementary lower triangular matrices $\mathbf{M_p}$ are established. In practice the matrix multiplications required to obtain $\mathbf{L}$ and $\mathbf{z}$ are not formally carried out; these arrays are established by directly modifying $\mathbf{K}$ and $\mathbf{f}$.

2. For brevity, outlines of direct solution algorithms have been omitted from the present development. Such information is available in most numerical general analysis texts, in books devoted to algorithms and their implementation [443], and in user's guides for subroutine libraries [14].

E.1.4 Operation Counts and Storage Requirements

Having introduced the basic variants to standard Gauss elimination, it is instructive to consider the approximate number of operations associated with the respective algorithms. In the subsequent discussion, $\mathbf{K}$ is assumed to have size $(n * n)$.

- Standard Gauss elimination (with one right-hand side vector):

$$\frac{n^3 + 3n^2 - n}{3} \text{ multiplications and divisions}$$

$$\frac{2n^3 + 3n^2 - 5n}{6} \text{ additions and subtractions}$$

To store the entries in $\mathbf{K}$ requires n^2 words of storage.

- Doolittle or Crout decomposition, with m right-hand side vectors:

$$\frac{n^3}{3} + mn^2 - \frac{n}{3} \text{ multiplications and divisions}$$

$$\frac{n^3}{3} + mn^2 - \frac{n^2}{2} - mn + \frac{n}{6} \quad \text{additions and subtractions}$$

To store the entries in **K** again requires n^2 words of storage.

- Cholesky decomposition (for symmetric **K**):

$$\frac{n^3 + 9n^2 + 2n}{6} \quad \text{multiplications and divisions}$$

$$\frac{n^3 + 6n^2 - 7n}{6} \quad \text{additions and subtractions}$$

and n square roots. To store the entries in the symmetric **K** requires $n(n-1)/2$ words of storage. Cholesky decomposition thus requires approximately one-half of the number of multiplications/divisions required for Gauss elimination, Doolittle decomposition or Crout decomposition. This is to be expected, since the algorithm takes advantage of the symmetry of the coefficient matrix. However, Cholesky decomposition tends to be slower than the aforementioned decomposition algorithms. This is because of the necessity to evaluate n square roots, an operation that can involve the computational effort of 5 to 7 multiplications [514, 266].

To increase efficiency through more optimal storage, direct methods are often implemented as "banded," "skyline" (or active column), or "frontal" solvers. General information regarding storage schemes for sparse matrices is given in [535].

E.1.5   Constant Bandwidth Solvers

In banded matrices, all of the non-zero coefficients are contained within a band; outside of the band all coefficients are zero. If the coefficient matrix is symmetric, only the diagonal coefficients and those above or below the diagonal need to be stored. Since only non-zero coefficients need to be considered, these are compactly stored in a reduced space of size $n(n_{bw})$, where n_{bw} is the half-bandwidth. As shown in Section 8.3, an optimal numbering of the nodes will minimize n_{bw}, and consequently the storage requirements. The numbering of elements in the mesh is immaterial.

For example, in the mesh shown in Figure 8.11(a), the following global **K** matrix was determined:

$$\mathbf{K} = \begin{bmatrix} K_{11}^{(1)} & 0 & 0 & K_{12}^{(1)} & 0 \\ 0 & K_{22}^{(4)} & 0 & 0 & K_{21}^{(4)} \\ 0 & 0 & K_{22}^{(2)} + K_{11}^{(3)} & K_{21}^{(2)} & K_{12}^{(3)} \\ K_{21}^{(1)} & 0 & K_{12}^{(2)} & K_{22}^{(1)} + K_{11}^{(2)} & 0 \\ 0 & K_{12}^{(4)} & K_{21}^{(3)} & 0 & K_{22}^{(3)} + K_{11}^{(4)} \end{bmatrix} \qquad (\text{E.19})$$

For this matrix, $n_{bw} = 4$. Assuming symmetric element arrays, $\mathbf{K}$ would be stored in banded, upper triangular form in the following manner:

$$\mathbf{K} = \begin{bmatrix} K_{11}^{(1)} & 0 & 0 & K_{12}^{(1)} \\ K_{22}^{(4)} & 0 & 0 & K_{21}^{(4)} \\ K_{22}^{(2)} + K_{11}^{(3)} & K_{21}^{(2)} & K_{12}^{(3)} & - \\ K_{22}^{(1)} + K_{11}^{(2)} & 0 & - & - \\ K_{22}^{(3)} + K_{11}^{(4)} & - & - & - \end{bmatrix} \qquad (\text{E.20})$$

The conversion of standard LU decomposition algorithms to incorporate banded storage is trivial. The associated factorization requires $n(n_{bw})^2/2$ multiplications and additions.

E.1.6 Skyline Solvers

An improvement in efficiency over constant bandwidth solvers can usually be realized through the use of algorithms that employ variable bandwidth techniques to store only the "profile" of $\mathbf{K}$. Such solvers take advantage of the fact that, in a given column, only the coefficients beginning with the first non-zero entry need to be stored. The hypothetical line separating the zeros at the top of a column from the first non-zero entry is typically referred to as the "skyline"; the columns of non-zero coefficients are referred to as "active columns"; these are typically stored in a single one-dimensional array (vector). The price paid for this compact storage is the requirement of maintaining a second "bookkeeping" vector that "points" to the diagonal entries in the one-dimensional array.

For a skyline solution of the mesh shown in Figure 8.11(a), the symmetric $\mathbf{K}$ and its corresponding "pointer" array $\mathbf{v}$ would be stored in the following manner:

$$
\mathbf{K} = \begin{bmatrix} K_{11}^{(1)} \\ K_{22}^{(4)} \\ K_{22}^{(2)} + K_{11}^{(3)} \\ K_{12}^{(1)} \\ 0 \\ K_{21}^{(2)} \\ K_{22}^{(1)} + K_{11}^{(2)} \\ K_{21}^{(4)} \\ K_{12}^{(3)} \\ 0 \\ K_{22}^{(3)} + K_{11}^{(4)} \end{bmatrix} \quad ; \quad \mathbf{v} = \begin{bmatrix} 1 \\ 2 \\ 3 \\ 7 \\ 11 \end{bmatrix} \qquad (E.21)
$$

Skyline solvers thus have the following advantages over other variants of Gauss elimination:

- They require less storage.

- Their storage requirements are not greatly affected by a few long columns.

- They can efficiently use scalar (dot) products to facilitate the reduction of $\mathbf{K}$ to LU form. This is particularly significant on modern vector processors.

E.1.7 Frontal Solvers

The "frontal" solution method was developed [388, 257, 246, 62] as a way to reduce the amount of fast-access memory required to solve large systems of equations. This direct method is closely connected to the finite element method, and has been successfully employed in certain finite element analysis programs. The frontal method is best suited for use with computers having a limited amount of fast-access memory, and in modified form for certain parallel architecture computers [430].

In a frontal procedure the entire global coefficient matrix is not considered. The element matrices are formed in the order that the elements are numbered (the optimal numbering of elements is thus essential; the nodes can be numbered in any fashion). Whenever the equations associated with a given node are completed (i.e., all the elements connecting at that node have been considered), then the unknowns associated with the node are eliminated.

E.2 Iterative Methods

For very large problems, the number of non-zero entries within the profile of K is *small* compared to the number of zero entries. This results in very large column heights within the matrix. Furthermore, for large enough systems of equations, a solution using a direct solver may be unattainable due to the excessive memory requirements and large number of operations required.

For this class of problems, *iterative* methods are generally be more efficient than direct ones.[1] The term "iterative method" refers to a wide range of techniques that use successive approximations to obtain more accurate solutions to a linear system at each step [56]. For relatively large systems of equations, experience shows that iterative solvers are significantly more efficient in that they offer the promise of substantially reduced operations (and thus reduced roundoff error) and storage requirements over direct methods. However, they have the potential disadvantage of slow convergence, if they converge at all. This is particularly true if K is poorly conditioned. Furthermore, in iterative methods the process of resolution is less flexible and more expensive than for direct techniques.

Iterative methods are commonly divided into two groups: *stationary* and *non-stationary*. Stationary methods derive their name from the fact that in proceeding from one iteration to the next, K does not need to be recomputed. This feature makes stationary methods relatively easy to understand and implement. However, they are not as effective as non-stationary methods [56]. The latter are a relatively recent development and, though they are typically more complex than stationary methods, can be highly effective for solving very large numbers of equations. Unlike stationary methods, in non-stationary methods K is recomputed at every iteration.

What follows is a brief overview of stationary and non-stationary iterative methods. Further details pertaining to such methods are available in texts on the subject such as [546, 579, 32, 56, 475].

E.2.1 Stationary Methods

Given a linear system of simultaneous equations, let $K = D + L + U$, where D is a diagonal matrix containing the diagonal entries in K, L is a strictly lower triangular matrix (zero diagonal entries), and U is a strictly upper triangular matrix (zero diagonal entries). If K is symmetric, then of course

---

[1] Iterative methods were used in certain early applications of the finite element method [58]. However, the speed and memory limitations of the computers of the day precluded efficient use of iterative methods and fostered the use of direct methods.

$\mathbf{U} = \mathbf{L}^T$. The above procedure is often referred to as the "splitting" of $\mathbf{K}$. In stationary methods we seek methods of the form

$$\hat{\phi}_n^{(k+1)} = \mathbf{T}\hat{\phi}_n^{(k)} + \mathbf{d} \quad ; \quad k = 0, 1, 2, \cdots \tag{E.22}$$

such that $\lim_{k \to \infty} \left(\mathbf{K}\hat{\phi}_n^{(k+1)} = \mathbf{f} \right)$. In the above expressions, k represents the iteration number, and $\mathbf{T}$ denotes an iteration matrix. Neither $\mathbf{T}$ nor the vector $\mathbf{d}$ depend on k.

The iterative solution is continued until:

1. A preset maximum number of iterations have been performed, or

2. The relative error in subsequent solutions is less than some preset error tolerance ε; for example, using the L2 norm,

$$\frac{\left\| \hat{\phi}_n^{(k+1)} - \hat{\phi}_n^{(k)} \right\|_2}{\left\| \hat{\phi}_n^{(k+1)} \right\|_2} < \varepsilon \tag{E.23}$$

Some details concerning four classic stationary methods are next presented.

Jacobi Method

The Jacobi method solves for each variable assuming that the other variables remain unchanged from their values computed in the previous iteration. That is, one iteration of the method corresponds to solving for every variable once. To achieve this, the original system of simultaneous equations is written as

$$(\mathbf{D} + \mathbf{L} + \mathbf{U})\, \hat{\phi}_n = \mathbf{f} \tag{E.24}$$

The equations are then rearranged into the following iterative scheme

$$\hat{\phi}_n^{(k+1)} = \mathbf{D}^{-1} \left[\mathbf{f} - (\mathbf{L} + \mathbf{U})\, \hat{\phi}_n^{(k)} \right] \tag{E.25}$$

where the computations required to determine $\mathbf{D}^{-1}$ are trivial. In light of equation (E.22), it follows that $\mathbf{T} = \mathbf{D}^{-1}(\mathbf{L} + \mathbf{U})$ and $\mathbf{d} = \mathbf{D}^{-1}\mathbf{f}$.

Equation (E.25) are solved *without* the need for reduction in the manner associated with direct solution techniques. Indeed, the solution involves only matrix multiplications and significantly reduces the associated round-off error. Furthermore, compared with the reduction process used in direct methods, the cost per iteration is quite small. Although the Jacobi method is easy to understand and implement, convergence is known to be slow [56].

Gauss-Seidel Method

In the Gauss-Seidel method [489] is commonly regarded as a logical improvement of the Jacobi method. Here the system of equations is rearranged into the following form:

$$\hat{\phi}_n^{(k+1)} = \mathbf{D}^{-1}\left(\mathbf{f} - \mathbf{L}\hat{\phi}_n^{(k+1)} - \mathbf{U}\hat{\phi}_n^{(k)}\right) \tag{E.26}$$

In light of equation (E.22), it follows that $\mathbf{T} = (\mathbf{D}+\mathbf{L})^{-1}\mathbf{U}$ and $\mathbf{d} = (\mathbf{D}+\mathbf{L})^{-1}\mathbf{f}$.

Equation (E.26) is solved by forward substitution – similar to the Jacobi method, no reduction is necessary. Unlike the Jacobi method, the Gauss-Seidel method uses *updated* values as soon as they become available via the product $\mathbf{L}\hat{\phi}_n^{(k+1)}$.

In general, if the Jacobi method converges, the Gauss-Seidel method will also converge, and at a faster rate. Both methods converge if the spectral radius (magnitude of the largest eigenvalue) of $\mathbf{T}$ is less than one [99].

The conditions for convergence are also related to the nature of $\mathbf{K}$. For example, if $\mathbf{K}$ is diagonally dominant and the system of equations is linear, then convergence is *guaranteed*. Also, if $\mathbf{K}$ is symmetric and positive definite, then both methods are known to converge. However, even if convergence is guaranteed, the *rate* at which convergence is realized is typically quite slow [56]. For this reason attempts have been made to accelerate the Gauss-Seidel method.

Successive Overrelaxation (SOR) Method

The Successive Overrelaxation (SOR) method can be derived from the Gauss-Seidel method by introducing a non-zero extrapolation parameter ω. More precisely, having "split" $\mathbf{K}$ in the usual fashion, multiply both sides of the equation by ω, and then add and subtract the product $\mathbf{D}\hat{\phi}_n$ from the left had side of the equation. After simple rearrangement, this gives the following iterative scheme:

$$\hat{\phi}_n^{(k+1)} = \omega \mathbf{D}^{-1} \left[\mathbf{f} - \mathbf{L}\hat{\phi}_n^{(k+1)} - \mathbf{U}\hat{\phi}_n^{(k)} \right] + (1 - \omega)\hat{\phi}_n^{(k)} \qquad (\text{E.27})$$

If $0 < \omega < 1$, the method is called *underrelaxation*. If $\omega = 1$, the standard Gauss-Seidel method is recovered. Finally if $\omega > 1$, an *overrelaxation* method is realized, which typically is called the Successive Overrelaxation method.

An optimal value of ω would be one that maximizes the asymptotic rate of convergence for the SOR method. Such an optimal value does not necessarily reduce the number of iterations or may not be the best initial choice. Although the optimal value of ω is not known a priori (and thus constitutes a disadvantage of the method), values in the range $1.3 < \omega < 1.9$ are typically used [421, 546]. For the optimal choice of ω, the SOR method may converge faster than Gauss-Seidel by an order of magnitude [56].

Symmetric Successive Overrelaxation (SSOR) Method

The Symmetric Successive Overrelaxation (SSOR) method involves two "sweeps" of the SOR method. It has no advantage over SOR as a stand-alone iterative method. However, it is a useful and easily derivable pre-conditioner for non-stationary methods [56]. Further details pertaining to preconditioners are deferred until Section E.2.3.

E.2.2   Non-Stationary Methods

As previously noted, non-stationary methods are typically more complex than stationary methods, but can be highly effective for solving very large numbers of equations. Some non-stationary methods are summarized below.

Multigrid Methods

The basic stationary methods are very effective at eliminating oscillatory (high frequency) components of error. The smooth (low frequency) components are, however, relatively unaffected. One way in which to improve a stationary method, at least in its early stages, is to use a good initial guess. A well known approach for obtaining a good initial guess is to perform some preliminary iterations on a coarse grid and then use the resulting approximation as an initial guess on the fine grid. This approach forms the basis of the multigrid methods, which evolved from attempts to improve upon limitations of the basic stationary iterative methods [211, 95]. These attempts have been relatively successful; used in a multigrid setting, non-stationary methods are competitive with the fast direct solvers when applied to many model problems [428, 429].

Conjugate Gradient Method

The Conjugate Gradient (CG) method is one of the oldest and best-known non-stationary methods. It derives its name from the fact that it generates a sequence of conjugate (or orthogonal) vectors [236]. These vectors are the residuals of the iterates. They are also the gradients of a quadratic functional, the minimization of which is equivalent to solving the linear system. CG is an extremely effective method when the coefficient matrix is symmetric positive definite (such as in many finite element applications), since storage for only a limited number of vectors is required [58, 605].

Minimum Residual and Symmetric LQ Methods

The Minimum Residual (MINRES) and Symmetric LQ (SYMMLQ) methods are computational alternatives for the conjugate gradient method for $\mathbf{K}$ that are symmetric but possibly indefinite. SYMMLQ will generate the same solution iterates as CG if the coefficient matrix is symmetric positive definite [56].

Generalized Minimal Residual Method

The Generalized Minimal Residual (GMRES) method computes a sequence of orthogonal vectors (like MINRES), and combines these through a least-squares solve and update [476, 475]. However, unlike MINRES (and CG) it requires storing the whole sequence, so that a large amount of storage is needed. For this reason, restarted versions of this method are used. In restarted versions, computational and storage costs are limited by specifying a fixed number of vectors to be generated. This method is useful for general non-symmetric matrices [56].

Bi-Conjugate Gradient Method

The Biconjugate Gradient (BiCG) method generates two CG-like sequences of vectors, one based on a system with the original coefficient matrix $\mathbf{K}$, and one on $\mathbf{K}^T$. Instead of orthogonalizing each sequence, they are made mutually orthogonal, or "bi-orthogonal." This method, like CG, uses limited storage. It is useful when the matrix is non-symmetric and non-singular; however, convergence may be irregular, and there is possibility that the method will break down. The BiCG method requires a multiplication with the coefficient matrix and with its transpose at each iteration [56].

Quasi-Minimal Residual Method

The Quasi-Minimal Residual (QMR) method applies a least-squares solve and update to the BiCG residuals, thereby smoothing out the irregular convergence behavior of BiCG, which may lead to more reliable approximations. In full glory, it has a look ahead strategy built in that avoids the BiCG breakdown. Even without look ahead, QMR largely avoids the breakdown that can occur in BiCG. On the other hand, it does not effect a true minimization of either the error or the residual, and while it converges smoothly, it often does not improve on the BiCG in terms of the number of iteration steps [56, 475].

E.2.3   Preconditioning

The rate of convergence of iterative methods depends on the spectral properties of $\mathbf{K}$ (see Section 3.5). Thus, improving the spectral properties of $\mathbf{K}$ should, in theory, improve the rate of convergence.

This improvement is usually realized through *preconditioning*. When preconditioned, $\mathbf{K}$ is premultiplied by a suitable *preconditioner* matrix that transforms $\mathbf{K}$ into a matrix with a more favorable spectrum. A good preconditioner improves the convergence of the iterative method sufficiently to overcome the extra cost of constructing and applying the preconditioner [56].

E.3   Solution Errors

The most significant loss of accuracy associated with finite element computations typically occurs in the solution of the system equations [376, 59]. The primary sources of error are *roundoff* and the *truncation* (chopping and rounding) of numbers associated with their finite precision representation.[2] However, experience and numerical experiments have indicated that in finite element analyses, roundoff errors may not be as severe as the truncation of significant digits [471].

Several measures are typically used in assessing the error associated with finite element analyses. Perhaps the simplest, though certainly not the most practical, is the estimation of the norm of the difference between the approximate solution obtained in absence of roundoff and truncation error $\hat{\phi}_n$ and the actual approximate solution $\hat{\phi}_n^*$, that is, $\left\| \hat{\phi}_n^* - \hat{\phi}_n \right\|$.

---

[2] Recall the discussion of Chapter 2.

Noting that

$$\mathbf{K}\left(\hat{\boldsymbol{\phi}}_n^* - \hat{\boldsymbol{\phi}}_n\right) = \boldsymbol{\Delta}\mathbf{f} \quad \Rightarrow \quad \left(\hat{\boldsymbol{\phi}}_n^* - \hat{\boldsymbol{\phi}}_n\right) = \mathbf{K}^{-1}\boldsymbol{\Delta}\mathbf{f} \tag{E.28}$$

From the definition of a vector norm it follows that

$$\left\|\hat{\boldsymbol{\phi}}_n^* - \hat{\boldsymbol{\phi}}_n\right\| = \left\|\mathbf{K}^{-1}\boldsymbol{\Delta}\mathbf{f}\right\| \leq \left\|\mathbf{K}^{-1}\right\|\left\|\boldsymbol{\Delta}\mathbf{f}\right\| \tag{E.29}$$

where a non-singular $\mathbf{K}$ is assumed, and where

$$\boldsymbol{\Delta}\mathbf{f} = \mathbf{f} - \mathbf{K}\hat{\boldsymbol{\phi}}_n^* \tag{E.30}$$

We next desire to estimate the relative error associated with the solution of the system equations. Noting that

$$\mathbf{K}\hat{\boldsymbol{\phi}}_n^* = \mathbf{f} \quad \Rightarrow \quad \|\mathbf{f}\| \leq \|\mathbf{K}\|\left\|\hat{\boldsymbol{\phi}}_n^*\right\| \quad \Rightarrow \quad \left\|\hat{\boldsymbol{\phi}}_n^*\right\| \geq \frac{\|\mathbf{f}\|}{\|\mathbf{K}\|} \tag{E.31}$$

In light of equation (E.29), the relative error is thus

$$\frac{\left\|\hat{\boldsymbol{\phi}}_n^* - \hat{\boldsymbol{\phi}}_n\right\|}{\left\|\hat{\boldsymbol{\phi}}_n^*\right\|} \leq \frac{\left\|\mathbf{K}^{-1}\right\|\left\|\boldsymbol{\Delta}\mathbf{f}\right\|}{\left\|\hat{\boldsymbol{\phi}}_n^*\right\|} \leq \frac{\|\mathbf{K}\|\left\|\mathbf{K}^{-1}\right\|\left\|\boldsymbol{\Delta}\mathbf{f}\right\|}{\|\mathbf{f}\|} \tag{E.32}$$

Defining the *matrix condition number* by

$$\text{cond}\,(\mathbf{K}) = \|\mathbf{K}\|\left\|\mathbf{K}^{-1}\right\| \tag{E.33}$$

we have the final desired form of the relative error, viz.,

$$\frac{\left\|\hat{\boldsymbol{\phi}}_n^* - \hat{\boldsymbol{\phi}}_n\right\|}{\left\|\hat{\boldsymbol{\phi}}_n^*\right\|} \leq \text{cond}\,(\mathbf{K})\frac{\left\|\boldsymbol{\Delta}\mathbf{f}\right\|}{\|\mathbf{f}\|} \tag{E.34}$$

The condition number can thus be viewed as an amplification factor between the relative changes in $\mathbf{f}$ and the solution $\hat{\boldsymbol{\phi}}_n$. If cond$(\mathbf{K})$ is small (a well-conditioned $\mathbf{K}$), then roundoff error will not have much of an effect on the solution. If cond$(\mathbf{K})$ is large (an ill-conditioned $\mathbf{K}$), then roundoff error becomes more significant. Finally, we note that $1 = \|\mathbf{I}\| = \|\mathbf{KK}^{-1}\| \leq \|\mathbf{K}\| \, \|\mathbf{K}^{-1}\| \equiv \text{cond}\,(\mathbf{K}) \quad \Rightarrow \quad \text{cond}\,(\mathbf{K}) \geq 1$.

To estimate the initial truncation error, assume that for a t-digit precision computer that [471]

$$\frac{\|\mathbf{\Delta f}\|}{\|\mathbf{f}\|} = 10^{-t} \tag{E.35}$$

Assuming s-digit precision in the solution, it follows that

$$\frac{\left\|\hat{\phi}_n^* - \hat{\phi}_n\right\|}{\left\|\hat{\phi}_n^*\right\|} = 10^{-s} \tag{E.36}$$

It follows that

$$10^{-s} \leq \text{cond}\,(\mathbf{K})10^{-t} \quad \text{or} \quad s \geq t\log_{10}\text{cond}\,(\mathbf{K}) \tag{E.37}$$

Equation (E.37) thus quantifies the degradation of the solution of a system of linear equations. It is evident that the larger the magnitude of the condition number, the less accurate the solution.

E.3.1   Sources of Ill-Conditioning

In general, an ill-conditioned system of equations possesses a greater likelihood of solution errors. In finite element analyses, the following sources contribute to such ill-conditioning [376].

For certain physically ill-conditioned problems, the solution is very sensitive to changes in the right-hand side vector. In elasticity applications, displacements may be extremely sensitive to applied forces. This occurs where combinations of structural elements have very different stiffness, and where structures or structural components undergo large rigid body (non-straining) displacements (e.g., a long cantilever beam). A rigid inclusion imbedded in a soft, nearly incompressible medium will typically cause some difficulties. Poor or incorrect idealization, such as sudden mesh variation or element thickness variation, or use of elongated two- or three-dimensional elements where one dimension is considerably smaller than the others, induces local ill-conditioning.

For well-conditioned physical problems, ill-conditioned equations may be caused by a correct, but very fine mesh idealization. No numerical problems are encountered when the problem is solved with coarse idealizations.

As the mesh is repeatedly subdivided, the condition number increases. Eventually the buildup of intermediate roundoff error in the calculations swamps any accuracy improvement due to the finer discretization. This deterioration is more pronounced when bending elements are used, since higher-order derivatives are included in the strain-energy expression. For example, the condition number for a prismatic beam element subdivided into N equal segments increases as N^4; whereas for an N-segment axial element, the condition number increases as N^2. As word lengths on digital computers increase in size, this source of error is less prevalent.

The condition of the system equations is influenced by the choice of the interpolation functions. These functions must be chosen in such a manner that the system remain well-conditioned as the element size goes to zero.

The present discussion of solution errors has, by design, been somewhat terse. Further details pertaining to this subject are available in numerous references such as [406, 565, 378, 567, 256, 471, 421].

Appendix F

Notes on Integration of Finite Elements

F.1 Introductory Remarks

The derivation of finite element equations involves integration. For straight-sided elements the integration can be performed exactly. However, for the more general case of mapped elements, the complexity of the associated integrals makes their exact evaluation impractical. In such cases the integration must be performed in an approximate (numerical) manner.

This appendix briefly reviews some techniques commonly used to integrate finite element equations for single and multi-dimensional problems. Although the emphasis is on numerical integration, some exact integration formulae are also presented.

Recall that the definition of the integral

$$I(f) \equiv \int_a^b f(x)\, dx \tag{F.1}$$

is the area under the curve of $f(x)$ from a to b. Herein the function f is assumed to be smooth and integrable and $[a, b]$ is a closed interval.

The integral $I(f)$ shall be approximated by a formula of the form

$$\hat{I}_n(f) \equiv \sum_{i=1}^n w_i f(x_i) \tag{F.2}$$

When employed as an approximation to an integral, a sum of this form is called a *numerical integration formula* or numerical *quadrature formula* [263].

The n distinct points x_i are called the *quadrature points* or *sampling points*, and the w_i are a set of *weighting coefficients* independent of $f(x)$. The associated *quadrature error*

$$E_n(f) = I(f) - \hat{I}_n(f) \tag{F.3}$$

is obviously a scalar. The degree of precision of a quadrature formula is defined in the following manner:

Definition. The *degree of precision* or *accuracy* of a quadrature formula is the positive integer n such that $E_n(P_k) = 0$ for all polynomials P_k of degree less than or equal to n, but for which $E_n(P_{n+1}) \neq 0$ for some polynomial of degree $(n + 1)$.

The goal is thus to approximate $I(f)$ as accurately as possible with the smallest number of function evaluations of the integrand. The analyst has the freedom to choose quadrature formulae of various *orders*, with higher order sometimes, but not always, giving higher accuracy [443].

F.2   Interpolatory Quadrature

One way to approach the problem of numerical integration is to approximate $f(x)$ by an interpolating polynomial, and to then integrate this polynomial. This requires *a priori* selection of a set of distinct points $\{x_0, x_1, \cdots, x_n\}$ from some interval $[c, d]$ that contains $[a, b]$. Let P_n be the interpolating polynomial

$$P_n(x) = \sum_{i=0}^{n} f(x_i)\Lambda_i^n(x) = \sum_{i=0}^{n} f(x_i) \prod_{\substack{j=0 \\ j \neq i}}^{n} \frac{(x - x_j)}{(x_i - x_j)} \tag{F.4}$$

where $\Lambda_i^n(x)$ represents the Lagrange polynomial[1] of degree n about the point $x = x_i$. Integrating equation (F.4) over $[a, b]$ gives the following quadrature formula

$$\hat{I}_{n+1}(f) = \int_a^b \left[\sum_{i=0}^{n} f(x_i)\Lambda_i^n(x) \right] dx = \sum_{i=0}^{n} f(x_i)w_i \tag{F.5}$$

---

[1] Named in honor of the French mathematician Joseph-Louis Lagrange (1736-1813).

where

$$w_i = \int_a^b \Lambda_i^n(x)\, dx = \int_a^b \left[\prod_{\substack{j=0 \\ j \neq i}}^n \frac{(x - x_j)}{(x_i - x_j)} \right] dx \quad ; \quad i = 0, 1, 2, \cdots \quad \text{(F.6)}$$

It is evident that the w_i are determined completely by the endpoints of the interval of integration and by the locations of the interpolation points x_i. As noted in conjunction with equation (F.2), the w_i are independent of $f(x)$. Formulae of the form of equations (F.5) and (F.6) are sometimes referred to as *interpolatory quadrature formulae* [263]. The error associated with such a quadrature formula of degree n is

$$E_{n+1}(f) = \frac{f^{(n+1)}(\mu)}{(n+1)!} \int_a^b \prod_{i=0}^n (x - x_i)\, dx \qquad \text{(F.7)}$$

where the specific location of the point $\mu \in [a, b]$ is unknown.

We next consider the specific case of Newton-Cotes[2] quadrature formulae. These employ *equally spaced* quadrature points x_i, where $x_i = x_0 + ih$ for each $i = 1, 2, \cdots, n$. To put the Newton-Cotes formulae in perspective, consider the following observations [443]:

> The classical formulas for integrating a function whose value is known at equally spaced steps have a certain elegance about them, and they are redolent with historical association. Through them, the modern numerical analyst communes with the spirits of his or her predecessors back across the centuries, as far as the time of Newton, if not farther. Alas, times *do* change; with the exception of two of the most modest formulas $\cdots$, the classical formulas are almost entirely useless. They are museum pieces, but beautiful ones.

F.2.1   Closed Newton-Cotes Formulae

The so-called *closed* Newton-Cotes formulae are obtained by setting $x_0 = a$, $x_n = b$ and $h = (b-a)/n$ in equation (F.5). The designation "closed" refers to the fact that the endpoints a and b are the extreme quadrature points in the formula.

---

[2] Named in honor of Sir Isaac Newton (1642-1727), who is best known for his discoveries of the calculus and the law of gravitation, and of Roger Cotes (1683-1717), an English mathematician and astronomer who is noted for his work in geometry.

To determine a general expression for the weighting coefficients w_i defined by equation (F.6), a suitable change of variable is required. The quadrature error associated with the closed Newton-Cotes formulae is given by the following theorem [263]:

Theorem. Let $x_0 = a$, $x_n = b$, $h = (b - a)/n$, and $t = (x - x_0)/h$. Let f have a continuous derivative of order $(n + 2)$ on $[a, b]$. If n is *even*, then there exists a number $\mu \in [a, b]$ for which

$$E_{n+1}(f) = \frac{h^{(n+3)} f^{(n+2)}(\mu)}{(n + 2)!} \int_0^n t^2 (t - 1) \cdots (t - n) \, dt \qquad (F.8)$$

If n is *odd*, then

$$E_{n+1}(f) = \frac{h^{(n+2)} f^{(n+1)}(\mu)}{(n + 1)!} \int_0^n t(t - 1) \cdots (t - n) \, dt \qquad (F.9)$$

The values of the integrals appearing in equations (F.8) and (F.9) are known to be less than zero [263].

In light of the above theorem, when n is an *even* integer, the degree of precision is $(n + 1)$, although the interpolation polynomial is of degree at most n. If n is *odd*, the degree of precision is only n. Thus, if n is even and more quadrature points are to be added, no accuracy is gained by adding odd numbers of points; points should be added in *multiples of two*.

The first four closed Newton-Cotes formulae, along with the associated error terms, are listed below. Additional formulae in the sequence are given in [263, 2, 519].

- $n = 1$ (Trapezoidal rule):

$$\int_{x_0}^{x_1} f(x) \, dx = \frac{h}{2} [f(x_0) + f(x_1)] - \frac{h^3}{12} f^{(2)}(\mu) \qquad (F.10)$$

- $n = 2$ (Simpson's rule[3]):

$$\int_{x_0}^{x_2} f(x) \, dx = \frac{h}{3} [f(x_0) + 4f(x_1) + f(x_2)] - \frac{h^5}{90} f^{(4)}(\mu) \qquad (F.11)$$

---

[3] Named in honor of the English mathematician Thomas Simpson (1710-1761). Apparently this formula was previously known by Newton and Cotes [519].

- $n = 3$ (Simpson's three-eighths rule):

$$\int_{x_0}^{x_3} f(x) \, dx = \frac{3h}{8} [f(x_0) + 3f(x_1) + 3f(x_2) + f(x_3)]$$

$$-\frac{3h^5}{80} f^{(4)}(\mu) \tag{F.12}$$

- $n = 4$ (Boole's rule):

$$\int_{x_0}^{x_4} f(x) \, dx = \frac{2h}{45} [7f(x_0) + 32f(x_1) + 12f(x_2) + 32f(x_3) + 7f(x_4)]$$

$$-\frac{8h^7}{945} f^{(6)}(\mu) \tag{F.13}$$

In the above formulae, $x_0 < \mu < x_n$.

F.2.2 Open Newton-Cotes Formulae

In the so-called *open* Newton-Cotes formulae the points $x_j = x_0 + jh$ are used for each $j = 0, 1, 2, \cdots, n$, where $h = (b-a)/(n+2)$ and $x_0 = a+h$. This implies that $x_n = b - h$. Labeling the endpoints by setting $x_{-1} = a$ and $x_{n+1} = b$, the open Newton-Cotes formulae are again described by equations (F.5) and (F.6). The values of $f(x)$ at the endpoints of $[a, b]$ are *not* used in the formulae.

The error associated with the open Newton-Cotes formulae is very similar to that for the closed formulae [263].

Theorem. Let $x_0 = a + h$, $x_n = b - h$, $h = (b-a)/(n+2)$, and $t = (x - x_0)/h$. Let f have a continuous derivative of order $(n+2)$ on $[a, b]$. If n is *even*, then there exists a number $\mu \in [a, b]$ for which

$$E_{n+1}(f) = \frac{h^{(n+3)} f^{(n+2)}(\mu)}{(n+2)!} \int_{-1}^{n+1} t^2(t-1) \cdots (t-n) \, dt \tag{F.14}$$

If n is *odd*, then

$$E_{n+1}(f) = \frac{h^{(n+2)} f^{(n+1)}(\mu)}{(n+1)!} \int_{-1}^{n+1} t(t-1) \cdots (t-n) \, dt \tag{F.15}$$

The values of the integrals appearing in equations (F.14) and (F.15) are known to be greater than zero [263].

The first four open Newton-Cotes formulae, along with the associated error terms, are listed below. Additional formulae in the sequence are given in [263, 2, 519].

- $n = 0$ (Midpoint rule):

$$\int_{x_0}^{x_2} f(x)\,dx = 2hf(x_1) + \frac{h^3}{3} f^{(2)}(\mu) \tag{F.16}$$

- $n = 1$:

$$\int_{x_0}^{x_3} f(x)\,dx = \frac{3h}{2} \left[f(x_1) + f(x_2) \right] + \frac{3h^3}{4} f^{(2)}(\mu) \tag{F.17}$$

- $n = 2$:

$$\int_{x_0}^{x_4} f(x)\,dx = \frac{4h}{3} \left[2f(x_1) - f(x_2) + 2f(x_3) \right] + \frac{14h^5}{45} f^{(4)}(\mu) \tag{F.18}$$

- $n = 3$:

$$\int_{x_0}^{x_5} f(x)\,dx = \frac{5h}{24} \left[11f(x_1) + f(x_2) + f(x_3) + 11f(x_4) \right]$$
$$+ \frac{95h^5}{144} f^{(4)}(\mu) \tag{F.19}$$

In the above formulae, $x_0 < \mu < x_{n+2}$.

F.3   Gaussian Quadrature

The Newton-Cotes formulae require that the values of the function $f(x)$ be known at *evenly spaced* points. Practically speaking, in the integration of the finite element equations, $f(x)$ is evaluated at a given quadrature point in the "parent" domain Ω^p by calling a suitable subroutine. Since such points may be located *anywhere* in Ω^p, no additional difficulties are introduced if the points are not equally spaced.

Gaussian quadrature[4] is concerned with choosing the quadrature points $x_1, x_2, \cdots, x_n$ and weighting coefficients $w_1, w_2, \cdots, w_n$ in an *optimal* manner. Since the values of the w_i are completely arbitrary and those of the x_i are restricted only in the sense that the function $f(x)$ must be defined at these points, there are at most $2n$ parameters involved. If the coefficients of a polynomial are also considered as parameters, the class of polynomials of degree less than or equal to $(2n - 1)$ contains at most $2n$ parameters. As such, this would be the largest class of polynomials for which it would be reasonable to expect Gaussian quadrature to be exact.[5]

The Gaussian quadrature formula with degree of precision of $(2n - 1)$ has the general form[6]

$$\int_a^b c(x)f(x)\,dx = \sum_{i=1}^n f(x_i)w_i + E_n(f) \qquad \text{(F.20)}$$

where $c(x)$ represents a continuous weighting function and the weighting coefficients w_i are again independent of the $f(x_i)$. The quadrature points x_i are chosen to be the n roots (zeros) of $Q_n(x)$, a polynomial of degree n that is *orthogonal* to the polynomial being integrated, that is to $P_n(x)$. More formally [519],

Definition
An nth degree polynomial $Q_n(x)$ is said to be orthogonal on the interval $[a, b]$ with respect to the weighting function $c(x)$ to all polynomials of degree less than or equal to $(n - 1)$ if

$$\int_a^b c(x)\,Q_n(x)\,P_k(x)\,dx = 0 \qquad \text{(F.21)}$$

for all polynomials $P_k(x)$ of degree less than or equal to $(n - 1)$.

The polynomial $Q_n(x)$ is part of a sequence $Q_0(x), Q_1(x), \ldots Q_n(x)$, where $Q_0(x), Q_1(x), \ldots, Q_{n-1}(x)$ are all known. Some theorems concerning orthogonal polynomials and their zeros, given here without proof, are particularly useful. Proofs for these can be found in [519].[7]

---

[4]Named in honor of the German mathematician Carl Friedrich Gauss (1777-1855).

[5]The formal proof that a Gaussian quadrature formula indeed has a degree of precision of $(2n - 1)$ is given in [519].

[6]It appears [240] that only the case in which $c(x) = 1$ was explicitly considered by Gauss [189], the generalization to other weighting functions being due to Christoffel [107].

[7]Further details pertaining to the theory of orthogonal polynomials can be found in [245, 575, 205].

Theorem

The polynomial $Q_n(x)$ that satisfies

$$\int_a^b c(x)\, Q_n(x)\, P_k(x)\, dx = 0 \qquad\qquad \text{(F.22)}$$

for all polynomials $P_k(x)$ of degree less than or equal to $(n-1)$, exists and is unique except for a multiplicative constant.

Theorem

Let $Q_n(x)$ be the orthogonal polynomial for the interval $[a,b]$ and weighting function $c(x)$. Then all the zeros of $Q_n(x)$ are *real* and *distinct* and lie in the open interval $[a,b]$

By a suitable choice of orthogonal polynomials $Q_n(x)$ and weighting function $c(x)$, Gaussian quadrature formulae can be derived that are very accurate for a wide range of functions over varying intervals $[a,b]$. Some examples of specific $Q_n(x)$ are: Chebyshev, Jacobi, Hermite, Laguerre, and Legendre.[8] A tabulation of these polynomials is given in [519, 443].

The weighting coefficients w_i are determined from the formula

$$w_i = \frac{\langle Q_{n-1}(x),\, Q_n'(x)\rangle}{nQ_{n-1}(x),\, Q_n'(x)} = \frac{\int_a^b [Q_{n-1}(x)\, Q_n'(x)\, c(x)]\, dx}{nQ_{n-1}(x),\, Q_n'(x)} \qquad\qquad \text{(F.23)}$$

and are always *positive* [263].

The quadrature error associated with general Gaussian formulae is [263]

$$E_n(f) = \frac{f^{(2n)}(\mu)}{(2n)!} \int_a^b c(x)\, [Q_n(x)]^2\, dx \qquad\qquad \text{(F.24)}$$

where μ is some point in $[a,b]$.

---

[8]Named in honor of the Russian mathematician P. L. Chebyshev (1821-1894), the German mathematician Karl Gustav Jacob Jacobi (1804-1851), and the French mathematicians Charles Hermite (1822-1901), E. N. Laguerre (1834-1886) and Adrien-Marie Legendre (1752-1833), respectively.

F.3.1 Gauss-Legendre Quadrature

The *Legendre* orthogonal polynomials are used in most finite element applications. These polynomials are orthogonal on the interval $[-1, 1]$ with respect to the weighting function $c(x) = 1$. Beginning with $Q_0(x) = 1$ and $Q_1(x) = x$, the Legendre polynomials for $k = 2, 3, \cdots$ are defined recursively by [99]:

$$Q_k(x) = (x - B_k) \, Q_{k-1}(x) - C_k \, Q_{k-2}(x) \qquad (\text{F.25})$$

where

$$B_k = \frac{\int_{-1}^{1} x \, [Q_{k-1}(x)]^2 \, dx}{\int_{-1}^{1} [Q_{k-1}(x)]^2 \, dx}, \quad C_k = \frac{\int_{-1}^{1} x \, Q_{k-1}(x) \, Q_{k-2}(x) \, dx}{\int_{-1}^{1} [Q_{k-2}(x)]^2 \, dx} \qquad (\text{F.26})$$

For Gauss-Legendre formulae, the quadrature error expression (F.24) becomes [263]

$$E_n(f) = \frac{2}{(2n + 1)!} \left[\frac{2^n (n!)^2}{(2n)!} \right]^2 f^{(2n)}(\mu) \qquad (\text{F.27})$$

for $-1 < \mu < 1$.

Provided $b \neq a$, the linear coordinate transformation

$$x = \frac{1}{2} \left[(b - a)\xi + (a + b) \right] \qquad (\text{F.28})$$

will map any interval $[a, b]$ into $[-1, 1]$. For finite element applications this transformation is, of course, not necessary provided that the actual element geometry is mapped to the parent domain $[-1, 1]$ in natural coordinates (see Chapter 8).

Some values of quadrature points x_i and weighting coefficients w_i for Gauss-Legendre formulae are listed in Table F.1. More extensive tabulations are given in a number of specialized references [132, 518, 133] and in most finite element textbooks.

Remark

1. It is interesting to note that for any order of Gauss-Legendre quadrature, $\sum_{i=1}^{n} w_i = 2.0$, which is the length of the interval $[-1, 1]$ (the one-dimensional "parent" domain Ω^p). This guarantees that constant functions will always be integrated exactly.

Table F.1: Locations and Weights for Gauss-Legendre Quadrature

$$\int_{-1}^{1} f(\xi) \, d\xi \cong \sum_{i=1}^{n} w_i f(\xi_i)$$

n	Precision	w_i	ξ_i
1	1	2.0	0.0
2	3	1.0	$\pm 1/\sqrt{3}$
3	5	5/9	$\pm\sqrt{3/5}$
		8/9	0.0
4	7	$\dfrac{18 - \sqrt{30}}{36}$	$\pm \left[\dfrac{15 + 2\sqrt{30}}{35} \right]^{0.5}$
		$\dfrac{18 + \sqrt{30}}{36}$	$\pm \left[\dfrac{15 - 2\sqrt{30}}{35} \right]^{0.5}$
5	9	$\dfrac{5103}{50\left(322 + 13\sqrt{70}\right)}$	$\pm\dfrac{\sqrt{35 + 2\sqrt{70}}}{\sqrt{63}}$
		$\dfrac{5103}{50\left(322 - 13\sqrt{70}\right)}$	$\pm\dfrac{\sqrt{35 - 2\sqrt{70}}}{\sqrt{63}}$
		$\dfrac{128}{225}$	0.0

F.3.2 Modifications to Gaussian Quadrature

The basic ideas underlying Gaussian Quadrature have been modified in a number of ways. Two of these, which are particularly pertinent to the finite element method are discussed below.

Constrained Gaussian Quadrature

The first important modification is the *advance assignment* of some of the quadrature point locations. The problem is then to choose the weighting coefficients and locations of the remaining points to maximize the degree of precision of the quadrature formula. For such approaches, the degree of precision would be no higher than that achieved by the corresponding Gaussian formula and no less than that achieved by the corresponding Newton-Cotes formula.

One of the most common examples of such constraints is *Gauss-Radau* quadrature [451], where one of the points is an endpoint of the interval, either a or b. Gauss-Radau formulae for use in finite element analyses were first presented by Irons [254]. However, due to a lack of symmetry, application of these formulae has been rather limited.

A second example of constrained Gaussian formulae is *Gauss-Lobatto* quadrature, in which two of the quadrature points are the endpoints a and b. The Lobatto quadrature formulae for $n = 2$ and $n = 3$ are identical to the trapezoidal and Simpson's rules, respectively. Although the degree of precision of Gauss-Lobatto formulae is no greater than the corresponding Gaussian formulae, the former prove useful in certain so-called "lumping" techniques employed in transient and dynamic finite element analyses. Quadrature points and weighting coefficients for Lobatto quadrature formulas for $n = 3$ to $n = 10$ are given in [132].

Gauss-Kronrod Quadrature

The second important modification is the *Gauss-Kronrod* formulae. If ordinary Gaussian formulae of orders n and $n + 1$ are used, none of the quadrature points coincide. Thus, if the quadrature error is to be estimated by successively increasing n, the previous function evaluations cannot be reused.[9]

Kronrod [296] proposed a method of searching for optimal sequences of formulae, each of which reuses all quadrature points of its predecessor. Thus, if one starts with n points and adds p new points, one has $(2p + n)$

---

[9]Romberg integration, which uses "composite" trapezoidal rule with progressively smaller h to give preliminary approximations and then applies the Richardson extrapolation process to improve the approximations, allows for such a reuse [99].

free parameters: the p new points and weighting coefficients, and n new weights for the n fixed previous points. The maximum expected degree of precision would thus be $(2p+n-1)$. The question is whether this maximum degree of precision can actually be achieved when the quadrature points are required to all lie inside $[a,b]$. The answer to this question is apparently not known in general [443], though similar ideas have been applied to other forms of Gaussian quadrature [192].

Closing Remarks

In using the Newton-Cotes formulae, $(n+1)$ function evaluations are required to exactly integrate a polynomial of (at best) degree $(n+1)$. On the other hand, if Gaussian quadrature is used, a polynomial of degree $(2n-1)$ is integrated exactly with only n function evaluations. Consequently, Gaussian quadrature is generally more efficient to use than Newton-Cotes formulae.[10]

F.4   Numerical Integration of Lines

Although the line elements developed in Chapter 7 were integrated exactly, such elements are typically integrated numerically using Gauss-Legendre formulae. When mapping such elements to Ω^p, the infinitesimal length dx_1 must be replaced by an equivalent expression in ξ-space. Thus for one-dimensional Gauss-Legendre quadrature:

$$\int f(x_1)\, dx_1 = \int_{-1}^{1} f(\xi)\, \det J\, d\xi \approx \sum_{i=1}^{N} w_i\, f(x_i)\, \det \mathbf{J}\,(\xi_i) \qquad \text{(F.29)}$$

Since $f(x_1)$ includes derivatives with respect to x_1, its transformation to ξ-space requires appropriate use of the chain rule of differentiation.

F.5   Numerical Integration of Rectangles

For numerical integration of rectangular elements, the two-dimensional counterpart of equation (F.29) is

$$\iint f(x_1,x_2)\, dx_1 dx_2 = \int_{-1}^{1} \int_{-1}^{1} f(\xi,\eta)\, \det \mathbf{J}\, d\xi\, d\eta \qquad \text{(F.30)}$$

---

[10]Newton-Cotes formulae may, however, be efficient in certain nonlinear applications [58].

which is approximated by integrating in the ξ- and η-directions independently. We begin by evaluating the inner integral

$$\int_{-1}^{1} \int_{-1}^{1} f(\xi, \eta) \det \mathbf{J} \, d\xi \, d\eta \approx \int_{-1}^{1} \left[\sum_{i=1}^{M} f(\xi_i, \eta) w_i \right] \det \mathbf{J}(\xi_i) \, d\eta \quad \text{(F.31)}$$

Integrate next in the η-direction gives

$$\int_{-1}^{1} \int_{-1}^{1} f(\xi, \eta) \det \mathbf{J} \, d\xi \, d\eta \approx \sum_{j=1}^{N} \sum_{i=1}^{M} w_i \, w_j f(\xi_i, \eta_j) \det \mathbf{J}(\xi_i, \eta_j) \quad \text{(F.32)}$$

If the integrals in the ξ and η directions separately are exact for polynomials of degree n, then equation (F.32) will exactly integrate all terms such as $\xi^p \eta^q$, where the integers p and q satisfy the equations $0 \le p, q \le n$, $p + q \le n$.

The double sum in equation (F.32) can easily be programmed as a single "do loop" with a range of 1 to $(M * N)$. For example, consider the 2 x 2 Gauss-Legendre quadrature for a square region ("parent" domain). The corresponding four quadrature points are numbered in the manner shown in Figure F.1 (a).

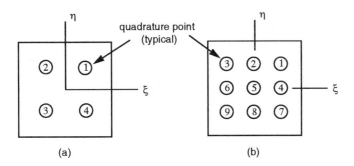

(a)   (b)

Figure F.1: Numbering of Quadrature Points for Hypothetical Rectangular Region: (a) 2 x 2 and (b) 3 x 3 Integration Formulae

Noting that the single loop index i ranges from 1 to 4, and forming appropriate products of weighting coefficients w_i obtained from Table F.1, the quadrature points and weights listed in Table F.2 would be used.

Next consider a 3 x 3 Gauss-Legendre formula. The corresponding nine quadrature points are numbered in the manner shown in Figure F.1 (b).

Table F.2: Locations and Weights for Single Loop 2 x 2 Quadrature

i	w_i	ξ_i	η_i
1	1.0	$1/\sqrt{3}$	$1/\sqrt{3}$
2	1.0	$-1/\sqrt{3}$	$1/\sqrt{3}$
3	1.0	$-1/\sqrt{3}$	$-1/\sqrt{3}$
4	1.0	$1/\sqrt{3}$	$-1/\sqrt{3}$

Table F.3: Locations and Weights for Single Loop 3 x 3 Quadrature

i	ξ_i	η_i	w_i
1	$\sqrt{3/5}$	$\sqrt{3/5}$	25/81
2	0	$\sqrt{3/5}$	40/81
3	$-\sqrt{3/5}$	$\sqrt{3/5}$	25/81
4	$\sqrt{3/5}$	0	40/81
5	0	0	64/81
6	$-\sqrt{3/5}$	0	40/81
7	$\sqrt{3/5}$	$-\sqrt{3/5}$	25/81
8	0	$-\sqrt{3/5}$	40/81
9	$-\sqrt{3/5}$	$-\sqrt{3/5}$	25/81

Noting that the single loop index i ranges from 1 to 9, and forming products of weighting coefficients w_i obtained from Table F.1, the quadrature points and weights listed in Table F.3 would be used.

Remark

1. For any order of Gauss-Legendre quadrature, $\sum\limits_{i=1}^{M*N} w_i = 4.0$, which is the area of the "parent" domain Ω^p $(-1 \leq \xi, \eta \leq 1)$.

F.6 Numerical Integration of Prisms

F.6.1 Right Prisms

The numerical integration of right prisms follows along similar lines, viz.,

$$\iiint f(x_1, x_2, x_3) \, dx_1 \, dx_2 \, dx_3$$

$$= \int_{-1}^{1} \int_{-1}^{1} \int_{-1}^{1} f(\xi, \eta, \zeta) \, \det \mathbf{J} \, d\xi \, d\eta \, d\zeta$$

$$\approx \sum_{k=1}^{P} \sum_{j=1}^{N} \sum_{i=1}^{M} w_i \, w_j w_k \, f(\xi_i, \eta_j, \zeta_k) \, \det \mathbf{J} \, (\xi_i, \eta_j, \zeta_k) \qquad (\text{F.33})$$

The triple sum can easily be programmed as a single "do loop" with a range of 1 to $(M * N * P)$. If the integrals in the ξ, η and ζ directions separately are exact for polynomials of degree n, then equation (F.33) will exactly integrate all terms such as $\xi^p \eta^q \zeta^r$, where the integers p, q and r satisfy the equations $0 \leq p, q, r \leq n$, $p + q + r \leq n$.

Remarks

1. For any order of Gauss-Legendre quadrature, $\sum_{i=1}^{M*N*P} w_i = 8.0$, which is the volume of the "parent" domain Ω^p $(-1 \leq \xi, \eta, \zeta \leq 1)$.

2. In multi-dimensional cases Gaussian quadrature formulae are no longer necessarily optimal [204, 252]. Three-dimensional examples include a six-point (non-Gaussian) formula [132] that is fourth-order accurate (the comparable Gaussian formula requires $2 \times 2 \times 2 = 8$ quadrature points), and a fourteen-point formula [216, 258, 219] that is sixth-order accurate (the comparable Gaussian formula requires $3 \times 3 \times 3 = 27$ quadrature points).[11] None the less, despite this level of inefficiency in multi-dimensional applications, Gaussian quadrature formulae are still widely used and are sufficient for a large number of typical applications.

---

[11] This fourteen-point formula accurately integrates any term h^5, where h is a linear combination of ξ, η and ζ. If this formula was truly as accurate as 27-point Gauss-Legendre quadrature, it would be quite efficient, as it requires slightly more than half as much computational effort. However, the fourteen-point formula does not accurately integrate the term $\xi^2 \eta^2 \zeta^2$ [370], which is exactly integrated by even the 2 x 2 x 2 Gauss-Legendre formula.

F.6.2   Triangular Prisms

Quadrature formulae for triangular prisms are constructed by combining triangular formulae in the $\xi - \eta$ plane with Gaussian type formulae in the ζ direction. Commonly used formulae for triangular prisms are the six-point (3 x 2) formula that is accurate to $O(h^2\zeta^3)$, and the twenty-one-point (7 x 3) formula that is accurate to $O(h^5\zeta^5)$ [370]. Here h is an arbitrary combination of ξ and η.

F.7   Integration of Triangles and Tetrahedra

F.7.1   Exact Integration

The integration of monomials over straight-edged triangles and flat-surfaced tetrahedra is facilitated by the following exact expressions:

For Triangles

$$\iint\limits_{\Omega} L_1^a L_2^b L_3^c \, d\Omega = \frac{a! \, b! \, c!}{(a + b + c + 2)!} \, 2A \tag{F.34}$$

where A, the area of Ω, is given by the following determinant

$$A = \frac{1}{2} \begin{vmatrix} 1 & 1 & 1 \\ x_1 & x_2 & x_3 \\ y_1 & y_2 & y_3 \end{vmatrix} \tag{F.35}$$

in which (x_i, y_i) are the coordinates of vertex i $(i = 1, 2, 3)$.

For Tetrahedra

$$\iiint\limits_{\Omega} L_1^a L_2^b L_3^c L_4^d \, d\Omega = \frac{a! \, b! \, c! \, d!}{(a + b + c + d + 3)!} \, 6V \tag{F.36}$$

where V, the volume of Ω, is given by the following determinant

$$V = \frac{1}{6} \begin{vmatrix} 1 & 1 & 1 & 1 \\ x_1 & x_2 & x_3 & x_4 \\ y_1 & y_2 & y_3 & y_4 \\ z_1 & z_2 & z_3 & z_4 \end{vmatrix} \tag{F.37}$$

in which (x_i, y_i, z_i) are the coordinates of vertex i ($i = 1, 2, 3, 4$).

In equations (F.34) and (F.36), L_1, L_2, L_3 and L_4 are the natural coordinates defined in Appendix C, and a, b, c, d are integers.[12]

F.7.2 Numerical Integration

In the case of *distorted* triangles and tetrahedra, the integrands will not consist solely of monomials. Consequently, numerical integration must be used. The derivation of an integration formula of a given degree of precision again involves the determination of weighting coefficients w_i, and the number and location of the corresponding quadrature points in the triangular or tetrahedral "parent" domain. Often, symmetry constraints are placed on the quadrature point locations to ensure invariance under element vertex reordering and thus avoid biasing one of the vertices. The efficiency of a given formula is based on the number of quadrature points required to give a certain degree of precision.

In theory, Gauss-Legendre quadrature could be extended to triangles by treating them as quadrilaterals with one degenerate edge of zero length [370]. However, such formulae lack the aforementioned symmetry with respect to the directions of the the the edges. Consequently, it is advantageous to use specially developed, symmetric formulae.

For Triangles

The first efficient (Gaussian type) and symmetric quadrature formulae for triangles and tetrahedra were given by Hammer and his collaborators [215, 216]. Each of the formulae was derived analytically, and although of relatively low order (up to fifth degree for triangles and up third degree for tetrahedra), they provided insight into the derivation of higher order formulae.

A different set of formulae, which make use of successive application of Gauss and Radau quadrature, was subsequently presented by Irons [254]. Although quite efficient and extending to relatively high orders, these employed sampling points that were not symmetrically arranged within the triangle.

As an alternative to the analytical approach employed by Hammer [215, 216], Silvester [495, 497] presented a Newton-Cotes type of procedure for deriving symmetric quadrature formulae for triangles and tetrahedra. This

---

[12]Equation (F.34) was first presented, without proof, by Stricklin [516]. Its extension to the volume integration described by equations (F.36) is attributed to Brebbia [93]. A formal derivation of both equations was subsequently given by Eisenberg and Malvern [147].

procedure is based on orthogonal polynomials and is easily extended to higher orders. However, being of the Newton-Cotes type, these formulae lacked the efficiency of the Gaussian quadrature.

Several authors have extended the work of Hammer [215, 216] to higher orders. Cowper [124] analytically determined sixth- and seventh-degree formulae for triangles.[13] Hillion [241] approximated integrals on a triangle by forming products of one-dimensional Gauss-Jacobi formulae. Laursen and Gellert [306] established a systematic criterion limiting the number of possible quadrature point configurations for triangles and presented some new high order formulae. Reddy [459] derived a three-point formula that provides exact integrals of some cubic terms; earlier three-point formulae [216, 334] exactly integrated only second-degree polynomials. Reddy and Shippy [460] presented some new fourth-, sixth-, and seventh-order formulae. Formulae for every degree up to twenty were presented by Dunavant [144], who solved the associated nonlinear equations numerically. A review of other quadrature formulae for triangles is presented by Cools and Rabinowitz [117].

Since they are of the Gaussian type and fully symmetric with respect to the vertices, the formulae of Hammer at al. [215] and Cowper [124] are commonly used when numerically integrating triangular finite elements. For numerically integrating a function f over a triangular area A, these formulae have the following general form:

$$\iint_A f \, dA = A \sum_{i=1}^{n} w_i \, f\left(L_{1i}, \, L_{2i}, \, L_{3i}\right) \qquad \text{(F.38)}$$

where L_{1i}, L_{2i} and L_{3i} are the natural (area) coordinates of the ith sampling point, and w_i is the associated weight. Values for L_{1i}, L_{2i}, L_{3i} and w_i are given in Table F.4. Due to the symmetry of the formulae the sampling points occur in groups of either one (at the centroid of the area), three (along bisectors of the vertices) or six (in pairs symmetrically offset from each bisector). The precision shown in the table indicates the highest degree polynomial that the formula integrates exactly.

Remark

1. Formulae that include *negative* w_i have been omitted since they tend to produce instabilities and thus cannot be used in developing element mass matrices for dynamic analyses [370].

---

[13]The four-point formula of Hammer et al. [215] and Cowper [124] was discussed by Lannoy [304], who derived different w_i for the same quadrature points.

Table F.4: Numerical Integration Formulae for Triangles

n	Precision	w_i	L_{1i}, L_{2i}, L_{3i}
1	1	1.0	1/3, 1/3, 1/3
3	2	1/3	1/2, 1/2, 0
		1/3	0, 1/2, 1/2
		1/3	1/2, 0, 1/2
3	2	1/3	2/3, 1/6, 1/6
		1/3	1/6, 2/3, 1/6
		1/3	1/6, 1/6, 2/3
6	4	β_1	$\alpha_1, \alpha_2, \alpha_2$
		β_1	$\alpha_2, \alpha_1, \alpha_2$
		β_1	$\alpha_2, \alpha_2, \alpha_1$
		β_2	$\alpha_3, \alpha_4, \alpha_4$
		β_2	$\alpha_4, \alpha_3, \alpha_4$
		β_2	$\alpha_4, \alpha_4, \alpha_3$
7	5	δ_1	1/3, 1/3, 1/3
		δ_2	$\gamma_1, \gamma_2, \gamma_2$
		δ_2	$\gamma_2, \gamma_1, \gamma_2$
		δ_2	$\gamma_2, \gamma_2, \gamma_1$
		δ_3	$\gamma_3, \gamma_4, \gamma_4$
		δ_3	$\gamma_4, \gamma_3, \gamma_4$
		δ_3	$\gamma_4, \gamma_4, \gamma_3$

$$\alpha_1 = 0.816847572980459, \ \alpha_2 = 0.091576213509771$$

$$\alpha_3 = 0.108103018168070, \ \alpha_4 = 0.445948490915965$$

$$\beta_1 = 0.109951743655322, \ \beta_2 = 0.223381589678011$$

$$\gamma_1 = 0.797426985353087, \ \gamma_2 = 0.101286507323456$$

$$\gamma_3 = 0.470142064105115, \ \gamma_4 = 0.059715871789770$$

$$\delta_1 = 0.225000000000000$$

$$\delta_2 = 0.125939180544827, \ \delta_3 = 0.132394152788506$$

Table F.5: Numerical Integration Formulae for Tetrahedra

n	Precision	w_i	$L_{1i}, L_{2i}, L_{3i}, L_{4i}$
1	1	1.0	1/4, 1/4, 1/4, 1/4
4	2	1/4	$\alpha, \beta, \beta, \beta$
		1/4	$\beta, \alpha, \beta, \beta$
		1/4	$\beta, \beta, \alpha, \beta$
		1/4	$\beta, \beta, \beta, \alpha$

$$\alpha = 0.5854101966249685, \ \beta = 0.138196601125015$$

For Tetrahedra

Quadrature formulae for tetrahedral domains have been developed using ideas similar to those discussed above for triangles. Although fully symmetric, the first sets of such formulae were either of a low order [215] or lacked the efficiency of Gaussian quadrature [495, 497]. The extension of Gaussian type formulae to higher orders was subsequently realized by Jinyun [268] and Keast [283]. Other formulae are given in [117]. Errors in one of Keast's formulae and in the formulae of Sunder and Cookson [521] have been noted [482].

For numerically integrating a function f over a tetrahedral volume V, the aforementioned formulae have the following general form:

$$\iint\limits_{V} f\left(L_1, L_2, L_3, L_4\right) dV \cong V \sum_{i=1}^{n} w_i \, f\left(L_{1i}, L_{2i}, L_{3i}, L_{4i}\right) \qquad \text{(F.39)}$$

where L_{1i}, L_{2i}, L_{3i} and L_{4i} are the natural (volume) coordinates of the i-th sampling point, and w_i is the associated weight. Values for L_{1i}, L_{2i}, L_{3i}, L_{4i} and w_i are given in Table F.5. Higher order formulas are available in the aforementioned papers by Jinyun [268] and Keast [283].

F.8 Order of Numerical Integration

The final issue requiring discussion is the order of the quadrature formula that should be used in integrating the finite element equations. Some general guidelines, applicable to one-, two- and three-dimensional elements are provided herein. The treatment is not meant to be exhaustive. Readers interested in further details pertaining to the subject should consult references such as [254, 161, 180, 181, 515, 100, 370].

The arrays requiring numerical integration are the element properties ("stiffness") matrix,

$$\mathbf{K}^{(e)} = \int_{\Omega^e} \mathbf{B}^T \mathbf{C} \mathbf{B} \, d\Omega \qquad (\text{F.40})$$

The element mass matrix (for dynamic analyses),

$$\mathbf{M}^{(e)} = \int_{\Omega^e} \mathbf{N}^T \rho \mathbf{N} \, d\Omega \qquad (\text{F.41})$$

The element "source" (body force) vector

$$\mathbf{f_b}^{(e)} = \int_{\Omega^e} \mathbf{N}^T \mathbf{b} \, d\Omega \qquad (\text{F.42})$$

and the element "surface traction" (flux) vector

$$\mathbf{f_t}^{(e)} = \int_{\Gamma_1^e} \mathbf{N}^T \bar{\mathbf{t}} \, d\Gamma \qquad (\text{F.43})$$

In addition, for stress-deformation analyses, the initial strain and initial stress vectors also require numerical integration. These two vectors, together with $\mathbf{f_b}^{(e)}$ and $\mathbf{f_t}^{(e)}$ are summed to form the element right-hand side vector $\mathbf{f}^{(e)}$.

F.8.1 General Observations

In selecting a quadrature formula, the primary concerns are efficiency, cost and accuracy. On the topic of efficiency, we have already noted in Section F.3.2 that for the same number of function evaluations, Gaussian quadrature formulae are more efficient than Newton-Cotes formulae.

When considering the cost of computations (i.e., the computational effort), we would like the order of the quadrature formula to be as low as possible. Assuming a symmetric $\mathbf{K}^{(e)}$, an element with N_{dof} degrees of freedom will require the evaluation of $N_{dof}\left(N_{dof} + 1\right)/2$ integrals. The cost of numerical integration is thus roughly proportional to the square of N_{dof}. The computational effort required to numerically integrate a given integrand is proportional to the number of quadrature points [370]. Consequently, if a particular quadrature formula can integrate an integrand exactly, it makes no sense to consider higher order formulae.

In summary, once a particular numerical quadrature scheme has been selected, the order of integration must be determined judiciously. In particular, we seek not only the *minimum* order of integration that guarantees convergence of the finite element approximation, but also the *optimal* order that ensures the highest possible rate of convergence while minimizing computational effort.

F.8.2   Minimum and Optimal Order of Integration

Consider first the lowest order formula that will merely preserve convergence in accordance with the criteria (which assume *exact* integration) discussed in Section 7.5 of Chapter 7. We recall that convergence will occur provided any arbitrary constant values of derivatives up to order s appearing in the element equations can be reproduced in the limit as the element size $h^{(e)} \to 0$. It has thus been argued [254, 515] that convergence should still occur if the volume (in three dimensions), area (in two dimensions) of length (in one dimension) can be integrated exactly.

To better understand the rationale behind this notion, consider the element properties ("stiffness") matrix $\mathbf{K}^{(e)}$ given by equation (F.40). In the limit as $h^{(e)} \to 0$, the integrand $\mathbf{B}^T\mathbf{C}\mathbf{B}$ approaches a constant and can be moved outside the integral. The numerical integration must thus integrate

$$\mathbf{B}^T\mathbf{C}\mathbf{B} \int_{\Omega^e} d\Omega$$

exactly. In natural coordinates, this would become

$$\mathbf{B}^T\mathbf{C}\mathbf{B} \int_{\Omega^e} \det \mathbf{J}\, d\xi d\eta d\zeta$$

with suitable simplification for two- and one-dimensional problems.

The "exact volume rule" tends to be unacceptable for three reasons [100]. First, since the order of integration tends to be low, the quadrature error can potentially exceed the discretization error associated with approximating the exact solution by a trial solution with a finite number of terms (this error was first discussed in Chapter 6). Second, even though it will preserve convergence, the "exact volume rule" will result in a decrease in the *rate* of convergence. Finally, the low order of integration frequently yields instabilities, commonly referred to as "zero-energy modes," that render $\mathbf{K}^{(e)}$ singular.[14].

Finally, even though it will preserve convergence, the "exact volume rule" will result in a decrease in the *rate* of convergence.

At the other extreme, one may use a high-order quadrature formula that is capable of evaluating the entire integrand exactly. This is only possible when the integrand is a polynomial, implying that the Jacobian determinant must be a constant. Such a condition is only true for undistorted elements such as rectangles (see Example 10.2), right prisms, etc. Besides being very costly, high-order quadrature formulas run the risk of producing another detrimental numerical effect called *mesh locking* [252]. This phenomenon is alleviated by using lower-order or "reduced" quadrature formulas [591, 404, 592], a subject that will be reviewed in Section F.8.3. In addition, if the order is taken too large, the quadrature error will be several orders of magnitude less than the discretization error. The additional accuracy in integration will thus be wasted, since the overall accuracy would be limited by the discretization error [100].

---

[14]An element instability is sometimes also referred to as a *spurious singular mode*. The term "zero-energy mode" is used to describe instabilities in structural mechanics applications (these may also be referred to as *mechanisms* or *hourglass modes*). In particular, it refers to a nodal displacement vector $\hat{\phi}_n^{(e)}$ that is not a rigid-body motion but none the less produces zero strain energy

$$\hat{U}^{(e)} = \frac{1}{2}\hat{\phi}_n^{(e)T} \mathbf{K}^{(e)} \hat{\phi}_n^{(e)} = \frac{1}{2} \int_{\Omega^e} \hat{\varepsilon}^{(e)T} \mathbf{C} \hat{\varepsilon}^{(e)} \, d\Omega$$

Instabilities occur because of irregularities in the element formation process, a prime example of which is the use of low order quadrature formulae. More precisely, when $\mathbf{K}^{(e)}$ is integrated numerically, it contains only the information at the quadrature points. If it happens that strains are zero at all quadrature points for a certain mode $\hat{\phi}_n^{(e)}$, then $\hat{U}^{(e)}$ will vanish for that $\hat{\phi}_n^{(e)}$. Since $\hat{U}^{(e)}$ is zero, it follows that $\mathbf{K}^{(e)}$ will be a zero stiffness matrix for this particular mode. Under normal circumstances, if $\hat{\phi}_n^{(e)}$ is a rigid body mode, $\hat{U}^{(e)} = 0$. If $\hat{U}^{(e)} = 0$ when $\hat{\phi}_n^{(e)}$ is *not* a rigid body mode, then an instability is present.

An element that displays a zero-energy mode is said to be *rank deficient*. That is, the rank of $\mathbf{K}^{(e)}$ is less than the number of element degrees of freedom minus the number of rigid body modes. An instability in $\mathbf{K}^{(e)}$ can be numerically assessed by means of an eigenvalue analysis. If the integration order is too low, $\mathbf{K}^{(e)}$ can have a larger number of zero eigenvalues than the number of rigid body modes.

For a more detailed discussion of element instabilities, the interested reader should consult the book by Cook et al. [116].

In light of the above discussion, we surmise that the desired optimal order of quadrature formula should lie somewhere between the two extremes of exact volume integration and high-order exact integration. More precisely, we would like the quadrature error to be comparable to the discretization error, and to at least preserve the *rate* of convergence due to the latter. To this end, Burnett [100] has proposed the following heuristic guidelines:

> To preserve the rate of convergence due to the discretization error, integrate $\mathbf{K}^{(e)}$ and $\mathbf{f}^{(e)}$ with a quadrature formula that has an accuracy $O(h^{2(p-s+1)})$. If Gauss-Legendre quadrature is used, a $(p - s + 1)$-point formula would be required.

In addition,

> For elements that are "considerably" distorted, accuracy can sometimes be improved by integrating $\mathbf{K}^{(e)}$ and $\mathbf{f}^{(e)}$ with a quadrature formula that has an accuracy $O(h^{2(p-s+1)+1})$. If Gauss-Legendre quadrature is used, a $(p - s + 2)$-point formula, or higher, should be tried.

In the above guidelines p is the highest degree of the *complete* polynomial in the element trial solution $\hat{\phi}^{(e)}$, and $2s$ is the order of the governing differential equation (recall the discussion of Section 1.3.4), which has been integrated by parts s times. Consequently, s is the order of the highest derivative of the element interpolation functions appearing in the integrand associated with $\mathbf{K}^{(e)}$.

Applying Burnett's guidelines to quadrilateral elements, the Gauss-Legendre quadrature formulas listed in Table F.6 would be specified. Not surprisingly, there are differences in opinion concerning these recommendations. For example, Bathe [58] recommends that even in the undistorted configuration, quadrilaterals should be integrated using the higher-order formulae listed in the third column of Table F.6.

A somewhat safer criterion, which is sufficient to ensure optimal rate of convergence for second-order elliptic C^0 problems requires that

> The squares of the first derivatives of all interpolation functions must be integrated exactly by the quadrature formula [218].

Thus, for elements employing *complete* interpolating polynomials of degree p (this would be the case for all one-dimensional, triangular and tetrahedral elements), the quadrature formula must have a degree of precision equal to at least $2(p - 1)$. If quadrilateral or hexahedral elements are used,

Table F.6: Requisite Number of Gauss-Legendre Quadrature Points from Guidelines of Burnett [100]

Element	Undistorted Geometry	Distorted Geometry
Linear ($p = 1$)	1 x 1	2 x 2
Quadratic ($p = 2$)	2 x 2	3 x 3
Cubic ($p = 3$)	3 x 3	4 x 4

higher-order quadrature formulae are typically required. This is because of the presence of additional terms of higher order than the complete polynomial contained in the interpolation functions. For such elements, at least $p + 1$ quadrature points are required in each coordinate direction.

Example F.1: Required Order of Integration

Consider first a linear (three-node) triangular element. Here $p = 1$ and, as shown in Chapter 9, the element interpolation functions are

$$N_i = \frac{1}{A^{(e)}} (a_i + b_i x + c_i y) \quad ; \quad i = 1, 2, 3 \tag{F.44}$$

where $A^{(e)}$ is the area of the element, and a_i, b_i and c_i are constants. The derivatives of the interpolation functions with respect to x and y, and thus their squares, are constants. The one-point quadrature formula given in Table F.4, which has a degree of precision of one, is thus sufficient.

Next consider a bi-linear (four-node) quadrilateral element (recall Example 9.1). Here again $p = 1$, however, the interpolating polynomial now contains a quadratic term, viz.,

$$\alpha_1 + \alpha_2 \xi + \alpha_3 \eta + \alpha_4 \xi \eta$$

where α_1, α_2, α_3 and α_4 are constants. The associated interpolation functions are

$$N_i = \frac{1}{4}(1 + \xi_i \xi)(1 + \eta_i \eta) \quad ; \quad i = 1, 2, 3, 4 \tag{F.45}$$

where the constants ξ_i and η_i are defined in Table 9.1. The derivative of N_i with respect to x is

$$\frac{\partial N_i}{\partial x} = \frac{\partial N_i}{\partial \xi}\frac{\partial \xi}{\partial x} + \frac{\partial N_i}{\partial \eta}\frac{\partial \eta}{\partial x} = \frac{\xi_i}{4}(1 + \eta_i\eta)\frac{\partial \xi}{\partial x} + \frac{\eta_i}{4}(1 + \xi_i\xi)\frac{\partial \eta}{\partial x} \qquad \text{(F.46)}$$

The first term is seen to be linear in η, while the second is linear in ξ. The square of $\partial N_i/\partial x$ thus contains expressions that are quadratic in ξ and quadratic in η. This finding likewise applies to the square of $\partial N_i/\partial y$. The evaluation of $\mathbf{K}^{(e)}$ thus involves quadratic (and lower order) terms. To integrate quadratic polynomials exactly, a two-point Gauss-Legendre formula, which has a degree of precision equal to three, is required (see Table F.1).

In a similar manner, it is easy to show that quadratic (six-node) triangular elements require a quadrature formula with degree of precision equal to two, and thus a three-point quadrature formula (see Table F.4). Biquadratic quadrilateral elements require a degree of precision of four, and thus a three-point Gauss-Legendre formula.

The extension of the above discussion to three-dimensional elements is straightforward.

We close this section with the following observations.

Remarks

1. Exceptions to the above guidelines for selecting the appropriate order of quadrature frequently arise, particularly for two- and three-dimensional elements.

2. Unlike the guidelines proposed by Burnett, the element right-hand side vector $\mathbf{f}^{(e)}$ and the element mass matrix $\mathbf{M}^{(e)}$ do not, in general, have to be integrated using the same order as $\mathbf{K}^{(e)}$. Whereas the selection of an integration order for $\mathbf{K}^{(e)}$ depends on more than just the considerations of precision and accuracy, in the case of $\mathbf{f}^{(e)}$ and $\mathbf{M}^{(e)}$, the order can be based strictly on precision [370]. Consequently, the order selected for $\mathbf{K}^{(e)}$ may be less than that selected for $\mathbf{f}^{(e)}$ and $\mathbf{M}^{(e)}$. This is summarized in Tables F.7 and F.8. In the former, the designations "Quad_4" and "Quad_8" refer to four- and eight-node quadrilateral elements, respectively. Eight- and twenty-node hexahedral elements are denoted by "Hex_8" and "Hex_20", respectively. In the latter, the designations "Tri_3", "Tri_6", "Tet_4" and "Tet_10" refer to three- and six-node triangular elements and to four- and ten-node tetrahedral elements, respectively. The quantity h is an arbitrary linear combination of the natural coordinates L_1, L_2 and L_3 (for triangles) or L_1, L_2, L_3 and L_4 (for tetrahedra).

3. Using the appropriate quadrature formula, the *nodal* values of the primary dependent variable $\hat{\phi}^{(e)}$ generally will be much more accurate than

Table F.7: Minimum Number of Points Required to Exactly Integrate Element Arrays for Quadrilateral and Hexahedral Elements (after Mac-Neal [370])

Element	Geometry	Loading	$\mathbf{K}^{(e)}$	$\mathbf{f}^{(e)}$	$\mathbf{M}^{(e)}$
Quad_4	Linear	Constant	2 x 2	1	2 x 2
	Bi-Linear	Constant	†	2 x 2	2 x 2
	Linear	Linear	2 x 2	2 x 2	2 x 2
	Bi-Linear	Linear	†	2 x 2	3 x 3
Quad_8	Linear	Constant	3 x 3	2 x 2	3 x 3
	Bi-Linear	Constant	†	2 x 2	3 x 3
	Bi-Linear	Linear	†	3 x 3	4 x 4
	General	Constant	†	3 x 3	4 x 4
Hex_8	Linear	Constant	2 x 2 x 2	1	2 x 2 x 2
	Tri-linear	Constant	†	2 x 2 x 2	3 x 3 x 3
Hex_20	Linear	Constant	3 x 3 x 3	2 x 2 x 2	3 x 3 x 3
	Tri-linear	Constant	†	3 x 3 x 3	4 x 4 x 4
	Tri-linear	Linear	†	3 x 3 x 3	4 x 4 x 4
	General	Constant	†	4 x 4 x 4	5 x 5 x 5

† - cannot be integrated exactly.

those computed elsewhere in the element. For the specific case of one-dimensional elements used to solve linear, second order $(s = 1)$, self-adjoint boundary-value problems, the $\hat{\phi}^{(e)}$ will converge at a rate of $O(h^{2p})$ at the interelement nodes (that is, at each end of the element) [417, 102]. This is *twice* the order of convergence for the approximation of the $\hat{\phi}^{(e)}$ elsewhere in the element. The interelement nodes are thus said to exhibit the quality known as *superconvergence*. That is, values sampled at these points show an error that decreases more rapidly than elsewhere in the element. We shall have more to say about superconvergence in Section F.8.4.

F.8.3 Reduced Integration

In the previous section we presented guidelines for selecting the appropriate order of integration. In the preferred approach, for complete polynomials of degree p, such as those for one-dimensional, triangular and tetrahedral elements, a degree of precision equal to at least $2(p - 1)$ guarantees an

Table F.8: Minimum Degree of Precision Required to Exactly Integrate Triangular and Tetrahedral Element Arrays (after MacNeal [370])

Element	Geometry	Loading	$\mathbf{K}^{(e)}$	$\mathbf{f}^{(e)}$	$\mathbf{M}^{(e)}$
Tri_3	Linear	Constant	1	h	h^2
	Linear	Linear	h	h^2	h^3
Tri_6	Linear	Constant	h^2	h^2	h^4
	Linear	Linear	h^3	h^3	h^5
	Quadratic	Constant	†	h^4	h^6
Tet_4	Linear	Constant	1	h	h^2
Tet_10	Linear	Constant	h^2	h^2	h^4
	Linear	Linear	h^3	h^3	h^5
	Quadratic	Constant	†	h^5	h^7

† - cannot be integrated exactly.

optimal rate of convergence. For quadrilateral and hexahedral elements, $p+1$ or more quadrature points are required in each coordinate direction. In both cases, we refer to the associated order of integration as *full integration*.

If an integration formula with degree of precision less than $2(p-1)$ is used for one-dimensional, triangular and tetrahedral elements, the approach is commonly referred to as *reduced integration*. For quadrilateral and hexahedral elements the corresponding integration formula typically uses p quadrature points in each coordinate direction.

For example, as shown in Example F.1, for a bilinear quadrilateral element, full integration requires a 2 x 2 formula. Reduced integration would thus use a one-point formula. The corresponding quadrature point locations in the "parent" element domain are shown in Figure F.2(a) and (b), respectively. In the case of a biquadratic serendipity element (eight or nine-node), 3 x 3 and 2 x 2 formulae constitute full and reduced integration, respectively. These are shown in Figure F.2(c) and (d), respectively.

The extension to hexahedral elements is straightforward. Full integration for the trilinear element requires a 2 x 2 x 2 formula; reduced integration is a one-point formula. For the triquadratic element (twenty or twenty-seven-node), 3 x 3 x 3 and 2 x 2 x 2 formulae constitute full and reduced integration, respectively.

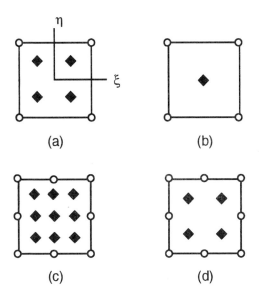

Figure F.2: Quadrature Point Locations for Quadrilateral Elements: (a) Full Integration for Linear Element, (b) Reduced Integration for Linear Element, (c) Full Integration for Quadratic Element, (d) Reduced Integration for Quadratic Element

Selective Reduced Integration

It is timely to note that different terms in the general expression for $\mathbf{K}^{(e)}$ can be integrated using different integration formulae. When this is done, the practice is referred to as *selective reduced integration*. In solid mechanics applications, this entails using reduced integration only on the volumetric contribution to the stiffness, while fully integrating the shear contribution. A beneficial characteristic of selective reduced integration is that it retains the correct rank of $\mathbf{K}^{(e)}$, thus ensuring that the global $\mathbf{K}$ will likewise possess correct rank [252].

Uniform Reduced Integration

When both the volumetric and shear contributions are integrated with a reduced formula, the practice is referred to as *uniform reduced integration*. The advantage of such an approach is the reduction in computational effort associated with the formation of the element arrays. In particular, since the number of multiplications required in the computation of $\mathbf{K}^{(e)}$ is proportional to the number of quadrature points, the ratio of computational effort for full and reduced integration of quadrilateral and hexahedral elements is

$$\left(\frac{p+1}{p}\right)^{sdim}$$

where *sdim* is the spatial dimension of the element. Thus, in the case of a bilinear quadrilateral element ($p = 1$, *sdim* $= 2$) the ratio is 4; for a trilinear hexahedral element ($p = 1$, *sdim* $= 3$) the ratio is 8.

The disadvantage of uniform reduced integration is that the rank of $\mathbf{K}^{(e)}$ may be reduced, resulting in the singularity or near singularity of the global $\mathbf{K}$. As previously mentioned, the associated element instabilities manifest themselves in the form of *mechanisms* or *hourglass modes*.

A relatively simple way in which to reintroduce the stiffness of the hourglass modes has, however, been proposed by Kosloff and Frazier [293]. The computational effort required to form the corresponding element is only slightly greater than that for one-point Gauss-Legendre quadrature. This computational economy motivated subsequent work to develop various uniformly underintegrated, but stabilized elements quadrilateral and hexahedral elements [163, 65, 66, 292] and to study their behavior [264].

F.8.4   Evaluating Fluxes

The accurate calculation of fluxes (derivatives of the primary dependent variables) has been of interest since the infancy of the finite element method. In heat transfer analyses, heat fluxes are computed from derivatives of the temperature. In solid mechanics applications, derivatives of displacements are used to compute strains and, through the constitutive relations, the stresses.

Although fluxes could be computed using the same quadrature points as the primary dependent variables, the results have been found to be less accurate than for approaches using a reduced order of integration. Barlow [52] appears to be the first to have noted that strains computed at the reduced order Gauss-Legendre quadrature points were more accurate than strains computed at other locations in the element. He subsequently [53] recommended that such points be used for computing strains and stresses, regardless of whether they are used for calculating $\mathbf{K}^{(e)}$.

It was subsequently shown [230, 397] that if the governing equations are linear and self-adjoint, then the finite element solution is equivalent to finding a weighted least squares fit of the s-th derivative of $\hat{\phi}^{(e)}$ to the s-th derivative of ϕ, in the limit as $h^{(e)} \to 0$. Here s is the highest derivative of $\hat{\phi}^{(e)}$ in the expression for the flux, which is also the highest derivative of the interpolation functions in the integrand for computing $\mathbf{K}^{(e)}$.

In one dimension the relation between the theory of orthogonal polynomials and approximation theory is well established [263]. It can be shown that if a complete one-dimensional polynomial of degree m forms a weighted least-squares fit to a one-dimensional polynomial of degree $m + 1$, then the two polynomials will equal each other at $m + 1$ points. These points are the zeros of the $(m + 1)$st degree polynomial in a family of orthogonal polynomials [245]. The orthogonal polynomials typically used in finite element analyses are Legendre polynomials. This leads to the following finding [397]:

> The flux converges at a rate $O(h^{p+1})$ at the zeros of the Legendre polynomial of degree $p - s + 1$. This is one order *higher* than the approximation of $\hat{\phi}^{(e)}$.

These special locations are often called the *optimal flux sampling points*, and the flux is said to be *superconvergent* at such points [515].

The existence of optimal flux points is supported by the observation [230, 515, 612] that the error in the flux does not have the same sign everywhere within the element, and thus must be zero at some locations. These locations are the optimal flux points.

More recently, the superconvergence phenomenon has been formalized by mathematicians, who have produced error analyses together with proofs of its existence. An excellent review of different approaches used to investigate the superconvergence phenomenon is given in reference [294].

In general, the above results are exact in one dimension and approximate in two and three dimensions [53]. None the less, for maximum accuracy in rectangular elements, the fluxes should be computed at the reduced Gauss-Legendre quadrature points. This translates into using one quadrature point for bilinear elements (Figure F.2b), 2 x 2 points for biquadratic elements (Figure F.2d), and so on. The extension to rectangular prisms is straightforward, and involves one and 2 x 2 x 2 quadrature points for trilinear and triquadratic elements, respectively.

When two- and three-dimensional rectangular elements are distorted isoparametrically, full superconvergence is lost. However, the accuracy of fluxes computed at reduced Gauss-Legendre quadrature points remains high. Consequently, use of these points for computing fluxes is recommended [605].

The situation for triangular and tetrahedral elements is not as straightforward as for rectangular elements. In general, superconvergent points do not exist in triangles, though certain points have been shown to be optimal [398].

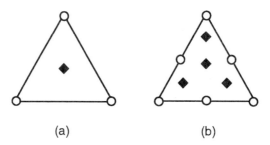

(a)   (b)

Figure F.3: Optimal Flux Sampling Points for Triangular Elements: (a) Linear, (b) Quadratic Element

For the linear triangular element, there appears to be only one optimal flux sampling point: at the centroid (Figure F.3a). There the accuracy is $O(h^2)$, whereas at other locations within the element it is $O(h)$.[15]

In the case of the quadratic triangular element, the optimal flux sampling points are located near the quadrature points corresponding to a four-point integration formula [398].[16] Consequently, these points are generally used for computing element fluxes (Figure F.3b).

F.8.5   Flux Recovery

Having shown that fluxes computed at optimal sampling points possess a higher order of accuracy than at other locations, we desire to "recover" similarly accurate fluxes elsewhere in the element. Typically, we would like to also compute *nodal* fluxes because these: 1) Provide a continuous representation of fluxes over the solution domain, 2) Are generally easier to work with since nodal locations are typically easier to identify than the locations of the optimal points, and 3) Most contouring programs expect nodal values of the variable to be plotted.

The simplest approach for computing nodal fluxes is to use smoothing and averaging procedures [100]. Other approaches involve extrapolat-

---

[15]The midpoints of the edges are apparently also superconvergent [16].

[16]Since it involves negative weighting coefficients w_i, this formulas was omitted from Table F.4. As such, we list the weighting coefficients and corresponding quadrature point locations:

$w_1 = -27/48$, $(L_1, L_2, L_3)_1 = (1/3, 1/3, 1/3)$
$w_2 = 25/48$, $(L_1, L_2, L_3)_2 = (3/5, 1/5, 1/5)$
$w_3 = 25/48$, $(L_1, L_2, L_3)_3 = (1/5, 3/5, 1/5)$
$w_4 = 25/48$, $(L_1, L_2, L_3)_4 = (1/5, 1/5, 3/5)$

ing quadrature point stresses to the nodes [242] and computing recovered stresses using a least squares approach [416, 243].

More recently, improved results have been obtained by direct polynomial smoothing of the superconvergent values using the so-called *superconvergent patch recovery* method [601, 602, 603] and its extensions [83, 562]. A second general approach, the so-called *recovery by equilibrium of patches* method [90, 91], has also proven successful. Further details pertaining to recovery techniques are beyond the scope of this introductory text. The interested reader is directed to the references cited above and to Chapter 14 in [605].

References

[1] Abbott, M. B., *An Introduction to the Method of Characteristics.* New York: American Elsevier Pub. (1966).

[2] Abramowitz, M. and I. A. Stegun, *Handbook of Mathematical Functions with Formulas, Graphs, and Mathematical Tables*, Applied Mathematics Series, **55**. Washington: National Bureau of Standards (1964).

[3] Adrain, R., "Research Concerning the Probabilities of the Errors Which Happen in Making Observations, Etc.," *Mathematical Museum*, **1**(4): 93-109 (1808).

[4] Ahmad, S., B. M. Irons, and O. C. Zienkiewicz, "Analysis of Thick and Thin Shell Structures by Curved Finite Elements, *International Journal for Numerical Methods in Engineering*, **2**(3): 419-451 (1970).

[5] Ainsworth, M. and Oden, J. T., "A Unified Approach to A-Posteriori Error Estimation Using Element Residual Methods," *Numerische Mathematik*, **65**(1): 23-50 (1993).

[6] Akhras, G. and G. Dhatt, "An Automatic Node Relabelling Scheme for Minimizing a Matrix or Network Bandwidth," *International Journal for Numerical Methods in Engineering*, **10**(4): 787-797 (1976).

[7] Akin, J. E., "The Generation of Elements with Singularities," *International Journal for Numerical Methods in Engineering*, **10**(6): 1249-1259 (1976).

[8] Akin, J. E., *Application and Implementation of Finite Element Methods.* London: Academic Press, Inc. LTD (1982).

[9] Akyuz, F. A. and S. Utku, "An Automatic Node-Relabeling Scheme for Bandwidth Minimization of Stiffness Matrices," *AIAA Journal*, **6**(4): 728-730 (1969).

[10] Alway, G. G. and D. W. Martin, "An Algorithm for Reducing the Bandwidth of a Matrix of Symmetrical Configuration," *Computer Journal*, **8**: 264-272 (1965).

[11] Ames, W. F., *Nonlinear Partial Differential Equations in Engineering*, Volume I. New York: Academic Press (1965).

[12] Ames, W. F., *Nonlinear Partial Differential Equations in Engineering*, Volume II. New York: Academic Press (1972).

[13] Anandarajah, A., "Time-Domain Radiation Boundary for Analysis of Plane Love-Wave Propagation Problem," *International Journal for Numerical Methods in Engineering*, **29**(5): 1049-1063 (1990).

[14] Anderson, E., Z. Bai, C. Bischof, J. Demmel, J. Dongarra, J. Du Croz, A. Greenbaum, S. Hammarling, A. McKenney, S. Ostrouchov, and D. Sorensen, *LAPACK User's Guide*, 2nd edition. Philadelphia: SIAM (1995).

[15] Anderssen, R. S. and A. R. Mitchell, "The Petrov Galerkin Method," *Mathematical Methods in Applied Science*. **1**: 3-15 (1979).

[16] Andreev, A. B., "Superconvergence of the Gradient for Linear Triangle Elements for Elliptic and Parabolic Equations," *C. R. Acad. Bulgare. Sci.*, **37**: 293-296 (1984).

[17] Anton, H., *Elementary Linear Algebra*, 3rd Edition. New York: J. Wiley and Sons (1981).

[18] Arantes e Oliveira, E. R. de, "Theoretical Foundations of the Finite Element Method," *International Journal of Solids and Structures*, **4**(10): 929-952 (1968).

[19] Arantes e Oliveira, E. R. de, "The Patch Test and the General Convergence Criteria of the Finite Element Method," *International Journal of Solids and Structures*, **13**(3): 159-178 (1977).

[20] Argyris, J. H., "Energy Theorems and Structural Analysis," *Aircraft Engineering*, **26** (Oct. - Nov.) (1954); 27 (Feb. - May) (1955). Reprinted in J. H. Argyris and S. Kelsey. *Energy Theorems and Structural Analysis*. London: Butterworth Scientific Publications (1960).

[21] Argyris, J. H., "The Matrix Analysis of Structures with Cut-Outs and Modifications," *Proc. 9th Int. Congress on App. Mech., Section II: Mechanics of Solids* (1956).

[22] Argyris, J. H., "The Matrix Theory of Statics," *Ing. Arch.*, **25** (1957).

[23] Argyris, J. H., "Continua and Discontinua," *Proceedings of the 1st Conference on Matrix Methods in Structural Mechanics*, Wright-Patterson AFB, Dayton, Ohio, 22 (1965).

[24] Argyris, J. H., "Triangular Elements with Linearly Varying Strain for the Matrix Displacement Method," *J. Roy. Aero. Soc. Tech. Note*, **69**: 711-713 (1965).

[25] Argyris, J. H., "Tetrahedron Elements with Linearly Varying Strain for the Matrix Displacement Method," *J. Roy. Aero. Soc.*, **69**: 877-880 (1965).

[26] Argyris, J. H., "Matrix Analysis of Three-Dimensional Elastic Media, Small and Large Deflections," *AIAA Journal*, **3**(1): 45-51 (1965).

[27] Argyris, J. H. and P. C. Patton, "Computer Oriented Research in a University Milieu," *Applied Mechanics Reviews*, **19**(12):1029-1039 (1966).

[28] Argyris, J. H., I. Fried, and D. W. Scharpf, "The TET 20 and the TEA 8 Elements for the Matrix Displacement Method," *Aeron. J. Roy. Aeron. Soc.*, **72**: 618-623 (1968).

[29] Argyris, J. H., K. E. Buck, D. W. Scharpf, H. M. Hilber, and G. Mareczek, "Some New Elements for the Matrix Displacement Method," *Proceedings of the 2nd Conference on Matrix Methods in Structural Mechanics*, Wright-Patterson AFB, Dayton, Ohio: 333-366 (1968).

[30] Atkinson, K. E., *An Introduction to Numerical Analysis*. New York: J. Wiley and Sons (1989).

[31] Atluri, S. N. and T. Zhu, "A New Meshless Local Petrov-Galerkin (MLPG) Approach in Computational Mechanics," *Computational Mechanics*, **22**(2): 117-127 (1998).

[32] Axelsson, O., *Iterative Solution Methods*. Cambridge: Cambridge University Press (1994).

[33] Azar, J. J., *Matrix Structural Analysis*. New York: Pergamon Press (1972).

[34] Babuška, I., "Stability of Domains of Definition with Respect to the Fundamental Problems in the Theory of Partial Differential Equations, Primarily in Relation to the Theory of Elasticity", (in Russian), *Czech Math Journal* **11**(86): 76-105 (Part I), 162-203 (Part II) (1961).

[35] Babuška, I., "The Rate of Convergence for the Finite Element Method," *SIAM Journal on Numerical Analysis*, **8**(2): 304-315 (1971).

[36] Babuška, I., "Error Bounds for Finite Element Method," *Numerische Mathematik*, **16**(4): 322-333 (1971).

[37] Babuška, I. and A. K. Aziz, eds., *The Mathematical Foundations of the Finite Element Method – with Applications to Partial Differential Equations.* New York: Academic Press (1973).

[38] Babuška, I. and A. K. Aziz, "On the Angle Condition in the Finite Element Method," *SIAM Journal on Numerical Analysis*, **13**(2): 214-226 (1976).

[39] Babuška, I. and W. C. Rheinboldt, "A-*posteriori* Error Estimates for the Finite Element Method," *International Journal for Numerical Methods in Engineering*, **12**(10): 1597-1615 (1978).

[40] Babuška, I. and W. C. Rheinboldt, "Adaptive Approaches and Reliability Estimations in Finite Element Analysis," *Computer Methods in Applied Mechanics and Engineering*, **17/18**(3): 519-540 (1979).

[41] Babuška, I., B. A. Szabó, and I. N. Katz, "The *p*-Version of the Finite Element Method," *SIAM Journal on Numerical Analysis*, **18**(3): 512-545 (1981).

[42] Babuška, I. and M. R. Dorr, "Error Estimates for the Combined *h* and *p* Versions of the Finite Element Method," *Numerische Mathematik*, **37**(2): 257-277 (1981).

[43] Babuška, I. and B. Q. Guo, "The *h-p* Version of the Finite Element Method for Domains with Curved Boundaries," *SIAM Journal on Numerical Analysis*, **25**(4): 837-861 (1988).

[44] Babuška, I. and M. Suri, "The *p*- and *h-p* Versions of the Finite Element Method, an Overview," *Computer Methods in Applied Mechanics and Engineering*, **80**(1-3): 5-26 (1990).

[45] Babuška, I., T. Strouboulis, C. S. Upadhyay, S. K. Gangaraj, and K. Copps, "An Objective Criterion for Assessing the Reliability of A-Posteriori Error Estimators in Finite Element Computations," *IACM Bulletin*, **9**(4): 27-37 (1994).

[46] Babuška, I. and B. Q. Guo, "Approximation Properties of the *h-p* Version of the Finite Element Method," *Computer Methods in Applied Mechanics and Engineering*, **133**(3-4): 319-349 (1996).

[47] Babuška, I. and J. M. Melenk, "The Partition of Unity Method," *International Journal for Numerical Methods in Engineering*, **40**(4): 727-758 (1997).

[48] Bäcklund, J., "On Isoparametric Elements," *International Journal for Numerical Methods in Engineering*, **12**(4): 731-732 (1978).

[49] Baker, T. J., "Automatic Mesh Generation for Complex Three-Dimensional Regions Using a Constrained Delaunay Triangulation," *Engineering with Computers*, **5**(3/4): 161-175 (1989).

[50] Ball, A. A., "The Interpolation Function of a General Serendipity Rectangular Element," *International Journal for Numerical Methods in Engineering*, **15**(5): 773-778 (1980).

[51] Banerjee, P. K. and R. Butterfield, *Boundary Element Methods in Engineering Science*. New York: McGraw-Hill (1981).

[52] Barlow, J., "A Stiffness MAtrix for a Curved Membrane Shell," *Conference on Recent Advances in Stress Analysis*, Royal Aeron. Society (1968).

[53] Barlow, J., "Optimal Stress Locations in Finite Element Models," *International Journal for Numerical Methods in Engineering*, **10**(2): 243-251 (1976).

[54] Barlow, J., "More on Optimal Stress Points – Reduced Integration, Element Distortions and Error Estimation," *International Journal for Numerical Methods in Engineering*, **28**(7): 1487-1504 (1989).

[55] Barrett, K. E., "Jacobians for Isoparametric Finite Elements," *Communications in Numerical Methods in Engineering*, **12**(11): 755-766 (1996).

[56] Barrett, R., M. Berry, T. F. Chan, J. Demmel, J. M. Donato, J. Dongarra, V. Eijkhout, R. Pozo, C. Romine, and H. Van der Vorst, *Templates for the Solution of Linear Systems: Building Blocks for*

Iterative Methods. Philadelphia: SIAM (1994). Also available at: $http : //www.netlib.org/templates/Templates.html.$

[57] Barsoum, R. S., "On the Use of Isoparametric Finite Elements in Linear Fracture Mechanics," *International Journal for Numerical Methods in Engineering,* **10**(1): 25-37 (1976).

[58] Bathe, K. J., *Finite Element Procedures.* Englewood Cliffs, NJ: Prentice Hall (1996).

[59] Bathe, K. J. and E. L. Wilson, *Numerical Methods in Finite Element Analysis.* Englewood Cliffs, NJ: Prentice Hall (1976).

[60] Bazeley, G. P., Y. K. Cheung, B. M. Irons, and O. C. Zienkiewicz, "Triangular Elements in Bending – Conforming and Nonconforming Solutions," *Proceedings of the 1st Conference on Matrix Methods in Structural Mechanics,* Wright-Patterson AFB, Dayton, Ohio (1965).

[61] Beer, G. and J. L. Meek, "Infinite Domain Elements," *International Journal for Numerical Methods in Engineering,* **17**(1): 43-52 (1981).

[62] Beer, G. and W. Haas, "A Partitioned Frontal Solver for Finite Element Analysis," *International Journal for Numerical Methods in Engineering,* **18**(11): 1623-1654 (1982).

[63] Beer, G., "An Isoparametric Joint/Interface Element for Finite Element Analysis," *International Journal for Numerical Methods in Engineering,* **12**: 585-600 (1985).

[64] Belytschko, T., "Are Finite Elements Passe," USACM Bulletin, **7**(3): 4-7 (1994).

[65] Belytschko, T. and C. S. Tsay, "A Stabilization Procedure for the Quadrilateral Plate Element with One-Point Quadrature," *International Journal for Numerical Methods in Engineering,* **19**(3): 405-419 (1983).

[66] Belytschko, T., J. S-J. Ong, W. K. Liu, and J. M. Kennedy, "Hourglass Control in Linear and Nonlinear Problems," *Computer Methods in Applied Mechanics and Engineering,* **43**(3): 251-276 (1984).

[67] Belytschko, T., J. Fish, and B. E. Engelmann, "A Finite Element with Embedded Localization Zones," *Computer Methods in Applied Mechanics and Engineering,* **70**(1): 59-89 (1988).

[68] Belytschko, T., Y. Y. Lu, and L. Gu, "Element-Free Galerkin Methods," *International Journal for Numerical Methods in Engineering*, **37**(2): 229-256 (1994).

[69] Belytschko, T., Y. Krongauz, D. Organ, M. Fleming, and P. Krysl, "Meshless Methods: An Overview and Recent Developments," *Computer Methods in Applied Mechanics and Engineering*, **139**(1-4): 3-47 (1996).

[70] Bergan, P. G. "Challenges in Computational Mechanics Applied to Offshore Engineering," *Proceedings of the 10th ASCE Engineering Mechanics Conference*, edited by S. Sture. New York: ASCE Press: 29-44 (1995).

[71] Besseling, J. F. "The Complete Analogy Between the Matrix Equations and the Continuous Field Equations of Structural Analysis," *Proceedings of the International Symposium on Analogue and Digital Techniques Applied to Aeronautics*, Liège, Belgium (1963).

[72] Bettess, P., "Infinite Elements," *International Journal for Numerical Methods in Engineering*, **11**(1): 53-64 (1977).

[73] Bettess, P., "More on Infinite Elements," *International Journal for Numerical Methods in Engineering*, **15**(11): 1613-1626 (1980).

[74] Bickerton, S., *Development of a Traslator from MSC-PATRAN to TWOD and Porflow*. Term Project for CIEG 801 (Advanced Topics in Finite Element Analysis), University of Delaware, Newark, DE (1996).

[75] Bickley, W. G., "Finite Difference Formulae for the Square Lattice, *The Quarterly Journal of Mechanics and Applied Mathematics*, **1**: 35-42 (1948).

[76] Biezeno, C. B., "Over een Vereenvoudiging en over een Uitbreiding van de Methode van Ritz," *Christiaan Huygens International Mathematisch Tijdschrift*. **3**: 69-75 (1923).

[77] Biezeno, C. B. and J. J. Koch, "Over een nieuwe Methode ter Berekening van vlakke Platen met Toepassing op Enkele voor de techniek Belangrijke belastings Gevallen," *De Ingenieur*. **38**: 25-36 (1923).

[78] Biezeno, C. B. and R. Grammel, *Technische Dynamik*. Berlin: Springer-Verlag (1939).

[79] Bird, R. B., R. C. Armstrong, and O. Hassager, *Dynamics of Polymeric Liquids, Volume 1, Fluid Mechanics*, New York: J. Wiley and Sons (1987).

[80] Birkhoff, G., M. H. Schultz, and R. S. Varga, "Piecewise Hermite Interpolation in One and Two Variables with Applications to Partial Differential Equations," *Numerische Mathematik*, **11**(3): 232-256 (1968).

[81] Black, A. N. and R. V. Southwell, "Relaxation Methods Applied to Engineering Problems. II. Basic Theory, with Application to Surveying and to Electrical Networks, and an Extension to Gyrostatic Systems," *Proc. Royal Soc.*, A164:447-467 (1938).

[82] Blacker, T. D. and R. J. Myers, "Seams and Wedges in Plastering: A 3-D Hexahedral Mesh Generation Algorithm," *Engineering With Computers*, **9**(2): 83-93 (1993).

[83] Blacker, T. D. and T. Belytschko, "Superconvergent Patch Recovery with Equilibrium and Conjoint Interpolant Enhancements," *International Journal for Numerical Methods in Engineering*, **37**(3): 517-536 (1994).

[84] Bliss, G. A., *Lectures on the Calculus of Variations*. Chicago: The University of Chicago Press (1946).

[85] Bolotin, V. V., *Nonconservative Problems of the Theory of Elastic Stability*. New York: Macmillan (1963).

[86] Bolza, O., *Lectures on the Calculus of Variations*. New York: Chelsea (1960).

[87] Bonet, J. and S. Kulasegaram, "Correction and Stabilization of Smooth Particle Hydrodynamics Methods with Applications in Metal Forming Simulations," *International Journal for Numerical Methods in Engineering*, **47**(6): 1189-1214 (2000).

[88] Boole, G., *A Treatise on the Calculus of Finite Differences*. London: Macmillan and Company, Ltd. (1860).

[89] Boresi, A. P., R. J. Schmidt, and O. M. Sidebottom, *Advanced Mechanics of Materials*, 5th edition. New York: J. Wiley and Sons (1993).

[90] Boroomand, B. and O. C. Zienkiewicz, "Recovery by Equilibrium in Patches (REP)," *International Journal for Numerical Methods in Engineering*, **40**(1): 137-164 (1997).

[91] Boroomand, B. and O. C. Zienkiewicz, "An Improved REP Recovery and the Effectivity Robustness Test', *International Journal for Numerical Methods in Engineering*, **40**(17): 3247-3277 (1997).

[92] Bossavit, A., *Computational Electromagnetism : Variational Formulations, Complementarity, Edge Elements*. San Diego, CA: Academic Press (1998).

[93] Brebbia, C. A., "Integration of Area and Volume Coordinates in the Finite-Element Method," *AIAA Journal*, **7**(6): 1212 (1969).

[94] Brebbia, C. A. and J. Dominguez, *Boundary Elements, An Introductory Course*, 2nd edition. Southampton: Computational Mechanics Publications, McGraw-Hill (1992).

[95] Briggs, W. L., *A Multigrid Tutorial*. Philadelphia, PA: SIAM (1987).

[96] Brodskii, M. L., "Asymptotic Estimates of the Errors in the Numerical Integration of Systems of Ordinary Differential Equations by Difference Methods," *Dokl. Akad. Nauk* SSSR (N.S.) 93:599-602 (1953).

[97] Bubnov, I. G., "Report on the works of Professor Timoshenko which were awarded the Zhuranskii prize," *Symposium of the Institute of Communication Engineers* (Sborn, inta inzh. putei soobshch.) **81**, All Union Special Planning Office (SPB) (1913).

[98] Buck, R. C., *Advanced Calculus*, Third Edition. New York: McGraw-Hill (1978).

[99] Burden, R. L., J. D. Faires, and A. C. Reynolds, *Numerical Analysis*. Boston: PWS Publishing (1980).

[100] Burnett, D. S., *Finite Element Analysis, from Concepts to Applications*. Reading, MA: Addison-Wesley (1988).

[101] Carey, G. F., *Computational Grids: Generation, Adaptation, and Solution Strategies*. Philadelphia: Taylor and Francis (1997).

[102] Carey, G. F., D. Humphrey, and M. F. Wheeler, "Galerkin and Collocation-Galerkin Methods with Superconvergence and Optimal Fluxes," *International Journal for Numerical Methods in Engineering*, **17**(6): 939-950 (1981).

[103] Carreau, P. J., "Rheological Equations from Molecular Network Theories," *Trans. Soc. Rheol.*, **16**(1): 99-127 (1972).

[104] Carroll, W. F., *A Primer for Finite Elements in Elastic Structures*. New York: J. Wiley and Sons (1999).

[105] Carslaw, H. S. and J. C. Jaeger, *Conduction of Heat in Solids*, 2nd edition. Oxford: The Clarendon Press (1959).

[106] Castigliano, A., *Theorie de l'equilibre des Systemes Elastiques*. Turin (1879).

[107] Christoffel, E. B., "Über die Gaissische Quadratur und eine Verallgemeinerung derselben," *J. Reine Angew. Math.*, **55**: 61-82 (1858).

[108] Ciarlet, P. G., M. H. Schultz, and R. S. Varga, "Numerical Methods of High-Order Accuracy for Nonlinear Boundary Value Problems, I. One-Dimensional Problem," *Numerische Mathematik*, **9**(5):394-430 (1967).

[109] Clough, R. W., "The Finite Element Method in Plane Stress Analysis," *Proceedings of the Second ASCE Conference on Electronic Computation*, Pittsburgh, PA, 345-378 (1960).

[110] Clough, R. W., "The Finite Element Method After Twenty-Five Years: A Personal View," *Computers and Structures*, **12**(4): 361-370 (1980).

[111] Collatz, L., "Eigenwertprobleme und ihre numerische Behandlung," *Mathematik und ihre Anwendungen in Physik und Technik*, Reihe A, B und 19. Akademische Verlags Gesellshaft, Leipzig (1945).

[112] Collatz, L., *Numerische Behandlung von Differentialgleichungen*. Berlin: Springer-Verlag (1951).

[113] Collatz, L., *The Numerical Treatment of Differential Equations*. Berlin: Springer (1960).

[114] Collins, R. J., "Bandwidth Reduction by Automatic Renumbering," *International Journal for Numerical Methods in Engineering*, **6**(3): 345-356 (1973).

[115] Cook, W., "The Effect of Geometric Shape on Two-Dimensional Finite Elements," *Nuclear Engineering and Design*, **70**(1): 13-26 (1982).

[116] Cook, R. D., D. S. Malkus, and M. E. Plesha, *Concepts and Applications of Finite Element Analysis*, 3rd edition. New York: J. Wiley and Sons (1989).

[117] Cools, R. and P. Rabinowitz, "Monomial Cubature Rules Since Stroud: A Compilation," *Journal of Computational Applied Math.*, **48**: 309-326 (1993).

[118] Coons, S. A., *Surfaces for Computer-Aided Design of Space Form*, MIT Project MAC, MAC-TR-41 (1967).

[119] Coulomb, C. A., "Histoire de lAcadémie," Paris, 229-269 (1787).

[120] Courant, R., "Über Randwertaufgaben bei partieller Differenzengleichungen," *Zeitschrift für Angewandte Mathematik und Mechanik*, 6:322-325 (1926).

[121] Courant, R., K. O. Friedrichs, and H. Lewy, "Über die Partiellen Differenzengleichungen der Mathematischen Physik," *Mathematische Annalen.* **100**: 32-74 (1928).

[122] Courant, R., "Variational Methods for the Solution of Problems of Equilibrium and Vibrations," *Bulletin of the American Mathematical Society* **49**:1-23 (1943).

[123] Courant, R. and D. Hilbert, *Methods of Mathematical Physics*, 1st English Edition. New York: Interscience Publishers (1953).

[124] Cowper, G. R., "Gaussian Quadrature Formulas for Triangles," *International Journal for Numerical Methods in Engineering*, **7**(3): 405-408 (1973).

[125] Crandall, S. H., *Engineering Analysis, A Survey of Numerical Procedures*. New York: McGraw-Hill Book Co. (1956).

[126] Crank, J. and P. Nicolson, "A Practical Method for Numerical Integration of Solutions of Partial Differential Equations of Heat Conduction Type," *Proc. Camb. Phil. Soc.*, **43**(50) (1947).

[127] Cui, L., V. N. Kaliakin, Y. Abousleiman, and A. H-D. Cheng, "Finite Element Formulation and Application of Poroelastic Generalized Plane Strain Problems," *International Journal for Rock Mechanics and Mining Sciences and Geomechanics Abstracts*, **6**: 953-962 (1997).

[128] Curnier, A., "A Static Infinite Element," *International Journal for Numerical Methods in Engineering*, **19**(10): 1479-1488 (1983).

[129] Cuthill, E. and J. McKee, "Reducing the Bandwidth of Sparse Symmetric Matrices," *Proceedings of the 24th National Conference, ACM*, San Francisco, CA, 157-172 (1969).

[130] Dahlquist, G., "Stability and Error Bounds in the Numerical Integration of Ordinary Differential Equations," *Trans. Royal Institute of Technology*, Stockholm, **130** (1959).

[131] Darcy, H., *Les Fontaines Publiques de la Ville de Dijon*. Paris: Dalmont (1856).

[132] Davis, P. J. and I. Polonsky, "Numerical Interpolation, Differentiation and Integration," Chapter 25 in *Handbook of Mathematical Functions with Formulas, Graphs, and Mathematical Tables*, Applied Mathematics Series, **55**, edited by M. Abramowitz and I. A. Stegun. Washington: National Bureau of Standards (1964).

[133] Davis, P. J. and P. Rabinowitz, *Methods of Numerical Integration*. Orlando: Academic Press (1984).

[134] De, S. and K. J. Bathe, "The Method of Finite Spheres," *Computational Mechanics*, **25**(4): 329-345 (2000).

[135] Delaunay, B. N., "Sur la Sphere Vide," *Izvestia Akademia Nauk SSSR, VII Seria, Otdelenie Matematicheskii i Estestvennyka Nauk*, **7**: 793-800 (1934).

[136] Demkowicz, L., J. T. Oden, W. Rachowicz, and O. Hardy, "Toward a Universal *h-p* Adaptive Finite Element Strategy, Part 1. Constrained Approximation and Data Structure," *Computer Methods in Applied Mechanics and Engineering*, **77**(1-2): 79-112 (1989).

[137] Desai, C. S., *Elementary Finite Element Method*. Englewood Cliffs, NJ: Prentice Hall (1979).

[138] Desai, C. S., H. V. Phan, and S. Sture, "Procedure, Selection and Application of Plasticity Models for a Soil," *International Journal for Numerical and Analytical Methods in Geomechanics*, **5**: 295-311 (1981).

[139] Desai, C. S. and H. J. Siriwardane, *Constitutive Laws for Engineering Materials with Emphasis on Geologic Materials*. Englewood Cliffs, NJ: Prentice Hall (1984).

[140] Desai, C. S., M. M. Zaman, J. G. Lightner, and H. J. Siriwardane, "Thin-Layer Element for Interfaces and Joints," *International Journal for Numerical and Analytical Methods in Geomechanics*, **8**: 19-43 (1984).

[141] Dirichlet, G. L., "Uber Die Reduction der Positiven Quadratischen Forment Mit Drei Underestimmten Ganzen Zahlen," *J. Reine Angew. Math.*, **40**(3): 209-227 (1850).

[142] Duarte, C. A. and J. T. Oden, "An h-p Adaptive Method Using Clouds," *Computer Methods in Applied Mechanics and Engineering*, **139**(1-4): 237-262 (1996).

[143] Duleau, A., "Essai théorique et expérimental sur la résistance du fer forgé," Paris (1820).

[144] Dunavant, D. A., "High Degree Efficient Symmetrical Gaussian Rules for the Triangle," *International Journal for Numerical Methods in Engineering*, **21**(6): 1129-1148 (1985).

[145] Dunne, P. C., "Complete Polynomial Displacement Fields for Finite Element Method," *Aeron. J. Roy. Aeron. Soc.*, **72**: 245-246 (1968).

[146] Eason, E. D., "A Review of Least-Squares Methods for Solving Partial Differential Equations," *International Journal for Numerical and Analytical Methods in Engineering*, **10**(5): 1021-1046 (1976).

[147] Eisenberg, M. A. and L.E. Malvern, "On Finite Element Integration in Natural Co-Ordinates," *International Journal for Numerical Methods in Engineering*, **7**(4): 574-575 (1973).

[148] El-Esnawy, N.A., A.Y. Akl, and A.S. Bazaraa, "A New Parametric Infinite Domain Element," *Finite Elements in Analysis and Design*, **19**(1-2): 103-114 (1995).

[149] Engineering with Computers, *Special Issue on Mesh Generation*, **12** (1996).

[150] Engineering with Computers, *Special Issue on Mesh Generation*, **15**(3) (1999).

[151] Ergatoudis, I., B. M. Irons, and O. C. Zienkiewicz, "Curved Isoparametric, 'Quadrilateral' Elements for Finite Element Analysis," *International Journal of Solids and Structures*, **4**(1): 31-42 (1968).

[152] Ergatoudis, J., B. M. Irons, and O. C. Zienkiewicz, "Three-Dimensional Analysis of Arch Dams and Their Foundations," *Proceedings of the Symposium on Arch Dams*. London: Institute of Civil Engineering, 37-50 (1970).

[153] Euler, L., *Institutiones Calculi Differentialis*. Academiae Imperialis Scientiarium Petropolitanae (1755).

[154] Everstine, G. C., "A Comparison of Three Resequencing Algorithms for the Reduction of Matrix Profile and Wavefront," *International Journal for Numerical Methods in Engineering*, **14**(6): 837-853 (1979).

[155] Farlow, S. J., *Partial Differential Equations for Scientists and Engineers*. New York: J. Wiley and Sons (1982).

[156] Felippa, C. A., *Refined Finite Element Analysis of Linear and Nonlinear Two-dimensional Structures*. Report UC SESM 66-22, Department of Civil Engineering, University of California, Berkeley (1966).

[157] Field, D. A., "Algorithms for Determining Invertible Two- and Three-Dimensional Quadratic Isoparametric Finite Element Transformations," *International Journal for Numerical Methods in Engineering*, **19**(6): 789-802 (1983).

[158] Finlayson, B. A., *The method of Weighted Residuals and Variational Principles, with Application in Fluid Mechanics, Heat and Mass Transfer*. New York: Academic Press (1972).

[159] Finlayson, B. A. and L. E. Scriven, "The Method of Weighted Residuals – A Review," *Applied Mechanics Reviews*, **19**(9): 735-748 (1966).

[160] Finlayson, B. A. and L. E. Scriven, "On The Search for Variational Principles," *International Journal of Heat and Mass Transfer*, **10**(6): 799-821 (1967).

[161] Fix, G., "On the Effect of Quadrature Errors in the Finite Element Method," in *The Mathematical Foundations of the Finite Element Method with Applications to Partial Differential Equations*, edited by A. K. Aziz. New York: Academic Press, 525-556 (1972).

[162] Fix, G. and N. Nassif, "On Finite Element Approximations on Time-Dependent Problems," *Numerische Mathematik*, **19**(2):127-135 (1972).

[163] Flanagan, D. P. and T. Belytschko, "A Uniform Strain Hexahedron and Quadrilateral with Orthogonal Hourglass Control," *International Journal for Numerical Methods in Engineering*, **17**(5): 679-706 (1981).

[164] Fletcher, C. A. J., *Computational Galerkin Methods*. New York: Springer-Verlag (1984).

[165] Foley, J. D. and A. Van Dam, *Fundamentals of Interactive Computer Graphics.* New York: Addison-Wesley Publishing Co. (1983).

[166] Forsythe, G. E. and W. R. Wascow, *Finite-Difference Methods for Partial Differential Equations.* New York: J. Wiley and Sons, Inc. (1960).

[167] Fort, T., *Finite differences and Difference Equations in the Real Domain.* Oxford: The Clarendon Press (1948).

[168] Fortner, B., *The Data Handbook, A Guide to Understanding the Organization and Visualization of Technical Data.* Champaign, IL.: Spyglass, Inc. (1992).

[169] Fourier, J. B., *Theorie Analytique de la Chaleur.* Paris (1822). English translation published by Dover, New York (1955).

[170] Fraeijs de Veubeke, B., "Upper and Lower Bounds in Matrix Structural Analysis," in *Matrix Methods of Structural Analysis*, AGARD, 72, edited by B. Fraeijs de Veubeke. London: Pergamon Press, 165-201 (1964).

[171] Fraeijs de Veubeke, B., "Displacement and Equilibrium Models in the Finite Element Method," Chapter 9 of *Stress Analysis*, edited by O. C. Zienkiewicz and G. S. Holister. New York: J. Wiley and Sons, Inc., 145-197 (1965).

[172] Fraeijs de Veubeke, B., "Variational Principles and the Patch Test," *International Journal for Numerical Methods in Engineering*, 8(4): 783-801 (1974).

[173] Fraeijs de Veubeke, B. and G. Sander, "Upper and Lower Bounds in Matrix Structural Analysis," *14th Meeting of the Structures and Materials Panel of AGARD*, Paris (1962).

[174] Frankel, S. P., "Convergence Rates of Iterative Treatments of Partial Differential Equations," *Mathematical Tables and other Aids to Computation*, 4:65-75 (1950).

[175] Frazer, R. A., W. P. Jones, and S. W. Skan, "Approximations to Functions and to the Solutions of Differential Equations," *Aeronautical Research Committee Report and Memo*, **1799** (1937).

[176] Frey, P. J. and George, P. L., *Mesh Generation.* Paris: Hermes Science Publishing (2000).

[177] Fried, J., "Some Aspects of the Natural Coordinate System in the Finite-Element Method," *AIAA Journal*, **7**(7): 1366-1368 (1969).

[178] Fried, I., "Discretization and Computational Errors in High-Order Finite Elements," *AIAA Journal*, **9**(10): 2071-2073 (1971).

[179] Fried, I., "Accuracy of Complex Finite Elements," *AIAA Journal*, **10**(3): 347-349 (1972).

[180] Fried, I., "Accuracy and Condition of Curved (Isoparametric) Finite Elements," *Journal of Sound and Vibration*, **31**(3): 345-355 (1973).

[181] Fried, I., "Numerical Integration in the Finite Element Method," *Computers and Structures*, **4**(5): 921-932 (1974).

[182] Friedrichs, K. O., "A Finite Difference Scheme for the Neumann and the Dirichlet Problem," *NYO-9760*. Courant Institute of Mathematical Sciences. New York: New York University (1962).

[183] Friedrichs, K. O. and H. B. Keller, "A Finite Difference Scheme for Generalized Neumann Problems," in *Numerical Solution of Partial Differential Equations*, edited by J. H. Bramble. New York: Academic Press (1966).

[184] Fung, Y. C., *Foundations of Solid Mechanics*. Englewood Cliffs, NJ: Prentice Hall (1965).

[185] Galerkin, B. G., "Rods and Plates. Series Occuring in Various Questions Concerning the Elastic Equilibrium of Rods and Plates" (Sterzhni i plastiny. Ryady v nekotorykh voprosakh uprogogo ravnovesia sterzhnei i plastin), *Vestnik Inzhenerov, Tech.* **19**:897-908 (1915).

[186] Gallagher, R. H., "Large-Scale Computer Programs for Structural Analysis," in *On General Purpose Finite Element Computer Programs*, edited by P. V. Marcal. New York: ASME, 1-14 (1970).

[187] Gallagher, R. H., *Finite Element Analysis Fundamentals*. Englewood Cliffs: Prentice Hall (1975).

[188] Gallagher, R. H., J. Padlog, and P. P. Bijlaard, "Stress Analysis of Heated Complex Shapes," *American Rocket Society Journal*, **32**(5): 700-707 (1962).

[189] Gauss, C. F., "Methodus nova integralium valores per approximationen inveniendi, *Commentationes Societatis Regiae Scientiarum Göttingensis Recentiores*, **3** (1814).

[190] Gauss, C. F., "Brief an Gerling," 26 December . *Werke.* 9:278-281 (1823) Translated by G. E. Forsythe, *Mathematical tables and other aids to computation.* 5:255-258 (1951).

[191] Gauss, C. F., *Carl Friedrich Gauss Werke,* von der Königlichen Gesellschaft der Wissenschaften zu Göttingen, 4 (1873).

[192] Gautschi, W. and S. Notaris, "Gauss Kronrod Quadrature Formulas for Weight Functions of the Bernstein-Szego Type," *J. Computat. Appl. Math.*, **29**: 199-224 (1989).

[193] Gens, A., I. Carol, and E. E. Alonso, "An Interface Element Formulation for the Analysis of Soil-Reinforcement Interaction," *Computers and Geotechnics*, **7**: 133-151 (1989).

[194] George, P. L., *Automatic Mesh Generation, Application to Finite Element Methods.* Paris: J. Wiley and Sons (1991).

[195] George, P. L., F. Hecht, and E. Saltel, "Automatic Mesh Generator with Specified Boundary," *Computer Methods in Applied Mechanics and Engineering*, **92**(3): 269-288 (1991).

[196] Ghaboussi, J., E. L. Wilson, and J. Isenberg, "Finite Element for Rock Joints and Interfaces," *Journal of Soil Mechanics and Foundations Division, ASCE*, **99**(SM10): 833-848 (1973).

[197] Gibbs, N. E., W. G. Poole, Jr., and P. K. Stockmeyer, "An Algorithm for Reducing the Bandwidth and Profile of a Sparse Matrix," *SIAM Journal on Numerical Analysis*, **13**(2): 236-250 (1976).

[198] Gibbs, N. E., W. G. Poole, Jr., and P. K. Stockmeyer, "A Comparison of Several Bandwidth and Profile Reduction Algorithms," *ACM Trans. Math. Software*, **2**: 322-330 (1976).

[199] Gifford, L. N., "More on Distorted Isoparametric Elements," *International Journal for Numerical Methods in Engineering*, **14**(2): 290-291 (1979).

[200] Glen, J. W. "The Creep of Polycrystalline Ice," *Proceedings of the Royal Society of London A*, **228**: 519-538 (1955).

[201] Goël, J. J., "Construction of Basic Functions for Numerical Utilisation of Ritz's Method," *Numerische Mathematik*, **12**(5): 435-447 (1968).

[202] Goodman, R. E., R. L. Taylor, and T. L. Brekke, "A model for the mechanics of jointed rock," *Journal of Soil Mechanics and Foundations Division, ASCE*, **94**(SM3): 637-659 (1968).

[203] Grace, E. M., Jr., *Coupled Numerical Modeling of a West Antarctic Ice-Sheet/Ice-Stream/Ice-Shelf Drainage*. PhD dissertation, Department of Geography, University of Delaware, Newark, DE (2001).

[204] Gray, W. G. and M. T. van Genuchten, "Economical Alternatives to Gaussian Quadrature over Isoparametric Quadrilaterals," *International Journal for Numerical Methods in Engineering*, **12**(9): 1478-1484 (1978).

[205] Greenberg, M. D., *Advanced Engineering Mathematics*, 2nd Edition. Upper Saddle River, N.J.: Prentice Hall (1998).

[206] Greenstadt, J., "On the Reduction of Continuous Problems to Discrete Form," *IBM J. Res. Div.* **3**: 355-363 (1959).

[207] Grooms, H. R., "Algorithm for Matrix Bandwidth Reduction," *Journal of the Structural Division, ASCE*, **98**(ST1): 203-214 (1972).

[208] Gui, W. and I. Babuška, "The h, p and h-p Version of the Finite Element Method in 1 Dimension. Part I: The Error Analysis of the p-Version. Part II: The Error Analysis of the h- and h-p Version. Part III: The Adaptive h-p Version," *Numerische Mathematik*, **49**(6): 557-683 (1986).

[209] Guo, B. and I. Babuška, "The h-p Version of the Finite Element Method. Part 1: The Basic Approximation Results. Part 2: General Results and Applications," *Computational Mechanics*, **1**: 21-41, 203-226 (1986).

[210] Gupta, A. K., "On Thick Superparametric Shell Element, *International Journal for Numerical Methods in Engineering*, **12**(12): 1883-1889 (1978).

[211] Hackbusch, W., *Multi-Grid Methods and Applications*. Berlin: Springer-Verlag (1985).

[212] Haftka, R. T. and J. C. Robinson, "Effect of Out-of-Planeness of Membrane Quadrilateral Finite Elements," *AIAA Journal*, **11**(5): 742-744 (1973).

[213] Hall, T., *Carl Friedrich Gauss*, (translated by A. Froderberg from *Gauss, Matematikernas Konung*, Stockholm, 1965), Cambridge, MA: MIT Press (1970).

[214] Halliday, D. and R. Resnick, *Physics, Part II*, 2nd Edition. New York: J. Wiley and Sons (1966).

[215] Hammer, P. C., O. P. Marlowe, and A. H. Stroud, "Numerical Integration over Simplexes and Cones," *Mathematical Tables and Aids to Computation*, **10**: 130-137 (1956).

[216] Hammer, P. C., and A. H. Stroud, "Numerical Evaluation of Multiple Integrals II," *Mathematical Tables and Other Aids to Computation*, **12**(64): 272-280 (1958).

[217] Harrington, R. F., "Origin and Development of the Method of Moments for Field Computation," in *Applications of the Method of Moments to Electromagnetic Fields*, edited by B. Strait. St. Cloud, FL: SCEEE Press (1981).

[218] Heinrich, J. C. and D. W. Pepper, *Intermediate Finite Element Method: Fluid Flow and Heat Transfer Applications*. Philadelphia: Taylor and Francis (1999).

[219] Hellen, T. K., "Effective Quadrature Rules for Quadratic Solid Isoparametric Finite Elements," *International Journal for Numerical Methods in Engineering*, **4**(4): 597-599 (1972).

[220] Henrici, P., *Error Propagation for Difference Methods*. New York: J. Wiley and Sons, Inc. (1963).

[221] Henshell, R. D. and K. G. Shaw, "Crack Tip Elements are Unnecessary," *International Journal for Numerical Methods in Engineering*, **9**(3): 495-507 (1975).

[222] Henshell, R. D., "Differences Between Isoparametric Assumptions and True Circles," *International Journal for Numerical Methods in Engineering*, **10**(5): 1193-1196 (1976).

[223] Heppler, G. R. and S. J. Hansen, "Timoshenko Beam Finite Elements Using Trigonometric Basis Functions," *AIAA Journal*, **26**(11):1378-1386 (1988).

[224] Herrera, I., "Boundary Methods for Fluids," Chapter 9 in *Finite Elements in Fluids*, edited by R. H. Gallagher, D. H. Norrie, J. T. Oden and O. C. Zienkiewicz. New York: Wiley, 403-432 (1982).

[225] Herrera, I. and H. Gourgeon, "Boundary Methods, C-Complete Systems for Stokes Problems," *Computer Methods in Applied Mechanics and Engineering*, **30**(2): 225-241 (1982).

[226] Herrera, I., "Trefftz Method," in *Progress in Boundary Element Methods*, Volume 3, edited by C. A. Brebbia. New York: Wiley (1983).

[227] Herrmann, L. R., "A Bending Analysis for Plates," *Proceedings of the 1st Conference on Matrix Methods in Structural Mechanics*, Wright-Patterson AFB, Dayton, Ohio (1965).

[228] Herrmann, L. R., "Elastic Torsional Analysis of Irregular Shapes," *Journal of the Engineering Mechanics Division, ASCE*, 91(EM6): 11-19 (1966).

[229] Herrmann, L. R., "Finite-Element Bending Analysis for Plates," *Journal of the Engineering Mechanics Division, ASCE*, 93(EM5): 13-25 (1967).

[230] Herrmann, L. R., "Interpretation of Finite Element Procedure as Stress Error Minimization Procedure," *Journal of the Engineering Mechanics Division, ASCE*, 98(EM5): 1330-1335 (1972).

[231] Herrmann, L. R., "Laplacian-Isoparametric Grid Generation Scheme," *Journal of the Engineering Mechanics Division, ASCE*, 102(EM5): 749-756 (1976).

[232] Herrmann, L. R., "Finite Element Analysis of Contact Problems," *Journal of Engineering Mechanics, ASCE*, 104(EM5): 1043-1057 (1978).

[233] Herrmann, L. R., Lecture Notes for CE 212B ("Finite Elements : Application to Structural Mechanics Problems"), Department of Civil Engineering, University of California, Davis (1981).

[234] Hersch, J., "Equations Differentielles et Functions de Cellules," *C. R. Acad. Sci*, Paris 240: 1602-1604 (1955).

[235] Hess, J. L. and A. M. O. Smith, *Journal of Ship Research*, 8: 22-42 (1964).

[236] Hestenes, M. R. and E. Stiefel, "Methods of Conjugate Gradients for Solving Linear Systems," *J. Res. Nat. Bur. Stand.*, 49: 409-435 (1965).

[237] Hetényi, M., *Beams on Elastic Foundation*. Ann Arbor, MI: The University of Michigan Press (1979).

[238] Hibbitt, H. D., "On General-Purpose Programs for Nonlinear Finite Element Analysis," in *On General Purpose Finite Element Computer Programs*, edited by P. V. Marcal. New York: ASME, 98-122 (1970).

[239] Higgins, T. J., "A survey of the Approximate Solution of Two-Dimensional Physical Problems by Variational Methods and Finite Difference Procedures," in *Numerical Methods of Analysis in Engineering*, edited by L. E. Grinter. New York: Macmillan Company, 169-198 (1949).

[240] Hildebrand, F. B., *Introduction to Numerical Analysis*. New York: McGraw-Hill (1974).

[241] Hillion, P. "Numerical Integration on a Triangle," *International Journal for Numerical Methods in Engineering*, **11**(5): 797-815 (1977).

[242] Hinton, E. and J. S. Campbell, "Local and Global Smoothing of Discontinuous Finite Element Functions Using a Least Squares Method," *International Journal for Numerical Methods in Engineering*, **8**(3): 461-480 (1974).

[243] Hinton, E., F. C. Scott, and R. E. Ricketts, "Local Least Squares Stress Smoothing for Parabolic Isoparametric Elements," *International Journal for Numerical Methods in Engineering*, **9**(1): 235-238 (1975).

[244] Hinton, E. and D. R. J. Owen, *Finite Element Programming*. London: Academic Press, Inc. LTD (1977).

[245] Hochstrasser, U. W., *Orthogonal Polynomials*, Chapter 22 in *Handbook of Mathematical Functions with Formulas, Graphs, and Mathematical Tables*, Applied Mathematics Series, **55**, edited by M. Abramowitz and I. A. Stegun. Washington: National Bureau of Standards (1964).

[246] Hood, P., "Frontal Solution Program for Unsymmetric Matrices," *International Journal for Numerical Methods in Engineering*, **10**(2): 379-400 (1976).

[247] Hooke, R., "Lectures de Potentia Restitutiva," (1678), or "Of Spring, Explaining the Power of Springing Bodies," pamphlet reproduced by R. T. Gunther, *Early Science in Oxford*, **8**: 119-152 (1931).

[248] Hooke, R. LeB., "Flow Law for Polycrystalline Ice in Glaciers: Comparison of Theoretical Predictions, Laboratory Data, and Field Measurements," *Reviews in Geophysics and Space Physics*, **19**(4): 664-672 (1981).

[249] Hornbeck, R. W. 1975. *Numerical Methods.* New York: Quantum Publishers, Inc.

[250] Hrennikoff, A., "Solution of Problems of Elasticity by the Framework Method," *Journal of Applied Mechanics,* Trans. ASME, **8**(4): A169-A175 (1941).

[251] Hughes, T. J. R. and J. E. Akin, "Techniques for Developing 'Special' Finite Element Shape Functions with Particular Reference to Singularities," *International Journal for Numerical Methods in Engineering,* **15**(5): 733-751 (1980).

[252] Hughes, T. J. R., *The Finite Element Method, Linear Static and Dynamic Finite Element Analysis.* Englewood Cliffs, NJ: Prentice Hall (1987).

[253] International Journal for Numerical Methods in Engineering, *Special Issue on Unstructured Mesh Generation,* **49**(1-2) (2000).

[254] Irons, B. M., "Engineering Applications of Numerical Integration in Stiffness Method," *AIAA Journal.* **4**(11): 2035-2037 (1966).

[255] Irons, B. M., "Numerical Integration Applied to Finite Element Methods," *Conference on the Use of Digital Computers in Structural Engineering,* University of Newcastle, England (1966).

[256] Irons, B. M., "Roundoff Criteria in Direct Stiffness Solutions," *AIAA Journal.* **6**(7): 1308-1312 (1968).

[257] Irons, B. M., "A Frontal Solution Program for Finite Element Analysis," *International Journal for Numerical Methods in Engineering,* **2**(1): 5-32 (1970).

[258] Irons, B. M., "Quadrature Rules for Brick Based Finite Elements," *International Journal for Numerical Methods in Engineering,* **3**(2): 293-294 (1971).

[259] Irons, B. M. and K. J. Draper, "Inadequacy of Nodal Connections in a Stiffness Solution for Plate Bending," *AIAA Journal,* **3**(5): 961 (1965).

[260] Irons, B. M. and O. C. Zienkiewicz, "The Isoparametric Finite Element System – a New Concept in Finite Element Analysis," *Proceedings of the Conference on Recent Advances in Stress Analysis,* Royal Aeronautical Society, London (1968).

[261] Irons, B. M. and A. Razzaque, A. "Experience with the Patch Test for Convergence of Finite Elements," in *The Mathematical Foundations of the Finite Element Method with Applications to Partial Differential Equations*, edited by A. K. Aziz. New York: Academic Press, 557-587 (1972).

[262] Irons, B. M. and M. Loikkanen, M. "An Engineer's Defense of the Patch Test," *International Journal for Numerical Methods in Engineering*, **19**(9): 1391-1401 (1983).

[263] Isaacson, E. and H. B. Keller, *Analysis of Numerical Methods*. New York: J. Wiley and Sons (1966).

[264] Jacquotte, O. -P. and J. T. Oden, "Analysis of Hourglass Instabilities and Control in Underintegrated Finite Element Methods," *Computer Methods in Applied Mechanics and Engineering*, **44**(3): 339-363 (1984).

[265] Jaluria, Y., *Computer Methods for Engineering*. Englewood Cliffs, NJ: Prentice Hall (1988).

[266] Jennings, A., *Matrix Computation for Engineers and Scientists*. New York: J. Wiley and Sons (1977).

[267] Jézéquel, J-M. and B. Meyer, "Design by Contract: The Lessons of Ariane," *Computer*, **30**(2): 129-130 (1997). Also available at: http://www.eiffel.com/doc/manuals/technology/contract/ariane.

[268] Jinyun, Y., "Symmetric Gaussian Quadrature Formulae for Tetrahedronal Regions," *Computer Methods in Applied Mechanics and Engineering*, **43**(3): 349-353 (1984).

[269] Johnson, W. M., Jr. and R. W. McLay, "Convergence of the Finite Element Method in the Theory of Elasticity," *Journal of Applied Mechanics*, Trans. ASME, **35**(2): 274-278 (1968).

[270] Jones, R. E., "A Generalization of the Direct-Stiffness Method of Structural Analysis," *AIAA Journal*, **2**(5): 821-826 (1964).

[271] Jordan, C., *Calculus of Finite Differences*. Budapest: Rottig and Romwalter (1939).

[272] Joshi, A. W., *Matrices and Tensors in Physics*, 2nd Edition. A Halsted Press Book, New York: J. Wiley and Sons (1984).

[273] Kaliakin, V. N., "A Simple Coordinate Determination Scheme for Two-Dimensional Mesh Generation," *Computers and Structures*, **43**(3): 505-516 (1992).

[274] Kaliakin, V. N., User Documentation for the APES (Analysis Program for Earth Structures) Computer Program, available at: http://www.ce.udel.edu/faculty/kaliakin/apes_doc.html.

[275] Kaliakin, V. N., User Documentation for SS1D, A Computer Program for One-Dimensional Steady State Finite Element Analyses, available at: http://www.ce.udel.edu/faculty/kaliakin/ss1d_doc.pdf.

[276] Kaliakin, V. N. and J. Li, "Insight Into Deficiencies Associated with Commonly Used Zero-Thickness Interface Elements," *Computers and Geotechnics*, **17**(2): 225-252 (1995).

[277] Kantorovich, L. V., "A Direct Method of Solving the Problem of the Minimum of a Double Integral" (in Russian), *Izvestiya Nauk SSSR, Mathematical Series*: 647-652 (1933).

[278] Kantorovich, L. V. and V. I. Krylov, *Approximate Methods of Higher Analysis*. New York: Wiley Interscience (1958).

[279] Kaplan, S., O. J. Marlowe, and J. Bewick, "Application of Synthesis Techniques to Problems Involving Time Dependence," *Nucl. Sci. Engng.*, **18**: 163-176 (1964).

[280] Kármán, T., "Über laminate und turbulente Reibung," *Zeitschrift für Angewandte Mathematik und Mechanik*, 1:233-252 (1921).

[281] Kármán, T. Von and M. A. Biot, *Mathematical Methods in Engineering, An Introduction to the Mathematical Treatment of Engineering Problems*. New York: McGraw-Hill (1940).

[282] Katona, M. G., "A Simple Contact Friction Interface Element with Applications to Buried Culverts," *International Journal for Numerical and Analytical Methods in Geomechanics*, **7**: 371-384 (1983).

[283] Keast, P., "Moderate-Degree Tetrahedral Quadrature Formulas," *Computer Methods in Applied Mechanics and Engineering*, **55**(3): 339-348 (1986).

[284] Kellison, S. G., *Fundamentals of Numerical Analysis*. Homewood, IL: Irwin, Inc. (1975).

[285] Kellogg, R. B., "Difference Equations on a Mesh Arising from a General Triangulation," *Math. Comp.* **18**: 203-210 (1964).

[286] Kern, D. Q. and A. D. Kraus, *Extended Surface Heat Transfer*. New York: McGraw-Hill Book Co. (1972).

[287] Kerr, A. D., "Elastic and Viscoelastic Foundation Models," *Journal of Applied Mechanics*, Trans. ASME, **31**(3): 491-498 (1964).

[288] Kerr, A. D., "An Extension of the Kantorovich Method," *Quarterly of Applied Mathematics*, **26**(2): 219-229 (1968).

[289] Kernighan, B. W. and P. J. Plauger, *Software Tools*. Menlo Park, CA: Addison-Wesley Publishing Co. (1976).

[290] Key, S. W., *A Convergence Investigation of the Direct Stiffness Method*. Ph.D. thesis, University of Washington, Seattle (1966).

[291] Knupp, P. M., "On the Invertibility of the Isoparametric Map," *Computer Methods in Applied Mechanics and Engineering*, **78**(3): 313-329 (1990).

[292] Koh, B. C. and N. Kikuchi, "New Improved Hourglass Control for Bilinear and Trilinear Elements in Anisotropic Linear Elasticity," *Computer Methods in Applied Mechanics and Engineering*, **65**(1): 1-46 (1987).

[293] Kosloff, D. and G. A. Frazier, "Treatment of Hourglass Patterns in Low Order Finite Element Codes," *International Journal for Numerical and Analytical Methods in Geomechanics*, **2**(1): 57-72 (1978).

[294] Křižek, M. and P. Neittaanmäki, "On Superconvergence Techniques," *Acta Applied Math.*, **9**: 75-198 (1987).

[295] Kron, G., *Tensor Analysis of Networks*. New York: J. Wiley and Sons (1939).

[296] Kronrod, A. S., *Nodes and Weights of Quadrature Formulas*, New York: Consultants Bureau (1965).

[297] Krahula, J. L. and J. F. Polhemus, "Use of Fourier Series in the Finite Element Method," *AIAA Journal*, **6**(4):726-728 (1968).

[298] Krauthammer, T., "Accuracy of the Finite Element Method Near a Curved Boundary," *Computers and Structures*, **10**(6): 921-929 (1979).

[299] Kravchuk, M. F., "Application of the Method of Moments to the Solution of Linear Differential and Integral Equations," (in Ukrainian), *Kiev Soobshch. Akad. Nauk UkSSR*, **1**:168 (1932).

[300] Krishnamoorthy, C. S., *Finite Element Analysis, Theory and Programming.* New Delhi: Tata McGraw-Hill Publishing Co. Ltd. (1994).

[301] Lacroix, F. S., *Traité des différences et séries.* Paris (1800).

[302] Lanczos, C., *The Variational Principles of Mechanics.* Toronto: University of Toronto Press, Toronto (1966).

[303] Langefors, B., "Analysis of Elastic Structures by Matrix Transformation, with Special Regard to Semimonocoque Structures," *Journal of Aeronautical Sciences*, **19**(10): 451-458 (1952).

[304] Lannoy, F. G., "Triangular Finite Elements and Numerical Integration," *Computers and Structures*, **7**(5): 613 (1977).

[305] Laplace, P. S., marquis de, *Théorie analytique des probabilités. Par M. le comte Laplace.* Paris: ve Courcier (1812).

[306] Laursen, M. E. and M. Gellert, "Some Criteria for Numerically Integrated Matrices and Quadrature Formulas for Triangles," *International Journal for Numerical Methods in Engineering*, **12**(1): 67-76 (1978).

[307] Lax, P. D. and R. D. Richtmyer, "Survey of the Stability of Finite Difference Equations," *Communications in Pure and Applied Mathematics*, **9**: 267-293 (1956).

[308] Lee, N-S. and K. J. Bathe, "Effects of Element Distortion on the Performance of Isoparametric Elements," *International Journal for Numerical Methods in Engineering*, **36**(20): 3553-3576 (1993).

[309] Legendre, A. M., "Nouvelles Méthodes pour la Détermination des Orbites des Cométes," Paris: Courcier (1805).

[310] Lekhnitskii, S. G., *Theory of Elasticity of an Anisotropic Body*, Moscow: Mir Publishers (1981).

[311] Levy, H., and F. Lessman, *Finite Difference Equations.* London: Sir Isaac Pitman and Sons, Ltd. (1959).

[312] Li, T.S., R.M. McKeag, and C.G. Armstrong, "Hexahedral Meshing Using Midpoint Subdivision and Integer Programming," *Computer Methods in Applied Mechanics and Engineering*, **124**(1-2): 171-193 (1995).

[313] Li, S. and W. K. Liu, "Moving Least-Square Reproducing Kernal Method, Part II: Fourier Analysis," *Computer Methods in Applied Mechanics and Engineering*, **139**(1-4): 159-193 (1996).

[314] Li, S. and W. K. Liu, "Reproducing Kernal Hierarchical Partition of Unity, Part I – Formulation and Theory," *International Journal for Numerical Methods in Engineering*, **45**(3): 251-288 (1999).

[315] Li, S. and W. K. Liu, "Reproducing Kernal Hierarchical Partition of Unity, Part II – Applications," *International Journal for Numerical Methods in Engineering*, **45**(3): 289-317 (1999).

[316] Liebmann, H., "Die angenäherte Ermittelung harmonischer Funktionen und Konformer Abbildung," *Sitzber. math-physik. Kl. bayer. Akad. Wiss. München* **3**:385-416 (1918).

[317] Lions, J. L., ARIANE 5, Flight 501 Failure, Report by the Inquiry Board, available at: http://www.esrin.esa.it/htdocs/tidc/Press/Press96/ariane5rep.html (1996).

[318] Liseikin, V.D., *Grid Generation Methods*. New York: Springer (1999).

[319] Liu, W-H. and A. H. Sherman, "Comparative Analysis of the Cuthill-McKee and the Reverse Cuthill-McKee Ordering Algorithms for Sparse Matrices," *SIAM Journal on Numerical Analysis*, **13**(2): 198-213 (1976).

[320] Liu, W. K., J. Adee, and S. Jun, "Reproducing Kernal Particle Methods for Elastic and Plastic Problems," in *Advanced Computational Methods for Material Modeling*, **AMD 180**, edited by D. J. Benson and R. A. Asaro. ASME, 175-190 (1993).

[321] Liu, W. K., S. Jun, S. Li, J. Adee, and T. Belytschko, "Reproducing Kernal Particle Methods for Structural Dynamics," *International Journal for Numerical Methods in Engineering*, **38**(10): 1655-1679 (1995).

[322] Liu, W. K., S. Jun, and Y. F. Zhang, "Reproducing Kernal Particle Methods," *International Journal for Numerical Methods in Fluids*, **20**(8/9): 1081-1106 (1995).

[323] Liu, W. K., Y. Chen, R. A. Uras, and C. T. Chang, "Generalized Multiple Scale Reproducing Kernal Particle Methods," *Computer Methods in Applied Mechanics and Engineering*, **139**(1-4): 91-157 (1996).

[324] Liu, W. K., S. Li, and T. Belytschko, "Moving Least-Square Repro-
      ducing Kernal Methods (I) Methodology and Convergence," *Com-
      puter Methods in Applied Mechanics and Engineering*, **143**(1-2): 113-
      154 (1997).

[325] Livesly, R. K., *Matrix Methods of Structural Analysis*. New York:
      Pergamon Press (1975).

[326] Lo, S. H., "A New Mesh Generation Scheme for Arbitrary Planar Do-
      mains," *International Journal for Numerical Methods in Engineering*,
      21(8): 1403-1426 (1985).

[327] Lo, S. H., "Volume Discretization into Tetrahedra - I. Verification
      and Orientation of Boundary Surfaces," *Computers and Structures*,
      39(5): 493-500 (1991).

[328] Lo, S. H., "Volume Discretization into Tetrahedra - II. 3D Trian-
      gulation by Advancing Front Approach," *Computers and Structures*,
      39(5): 501-511 (1991).

[329] Logan, D. L., *A First Course in the Finite Element Method Using
      Algor^{TM}*. Boston, MA: PWS Publishing Co. (1997).

[330] Löhner, R., "Progress in Grid Generation via the Advancing Front
      Technique," *Engineering with Computers*, **12**(3-4): 186-210 (1996).

[331] Love, A. E. H., *A Treatise on the Mathematical Theory of Elasticity*,
      4th edition. Cambridge: Cambridge Univ. Press (1927).

[332] Lu, Y. Y., T. Belytschko, and L. Gu, "A New Implementation of
      the Element Free Galerkin Method," *Computer Methods in Applied
      Mechanics and Engineering*, **113**(3-4): 397-414 (1994).

[333] Lusternik, L. A., "Über einige Anwendungen der direkten Methoden
      in Variationsrechnung," *Mathematicheski Sbornik*. **23**:173-201 (1926).

[334] Lyness, J. N. and D. Jespersen, "Moderate Degree Symmetric
      Quadrature Rules for the Triangle," *J. Inst. Math. Applic.*, **15**: 19-32
      (1975).

[335] Mackerle, J., "Finite Element Methods: A Guide to Information
      Sources," *Finite Elements in Analysis and Design*, **8**(1-4):5-371
      (1990).

[336] Mackerle, J., "Finite and Boundary Elements and Supercomputing
      – A Bibliography (1989 – 1991)," *Finite Elements in Analysis and
      Design*, **12**(2): 151-159 (1992).

[337] Mackerle, J., "Finite and BE Analysis of Shells – A Bibliography (1990 – 1992)," *Finite Elements in Analysis and Design*, **14**(1): 73-83 (1993).

[338] Mackerle, J., "Mesh Generation and Refinement for FEM and BEM – A bibliography (1990 – 1993)," *Finite Elements in Analysis and Design*, **15**(2): 177-188 (1993).

[339] Mackerle, J., "Finite and Boundary Element Techniques in Acoustics – A Bibliography (1990 – 1992)," *Finite Elements in Analysis and Design*, **15**(3): 263-272 (1994).

[340] Mackerle, J., "Finite and Boundary Element Analyses of Concrete and Concrete Structures – A Bibliography (1991 – 1993)," *Finite Elements in Analysis and Design*, **16**(1): 71-83 (1994).

[341] Mackerle, J., "Finite and Boundary Element Techniques in Biomechanics – A Bibliography (1991 – 1993)," *Finite Elements in Analysis and Design*, **16**(2): 163-174 (1994).

[342] Mackerle, J., "Finite Element and Boundary Element Library for Composites – A Bibliography (1991 – 1993)," *Finite Elements in Analysis and Design*, **17**(2): 155-165 (1994).

[343] Mackerle, J., "Error Analysis, Adaptive Techniques, and Finite and Boundary Elements – A Bibliography (1992 – 1993)," *Finite Elements in Analysis and Design*, **17**(3): 231-246 (1994).

[344] Mackerle, J., "Finite Element and Boundary Element Methods in Probabilistic Engineering Mechanics – A Bibliography (1992 – 1994)," *Finite Elements in Analysis and Design*, **17**(4): 339-345 (1994).

[345] Mackerle, J., "Fracture Mechanics Parameters and Finite Element and Boundary Element Methods – A Bibliography (1992 – 1994)," *Finite Elements in Analysis and Design*, **19**(3): 209-223 (1995).

[346] Mackerle, J., "Thermomechanical Analysis with Finite and Boundary Element Methods – A Bibliography (1992 – 1994)," *Finite Elements in Analysis and Design*, **20**(1): 71-82 (1995).

[347] Mackerle, J., "Static and Dynamic Analysis of Plates Using FE and BE Techniques – A Bibliography (1992 – 1994)," *Finite Elements in Analysis and Design*, **20**(2): 139-154 (1995).

[348] Mackerle, J., "Fastening and joining : Finite Element and Boundary Element Analyses – A Bibliography (1992 – 1994)," *Finite Elements in Analysis and Design*, **20**(3): 205-215 (1995).

[349] Mackerle, J., "FEM and BEM in Geomechanics : Foundations and Soil-Structure Interaction – A Bibliography (1992 – 1994)," *Finite Elements in Analysis and Design*, **22**(3): 249-264 (1996).

[350] Mackerle, J., "Finite Element and Boundary Element Analyses of Beams and Thin-Wall Structures – A Bibliography (1994 – 1995)," *Finite Elements in Analysis and Design*, **23**(1): 77-89 (1996).

[351] Mackerle, J., "Structural Response to Impact, Blast, and Shock Loadings – A FE/BE Bibliography (1993 – 1995)," *Finite Elements in Analysis and Design*, **24**(2): 95-110 (1996).

[352] Mackerle, J., "Fatigue Analysis with Finite and Boundary Element Methods – A Bibliography (1993 – 1995)," *Finite Elements in Analysis and Design*, **24**(3): 187-196 (1997).

[353] Mackerle, J., "Stability Problems Analysed by Finite Element and Boundary Element Techniques – A Bibliography (1994 – 1996)," *Finite Elements in Analysis and Design*, **26**(4): 337-353 (1997).

[354] Mackerle, J., "Finite Element Analysis of Electrical Machines/Motors – A Bibliography (1994 – 1996)," *Finite Elements in Analysis and Design*, **27**(2): 215-224 (1997).

[355] Mackerle, J., "Finite Element/Boundary Element Analysis of Viscoelastic and Viscoplastic Problems – A Bibliography (1994 – 1996)," *Finite Elements in Analysis and Design*, **27**(3): 273-288 (1997).

[356] Mackerle, J., "Finite Element and Boundary Element Analysis of Bridges, Roads and Pavements – A Bibliography (1994 – 1997)," *Finite Elements in Analysis and Design*, **29**(1-4): 65-74 (1998).

[357] Mackerle, J., "Finite Element and Boundary Element Analyses of Eddy Current Problems – A bibliography (1995 – 1997)," *Finite Elements in Analysis and Design*, **31**(1): 73-84 (1998).

[358] Mackerle, J., "Fluid-Structure Interaction Problems, Finite Element and Boundary Element Approaches – A Bibliography (1995 – 1998)," *Finite Elements in Analysis and Design*, **31**(3): 231-240 (1999).

[359] Mackerle, J., "Geometric-Nonlinear Analysis by Finite Element and Boundary Element Methods – A Bibliography (1997 – 1998)," *Finite Elements in Analysis and Design*, **32**(1): 51-62 (1999).

[360] Mackerle, J., "Sensors and Actuators: Finite Element and Boundary Element Analyses and Simulations – A Bibliography (1997 – 1998)," *Finite Elements in Analysis and Design*, **33**(3): 209-220 (1999).

[361] Mackerle, J., "Finite Element Crash Simulations and Impact-Induced Injuries – A Bibliography (1980-1998)," *Shock and Vibration*, **6**(5-6):321 (1999).

[362] Mackerle, J., "Finite Element and Boundary Element Modeling of Surface Engineering Systems – A Bibliography (1996 – 1998)," *Finite Elements in Analysis and Design*, **34**(1): 113-124 (2000).

[363] Mackerle, J., "Heat Transfer Analyses by Finite Element and Boundary Element Methods – A Bibliography (1996 – 1998)," *Finite Elements in Analysis and Design*, **34**(3-4): 309-320 (2000).

[364] Mackerle, J., "Parallel Finite Element and Boundary Element Analysis: Theory and Applications – A Bibliography (1997-1999)," *Finite Elements in Analysis and Design*, **35**(3): 283-296 (2000).

[365] Mackerle, J., "Parallel finite element and boundary Object-oriented techniques in FEM and BEM - A bibliography (1996-1999)," *Finite Elements in Analysis and Design*, **36**(2): 189-196 (2000).

[366] Mackerle, J., "Finite Element Vibration and Dynamic Response Analysis of Engineering Structures – A Bibliography (1994-1998)," *Shock and Vibration*, **7**(1):39 (2000).

[367] Mackerle, J., "Smart Materials and Structures: FEM and BEM Simulations – A Bibliography (1997-1999)," *Finite Elements in Analysis and Design*, **37**(1): 71-83 (2001).

[368] Mackerle, J., "FEM and BEM in electronic packaging – A Bibliography (1998-1999)," *Finite Elements in Analysis and Design*, **37**(2): 159-171 (2001).

[369] Mackerle, J., "FEM and BEM analysis and Modelling of Residual Stresses – A Bibliography (1998-1999)," *Finite Elements in Analysis and Design*, **37**(3): 253-262 (2001).

[370] MacNeal, R. H., *Finite Elements: Their Design and Performance*. New York: Marcel Dekker (1994).

[371] MacNeal, R. H. and R. L. Harder, "A Proposed Standard Set of Problems to Test Finite Element Accuracy," *Finite Elements in Analysis and Design*, **1**(1): 3-20 (1985).

[372] Maday, Y. and A. T. Patera, "Spectral Element Methods for the Incompressible Navier-Stokes Equations," in *State-of-the-Art Surveys on Computational Mechanics*, edited by A. K. Noor and J. T. Oden. New York: ASME, 71-143 (1989).

[373] Markoff, A. A., *Differenzenrechnung*. Leipzig (1896).

[374] Marques, J.M.M.C. and D.R.J. Owen, "Infinite Elements in Quasi-Static Materially Nonlinear Problems," *Computers and Structures*, 18(4): 739-751 (1984).

[375] Martin, H. C., *Introduction to Matrix Methods of Structural Analysis*. New York: McGraw-Hill (1966).

[376] Martin, H. C. and G. F. Carey, *Introduction to Finite Element Analysis, Theory and Application*. New York: McGraw-Hill (1973).

[377] Martin, R. S., G. Peters, and J. H. Wilkinson, "Symmetric Decomposition of a Positive Definite Matrix," *Numerische Mathematik*, 7(5): 362-383 (1965).

[378] Martin, R. S., G. Peters, and J. H. Wilkinson, "Iterative Refinement of the Solution of a Positive Definite Matrix," *Numerische Mathematik*, 8(3): 203-216 (1966).

[379] Mase, G. E., *Theory and Problems of Continuum Mechanics*, Schaum's Outline Series. New York: McGraw-Hill (1970).

[380] Maxwell, J. C., "On the Calculations of the Equilibrium and Stiffness of Frames," *Phil. Mag.* 27(4): 294 (1864).

[381] McGuire, W. and R. H. Gallagher, *Matrix Structural Analysis*. New York: J. Wiley and Sons (1979).

[382] McHenry, D., "A Lattice Analogy for the Solution of Plane Stress Problems," *J. Inst. Civ. Eng*, 21: 59-82 (1943).

[383] McMahon, J., "Lower Bounds for the Electrostatic Capacity of a Cube," *Proc. Roy. Irish Acad.*, 55(A): 133-167 (1953).

[384] Meek, J. L., *Matrix Structural Analysis*. New York: McGraw-Hill (1971).

[385] Melenk, J. M. and I. Babuška, "The Partition of Unity Finite Element Method: Basic Theory and Applications," *Computer Methods in Applied Mechanics and Engineering*, 139(1-4): 289-314 (1996).

[386] Melosh, R. J., "Basis for Derivation of Matrices for the Direct Stiffness Method," *AIAA Journal*, **1**(7): 1631-1637 (1963).

[387] Melosh, R. J., "Structural Analysis of Solids," *Journal of the Structural Division, ASCE*, **89**(ST4): 205-223 (1963).

[388] Melosh, R. J. and R. M. Bamford, "Efficient Solution of Load-Deflection Equations," *Journal of the Structural Division, ASCE*, **95**: 661-676 (1969).

[389] Mikeladze, S. E., *Numerical Methods for the Solution of Partial Differential Equations*. Moscow (1936).

[390] Mikhlin, S. G., "The Convergence of Galerkin's Method," sl Dokladi Akademie Nauk SSSR, 61(2) (1948).

[391] Mikhlin, S. G., "Some Sufficient Conditions for the Convergence of Galerkin's Method," *Uch. zap. Len. gos. un-ta*, 135, Ser. Matem. Nauk. **21** (1950).

[392] Mikhlin, S. G., *Variational Methods in Mathematical Physics*. Oxford: Pergamon Press, LTD. (1964).

[393] Mikhlin, S. G., *The Problem of the Minimum of a Quadratic Functional*. San Francisco: Holden-Day (1965).

[394] Milne, W. E., *Numerical Solution of Differential equations*. New York: J. Wiley and Sons (1953).

[395] Milne-Thomson, L. M., *The Calculus of Finite Differences*. London: Macmillan and Co., Ltd. (1933).

[396] Mitchell, A. R., and Griffiths, D. E. 1980. *The Finite Difference Method in Partial Differential Equations*. New York: John Wiley.

[397] Moan, T., "On the Local Distribution of Errors by the Finite Element Approximation," in *Theory and Practice in Finite Element Standard Analysis*, edited by Y. Yamada and R. H. Gallagher. Tokyo: University of Tokyo Press, 138-168 (1973).

[398] Moan, T., "Experiences with Orthogonal Polynomials and 'Best' Numerical Integration Formulas on a Triangle; with Particular Reference to Finite Element Approximations," *Zeitschrift für Angewandte Mathematik und Mechanik*, **54**(8): 501-508 (1974).

[399] Mohr, O., "Beitrag zur Theorie der Holz und Eisenkonstruktion," *Zeitschrift des Acrhitekten–und Ingenieur–Vereins Hannover* (1868).

[400] Monaghan, J. J., "An Introduction to SPH," *Computer Physics Communications*, **48**(1): 89-96 (1988).

[401] Morse, P. M. and H. Feshbach, *Methods of Theoretical Physics*. New York: McGraw-Hill (1953).

[402] Nagtegaal, J. C., D. M. Parks, and J. R. Rice, "On Numerically Accurate Finite Element Solutions in the Fully Plastic Range," *Computer Methods in Applied Mechanics and Engineering*, **4**(2): 153-177 (1974).

[403] Navier, "Résumé des Leçons sur l'Application de la Méchanique," 3rd edition, edited by B. Saint-Venant. Paris: (1864).

[404] Naylor, D. J., "Stresses in Nearly Incompressible Materials for Finite Elements with Application to the Calculation of Excess Pore Pressure," *International Journal for Numerical Methods in Engineering*, **8**(3): 443-460 (1974).

[405] Nayroles, B., G. Touzot, and P. Villon, "Generalizing the Finite Element Method: Diffuse Approximation and Diffuse Elements," *Computational Mechanics*, **10**: 307-318 (1992).

[406] Neumann, J. Von and H. H. Goldstine, "Numerical Inverting of Matrices of High Order," *Bulletin of the Americam Mathematical Society*, **53**(1):1021-1099 (1947).

[407] Newmark, N. M., "Numerical Analysis of Bars, Plates and Elastic Bodies," in *Numerical Methods of Analysis in Engineering*, edited by L. E. Grinter. New York: Macmillan Company, 138-168 (1949).

[408] Newton, R. E., "Degeneration of Brick-Type Isoparametric Elements," *International Journal for Numerical Methods in Engineering*, **7**(4): 579-581 (1973).

[409] Nielan, P. E., K. J. Perano, and W. E. Mason, *ANTIPASTO: An Interactive Mesh Generator and Preprocessor for Two-Dimensional Analysis Programs*. Sandia Report SAND90-8203, Livermore, CA: Sandia National Laboratories (1990).

[410] Noble, B., *Applied Linear Algebra*. Englewood Cliffs, NJ: Prentice Hall (1969).

[411] Noor, A. K., "Bibliography of Books and Monographs on Finite Element Technology," *Applied Mechanics Reviews*, **44**(6): 307-317 (1991).

[412] Nörlund, N. E., *Differenzenrechnung*. Berlin: Springer (1924).

[413] Norrie, D. and G. deVries, *Finite Element Bibliography*. New York: IFI/Plenum Data Co. (1976).

[414] Oden, J. T., *Finite Elements of Nonlinear Continua*. New York: McGraw-Hill (1972).

[415] Oden, J. T., "Finite Elements: An Introduction," in *Handbook of Numerical Analysis*, Vol. II, in Finite Element Methods (Part 1), edited by P. G. Ciarlet and J. L. Lyons, North-Holland: Elsevier Science, 3-15 (1991).

[416] J. T. Oden and H. J. Brauchli, "On the Calculation of Consistent Stress Distributions in Finite Element Approximations," *International Journal for Numerical Methods in Engineering*, **3**(3): 317-325 (1971).

[417] Oden, J. T. and J. N. Reddy, *An Introduction to the Mathematical Theory of Finite Elements*. New York: J. Wiley and Sons (1976).

[418] Oden, J. T. and G. F. Carey, *Finite Elements, Volume IV, Mathematical Aspects*. Englewood Cliffs, NJ: Prentice Hall (1983).

[419] Oden, J. T., L. Demkowicz, W. Rachowicz, and T. A. Westermann, "Toward a Universal h-p Adaptive Finite Element Strategy, Part 2. A Posteriori Error Estimation," *Computer Methods in Applied Mechanics and Engineering*, **77**(1-2): 113-180 (1989).

[420] Oñate, E., S. Idelsohn, O. C. Zienkiewicz, and R. L. Taylor, "A Finite Point Method in Computational Mechanics. Applications to Convective Transport and Fluid Flow," *International Journal for Numerical Methods in Engineering*, **39**(22): 3839-3866 (1996).

[421] Ortega, J. M., *Numerical Analysis, A Second Course*. New York: Academic Press (1972).

[422] Ortiz, M., Y. Leroy, and A. Needleman, "A Finite Element Method for Localized Failure Analysis," *Computer Methods in Applied Mechanics and Engineering*, **61**(2): 189-214 (1987).

[423] Owen, S. " Mesh Generation: A Quick Introduction," available at: http://www.andrew.cmu.edu/user/sowen/mintro.html (2000).

[424] Owen, S. "A Survey of Unstructured Mesh Generation Technology," available at: http://www.andrew.cmu.edu/user/sowen/survey/index.html (2000).

[425] Özişik, M. N., *Basic Heat Transfer*. New York: McGraw-Hill (1985).

[426] Panov, D. Y., *Handbook on Numerical Solution of Partial Differential Equations*. Moscow (1951).

[427] Pardhanani, A. and G. F. Carey, "Optimization of Computational Grids," *Numerical Methods for Partial Differential Equations*, **4**(2): 95-117 (1988).

[428] Parsons, D. and J. F. Hall, "The Multigrid Method in Solid Mechanics: Part I – Algorithm Description and Behaviour," *International Journal for Numerical Methods in Engineering*, **29**(4): 719-737 (1990).

[429] Parsons, D. and J. F. Hall, "The Multigrid Method in Solid Mechanics: Part II – Practical Applications," *International Journal for Numerical Methods in Engineering*, **29**(4): 739-753 (1990).

[430] Peterson, A. F., S. L. Ray, and R. Mittra, *Computational Methods for Electromagnetics*. New York: IEEE Press (1998).

[431] Petrov, G. I., "Application of the Method of Galerkin to a Problem Involving the Stationary Flow of a Viscous Fluid," *Prikladnaya Matematika i Mekhanika* **4**(3) (1940).

[432] Petrovsky, I. G., *Uspekhi Matematicheskikh Nauk*. **8**:161 (1941).

[433] Pian, T. H. H., "Derivation of Element Stiffness Matrices by Assumed Stress Distributions," *AIAA Journal*, **2**(7): 1333-1336 (1964).

[434] Pian, T. H. H., "State-of-the-Art Development of Hybrid-Mixed Finite Element Method, *Finite Elements in Analysis and Design*, **21**(1-2): 5-20 (1995).

[435] Pian, T. H. H. and P. Tong, "Basis of Finite Element Methods for Solid Continua," *International Journal for Numerical Methods in Engineering* **1**(1): 3-28 (1969).

[436] Picone, M."Sul metodo delle minime potenze ponderate e sul metodo di Ritz per il calcolo approssimato nei problemi della fisica-matematica," *Rend. Circ. Mat. Palermo*, **52**: 225-253 (1928).

[437] Pohlhausen, K. "Zur näherungsweisen Integration der Differentialgleichung der laminar en Grenzschicht," *Zeitschrift für Angewandte Mathematik und Mechanik*, **1**:252-268 (1921).

[438] Polya, G., "Sur une Interprétation de la Méthode des Différences Finies Qui Peut Fournir des Bornes supérieures ou Inférieures," *Compt. Rend. Acad. Sci*, Paris, **235**: 995-997 (1952).

[439] Polya, G., "Estimates for Eigenvalues," *Studies Presented to Richard von Mises*, New York: Academic Press, 200-207 (1954).

[440] Popov, E. P., *Introduction to Mechanics of Solids*. Englewood Cliffs, NJ: Prentice Hall (1968).

[441] Prager, W. and J. L. Synge, "Approximations in Elasticity Based on the Concept of Function Space," *Quarterly of Applied Mathematics*, **5**(3): 241-269 (1947).

[442] Prandtl, L., "Zur Torsion von Prisma Atischen Stabben," *Physik. Zeit*, **4**: 758-770 (1903).

[443] Press, W. H., S. A. Teukolsky, W. T. Vetterling, and B. P. Flannery, *Numerical Recipes in C, The Art of Scientific Computing*, 2nd Edition. New York: Cambridge University Press (1992).

[444] Price, M.A., and C.G. Armstrong, "Hexahedral Mesh Generation by Medial Surface Subdivision: Part I.," *International Journal for Numerical Methods in Engineering*, **38**(19): 3335-3359 (1995).

[445] Price, M.A., and C.G. Armstrong, "Hexahedral Mesh Generation by Medial Surface Subdivision: Part II. Solids with Flat and Concave Edges," *International Journal for Numerical Methods in Engineering*, **40**(1): 111-136 (1997).

[446] Prigogine, I., *Thermodynamics of Irreversible Processes*. New York: Wiley-Interscience (1961).

[447] Przemieniecki, J. S., "Matrix Structural Analysis of Substructures," *AIAA Journal*, **1**(1):138-147 (1963).

[448] Przemieniecki, J. S., *Theory of Matrix Structural Analysis*. New York: McGraw-Hill (1968).

[449] Puttonen, J., "Simple and Effective Bandwidth Reduction Algorithm," *International Journal for Numerical Methods in Engineering*, **19**(8): 1139-1152 (1983).

[450] Rachowicz, W., J. T. Oden,and L. Demkowicz, "Toward a Universal h-p Adaptive Finite Element Strategy. Part 3: Design of h-p Meshes," *Computer Methods in Applied Mechanics and Engineering*, **77**(1-2): 181-211 (1989).

[451] Radau, R., "Sur les formules de quadrature à coefficients égaux," *C. R. Acad. Sci. Paris*, **90**: 520-529 (1880).

[452] Ralston, A. 1965. *A First Course in Numerical Analysis*. New York: McGraw-Hill.

[453] Randles, P. W. and L. D. Libersky, "Smoothed Particle Hydrodynamics: Some Recent Improvements and Applications," *Computer Methods in Applied Mechanics and Engineering*, **139**(1-4): 375-408 (1996).

[454] Rao, A.K. and K. Rajaiah, "Polygon-Circle Paradox of Simply Supported Thin Plates Under Uniform Pressure, " *AIAA Journal*, **6**(1): 155-156 (1968).

[455] Rauch, R. D., J. T. Batina, and H. T. Y. Yang, "Spatial Adaptation Procedure on Tetrahedral Meshes for Unsteady Aerodynamic Flow Calculations," *Proceedings of the AIAA 31st Aerospace Sciences Meeting*, AIAA Paper 93-0670, Reno, Nevada (1993).

[456] Rayleigh, Lord (Sir John William Strutt), *Theory of Sound*, First edition. London (1877). Second edition revised in 1945, Dover Publications, New York.

[457] Razzaque, A., "Automatic Reduction of Frontwidth for Finite Element Analysis," *International Journal for Numerical Methods in Engineering*, **15**(9): 1315-1324 (1980).

[458] Rebay, S., "Efficient Unstructured Mesh Generation by Means of Delaunay Triangulation and Bowyer-Watson Algorith," *Journal of Computational Physics*, **106**: 125-138 (1993).

[459] Reddy, C. T., "Improved Three Point Integration Schemes for Triangular Finite Elements," *International Journal for Numerical Methods in Engineering*, **12**(12): 1890-1896 (1978).

[460] Reddy, C. T. and D. J. Shippy, "Alternative Integration Formulae for Triangular Finite Elements," *International Journal for Numerical Methods in Engineering*, **17**(1): 133-139 (1981).

[461] Reddy, J. N., *An Introduction to the Finite Element Method*, 2nd Edition. New York: McGraw-Hill (1993).

[462] Repman, Y. V., "A Problem in the Mathematical Bases of Galerkins Method for Solving Problems on the Stability of Elastic Systems," *Prikladnaya Matematika i Mekhanika*, **4**(2) (1940).

[463] Richardson, L. F., "The Approximate Arithmetical Solution by Finite Differences of Physical Problems Involving Differential Equations with an Application to the Stresses in a Masonary Dam," *Trans. Roy. Soc.* (London) ser. A(210):307-357 (1910).

[464] Riedel, W., "Beiträge zur Lösung des ebenen Problems eines elastichen Körpers mittels der Airyschen Spannungsfunktion," *Zeitschrift für Angewandte Mathematik und Mechanik*, **7**(3):169-188 (1927).

[465] Ritz, W., "Über eine neue Methode zur Lösung gewisser Variationsprobleme der mathematischen Physik," *J. Reine Angew, Mathematik*, **135**(1): 1-61 (1908).

[466] Ritz, W., "Theorie der Transversalschwingungen einer quadratischen Platte mit freien Raendern," *Annalen der Physik*, **38** (1909).

[467] Robinson, J., "Some New Distortion Measures for Quadrilaterals," *Finite Elements in Analysis and Design*, **3**(3): 183-197 (1987).

[468] Robinson, J., "Distortion Measures for Quadrilaterals with Curved Boundaries," *Finite Elements in Analysis and Design*, **4**(2): 115-131 (1988).

[469] Robinson, J., "Validity of Aspect Ratio Sensitivity Testing – An Analytical Investigation," *Finite Elements in Analysis and Design*, **9**(2): 125-132 (1991).

[470] Rosen, S. *Programming Systems and Languages*. New York: McGraw-Hill (1967).

[471] Roy, J. R., "Numerical Error in Structural Solutions," *Journal of the Structural Division, ASCE*, **97**(ST4): 1039-1054 (1971).

[472] Rubbert, P. E. and G. R. Saaris, "Review and Evaluation of a Three-Dimensional Lifting Potential Flow Analysis Method for Arbitrary Configurations," *AIAA Paper* : 72-188 (1972).

[473] Rubinstein, M. F., *Matrix Computer Analysis of Structures*. Englewood Cliffs, NJ: Prentice Hall (1966).

[474] Runge, C., "Über eine Methode die partielle Differentialgleichung Δu = constans numerisch zu integrieren," *Zeitschrift für Mathematik und Physik* **56**: 225-232 (1908).

[475] Saad, Y., *Iterative Methods for Sparse Linear Systems*. Boston: PWS Publishing (1996).

[476] Saad, Y. and M. H. Schultz, "GMRES: A Generalized Minimal Residual Algorithm for Solving Non-Symmetric Linear Systems," *SIAM J. Sci. Statist. Comput.*, **7**: 856-869 (1986).

[477] Sack, R. L., *Matrix Structural Analysis*. Boston: PWS Publishing (1989).

[478] Saint-Venant, B., "Mémoires Savants Etrangers," **14**: 233-560 (1855).

[479] Salem, A. Z. I., S. A. Canann, and S. Saigal "Mid-Node Admissible Spaces for Quadratic Triangular 2D Finite Elements with One Edge Curved," *International Journal for Numerical Methods in Engineering*, **50**(1): 181-197 (2001).

[480] Salem, A. Z. I., S. A. Canann, and S. Saigal "Mid-Node Admissible Spaces for Quadratic Triangular Arbitrarily Curved 2D Finite Elements," *International Journal for Numerical Methods in Engineering*, **50**(2): 253-272 (2001).

[481] Salvadori, M. G., and Baron, M. L. 1961. *Numerical Methods in Engineering*, 2nd ed., Englewood Cliffs: Prentice Hall.

[482] Savage, J. S., and A. F. Peterson, "Quadrature Rules for Numerical Integration over Triangles and Tetrahedra," *IEEE Antennas and Propagation Magazine*, **38**(3): 100-102 (1996).

[483] Scarborough, J. B., *Numerical Mathematical Analysis*, 6th edition. Baltimore, MD: Johns Hopkins Press (1966).

[484] Schellbach, K., "Probleme der Variationsrechnung," *J. Reine Angew. Math.*, **41**: 293-363 (1851).

[485] Schellekens, J. C. J. and R. De Borst, "On the Numerical Integration of Interface Elements," *International Journal for Numerical Methods in Engineering*, **36**: 36-43 (1993).

[486] Schneiders, R., "A Grid-Based Algorithm for the Generation of Hexahedral Element Meshes," *Engineering With Computers*, **12**(3-4): 168-177 (1996).

[487] Schoenberg, I. J., "Contribution to the Problem of Approximation of Equidistant Data by Analytic Functions. Part A. On the Problem of Smoothing of Graduation," A First Class of Analytic Approximation Formulae, *Quart. Appl. Math.* **4**: 45-99 (1946).

[488] Segerlind, L. J., *Applied Finite Element Analysis*. New York: J. Wiley and Sons (1984).

[489] Seidel, L., "Ueber ein Verfahren, die Gleichungen, auf welche die Methode der kleinsten Quadrate führt, sowie lineäre Gleichungen überhaupt, durch successive Annäherung aufzulösen," *Abhandlungen mathematisch. physischen Klasse, Bayrische Akademie der Wissenschaften (München).* **11**: 81-108 (1874).

[490] Shames, I. H., *Mechanics of Fluids*, New York: McGraw-Hill (1962).

[491] Shelly, G. B. and T. J. Cashman, *Introduction to Computers and Data Processing.* Brea, CA: Anaheim Pub. Co. (1980).

[492] Shephard, M. S. and M. K. Georges, "Automatic Three-Dimensional Mesh Generation by Finite Octree Technique," *International Journal for Numerical Methods in Engineering*, **32**(4): 709-749 (1991).

[493] Sheppard, W. F., *Proc. London Mathematical Society* **31**:449-488 (1899).

[494] Shuleshko, P., "A New Method of Solving Boundary-Value Problems of Mathematical Physics," *Australian Journal of Applied Science*, **10**(1-7): 8-16.

[495] Silvester, P., "Newton-Cites Quadrature Formulae for N-dimensional Simplexes," *Proceedings of the 2nd Canadian Congress on Applied Mechanics*, Waterloo, Canada, 361-362 (1969).

[496] Silvester, P., "Higher-Order Polynomial Triangular Finite Elements for Potential Problems," *International Journal of Engineering Science*, **7**(8): 849-861 (1969).

[497] Silvester, P., "Symmetric Quadrature Formulae for Simplexes," *Mathematics of Computation*, **24**(109): 95-100 (1970).

[498] Silvester, P. P. and R. L. Ferrari, *Finite Elements for Electrical Engineers*, 3rd edition. Cambridge, UK: Cambridge University Press (1996).

[499] Sloan, S. W., "An Algorithm for Profile and Wavefront Reduction of Sparse Matrices," *International Journal for Numerical Methods in Engineering*, **23**(2): 239-251 (1986).

[500] Sloan, S. W. and M. F. Randolph, "Numerical Prediction of Collapse Loads Using Finite Element Methods," *International Journal for Numerical and Analytical Methods in Geomechanics*, **6**(1): 47-76 (1982).

[501] Sloan, S. W. and M. F. Randolph, "Automatic Element Reordering for Finite Element Analysis with Frontal Schemes," *International Journal for Numerical Methods in Engineering*, **19**(8): 1153-1181 (1983).

[502] Smith, I. M. and D. V. Griffiths, *Programming the Finite Element Method*, 3rd Edition. Chichester: J. Wiley and Sons (1998).

[503] Sobolev, S. L., *Partial Differential Equations of Mathematical Physics*, First Edition. Gostekhizdat (1947). English translation published by Pergamon Press, Oxford (1964).

[504] Sokolnikoff, I. S., *Mathematical Theory of Elasticity*. Malabar, FL: R. E. Krieger Pub. Co. (1983).

[505] Southwell, R. V., *Relaxation Methods in Engineering Science, a Treatise on Approximate Computation*. London: Oxford University Press (1940).

[506] Southwell, R. V., *Relaxation Methods in Theoretical Physics, a Continuation of the Treatise Relaxation Methods in Engineering Science.* Oxford: Clarendon Press (1946).

[507] Steffensen, J. F., *Interpolation*. London (1927).

[508] Stern, M., "Families of Consistent Conforming Elements with Singular Derivative Fields," *International Journal for Numerical Methods in Engineering*, **14**(3): 409-421 (1979).

[509] Stewart, G. W., *Introduction to Matrix Computations*. New York: Academic Press (1973).

[510] Stirling, J., *Methodus Differentialis*. London (1730).

[511] Sterbenz, P. H., *Floating Point Computation*. Englewood Cliffs, NJ: Prentice Hall (1974).

[512] Stoer, J. and R. Bulirsch, *Introduction to Numerical Analysis*. New York: Springer-Verlag (1980).

[513] Strang, G. "Variational Crimes in the Finite Element Method," in *The Mathematical Foundations of the Finite Element Method with Applications to Partial Differential Equations*, edited by A. K. Aziz. New York: Academic Press, 689-710 (1972).

[514] Strang, G., *Linear Algebra and Its Applications*. New York: Academic Press (1976).

[515] Strang, G. and G. J. Fix, *An Analysis of the Finite Element Method*. Englewood Cliffs, NJ: Prentice Hall (1973).

[516] Stricklin, J. A., "Integration of Area Coordinates in Matrix Structural Analysis," *AIAA Journal*, **6**(10):2023 (1968).

[517] Stricklin, J. A., W. S. Ho, E. Q. Richardson, and W. E. Haisler, "On Isoparametric vs. Linear Strain Triangular Elements," *International Journal for Numerical Methods in Engineering*, **11**(6): 1041-1043 (1977).

[518] Stroud, A. H. and D. Secrest, *Gaussian Quadrature Formulas*, Englewood Cliffs, NJ: Prentice-Hall (1966).

[519] Stroud, A. H., *Numerical Quadrature and Solution of Ordinary Differential Equations, A Textbook For a Beginning Course in Numerical Analysis*. New York: Springer-Verlag (1974).

[520] Stummel, F., "The Limitations of the Patch Test," *International Journal for Numerical Methods in Engineering*, **15**(2): 177-188 (1980).

[521] Sunder, K. S. and R. A. Cookson, "Integration Points for Triangles and Tetrahedrons Obtained from Gaussian Quadrature," *Computers and Structures*, **21**(5): 881-885 (1985).

[522] Synge, J. L., "Triangulation in the Hypercircle Method for Plane Problems," *Proc. Roy. Irish Acad.* **54**(A): 341-367 (1952).

[523] Synge, J. L., *The Hypercircle in Mathematical Physics*. London: Cambridge University Press (1957).

[524] Szabó, B. A. and A. K. Mehta, "*p*-Convergent Finite Element Approximations in Fracture Mechanics," *International Journal for Numerical Methods in Engineering*, **12**(3): 551-560 (1978).

[525] Szabó, B. A. and I. Babuška, *Finite Element Analysis*. New York: J. Wiley and Sons (1991).

[526] Taig, I. C., *Structural Analysis by the Matrix Displacement Method.* English Electric Aviation Report S017 (1961).

[527] Taig, I. C. and R. I. Kerr, "Some Problems in the Discrete Element Representation of Aircraft Structures," in *Matrix Methods of Structural Analysis*, edited by B. Fraeijis de Veubeke. New York: The Macmillan Co., 267-315 (1964).

[528] Tashiro, K. and M. Kobayashi, "Theoretical Evaluation of Three-Dimensional Elastic Constants of Native and Regenerated Celluloses : Role of Hydrogen Bonds," *Polymer*, **32**(8): 1516-1526 (1991).

[529] Tauchert, T. R., *Energy Principles in Structural Mechanics.* New York: McGraw-Hill (1974).

[530] Tautges, T. J., T. Blacker, and S. Mitchell, "The Whisker-Weaving Connectivity Based Method for Constructing All-Hexahedral Finite Element Meshes," *International Journal for Numerical Methods in Engineering*, **39**: 3327-3349 (1996).

[531] Taylor, B., *Methodus Incrementorum.* London (1717).

[532] Taylor, R. L., "On Completeness of Shape Functions for Finite Element Analysis," *International Journal for Numerical Methods in Engineering*, **4**(1): 17-22 (1972).

[533] Taylor, R. L., J. C. Simo, O. C. Zienkiewicz, and A. C. Chan, "The Patch Test: A Condition for Assessing Finite Element Convergence," *International Journal for Numerical Methods in Engineering*, **22**(1): 39-62 (1986).

[534] Temple, G., "The General Theory of Relaxation Methods Applied to Linear Systems," *Proc. Royal Soc.* **A169**: 476-500 (1938).

[535] Tewarson, R. P., *Sparse Matrices.* New York: Academic Press (1973).

[536] Thompson, J. F., Z. U. A. Warsi, and C. W. Mastin, *Numerical Grid Generation: Foundation and Applications.* Amsterdam: North Holland (1985).

[537] Timoshenko, S. P., *Elasticity Theory* (Theoria uprugosti). ONTI (1937).

[538] Timoshenko, S. P. and J. N. Goodier, *Theory of Elasticity*, Third Edition. New York: McGraw-Hill Book Co (1970).

[539] Timoshenko, S. P., *History of Strength of Materials*. New York: Dover Publications, Inc. (1983). Original publication by McGraw-Hill Book Co. (1953).

[540] Todhunter, I. and K. Pearson, *A History of the Theory of Elasticity and of the Strength of Materials: from Galilei to the Present Time*. Cambridge: University Press (1886 – 1893).

[541] Tong, P. "A Hybrid Crack Element for Rectilinear Anisotropic Material," *International Journal for Numerical Methods in Engineering*, 11(2): 377-403 (1977).

[542] Tong, P. and T. H. H. Pian, "The Convergence of Finite Element Method in Solving Linear Elastic Problems," *International Journal of Solids and Structures*, 3(5): 865-879 (1967).

[543] Trefftz, E., "Gegenstück zum Ritzschen Verfahren," *Proceedings of the 2nd International Congress on Applied Mechanics*, Zurich, 131-137 (1926).

[544] Truesdell, C., "The Mechanical Foundations of Elasticity and Fluid Dynamics," *Journal of Rational Mechanics and Analysis*, 1: 125-300 (1952).

[545] Turner, M. J., R. W. Clough, H. C. Martin, and L. J. Topp, "Stiffness and Deflection Analysis of Complex Structures," *Journal of Aeronautical Sciences*, 23(9): 805-823 (1956).

[546] Varga, R. S., *Matrix Iterative Analysis*. Englewood Cliffs, NJ: Prentice-Hall (1962).

[547] Varga, R. S., "Hermite Interpolation-Type Ritz Methods for Two-Point Boundary Value Problems," in *Numerical Solution of Partial Differential Equations*, edited by J. H. Bramble, New York: Academic Press, 365-373 (1966).

[548] Varga, R. S., "Functional Analysis and Approximation Theory in Numerical Analysis," *SIAM Regular Conference Series in Applied Mathematics*, Philadelphia, PA (1971).

[549] Visser, W., "A Finite Element Method for Determination of Non-Stationary Temperature Distribution and Thermal Deformations," *Proceedings of the 1st Conference on Matrix Methods in Structural Mechanics*, Wright-Patterson AFB, Dayton, Ohio (1965).

[550] Voronoï, G., "Novelles Applicatons des Parametre Continus a la Theorie des Formes Quadratiques: Recherches Sur les Paralleloedres
Primitifs," *J. Reine Angew. Math.*, **134** (1908).

[551] Washizu, K., *Variational Methods in Elasticity and Plasticity.* New
York: Pergamon (1975).

[552] Watson, D. F., "Computing the Delaunay Tesselation with Application to Voronoi Polytopes," *The Computer Journal*, **24**(2): 167-172
(1981).

[553] Weatherill, N. P. and O. Hassan, "Efficient Three-Dimensional Delaunay Triangulation with Automatic Point Creation and Imposed
Boundary Constraints," *International Journal for Numerical Methods in Engineering*, **37**(12): 2005-2039 (1994).

[554] Weaver, W., Jr. and P. R. Johnston, *Finite Elements for Structural
Analysis.* Englewood Cliffs, NJ: Prentice-Hall (1984).

[555] Weaver, W., Jr. and J. M. Gere, *Matrix Analysis of Framed Structures*, 3rd edition. New York: Van Nostrand Reinhold (1990).

[556] Weinberger, H. F., "Upper and Lower Bounds for Eigenvalues by
Finite Difference Methods," *Commun. Pure Appl. Math.*, **9**: 613-623
(1956).

[557] Weinberger, H. F., "Lower Bounds for Higher Eigenvalues by Finite
Difference Methods," *Pacific Journal of Mathematics*, **8**(2): 339-368
(1958).

[558] White, F. M., *Heat and Mass Transfer.* Reading, MA: Addison Wes-
ley (1988).

[559] White, G. N., *Difference Equations for Plane Thermal Elasticity.* Los
Alamos Scientific Laboratory Report LAMS-2745. Los Alamos, NM
(1962).

[560] Whiteman, J., ed. *The Mathematics of Finite Elements and Applications.* New York: Academic Press.

[561] Whiteman, J. R. 1975. *A Bibliography for Finite Elements.* New York:
Academic Press.

[562] Wiberg, N. E., F. Abdulwahab, and S. Ziukas, "Enhanced Superconvergent Patch Recovery Incorporating Equilibrium and Boundary
Conditions," *International Journal for Numerical Methods in Engineering*, **37**(20): 3417-3440 (1994).

[563] Wieghardt, K., "Über einen Grenzübergang der Elastizitätslehre und seine Andwendung auf die Statik hochgradig statisch unbestimmter Fachwerke," *Verhandlungen des Vereins z. Beförderung des Gewerbefleisses, Abhandlungen*, **85**:139-176 (1906).

[564] Wilkinson, J. H., "Error Analysis of Floating-Point Computation," *Numerische Mathematik*, **2**: 319-340 (1960).

[565] Wilkinson, J. H., *Rounding Errors in Algebraic Processes*, Englewood Cliffs, NJ: Prentice Hall (1962).

[566] Wilkinson, J. H., *The Algebraic Eigenvalue Problem*, London: Oxford University Press, Inc. (1965).

[567] Wilkinson, J. H., "The Solution of Ill-Conditioned Linear Equations," in *Mathematical Methods for Digital Computers*, Vol. 2, edited by A. Ralston and H. S. Wilf. New York: J. Wiley and Sons (1967).

[568] Williamson, F., Jr., "An Historical Note on the Finite Element Method," *International Journal for Numerical Methods in Engineering*, **15**(6): 930-934 (1980).

[569] Wilson, E. L., "Structural Analysis of Axisymmetric Solids," *AIAA Journal*, **3**(12): 2269-2274 (1965).

[570] Wilson, E. L., "The Static Condensation Algorithm," *International Journal for Numerical Methods in Engineering*, **8**(1): 198-203 (1974).

[571] Wilson, E. L. and R. E. Nickell, "Application of the Finite Element Method to Heat Conduction Analysis," *Nuclear Engineering and Design*, **4**(3): 276-286 (1966).

[572] Winkler, E. , "Die Lehre von der Elasticitaet und Festigkeit," *Prag. Dominicus* (1867).

[573] Winslow, A. M., "Numerical solutions of the quasi-linear Poisson equation in a non-uniform triangle mesh," *Journal of Computational Physics*, 1:149-172 (1967).

[574] Wright, K., "Chebyshev Collocation Methods for Ordinary Differential Equations," *Comput. Journal*, **6**: 358-365 (1964).

[575] Wylie, C. R., *Advanced Engineering Mathematics*, Fourth Edition. New York: McGraw-Hill (1975).

[576] Yamamoto, Y. and N. Tokuda, "A Note on the Convergence of Finite Element Solutions," *International Journal for Numerical Methods in Engineering*, **3**(4): 485-494 (1971).

[577] Yarry, M. A. and M. S. Shephard, "A Modified Quadtree Approach to Finite Element Mesh Generation," *IEEE Comput. Graph. Applic.*, **3**: 39-46 (1983).

[578] Yarry, M. A. and M. S. Shephard, "Automatic Three-Dimensional Mesh Generation by Modified Octree Technique," *International Journal for Numerical Methods in Engineering*, **20**(11): 1965-1990 (1984).

[579] Young, D. M., *Iterative Solution of Large Linear Systems*. New York: Academic Press (1971).

[580] Ženíšek, A., "Interpolation Polynomials on the Triangle," *Numerische Mathematik*, **15**(4): 283-296 (1970).

[581] Zhu, J. Z., E. Hinton, and O. C. Zienkiewicz, "Mesh Enrichment Against Mesh Regeneration using Quadrilateral Elements," *Communications in Numerical Methods in Engineering*, **9**(7): 547-554 (1993).

[582] Zhu, T., J. D. Zhang, and S. N. Atluri, "A Local Boundary Integral Equation (LBIE) Method in Computational Mechanics, and a Meshless Discretization Approach," *Computational Mechanics*, **21**(3): 223-235 (1998).

[583] Zienkiewicz, O. C., "The Finite Element Method: From Intuition to Generality," *Applied Mechanics Reviews*, **23**(3): 249-256 (1970).

[584] Zienkiewicz, O. C., "The Generalized Finite Element Method – State of the Art and Future Directions," *Journal of Applied Mechanics*, Trans. ASME, **50**: 1210-1217 (1983).

[585] Zienkiewicz, O. C. and Y. K. Cheung, "Finite Elements in the Solution of Field Problems," *Engineer*, **220**: 507-510 (1965).

[586] Zienkiewicz, O. C., P. Mayer, and Y. K. Cheung, "Solution of Anisotropic Seepage Problems by Finite Elements," *Journal of the Engineering Mechanics Division, ASCE*, **92**(EM1): 111-120 (1966).

[587] Zienkiewicz, O. C. and Y. K. Cheung, *The Finite Element Method in Structural and Continuum Mechanics*. London: McGraw-Hill (1967).

[588] Zienkiewicz, O. C., P. L. Arlett, and A. K. Bahrani, "Solution of Three-Dimensional Field Problems by the Finite Element Method," *Engineer*, **224**: 547-550 (1967).

[589] Zienkiewicz, O. C., B. M. Irons, J. Ergatoudis, S. Ahmad, and F. C. Scott, "Isoparametric and Associated Element Families for Two and Three Dimensional Analysis," Chapter 13 in *Finite Element Methods in Stress Analysis*, edited by I. Holland and K. Bell. Trondheim, Technical University of Norway: Tapir Press (1969).

[590] Zienkiewicz, O. C., B. M. Irons, J. Campbell, and F. C. Scott, "Three-Dimensional Stress Analysis," *Proceedings of the IUTAM Symposium on High Speed Computing in Elasticity*, Liége (1970).

[591] Zienkiewicz, O. C., R. L. Taylor, and J. M. Too, "Reduced Integration Technique in General Analysis of Plates and Shells," *International Journal for Numerical Methods in Engineering*, **3**(2): 275-290 (1971).

[592] Zienkiewicz, O. C. and E. Hinton, "Reduced Integration, Function Smoothing and Non-Conformity in Finite Element Analysis (with Special Reference to Thick Plates)," *Journal of the Franklin Institute*, **302**: 443-461 (1976).

[593] Zienkiewicz, O. C., D. W. Kelly, and P. Bettess, "The Coupling of the Finite Element Method and Boundary Solution Procedures," *International Journal for Numerical Methods in Engineering*, **11**(2): 355-375 (1977).

[594] Zienkiewicz, O. C. and K. Morgan, *Finite Elements and Approximation*. New York: J. Wiley and Sons (1983).

[595] Zienkiewicz, O. C., C. Emson, and P. Bettess, "A Novel Boundary Infinite Element," *International Journal for Numerical Methods in Engineering*, **19**(3): 393-404 (1983).

[596] Zienkiewicz, O. C., J. P. de S. R. Gago, and D. W. Kelly, "The Hierarchical Concept in Finite Element Analysis," *Computers and Structures*, **16**(1-4): 53-65 (1983).

[597] Zienkiewicz, O. C. and A. W. Craig, "A-posteriori Error Estimation and Adaptive Mesh Refinement in the Finite Element Method," *The Mathematical Basis of Finite Element Methods, The Institute of Mathematics and its Applications Conference Series*, **2**: 71-89 (1984).

[598] Zienkiewicz, O. C., S. Qu, R. L. Taylor, and S. Nakazawa, "The Patch Test for Mixed Formulations," *International Journal for Numerical Methods in Engineering*, **23**(10): 1873-1883 (1986).

[599] Zienkiewicz, O. C. and J. Z. Zhu, "A Simple Error Estimator and Adaptive Procedure for Practical Engineering Analysis," *International Journal for Numerical Methods in Engineering*, **24**(2): 337-357 (1987).

[600] Zienkiewicz, O. C., J. Z. Zhu, and N. G. Gong, "Effective and Practical *h-p* Version Adaptive Analysis Procedures for the Finite Element Method," *International Journal for Numerical Methods in Engineering*, **28**(4): 879-891 (1989).

[601] Zienkiewicz, O. C. and J. Z. Zhu, "The Superconvergent Patch Recovery and A Posteriori Error Estimates. Part 1: The Recovery Technique," *International Journal for Numerical Methods in Engineering*, **33**(7): 1331-1364 (1992).

[602] Zienkiewicz, O. C. and J. Z. Zhu, "The Superconvergent Patch Recovery and A Posteriori Error Estimates. Part 2: Error Estimates and Adaptivity," *International Journal for Numerical Methods in Engineering*, **33**(7): 1365-1382 (1992).

[603] Zienkiewicz, O. C. and J. Z. Zhu, "The Superconvergent Patch Recovery (SPR) and Adaptive Finite Element Refinement," *Computer Methods in Applied Mechanics and Engineering*, **101**(1-3): 207-224 (1992).

[604] Zienkiewicz, O. C. and R. L. Taylor, "The Finite Element Patch Test Revisited. A Computer Test for Convergence, Validation and Error Estimates," *Computer Methods in Applied Mechanics and Engineering*, **149**(1): 223-254 (1997).

[605] Zienkiewicz, O. C. and R. L. Taylor, *The Finite Element Method, Fifth Edition, Volume 1: The Basis*. Oxford: Butterworth Heinemann (2000).

[606] Zlámal, M., "On the Finite Element Method," *Numerische Mathematik*, **12**(5):394-409 (1968).

[607] Zlámal, M., "A Finite Element Procedure of the Second Order of Accuracy," *Numerische Mathematik*, **14**(4):394-402 (1970).

[608] Zlámal, M., "The Finite Element Method in Domains with Curved Boundaries," *International Journal for Numerical Methods in Engineering*, **5**(3): 367-373 (1973).

[609] Zlámal, M., "A Remark on the 'Serendipity' Family," *International Journal for Numerical Methods in Engineering*, **7**(1): 98-100 (1973).

[610] Zlámal, M., "Curved Elements in the Finite Element Method, I.," *SIAM Journal on Numerical Analysis*, **10**(1): 229-240 (1973).

[611] Zlámal, M., "Curved Elements in the Finite Element Method, II.," *SIAM Journal on Numerical Analysis*, **11**(2): 347-362 (1974).

[612] Zlámal, M., "Superconvergence and Reduced Integration in the Finite Element Method" *Math. Computations*, **32**: 663-685 (1978).

Index